Nanostructured Electrodes for High-Performance Supercapacitors and Batteries

Nanostructured Electrodes for High-Performance Supercapacitors and Batteries

Editor

Xiang Wu

Basel • Beijing • Wuhan • Barcelona • Belgrade • Novi Sad • Cluj • Manchester

Editor
Xiang Wu
School of Materials Science
and Engineering
Shenyang University of Technology
Shenyang
China

Editorial Office
MDPI
St. Alban-Anlage 66
4052 Basel, Switzerland

This is a reprint of articles from the Special Issue published online in the open access journal *Nanomaterials* (ISSN 2079-4991) (available at: www.mdpi.com/journal/nanomaterials/special_issues/supercapacitors_batteries).

For citation purposes, cite each article independently as indicated on the article page online and as indicated below:

Lastname, A.A.; Lastname, B.B. Article Title. *Journal Name* **Year**, *Volume Number*, Page Range.

ISBN 978-3-0365-9581-8 (Hbk)
ISBN 978-3-0365-9580-1 (PDF)
doi.org/10.3390/books978-3-0365-9580-1

Contents

Preface

Emerging renewable energy sources have received extensive attention in the past few decades. Energy storage has been one of the hottest research fields for more than half a century. Recently, low-dimensional nanostructured materials have been widely investigated due to their fascinating performances in energy storage fields. Great efforts have been devoted to studying their synthesis strategies, unique properties, and potential applications in various electrochemical devices. Nevertheless, challenges still exist, and many energy devices are still highly demanded for practical applications. At the same time, the ever-increasing demand for alternative energy strategies to fossil fuels and electrochemical cells has initiated considerable efforts to develop novel and renewable energy storage systems. Considering the synergetic effects between different components, hybrid nanomaterials may demonstrate dramatically enhanced performance compared to their single components. It is therefore urgent and significant to have a Special Issue to appreciate updated advances and to review recent progress regarding nanostructured electrodes for high-performance supercapacitors and batteries. In this Special Issue, we collected 21 papers, including 17 research articles and 4 review papers. It covers supercapacitors, sensors, catalysts, metal-ion batteries, solar cells, and so on.

Xiang Wu
Editor

Editorial

Nanostructured Electrodes for High-Performance Supercapacitors and Batteries

Xiang Wu

School of Materials Science and Engineering, Shenyang University of Technology, Shenyang 110870, China; wuxiang05@sut.edu.cn

Citation: Wu, X. Nanostructured Electrodes for High-Performance Supercapacitors and Batteries. *Nanomaterials* **2023**, *13*, 2807. https://doi.org/10.3390/nano13202807

Received: 10 October 2023
Accepted: 13 October 2023
Published: 23 October 2023

Emerging renewable energy sources have received extensive attention in the past few decades. Energy storage has been one of the hottest research fields for over half a century. Recently, low-dimensional nanostructured materials have been widely investigated due to their fascinating performances in energy storage fields. Great efforts have been devoted to studying their synthesis strategies, unique properties, and potential applications in various electrochemical devices. Nevertheless, challenges still exist, and many energy devices are in high demand for practical applications. At the same time, the ever-increasing demand for alternative energy strategies to fossil fuels/electrochemical cells has initiated considerable efforts to develop novel and renewable energy storage systems. Considering the synergetic effects between different components, hybrid nanomaterials may demonstrate dramatically enhanced performance compared to their single components. Therefore, It is urgent and significant to have a Special Issue to appreciate updated advances and review recent progress regarding nanostructured electrodes for high-performance supercapacitors and batteries. In this Special Issue, we collected 21 papers, including 17 research articles and four review papers. It covers supercapacitors, sensors, catalysts, metal ion batteries, solar cells, etc.

Supercapacitors show an important development perspective owning to their high-power density and excellent cycling performance [1–4]. Obaidat et al. reported hierarchical $CuMn_2O_4$ nanosheet arrays nanostructures using a one-step hydrothermal route on a nickel foam substrate. The obtained samples were utilized as battery-type electrode material, which delivered a specific capacity of 125.56 mA h g^{-1} at 1 A g^{-1} with a rate capability of 84.1% and a cycling stability of 92.15% over 5000 cycles [5]. Niu and coworkers prepared α-Fe_2O_3@MnO_2 electrode materials on carbon cloth using hydrothermal strategies and subsequent electrochemical deposition. The specific capacitance of the as-obtained product is 615 mF cm^{-2} at 2 mA cm^{-2}. Moreover, a flexible supercapacitor presents an energy density of 0.102 mWh cm^{-3} at 4.2 W cm^{-2}. Bending tests of the device at different angles show excellent mechanical flexibility [6]. Nitrogen/oxygen-doped porous carbon materials were also fabricated by calcining and activating an organic crosslinked polymer. The optimized porous carbon material showed a specific capacitance of 522 F g^{-1} at 0.5 A g^{-1} in a three-electrode system. Furthermore, an energy density of 18.04 Wh kg^{-1} was obtained at a power density of 200.0 W kg^{-1} in a two-electrode system [7]. $NiMoO_4$ is a very suitable electrode material for SCs because of its advantages of outstanding electrochemical performance and low price. Zhao's group synthesized $NiMoO_4$@$MnCo_2O_4$ composite electrodes based on a two-step hydrothermal method. The sample reaches 3000 mF/cm^2 at 1 mA/cm^2. The asymmetric supercapacitor was constructed with activated carbon as the negative electrode, which showed a maximum energy density of 90.89 mWh/cm^3 at a power density of 3726.7 mW/cm^3, and the capacitance retention can achieve 78.4% after 10,000 cycles [8]. In addition, Svetukhin et al. prepared PANI/VA-MWCNT pseudo-capacitors and studied the device's temperature-dependent charging–discharging dynamics [9]. Wang's group summarized the research progress of cobalt-based nanomaterials as electrode materials for supercapacitors and focused on the strategies to improve the electrochemical properties of these materials [10]. Polyaniline (PANI) is thought to be an excellent candidate for energy-storage applications owing to its tunable structure, multiple oxidation/reduction reaction, and environmental stability. Scientists from different countries summarized updated progress about polyaniline/metal-organic framework composite electrodes for supercapacitor applications [11]. Zhang and coworkers fabricated the n*AlN/n*ScN superlattices by the epitaxial growth technology of controlling layer interface in an atom resolution. Their energy storage characteristics were studied using first principle calculation of the band-structure and dielectric polarizability dependent on the electrical field and superlattice configuration [12].

Besides the above supercapacitor reports, we collected some work about metal ion batteries, such as Li-ion, Zn ion, and sodium batteries. Zhang and You synthesized an environmentally friendly and cost-effective CoO@rGO flexible membrane with excellent electrochemical properties as anode material. It showed that the hollow material is much more favorable for lithium-ion transport and storage [13]. Cui et al. prepared fibrous red phosphorus as an anode material for LIBs using chemical vapor transport. The obtained composite material showed a reversible specific capacity of 1621 mAh/g and a capacity of 742.4 mAh/g after 700 cycles at a current density of 2 A/g. The Coulombic efficiencies reached almost 100% for each cycle [14]. It is known that Li-rich oxides are promising cathode materials for Li-ion batteries. Scientists from Russia studied different compositions of Li-rich materials and various electrochemical testing modes. The results showed that the $Li_{1.149}Ni_{0.184}Mn_{0.482}Co_{0.184}O_2$ cathode material demonstrated the best functional properties [15]. SiO_2 is also used as an anode materials for lithium battery. Qin et al. prepared SiO_2 aerogels using a Sol-Gel method. The results showed that Ketjen Black provides superior cycling and rate performance with a reversible specific capacity of 351.4 mA h g^{-1} at 0.2 A g^{-1} after 200 cycles and 311.7 mA h g^{-1} at 1.0 A g^{-1} after 500 cycles [16]. Hierarchical Si@MnO_2@reduced graphene oxide (rGO) can effectively reduce the volume change of Si and increase the lithium-ion battery capacity due to the dual protection of MnO_2 and rGO. It showed a discharge-specific capacity of 1282.72 mAh g^{-1} at 1 A g^{-1} after 1000 cycles. Moreover, the volume expansion of the anode material is 50% after 150 cycles, which is much less than that of Si (300%) [17].

Aqueous zinc ion batteries (AZIBs) have attracted much attention owing to their low cost, high capacity, and non-toxic characteristics. Therefore, transition metal chalcogenides with a layered structure are considered suitable electrode materials. The large layer spacing facilitates the intercalation/de-intercalation of Zn^{2+} between the layers. Wu's group summarized many design strategies for modifying cathodes and specifically emphasized the zinc storage capacity of the optimized electrodes. They also proposed the challenges and prospects of cathode materials for high-energy AZIBs [18]. Gong et al. investigated the PANI cathode's electrochemical performance and ion transport kinetics to further understand Zn^{2+} storage mechanisms. The assembled PANI/Zn cell achieves a capacity of 74 mAh g^{-1} a t 0.3 A g^{-1} and maintains 48.4% of its initial discharge capacity after 1000 cycles [19]. Additionally, biomass-derived hard carbon as anode material for sodium-ion batteries has attracted attention because of its renewable nature and low cost. Qin's group employed a two-step method to prepare three different structures of hard carbon materials from sisal fibers. It showed the best electrochemical performance, with an initial Coulomb efficiency of 76.7% [20].

Indeed, graphitic carbon nitride (g-C_3N_4) is extensively used as an electron transport layer or interfacial buffer layer for realizing photoelectric performance improvement in perovskite solar cells (PSCs). Chu and Li overviewed different g-C_3N_4 nanostructures as an additive and surface modifier layers applied to PSCs. They emphasized the mechanism of reducing the defect state in PSCs and proposed the potential challenges and perspectives of g-C_3N_4 incorporated into perovskite-based optoelectronic devices [21]. Zhu et al. from Northeastern University investigated the durability of proton exchange membrane fuel cells (PEMFCs) by 300 h accelerated stress test under vibration and non-vibration conditions.

The voltage under vibration slightly declines at the current density of 400 mA cm^{-2} and decreases quickly over time in high current density [22].

Hydrogen is regarded as a promising clean energy source in the future due to its high heat value and environmentally friendly features. Sodium borohydride (NaBH$_4$) is a good candidate for hydrogen generation from hydrolysis because of its high hydrogen storage capacity and hydrolysis products. However, due to its sluggish hydrogen generation (HG) rate in the water, it usually needs an efficient catalyst to enhance the HG rate. Sun's group reported graphene oxide (GO)-modified Co-B-P catalysts by a chemical in situ reduction route. The results showed that the as-prepared catalyst with a GO content of 75 mg possesses an optimal catalytic efficiency with an HG rate of 12,087.8 mL min^{-1} g^{-1} at 25 °C [23]. They also synthesized a porous titanium oxide cage (PTOC) using a one-step hydrothermal method using NH$_2$-MIL-125 as the template and L-alanine as the coordination agent. Due to the synergistic effect between the PTOC and PtNi alloy particles, the catalysts present a hydrogen generation rate of 10,164.3 mL min^{-1} g^{-1}) and activation energy of 28.7 kJ mol^{-1} [24].

Moreover, wearable motion-monitoring systems have been widely studied in recent years. However, traditional wearable devices' battery energy storage problem limits the development of human sports training applications. Mao and coworkers reported a self-powered, portable micro-structure triboelectric nanogenerator (MS-TENG). It provides a maximum output voltage of 74 V, angular sensitivity of 1.016 V/degree, and high signal-to-noise ratio and can a power electronic calculator and electronic watch. In addition, as a flexible electrode hydrogel, it can readily stretch over 1300%, which can help improve the service life and work stability of MS-TENG [25].

Funding: This research received no external funding.

Acknowledgments: I am grateful to all the authors who contributed to this Special Issue and the referees for reviewing the manuscripts.

Conflicts of Interest: The authors declare no conflict of interest.

References

1. Dai, M.Z.; Zhao, D.P.; Wu, X. Research progress on transition metal oxide based electrode materials for asymmetric hybrid capacitors. *Chin. Chem. Lett.* **2020**, *31*, 2177–2188.
2. Liu, H.Q.; Dai, M.Z.; Zhao, D.P.; Wu, X.; Wang, B. Realizing superior electrochemical performance of asymmetric capacitors through tailoring electrode architectures. *ACS Appl. Energy Mater.* **2020**, *3*, 7004–7010. [CrossRef]
3. Wang, X.W.; Sun, Y.C.; Zhang, W.C.; Wu, X. Flexible CuCo$_2$O$_4$@Ni-Co-S hybrids as electrode materials for high-performance energy storage devices. *Chin. Chem. Lett.* **2023**, *34*, 107593.
4. Sun, Y.C.; Wang, X.W.; Wu, X. High-performance flexible hybrid capacitors by regulating NiCoMoS@Mo$_{0.75}$-LDH electrode structure. *Mater. Res. Bull.* **2023**, *158*, 112073.
5. Gopi, C.V.V.M.; Ramesh, R.; Vinodh, R.; Alzahmi, S.; Obaidat, I.M. Facile Synthesis of Battery-Type CuMn$_2$O$_4$ Nanosheet Arrays on Ni Foam as an Efficient Binder-Free Electrode Material for High-Rate Supercapacitors. *Nanomaterials* **2023**, *13*, 1125.
6. Niu, Y.Q.; Shang, D.Y.; Li, Z.P. Micro/Nano Energy Storage Devices Based on Composite Electrode Materials. *Nanomaterials* **2022**, *12*, 2202. [CrossRef]
7. Lao, J.H.; Lu, Y.; Fang, S.W.; Xu, F.; Sun, L.X.; Wang, Y.; Zhou, T.H.; Liao, L.M.; Guan, Y.X.; Wei, X.Y.; et al. Organic Crosslinked Polymer-Derived N/O-Doped Porous Carbons for High-Performance Supercapacitor. *Nanomaterials* **2022**, *12*, 2186. [CrossRef] [PubMed]
8. Di, Y.F.; Xiang, J.; Bu, N.; Loy, S.; Yang, W.D.; Zhao, R.D.; Wu, F.F.; Sun, X.B.; Wu, Z.H. Sophisticated Structural Tuning of NiMoO$_4$@MnCo$_2$O$_4$ Nanomaterials for High Performance Hybrid Capacitors. *Nanomaterials* **2022**, *12*, 1674. [CrossRef] [PubMed]
9. Yavtushenko, I.O.; Makhmud-Akhunov, M.Y.; Sibatov, R.T.; Kitsyuk, E.P.; Svetukhin, V.V. Temperature-Dependent Fractional Dynamics in Pseudo-Capacitors with Carbon Nanotube Array/Polyaniline Electrodes. *Nanomaterials* **2022**, *12*, 739. [CrossRef] [PubMed]
10. Yang, L.; Zhu, Q.H.; Yang, K.; Xu, X.K.; Huang, J.C.; Chen, H.F.; Wang, H.W. A Review on the Application of Cobalt-Based Nanomaterials in Supercapacitors. *Nanomaterials* **2022**, *12*, 4065.
11. Vinodh, R.; Babu, R.S.; Sambasivam, S.; Gopi, C.V.V.M.; Alzahmi, S.; Kim, H.J.; de Barros, A.L.F.; Obaidat, I.M. Recent Advancements of Polyaniline/Metal Organic Framework (PANI/MOF) Composite Electrodes for Supercapacitor Applications: A Critical Review. *Nanomaterials* **2022**, *12*, 1511. [PubMed]

12. Zhang, W.C.; Wu, H.; Sun, W.F.; Zhang, Z.P. First-Principles Study of n*AlN/n*ScN Superlattices with High Dielectric Capacity for Energy Storage. *Nanomaterials* **2022**, *12*, 1966. [CrossRef] [PubMed]
13. Zhang, J.X.; You, J.; Wei, Q.; Han, J.I.; Liu, Z.M. Hollow Porous CoO@Reduced Graphene Oxide Self-Supporting Flexible Membrane for High Performance Lithium-Ion Storage. *Nanomaterials* **2022**, *13*, 1986.
14. Liu, L.; Gao, X.; Cui, X.M.; Wang, B.F.; Hu, F.Z.; Yuan, T.H.; Li, J.H.; Zu, L.; Lian, H.Q.; Cui, X.G. Chemical Vapor Transport Synthesis of Fibrous Red Phosphorus Crystal as Anodes for Lithium-Ion Batteries. *Nanomaterials* **2023**, *13*, 1060. [PubMed]
15. Pechen, L.; Makhonina, E.; Medvedeva, A.; Politov, Y.; Rumyantsev, A.; Koshtyal, Y.; Goloveshkin, A.; Eremenko, I. Influence of the Composition and Testing Modes on the Electrochemical Performance of Li-Rich Cathode Materials. *Nanomaterials* **2022**, *12*, 4054.
16. Hu, G.B.; Sun, X.H.; Liu, H.G.; Xu, Y.Y.; Liao, L.; Guo, D.L.; Liu, X.M.; Qin, A.M. Improvement of Lithium Storage Performance of Silica Anode by Using Ketjen Black as Functional Conductive Agent. *Nanomaterials* **2022**, *12*, 692.
17. Liu, J.J.; Wang, M.; Wang, Q.; Zhao, X.S.; Song, Y.T.; Zhao, T.M.; Sun, J. Sea Urchin-like Si@MnO$_2$@rGO as Anodes for High-Performance Lithium-Ion Batteries. *Nanomaterials* **2022**, *12*, 285. [CrossRef]
18. Liu, Y.; Wu, X. Recent Advances of Transition Metal Chalcogenides as Cathode Materials for Aqueous Zinc-Ion Batteries. *Nanomaterials* **2022**, *12*, 3298. [CrossRef]
19. Gong, J.F.; Li, H.; Zhang, K.X.; Zhang, Z.P.; Cao, J.; Shao, Z.B.; Tang, C.M.; Fu, S.J.; Wang, Q.J.; Wu, X. Zinc-Ion Storage Mechanism of Polyaniline for Rechargeable Aqueous Zinc-Ion Batteries. *Nanomaterials* **2022**, *12*, 1438.
20. Luo, Y.; Xu, Y.Y.; Li, X.N.; Zhang, K.Y.; Pang, Q.; Qin, A.M. Boosting the Initial Coulomb Efficiency of Sisal Fiber-Derived Carbon Anode for Sodium Ion Batteries by Microstructure Controlling. *Nanomaterials* **2023**, *13*, 881.
21. Yang, J.; Ma, Y.H.; Yang, J.P.; Liu, W.; Li, X.A. Recent Advances in g-C3N4 for the Application of Perovskite Solar Cells. *Nanomaterials* **2022**, *12*, 3625. [PubMed]
22. Chen, S.T.; Wang, X.K.; Zhu, T. Effect of Mechanical Vibration on the Durability of Proton Exchange Membrane Fuel Cells. *Nanomaterials* **2023**, *13*, 2191. [PubMed]
23. Jia, X.L.; Sang, Z.; Sun, L.X.; Xu, F.; Pan, H.G.; Zhang, C.C.; Cheng, R.G.; Yu, Y.Q.; Hu, H.P.; Kang, L.; et al. Graphene-Modified Co-B-P Catalysts for Hydrogen Generation from Sodium Borohydride Hydrolysis. *Nanomaterials* **2022**, *12*, 2732. [PubMed]
24. Yu, Y.Q.; Kang, L.; Sun, L.X.; Xu, F.; Pan, H.G.; Sang, Z.; Zhang, C.C.; Jia, X.L.; Sui, Q.L.; Bu, Y.T.; et al. Bimetallic Pt-Ni Nanoparticles Confined in Porous Titanium Oxide Cage for Hydrogen Generation from NaBH4 Hydrolysis. *Nanomaterials* **2022**, *12*, 2550. [CrossRef]
25. Lu, Z.; Jia, C.J.; Yang, X.; Zhu, Y.S.; Sun, F.X.; Zhao, T.M.; Zhang, S.W.; Mao, Y.P. A Flexible TENG Based on Micro-Structure Film for Speed Skating Techniques Monitoring and Biomechanical Energy Harvesting. *Nanomaterials* **2022**, *12*, 1576.

Article

Effect of Mechanical Vibration on the Durability of Proton Exchange Membrane Fuel Cells

Sitong Chen [1], Xueke Wang [2] and Tong Zhu [3],*

1   School of Mechanical Engineering, Shenyang University of Technology, Shenyang 110870, China; chensitong0525@126.com
2   Beijing Institute of Space Launch Technology, Beijing 100076, China; 1810135@stu.neu.edu.cn
3   School of Mechanical Engineering and Automation, Northeastern University, Shenyang 110819, China
*   Correspondence: tongzhu_neu@126.com

Abstract: To study the durability of proton exchange membrane fuel cells (PEMFCs), the experiments were performed by using a 300 h accelerated stress test under vibration and non-vibration conditions. Before and after chronic operation, the polarization curve, impedance spectra and cyclic voltam-mogram were measured at regular intervals. The voltage under vibration shows a small decline at the current density of 400 mA cm^{-2} and decreases quickly along the time in high current density. Meanwhile, the pavement vibration dramatically impacts the contact resistance of the membrane electrode assembly to the bipolar plates and the clamping screws of the fuel cell easily loosen under vibration. The calculations from X-ray diffraction patterns indicate that the average diameters of Pt particles under vibration are smaller than those under no-vibration conditions. It increases from 3.17 nm in the pristine state to 3.43 nm and 4.62 nm, respectively. Moreover, much more platinum that dissolved from the catalyst layer and redeposited was detected inside the polymer membrane under vibration conditions.

Keywords: proton exchange membrane fuel cell; mechanical vibration; durability; platinum migration; accelerated stress test

Citation: Chen, S.; Wang, X.; Zhu, T. Effect of Mechanical Vibration on the Durability of Proton Exchange Membrane Fuel Cells. *Nanomaterials* **2023**, *13*, 2191. https://doi.org/10.3390/nano13152191

Academic Editor: Ioannis V. Yentekakis

Received: 23 April 2023
Revised: 30 June 2023
Accepted: 7 July 2023
Published: 27 July 2023

1. Introduction

The proton exchange membrane fuel cell (PEMFC) have become a new power option which can supersede the engines powered by fossil fuels in passenger vehicles owing to their characteristics of zero emissions, high efficiency and power density [1,2]. However, there are still challenges to fulfilling the target of 5000 hours' steady operation for passenger vehicles presented by the US Department of Energy (DOE) [3].

In order to ensure better forecast accuracy of the fuel cell's lifetime and analyze its probable degradation mechanism, the accelerated stress tests (ASTs) have been implemented by many researchers in experimental studies with regard to the contamination in hydrogen and air [4,5], start-up at subzero temperature [6–9] and dynamic response under driving cycles [10]. For the on-board fuel cells in vehicles, the performance degradation consists of membrane degradation, Pt/C catalyst degradation and gas diffusion layer degradation [11–19]. More specifically, the typical membrane degradation includes cracks, punctures and pinholes, resulting from a harsh operating environment accompanied by improper temperature, relative humidity and mechanical conditions. In the long run, rupture, delamination, agglomeration and migration of Pt or carbon corrosion will all result in the Pt/C catalyst degradation. All of these negative effects, which arise from either the changes in the microstructure of carbon-supported platinum or the decline of electrons and ions involved in the electrochemical reaction, lead to an obvious loss of catalyst activity. Additionally, the gas diffusion layer degradation due to the poly-tetrafluoroethylene (PTFE) decomposition, the movement of micropores and the loss of microporous layer (MPL) hydrophobicity caused by the carbon corrosion-based degradation process often cause the variation of water content.

Even though the degradation phenomenon of PEMFC has been investigated, research on the performance degradation of vehicle fuel cell stacks subjected to vibration situations is in progress. Among the few open literatures, Imen et al. [20] evaluated the effect of mechanical loads and vibrations on an open-cathode PEMFC in operating state as well as a non-operating PEMFC; the experimental results reveal that these external factors can change the performance and reliability of the fuel cell by causing physical damage to the fuel cell components. Rajalakshmi et al. [21] performed vibration test analysis on a 500 W PEM fuel cell stack developed by simulating some application situations in the stack and evaluated the robustness of the stack; the fuel cell performance was in good agreement before and after the vibration and shock tests, showing the mechanical integrity of the system. Hou et al. [22] studied the performance of a fuel cell stack through long-term strengthened road vibration tests, the individual cell voltage uniformity became distinctly worse. With the increase of vibration duration, the ohmic resistance obtained from AC impedance diagnosis ascended approximately linearly and presented a growth of 5.36% ultimately. Wu et al. [23] numerically analyzed the mechanical response of a large fuel cell stack clamped by steel belts to a violent impact. The results indicate that the stack may give rise to interface slippage between cells when subjected to a large impact in the direction parallel with the cells, showing a downward bowing phenomenon. Wang et al. [24] experimentally investigated the effect of mechanical vibration on the dynamic response of PEMFC. Diloyan et al. [25] discussed the effect of mechanical vibration on platinum particle agglomeration and growth in the catalyst layer of PEMFC; it was observed that the average diameter of Pt particles under vibration was 10% smaller than the ones that were under no-vibration conditions.

In the current research, the influence of vibration on the durability of PEMFCs was studied, presenting the results of the performance variation in PEMFCs. A 300 h vibration test was accomplished to assess the effect of vibration as for the endurance of MEA and the results were characterized clearly by electrochemical and physical methods. The polarization curve, impedance spectra and cyclic voltammogram were measured at regular intervals. Meanwhile, the changes in the microstructure of MEAs were observed and measured by X-ray Diffraction (XRD), Scanning Electron Microscopy (SEM) and Transmission Electron Microscopy (TEM) before and after the experiments.

2. Experimental Section

2.1. Experimental Setup

In Figure 1, the experimental bench is composed of the vibration generator, the electronic load, the data acquisition system, the bubble humidifier and the gas supply system. The vibration tests of PEMFC were performed on the horizontal and vertical vibration generator controlled by computer software, which could produce excitations of multiple waveforms within the frequency scope of 1 Hz to 600 Hz with the maximum displacement of 5 mm. Its acceleration amplitude is up to 20 g and the maximum load is 100 kg.

As to PEMFC, the parallel serpentine flow channel of 25 cm^2 was used for the experiments. For MEA, a commercially available Nafion® 211 membrane and SGL-25BC carbon papers, of which the porosity was 80% and the air permeability was 1.0 cm^3/(cm^2·s), were chosen. The Pt catalyst loading on the cathode side and anode side were, respectively, 0.48 mg cm^{-2} and 0.28 mg cm^{-2}.

2.2. Test Procedure

Based on the typical vibration feature of running vehicles and the actual road conditions, the accelerated stress test was designed to research the durability of PEMFCs under both the cases of vibration and non-vibration. The vibrational frequency of vehicles on the road is close to 17–40 Hz and the maximum vibration acceleration of the vehicles in the horizontal and vertical directions are usually less than 2.06 g and 4.665 g, respectively [26].

Therefore, the vibration experiments were conducted at the frequency of 20 Hz and the horizontal acceleration and vertical acceleration of 2.0 g. During an experimental cycle, the vibration generator first provided the horizontal excitation for 20 min with a rest for 5 min, and then produced the vertical excitation for 20 min. For the purpose of analyzing the influence of mechanical vibration on the durability of PEMFC, the PEMFC was operated under vibrational and static conditions, along with the operating parameters summarized in Table 1.

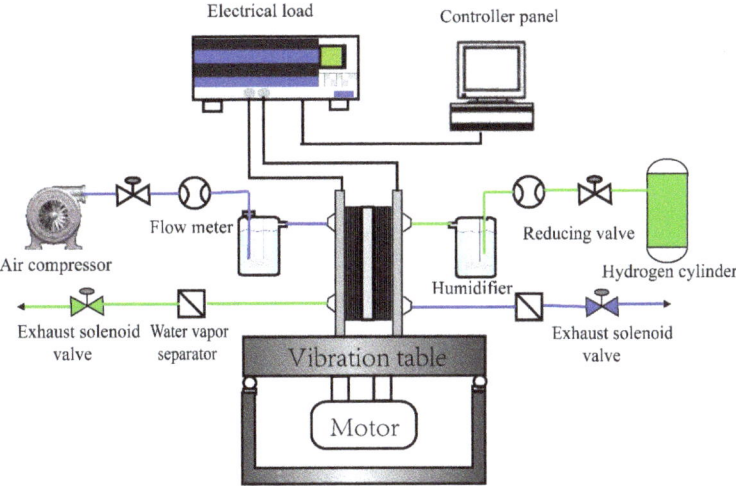

Figure 1. The schematic illustration of the test rig.

Table 1. The test conditions of the fuel cell.

Test Conditions	
Fuel/Oxidant	Hydrogen/Air
Constant current density	400 mA/cm^2
PEMFC temperature	65 °C
Anode/Cathode humidity	100%/100%
Anode/Cathode stoic	1.5/2.5
Test length	300 h

Throughout the vibration tests, the performance of PEMFC was evaluated several times at regular intervals by measuring the polarization curves, electrochemical impedance spectroscopy (EIS) and cyclic voltammogram (CV). For the EIS measurements, the test frequency was chosen in the range of 10 kHz to 10 mHz while the amplitude was kept at 5 mV. Moreover, the scanning rate of CV was recorded at 50 mV/s.

2.3. Characterization of MEA

The micromorphology characteristics and the changes in the microstructure of MEAs were observed and measured by X-ray Diffraction (XRD), Scanning Electron Microscopy (SEM) and Transmission Electron Microscopy (TEM) before and after the experiments. In order to compare the Pt grain diameter of the cathode side with TEM, we first scraped some Pt/C catalyst off the MEA and dissolved it in the ethanol–water mixture, and then split the Pt particles from the carbon support with ultrasonic shaking for 2 h. Afterwards, 1 mL mixture was extracted and dripped onto the copper mesh for observing.

3. Results and Discussion

The degradation of the PEMFC running continuously for 300 h under vibrational and static conditions is exhibited in Figure 2. Under vibration conditions, the cell voltage drops with a faster rate during the 300 h operation, the average performance of the PEMFC decreases from 0.643 V to 0.612 V and the degradation rate reaches 103 μVh^{-1}. While running at no-vibration conditions, a voltage difference from 0.658 V to 0.642 V is indicated and the voltage decay rate is approximately 53 μVh^{-1}. The U.S. Department of Energy sets the stack voltage at the end of lifetime as 90% of the initial voltage at the rated power output [3], and the degradation rate is assumed to be constant based on the results of relevant studies [22,27]. It can be inferred that the durability of the MEA under static condition can reach 1500 h, which only reaches 700 h under vibration conditions. The voltage drop is one of the manifestations of the vehicular PEMFCs' performance degradation caused by chronic pavement vibration. In addition, the voltage fluctuates greatly under vibration conditions, which is an interruption phenomenon caused by instability in the PEM fuel cell operation under vibration.

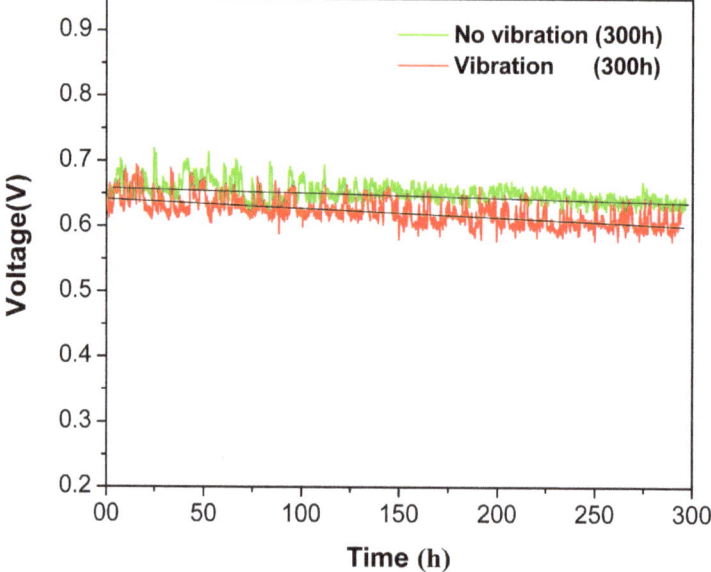

Figure 2. The effect of vibration on voltage degradation of the PEMFC operated at 400 mA cm^{-2}.

Figure 3a,b present the polarization curves of the PEMFC operating at different duration under vibration and no-vibration conditions, respectively. The kinetics region remains almost unaltered, and the ohmic region seems to slightly degrade compared to the non-vibration experiments. With the growth of current density, the attenuation ratio of voltage that represents the slope of the voltage value before and after the decay under the increase of the unit current density becomes larger under both cases. Particularly when running for 240 h at no-vibration conditions, the performance of the cell remains stable with the current density less than 400 mA cm^{-2}, but in the mass-transfer region density the voltage gradually falls off; the mass-transfer region is the most affected section of the polarization curves, indicative of a degradation of the electrode structure. Compared with above results, the voltage under vibration has shown a small decline at the current density of 400 mA cm^{-2} and decreases faster along the time in the high-current-density region.

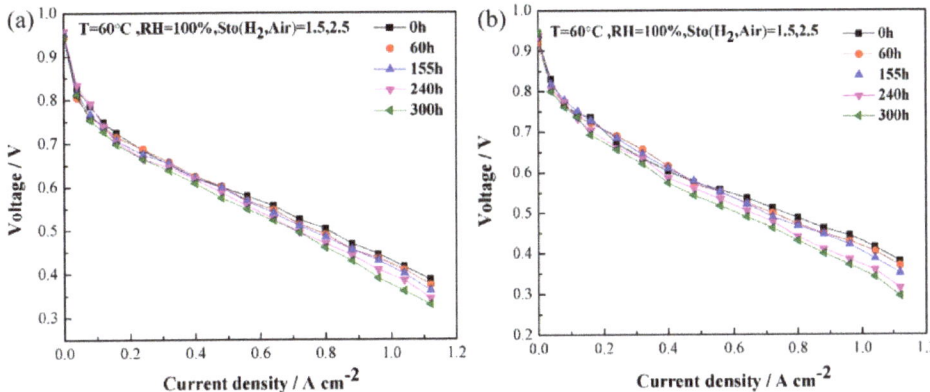

Figure 3. The polarization curves of a 25 cm^2 single cell under (**a**) no-vibration and (**b**) vibration conditions.

For all the single PEMFC, the high frequency intercept denotes the sum of interfacial contact and material bulk resistance (Rohm). The medium frequency arc reflects the combination of a charge transfer resistance (Rct) and a double layer capacitance (C) within the catalyst layer. The low frequency arc represents the mass transport process [28]. As is revealed in Figure 4a,b, the impedance loop of the PEMFC increases with the extension of discharge time, the ohmic resistance of PEMFC at no-vibration conditions increases from 0.09 Ω cm^{-1} to 0.102 Ω cm^{-1}, while the ohmic resistance under vibration enhances from 0.09 Ω cm^{-1} to 0.114 Ω cm^{-1}. This indicates that the pavement vibration dramatically impact the contact resistance of the membrane electrode assembly to the bipolar plates and the clamping screws of the fuel cell easily loosen under vibration conditions.

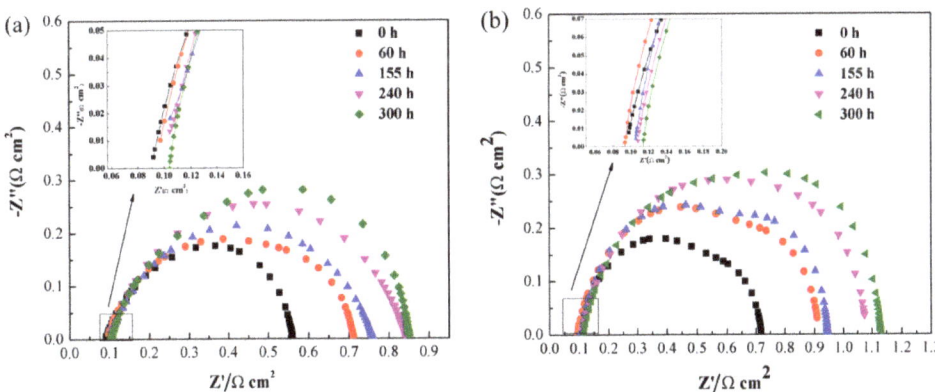

Figure 4. The Nyquist plots of a 25 cm^2 single cell measured at 400 mA cm^{-2} under (**a**) no-vibration and (**b**) vibration conditions.

The cyclic voltammograms measured after running at different test duration under vibrational and static conditions are presented in Figure 5. Obviously, the oxidation desorption peak of hydrogen is observed at about 0.2 V and the peak area decreases with the increase of operating time under both cases. The electrochemical active surface area (ECSA) can be calculated with the following equation [29]:

$$ECSA = \frac{[\text{Charge area}/V] \times 1000}{0.21 \times \text{Catalyst loading} \times 10000} \qquad (1)$$

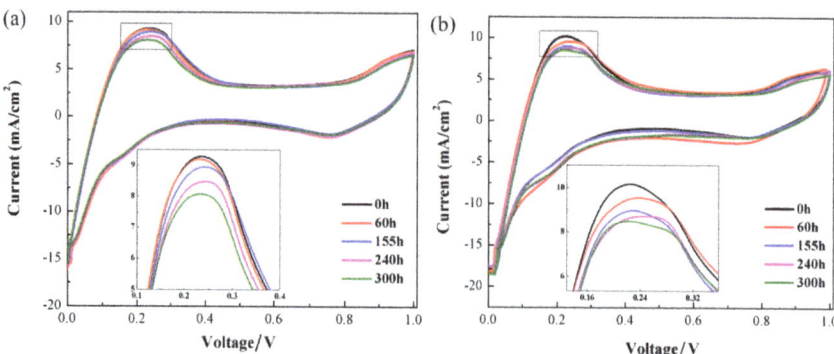

Figure 5. The cyclic voltammograms of a 25 cm^2 single cell operated under (**a**) no-vibration and (**b**) vibration conditions.

As is shown in Table 2, ECSA reduces with the increase of operating time, which changes between 60.1 and 49.6 m^2 g^{-1}, respectively, under vibration conditions and falls from 57.5 to 50 m^2 g^{-1} under static conditions. In Figure 6, the ratio between the effective active area and the initial active area after different running times is plotted. It can be observed that the effective active area has decreased greatly in the vibrational environment since the operation time of 150 h and the drop degree of the ratio under vibration is always greater than that under no-vibration conditions. The decline of ECSA to the catalyst may be on account of the agglomeration and dissolution of Pt nanoparticles, as well as the detachment of Pt nanoparticles from the carbon support.

Table 2. The ECSA of MEA with operating time under vibration and no-vibration conditions.

Operating Time	Under Vibration		Under No Vibration	
	Peak Area (mC cm^{-2})	ECSA (m^2 g^{-1})	Peak Area (mC cm^{-2})	ECSA (m^2 g^{-1})
0 h	60.6	60.1	58	57.5
60 h	60.2	59.7	56.4	55.9
155 h	54.4	53.9	55.6	55.1
240 h	52	51.5	52.6	52.1
300 h	50	49.6	50.4	50

The microscopic state of Pt particles on the cathode side of MEA was analyzed with TEM in terms of the state before the test, and under no-vibration and vibration conditions after 300 h accelerated test, as shown in Figure 7a–c. From the obtained TEM images, we can see that after the static and vibrational tests, the crystal particles of Pt become larger and agglomerated, and then the overall uniformity of the particles are worse than that before the test. In order to confirm and quantify the particle size distribution, X-ray scattering measurements were executed to calculate the average particle size. As can be seen from X-ray diffraction patterns, the characteristic diffraction peaks of Pt at 2θ = 40°, 46°, 68° and 81°, respectively, belong to the (111), (200), (220) and (311) crystal planes. Before the experiments, the diffraction peaks of the catalyst showed a wide broadened full-width at half of maximum, indicating that the grain diameter of Pt is smaller compared with under vibration and no-vibration conditions. Due to the diffraction intensity of Pt peaks, (111), (200) and (311) are interfered by the carbon-10 and carbon-11 characteristic peaks, and the Pt (220) crystal plane was used to calculate the average grain size of the catalyst with Sherrer formula [30]:

$$d = \frac{0.9\lambda_{ka}}{B_{2\theta} \cos \theta_B} \tag{2}$$

where d represents the average grain diameter of the catalyst, λ_{ka} is 1.54056 Å, θ_B represents the diffraction angle of the Pt-220 crystal plane and $B_{2\theta}$ represents the full width at half of maximum.

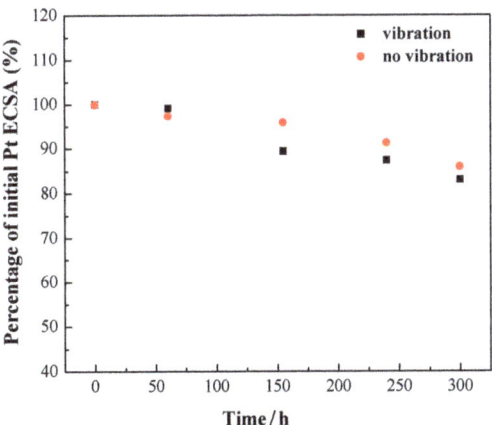

Figure 6. The variation of the effective active area of catalyst layer with running time.

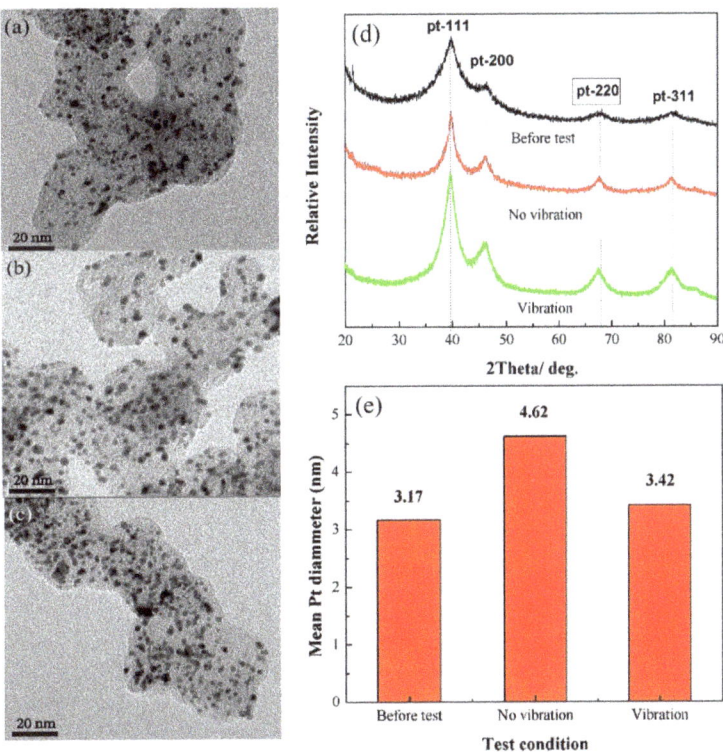

Figure 7. TEM micrographs of the carbon-supported Pt catalyst layer on the cathode side (**a**) before test, under (**b**) no-vibration and (**c**) vibration conditions after 300 h accelerated test. (**d**) X-ray diffraction patterns of Pt/C on the cathode side for the three cases; (**e**) the average particle size of Pt/C calculated from XRD analysis.

The average grain diameter of the catalyst is summarized in Figure 7e. The grain diameter before the test was 3.17 nm; after 300 h of operation under vibration and no-vibration conditions, the grain diameter of catalyst enhanced to 3.43 nm and 4.62 nm, respectively. It can be observed that the Pt particles on the cathode side have different extents of agglomeration after chronic operation under both cases in comparison to the state before the test. The catalyst agglomeration will bring about the drain in the ECSA, leading to degradation in the performance of PEMFC [15,31], but the grain diameter of a catalyst under vibration is smaller than that in a static environment, and so the Pt particles' agglomeration is not the only reason for the vehicular PEMFCs' performance degradation after chronic pavement vibration.

To further explain the attenuation mechanism of MEA in a vibration environment, EDX element analysis is performed on the cross-section of MEA at the end of 300 h accelerated tests in vibration and no-vibration environments. From Figure 8a,b, the cross-section of MEA shows that the interfaces of the catalyst layer to the membrane are very flat and without delamination under vibration and static conditions. Furthermore, it can be seen in Figure 8c,d that after a long time of operation, the polymer membrane which contains no platinum element shows the existence of platinum under both cases. This illustrates that the dissolved Pt ions from the cathode side transfer into the polymer membrane along the hydrophilic channels and redeposit in it, which also give rise to the drain of ECSA [32–34]. Meanwhile, much more platinum is detected inside the proton exchange membrane under vibration conditions than static conditions. The Pt catalyst loading of both cases before the tests are equivalent, so that more Pt particles drain from the catalyst layer, resulting in greater loss of the electrochemical active area and the performance of PEMFC.

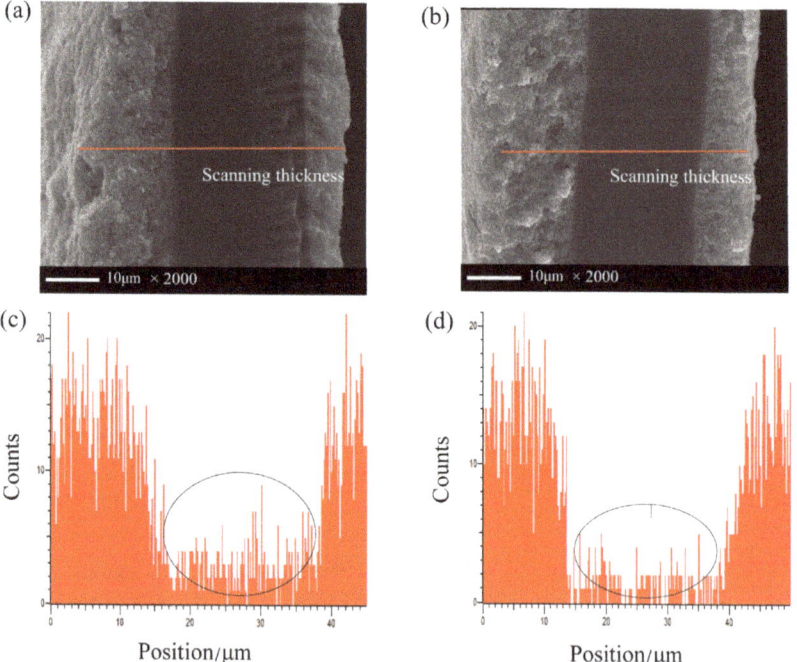

Figure 8. EDX element analysis of Pt after 300 h accelerated tests: (**a**) the cross-section of MEA and (**c**) EDX element analysis under vibration; (**b**) the cross-section of MEA and (**d**) EDX element analysis under no-vibration conditions.

4. Conclusions

In the present study, we demonstrate that the voltage degradation rate is 103 μV h^{-1} under vibration conditions during a 300 h accelerated test, which is higher than the 53 μV h^{-1} degradation rate under no-vibration conditions. The high frequency resistance under vibration conditions increases faster in the test time. This is by reason of the clamping screws of the PEMFC easily loosen under vibration conditions. In addition, according to TEM images and XRD patterns, the Pt particles on the cathode side show different extents of agglomeration after chronic operation under both cases, which lead to degrade in the performance of PEMFC. To explain why the performance and the electrochemical active area of the PEMFC under vibration operation remain a faster decrease than those of the PEMFC under static environment, EDX element analysis was performed. Much more platinum that dissolved from the catalyst layer and redeposited was detected inside the polymer membrane under vibration conditions, resulting in greater loss of the electrochemical active area and the performance of PEMFC.

Author Contributions: Conceptualization, S.C. and X.W.; methodology, S.C. and X.W.; software, S.C.; validation, S.C. and X.W.; formal analysis, X.W.; investigation, S.C.; resources, X.W.; data curation, S.C.; writing—original draft preparation, X.W.; writing—review and editing, S.C.; visualization, S.C.; supervision, T.Z.; project administration, T.Z.; funding acquisition, T.Z. All authors have read and agreed to the published version of the manuscript.

Funding: This work was supported by the National Key Research and Development Program of China (Grant No. 2020YFC1806402).

Data Availability Statement: The data presented in this study are available in this article.

Conflicts of Interest: The authors declare no conflict of interest.

References

1. Wang, X.; Ma, Y.; Gao, J.; Li, T.; Jiang, G.Z.; Sun, Z.Y. Review on water management methods for proton exchange membrane fuel cells. *Int. J. Hydrogen Energy* **2021**, *46*, 12206–12229. [CrossRef]
2. Xu, J.M.; Zhang, C.Z.; Wan, Z.M.; Chen, X.; Chan, S.H.; Tu, Z.K. Progress and perspectives of integrated thermal management systems in PEM fuel cell vehicles: A review. *Renew. Sust. Energ. Rev.* **2022**, *155*, 111908.1–111908.23. [CrossRef]
3. U.S. Department of Energy. Hydrogen and Fuel Cells Program Overview. Available online: http://www.eere.energy.gov/hydrogenandfuelcells/presentations.html (accessed on 20 December 2018).
4. Zhao, Y.; Mao, Y.; Zhang, W.; Tang, Y.; Wang, P. Reviews on the effects of contaminations and research methodologies for PEMFC. *Int. J. Hydrogen Energy* **2020**, *45*, 23174–23200. [CrossRef]
5. Cheng, X.; Shi, Z.; Glass, N.; Zhang, L.; Zhang, J.J.; Song, D. Poisoning of proton exchange membrane fuel cells by contaminants and impurities: Review of mechanisms, effects, and mitigation strategies. *J. Power Sources* **2019**, *427*, 21–48.
6. Wang, P.; Li, L.J. Cold start optimization of the proton-exchange membrane fuel cell by penetrating holes in the cathode micro-diffusion layer. *Int. J. Hydrogen Energy* **2022**, *47*, 36650–36658. [CrossRef]
7. Xie, X.; Zhang, G.; Zhou, J.; Jiao, K. Experimental and theoretical analysis of ionomer/carbon ratio effect on PEM fuel cell cold start operation. *Int. J. Hydrogen Energy* **2017**, *42*, 12521–12530. [CrossRef]
8. Chen, S.T.; Wang, S.B.; Wang, X.K.; Zhu, T. Microencapsulated Phase Change Material Suspension for Cold Start of PEMFC. *Materials* **2021**, *14*, 1514. [CrossRef]
9. Yang, X.K.; Sun, J.Q.; Sun, S.C.; Shao, Z.G. An efficient cold start strategy for proton exchange membrane fuel cell stacks. *J. Power Sources* **2022**, *542*, 231492–231503. [CrossRef]
10. Cheng, S.; Hu, D.H.; Hao, D. Investigation and analysis of proton exchange membrane fuel cell dynamic response characteristics on hydrogen consumption of fuel cell vehicle. *Int. J. Hydrogen Energy* **2022**, *47*, 15845–15864. [CrossRef]
11. Bajaj, A.; Liu, F.; Kulik, H.J. Uncovering Alternate Pathways to Nafion Membrane Degradation in Fuel Cells with First-Principles Modeling. *J. Phys. Chem. C* **2020**, *124*, 15094–15106. [CrossRef]
12. Robert, M.; Kaddouri, A.E.; Perrin, J.C.; Raya, J.; Lottin, O. Time-resolved monitoring of composite Nafion (TM) XL membrane degradation induced by Fenton's reaction. *J. Membr. Sci.* **2021**, *621*, 118977–118988. [CrossRef]
13. Madhav, D.; Shao, C.Y.; Mus, J. The Effect of Salty Environments on the Degradation Behavior and Mechanical Properties of Nafion Membranes. *Energies* **2023**, *16*, 2256. [CrossRef]
14. Fan, L.; Zhao, J.; Luo, X.; Tu, Z. Comparison of the performance and degradation mechanism of PEMFC with Pt/C and Pt black catalyst. *Int. J. Hydrogen Energy* **2022**, *47*, 5418–5428. [CrossRef]
15. Ohyagi, S.; Matsuda, T.; Iseki, Y.; Sasaki, T.; Kaito, C. Effects of operating conditions on durability of polymer electrolyte membrane fuel cell Pt cathode catalyst layer. *J. Power Sources* **2011**, *196*, 3743–3749. [CrossRef]

16. Kang, M.S.; Sim, J.; Min, K.Y.D. Analysis of surface and interior degradation of gas diffusion layer with accelerated stress tests for polymer electrolyte membrane fuel cell. *Int. J. Hydrogen Energy* **2022**, *47*, 29467–29480. [CrossRef]
17. Li, B.; Wan, K.C.; Xie, M.; Chu, T.K.; Zhang, C.M. Durability degradation mechanism and consistency analysis for proton exchange membrane fuel cell stack. *Appl. Energy* **2022**, *314*, 119020–119031. [CrossRef]
18. Liu, H.; George, M.G.; Ge, N.; Muirhead, D. Microporous Layer Degradation in Polymer Electrolyte Membrane Fuel Cells. *J. Electrochem. Soc.* **2018**, *165*, 3271–3279. [CrossRef]
19. Liu, H.; George, M.G.; Zeis, R.; Messerschmidt, M. The Impacts of Microporous Layer Degradation on Liquid Water Distributions in Polymer Electrolyte Membrane Fuel Cells Using Synchrotron Imaging. *ECS Trans.* **2017**, *80*, 155–164. [CrossRef]
20. Imen, S.J.; Shakeri, M. Reliability Evaluation of an Open-Cathode PEMFC at Operating State and Longtime Vibration by Mechanical Loads. *Fuel Cells* **2016**, *16*, 126–134. [CrossRef]
21. Rajalakshmi, N.; Pandian, S.; Dhathathreyan, K.S. Vibration tests on a PEM fuel cell stack usable in transportation application. *Int. J. Hydrogen Energy* **2009**, *34*, 3833–3837. [CrossRef]
22. Hou, Y.P.; Hao, D.; Shen, J.P.; Li, P.; Zhang, T.; Wang, H. Effect of strengthened road vibration on performance degradation of PEM fuel cell stack. *Int. J. Hydrogen Energy* **2016**, *41*, 5123–5134. [CrossRef]
23. Wu, C.W.; Liu, B.; Wei, M.Y.; Zhang, W. Mechanical response of a large fuel cell stack to impact: A numerical analysis. *Fuel Cells* **2015**, *15*, 344–351. [CrossRef]
24. Wang, X.K.; Wang, S.B.; Chen, S.T.; Zhu, T. Dynamic response of proton exchange membrane fuel cell under mechanical vibration. *Int. J. Hydrogen Energy* **2016**, *41*, 16287–16295. [CrossRef]
25. Diloyan, G.; Sobel, M.; Das, K.; Hutapea, P. Effect of mechanical vibration on platinum particle agglomeration and growth in Polymer Electrolyte Membrane Fuel Cell catalyst layers. *J. Power Sources* **2012**, *214*, 59–67. [CrossRef]
26. Yildirim, S.; Erkaya, S.; Eski, I.; Uzmay, I. Noise and Vibration Analysis of Car Engines using Proposed Neural Network. *J. Vib. Control* **2009**, *15*, 133–156. [CrossRef]
27. Hou, Y.; Zhou, B.; Zhou, W.; Shen, C.; He, Y. An investigation of characteristic parameter variations of the polarization curve of a proton exchange membrane fuel cell stack under strengthened road vibrating conditions. *Int. J. Hydrogen Energy* **2012**, *37*, 11887–11893. [CrossRef]
28. Hou, J.B.; Yu, H.M.; Zhang, S.S.; Sun, S.C.; Wang, H.W.; Yi, B.L.; Ming, P.W. Analysis of PEMFC freeze degradation at $-20\,^{\circ}$C after gas purging. *J. Power Sources* **2006**, *162*, 513–520. [CrossRef]
29. Ralph, T.R.; Hards, G.A.; Keating, J.E.; Campbell, S.A.; Wilkinson, D.P. ChemInform Abstract: Low Cost Electrodes for Proton Exchange Membrane Fuel Cells Performance in Single Cells and Ballard Stacks. *J. Electrochem. Soc.* **1998**, *144*, 3845–3857. [CrossRef]
30. Radmilovic, V.; Gasteiger, H.A.; Ross, P.N. Structure and Chemical Composition of a Supported Pt-Ru Electrocatalyst for Methanol Oxidation. *J. Catal.* **1995**, *154*, 98–106. [CrossRef]
31. Zhai, Y.F.; Zhang, H.M.; Xing, D.M.; Shao, Z.G. The stability of Pt/C catalyst in H_3PO_4/PBI PEMFC during high temperature life test. *J. Power Sources* **2007**, *164*, 126–133. [CrossRef]
32. Akita, T.; Taniguchi, A.; Maekawa, J.; Sirorna, Z.; Tanaka, K.; Kohyama, M. Analytical TEM study of Pt particle deposition in the proton-exchange membrane of a membrane-electrode-assembly. *J. Power Sources* **2006**, *159*, 461–467. [CrossRef]
33. Shao, Y.Y.; Yin, G.P.; Gao, Y.Z. Understanding and approaches for the durability issues of Pt-based catalysts for PEM fuel cell. *J. Power Sources* **2007**, *171*, 558–566. [CrossRef]
34. Ettingshausen, F.; Kleemann, J.; Marcu, A.; Toth, G.; Fuess, H.; Roth, C. Dissolution and Migration of Platinum in PEMFCs Investigated for Start/Stop Cycling and High Potential Degradation. *Fuel Cells* **2011**, *11*, 238–245. [CrossRef]

nanomaterials

MDPI

Article

Hollow Porous CoO@Reduced Graphene Oxide Self-Supporting Flexible Membrane for High Performance Lithium-Ion Storage

Junxuan Zhang [1,†], Jie You [2,†], Qing Wei [2], Jeong-In Han [1,*] and Zhiming Liu [2,3,*]

1. Flexible Display and Printed Electronics Laboratory, Department of Chemical and Biochemical Engineering, Dongguk University-Seoul, Seoul 04620, Republic of Korea; zjx950518@gmail.com
2. Shandong Engineering Laboratory for Preparation and Application of High-Performance Carbon-Materials, College of Electromechanical Engineering, Qingdao University of Science & Technology, Qingdao 266061, China; yj1048154827@163.com (J.Y.); qing3303@163.com (Q.W.)
3. Qingdao Institute of Bioenergy and Bioprocess Technology, Qingdao Industrial Energy Storage Research Institute, Chinese Academy of Sciences, Qingdao 266101, China
* Correspondence: hanji@dongguk.edu (J.-I.H.); zmliu@qust.edu.cn (Z.L.)
† These authors contributed equally to this work.

Abstract: We report an environment-friendly preparation method of rGO-based flexible self-supporting membrane electrodes, combining Co-MOF with graphene oxide and quickly preparing a hollow CoO@rGO flexible self-supporting membrane composite with a porous structure. This unique hollow porous structure can shorten the ion transport path and provide more active sites for lithium ions. The high conductivity of reduced graphene oxide further facilitates the rapid charge transfer and provides sufficient buffer space for the hollow Co-MOF nanocubes during the charging process. We evaluated its electrochemical performance in a coin cell, which showed good rate capability and cycling stability. The CoO@rGO flexible electrode maintains a high specific capacity of 1103 mAh g^{-1} after 600 cycles at 1.0 A g^{-1}. The high capacity of prepared material is attributed to the synergistic effect of the hollow porous structure and the 3D reduced graphene oxide network. This would be considered a promising new strategy for synthesizing hollow porous-structured rGO-based self-supported flexible electrodes.

Keywords: CoO@rGO; graphene; MOF; flexible electrodes; hollow structure; ultrafast integration; lithium-ion batteries

Citation: Zhang, J.; You, J.; Wei, Q.; Han, J.-I.; Liu, Z. Hollow Porous CoO@Reduced Graphene Oxide Self-Supporting Flexible Membrane for High Performance Lithium-Ion Storage. *Nanomaterials* **2023**, *13*, 1986. https://doi.org/10.3390/nano13131986

Academic Editor: Carlos Miguel Costa

Received: 6 June 2023
Revised: 27 June 2023
Accepted: 29 June 2023
Published: 30 June 2023

1. Introduction

Due to the increasing demand for higher capacity, higher power density, longer cycle life of energy storage devices for portable electronic devices, wearable electronic devices, and energy-consuming devices such as electric vehicles, research on new lithium-ion batteries with higher performance and more excellent safety has become more and more urgent [1–4]. Lithium-ion batteries have significant advantages and have become the first choice for the modern day owing to their high energy density, high cycle life, and high efficiency [5–7]. The low theoretical capacity of conventional graphite anode materials contradicts the need for higher-capacity Li-ion batteries [8–10].

Transition metal oxides (TMOs), the most promising active anode materials for the new generation of Li-ion batteries, have received much attention from researchers due to their high specific capacity and exemplary safety [11,12]. Among them, CoO has been further investigated due to its high theoretical capacity (716 mAh g^{-1}), relatively low cost, and fully reversible electrochemical reaction. However, transition metal oxides, including CoO, are conversion-type anode materials that generate large volume expansion during cycling, leading to the chalking of active material particles and greatly reduced cycle life. Thus, the practical application of metal compounds for lithium-ion batteries is severely limited [13–16]. Building nanostructures is an effective solution to improve their performance [17,18]. Metal–organic backbone (MOF) derivatives can effectively maintain the morphology of MOF precursors due to their tunability at the molecular level and unique porous backbone structure [19,20]. Therefore, MOF is widely used as a template [21,22], and TMO nanostructures such as CoO can be fabricated by a simple process [23]. However, the low electrical conductivity and slow reaction kinetics of CoO nanomaterials still limit their performance in lithium-ion battery electrode materials.

Graphene is often used as an ideal substrate material due to its high electrical conductivity, huge specific surface area, and excellent physical and mechanical flexibility [24–26]. The composite of graphene with TMOs is considered a reasonable and effective solution at present. On the other hand, it can also effectively alleviate the chalking of the active material caused by the volume expansion during cycling, thus greatly improving the battery cycling stability [27]. To date, various composites of TMOs with graphene, including Fe_3O_4 nanoflakes/RGO composites [28], MnO_2/reduced Graphene Oxide Nanosheet [29], hollow Fe_3O_4/graphene hybrid films [30], and Co_3O_4@rGO nanocomposites [31], etc., have been successfully reported for lithium-ion battery electrodes. However, these reported TMOs/graphene composites usually have more preparation steps, high preparation cost, and complicated processing, which seriously hinder their industrial application in Li-ion batteries.

Herein, we designed and synthesized a simple, environmentally friendly, cost-effective CoO@rGO flexible membrane electrode. Co-MOF is known as a metal–organic skeleton and its derivatives can effectively maintain its intrinsic morphology, and its composite with a three-dimensional (3D) graphene network has also exhibited excellent electrochemical properties. In addition, it has been shown that ammonia has an etching effect on some MOFs, which can be etched into hollow materials [28], and at the same time, graphene oxide is rapidly reduced. The hollow material is more favorable for lithium ion transport and storage, so the composite of hollow structured CoO with graphene oxide would be an excellent self-supporting anode material.

2. Material and Methods

2.1. Synthesis of Co-MOF

In the typical synthesis of Co-MOF, 0.6 mmol cobalt acetate tetrahydrate and 0.9 mmol sodium citrate were dissolved in 20 mL of deionized (DI) water to obtain solution A. In addition, 0.4 mmol potassium cobalt cyanide was dissolved in 20 mL of deionized water to obtain solution B. The solution A and solution B were then quickly mixed together and stirred for 12 h [32]. The precipitated product was collected by centrifugation and washed three times with deionized water and alcohol, followed by vacuum drying overnight at 70 °C to obtain the Co-MOF precursor material.

2.2. Synthesis of Co-MOF@rGO and CoO@rGO Flexible Membranes

A total of 20 mg of GO and 60 mg of Co-MOF precursor were dissolved into 5 mL DI water. The precursor solution was mixed well by stirring for 10 min, poured into disposable Petri dishes, and then freeze-dried to obtain Co-MOF@GO membrane material. Then, hot $(NH_4)_2S$ solution was added dropwise to the above membrane for rapid reduction. After 1 min, the excess $(NH_4)_2S$ was quickly washed off with deionized water and freeze-dried to obtain Co-MOF@rGO flexible membrane finally.

The above-obtained Co-MOF@rGO flexible film was heat treated at 550 °C for 2 h under Ar atmosphere with a heating rate of 2 °C min^{-1}, and the CoO@rGO flexible film could be directly used as an electrode after natural cooling.

2.3. Material Characterization

The morphology of the products was characterized by field emission scanning electron microscopy (SU8010, HITACHI, Japan), transmission electron microscopy (TEM), and high-resolution TEM (HRTEM). Energy dispersive X-ray spectroscopy (EDX) analysis and corresponding elemental mapping were performed using an X-ray spectrometer attached to the TEM instrument. The crystalline phases of the products were collected by X-ray diffraction (MiniFlex600, Rigaku, Japan) using monochromatic Cu-Kα lines as a radiation source. The thermal stability of the samples in the air was evaluated by thermogravimetric analysis (209-F3, NETZSCH, Germany) in the temperature range of 40–800 °C, with a ramp rate of 10 °C min^{-1}. Raman spectra were obtained by testing on a Raman spectrometer (1PQG99, Renishaw, Korea) at a laser wavelength of 532 nm.

2.4. Electrochemical Measurements

CoO@rGO and rGO flexible films were cut into small 10 mm diameter discs and used directly for an electrochemical evaluation. As a comparison, CoO electrodes were prepared by mixing 80 wt% of active electrode material (CoO), 10 wt% of carbon black, and 10 wt% of polyvinylidene fluoride binder to prepare a slurry. Then, the slurry was uniformly coated on the copper foil and dried overnight in a vacuum oven at 60 °C. The CR2032 half-cells were assembled in an argon-filled glove box by cutting into 10 mm-diameter disc electrodes by a button cell slicer, using lithium sheets as counter electrodes and a mixture of vinyl carbonate (EC) and diethyl carbonate (DEC) with 5% fluoroethylene carbonate (FEC) as the electrolyte. Cycling performance and rate performance tests were performed on a cell test system (LANHE, CT2001A, Wuhan, China) with a voltage range of 0.01–3.0 V. Cyclic voltammetry tests with a voltage window of 0.01–3.0 V (vs. Li$^+$/Li, 0.1 mV s^{-1}) were performed on an electrochemical workstation (CHI, 760E, Shanghai, China).

3. Results and Discussion

3.1. Characterization

As shown in Scheme 1, the Co-MOF nanocubes (Figure S1) prepared in advance were co-dispersed with GO in deionized water and later freeze-dried to form a film directly. Then, using the strong reduction effect of (NH$_4$)$_2$S, GO was rapidly reduced to rGO, and the Co-MOF@rGO flexible film was obtained. It is noteworthy that Co-MOF is a tunable and unique porous backbone structure at the molecular level, and in addition, it can be etched by NH$_4$$^+$ to form a hollow structure. This is because ammonia will coordinate with cobalt cations to form complex ions, which cause the etching of cobalt [33]. Therefore, while the three-dimensional reduced graphene oxide network improves the electrical conductivity of the material, its hollow structure further shortens the ion transport path; thereby, the ion transport efficiency can be effectively improved. Meanwhile, the buffering effect of the 3D reduced graphene oxide network can effectively adapt to the volume change during the cycling process. Thus, it will be a promising new anode material for energy storage devices.

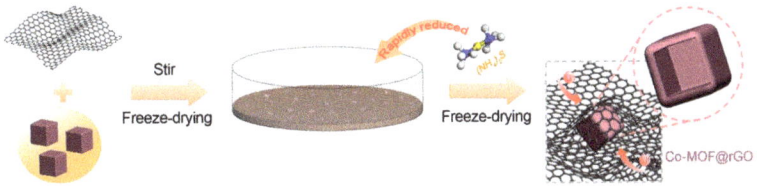

Scheme 1. Schematic illustration for the synthesis and construction of the Co-MOF@rGO flexible film.

The flexible membrane electrodes prepared by this promising synthetic strategy are shown in Figures 1a and S2, which can well meet the requirements of bendability and wearability. Figure S1 shows the SEM images of the prepared Co-MOF nanocubes, and all the cubes show smooth surface structures with a size of about 1 μm. The SEM images of the Co-MOF@GO composites were obtained after compounding with graphene oxide and are shown in Figure 1b,c. The Co-MOF nanocubes are uniformly wrapped in a 3D network of graphene oxide, showing a good 3D buffer structure. In addition, it can be seen that the Co-MOF@GO composite keeps its initial morphology intact and is successfully compounded with graphene oxide before being etched by ammonium sulfide.

Figure 1. (a) Digital images of CoO@rGO films; SEM images of (**b,c**) Co-MOF@GO composites, (**d–g**) etched Co-MOF@rGO composites, (**h,i**) CoO@rGO flexible film.

When ammonium sulfide is added, GO is rapidly reduced to rGO, as shown in Figure 1d,e, and the original oxygen-containing functional groups in graphene oxide are destroyed, thus contracting and cross-linking each other and tightly wrapping the Co-MOF nanocubes. It is noteworthy that the Co-MOF in the obtained Co-MOF@rGO flexible composite film produced a slight deformation due to the NH_4^+ in ammonium sulfide, which simultaneously had a strong etching effect on the Co-MOF nanocubes; it was etched into a hollow structure in a very short period of time. As can be seen from the SEM images at high magnification in Figure 1e, the etched Co-MOF nanocubes exhibit a different contrast around the etched area from the central position, indicating that they may have been etched into hollow structures. This judgment is confirmed by the subsequent SEM of individual broken Co-MOF nanocubes, as shown in Figure 1f,g. It can be seen from the broken Co-MOF nanocubes that the Co-MOF nanocubes are hollow inside, leaving a layer of outer wall. This proves the strong etching effect of ammonium sulfide on Co-MOF nanocubes and illustrates the successful synthesis of hollow Co-MOF nanocubes with rGO-based flexible composites.

The Co-MOF@rGO flexible composite film was heat-treated, and the Co-MOF was heat-treated to obtain its corresponding derivatives, which were then used as electrode materials. The SEM images of the obtained CoO@rGO flexible composite films are shown in Figures 1h,i and S3, from which it can be seen that the CoO nanocubes derived from the Co-MOF nanocubes obtained after heat treatment still maintain the complete cubic structure with a size of about 1 μm. Notably, the CoO nanocubes obtained after heat treatment have a hazy, gauzy texture and exhibit a more sparse and porous surface structure. This may be attributed to the porous structure caused by the escape of gases generated during the heat treatment process. This porous structure can further reduce the ion transport path and promote rapid lithium-ion transport, which is conducive to better lithium-ion insertion and extraction, and synergize with the high electrical conductivity of the 3D reduced graphene oxide network and the excellent buffer structure, thus effectively improving the cycling stability while also effectively improving the composite multiplicity performance. Figure S4 shows the EDS analysis of the CoO@rGO flexible composite film. As expected, the Co and O elements in the obtained CoO@rGO flexible composite film are uniformly distributed and correspond to the CoO nanocubes. At the same time, the C elements corresponding to the reduced graphene oxide around the CoO nanocubes were also uniformly distributed around the CoO nanocubes, and they wrapped the CoO nanocubes tightly. In addition, the S and N heteroatom doping introduced by adding ammonium sulfide is also present in the heat-treated CoO nanocubes, further demonstrating the successful composite of CoO@rGO flexible composite film materials.

The internal structure of Co-MOF nanocubes in the Co-MOF@rGO flexible composite film was further investigated by TEM analysis, as shown in Figure 2a,b. It is obvious that the Co-MOF nanocubes etched by ammonium sulfide exhibit a distinct hollow structure with an empty internal structure and an outer wall thickness of about 155 nm (Figure 2c). Figure S6 shows the tight junctions between the rGO and CoO crystals, and the 0.24 nm lattice stripe labeled in the figure is attributed to the CoO(111) crystal plane. And the elemental distribution in the Co-MOF@rGO flexible composite film was analyzed in Figure 2d, from which the hollow structure of the hollow Co-MOF nanocubes could be seen, (e), (f), (g), (h), (i) are the elemental mapping of C, Co, O, S and N, respectively.

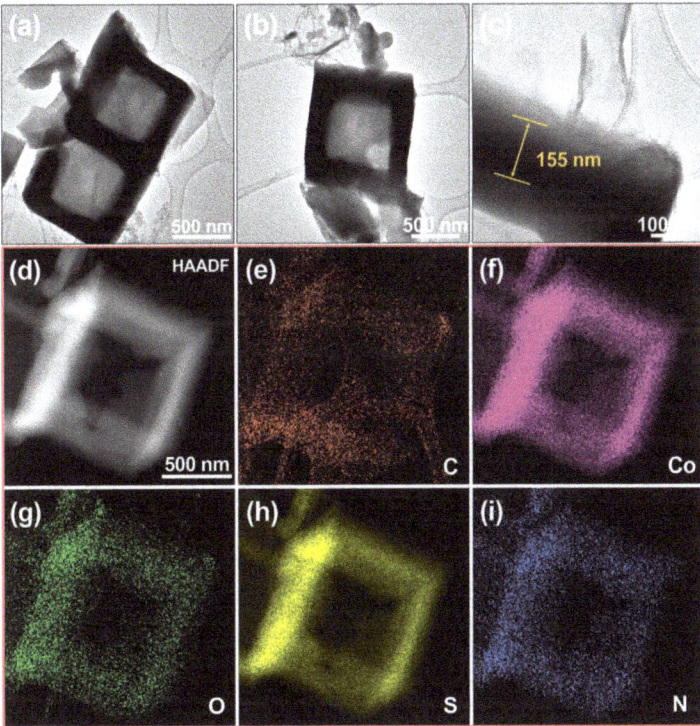

Figure 2. (**a–c**) TEM image of Co-MOF@rGO, (**d**) HRTEM images of EDX mapping of hollow Co-MOF@rGO flexible films, (**e–i**) are the elemental mapping of C, Co, O, S and N, respectively.

The elemental distribution shows that Co and O elements are uniformly distributed on the outer wall of the hollow Co-MOF nanocubes, and C elements can also be seen around the hollow Co-MOF nanocubes due to the peripheral wrapping of the reduced graphene oxide. In addition, the introduction of S and N elemental heteroatoms can also be seen on the elemental distribution map due to the reduction and etching of ammonium sulfide, which is beneficial for the storage of lithium ions. From this elemental mapping analysis, it can be well illustrated that the Co-MOF@rGO flexible composite film material is successfully compounded.

To further determine the composition of the hollow porous CoO@rGO flexible composite films, the resulting composites were characterized by XRD. As shown in Figure 3a, all major diffraction peaks of the hollow porous CoO@rGO flexible composite film can be successfully indexed to the corresponding standard card (JCPDS 48-1719). The hollow porous CoO@rGO flexible composite membrane shows very distinct diffraction peaks at 36.74, 42.61, 61.62, 73.97, and 77.65°, corresponding to the (111), (200), (220), (311), and (222) crystal planes of cobalt oxide, respectively. The broad diffraction peaks centered at $2\theta \approx 26°$ in the diffraction pattern of the hollow porous CoO@rGO flexible composite film is be attributed to the typical (002) crystal plane of rGO [34], and the absence of other impurity peaks is a good proof of the successful synthesis of the hollow porous CoO@rGO flexible composite film.

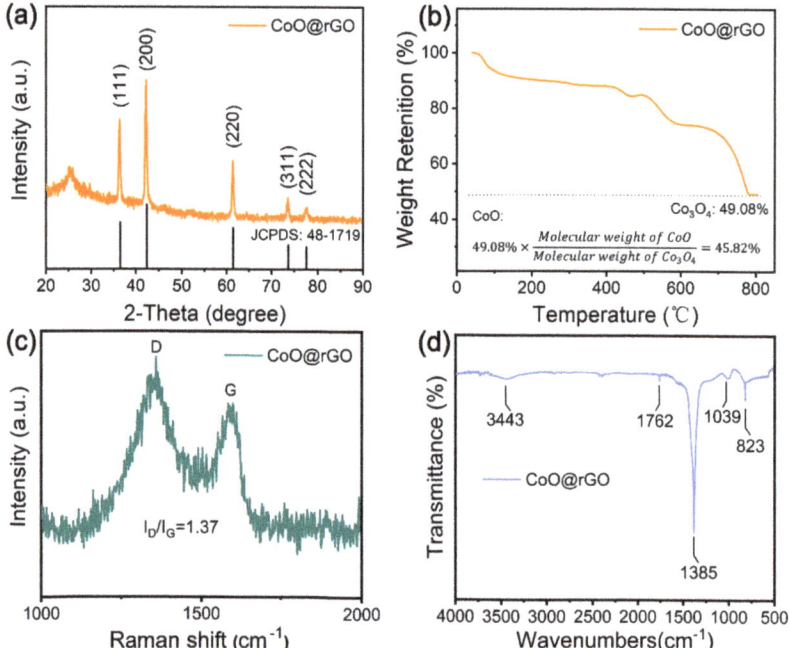

Figure 3. (**a**) XRD patterns; (**b**) TG curve; (**c**) Raman spectra; (**d**) FT-IR spectra of the CoO@rGO flexible film.

In order to obtain the carbon content in the hollow porous CoO@rGO flexible composite film, thermogravimetric analysis was also performed and represented in Figure 3b, and the mass content of rGO in the hollow porous CoO@rGO flexible composite film can be calculated from the thermogravimetric analysis results. During the heating process from 40–800 °C under air atmosphere, carbon is gradually decomposed to form carbon dioxide and expelled, while CoO nanocubes are gradually oxidized to Co_3O_4 when heated in the air [35]. It can be seen that at 800 °C, 49.08 wt% content of Co_3O_4 is finally obtained, so it can be calculated that the hollow porous CoO@rGO flexible composite film has a mass content of CoO of about 45.82%, which leads to the further calculation that the mass content of rGO is about 54.18%.

As shown in Figure 3c, the Raman spectra of CoO@rGO 3D mesh composites have two main peaks at 1349 cm^{-1} and 1574 cm^{-1} for the D-band of the A_{1g} vibrational mode used for disordered carbon and the G-band of the E_{2g} vibrational mode for ordered graphitic carbon. The D to G band's intensity ratio (I_D/I_G) usually reflects the degree of defects and disorder in carbon materials [36]. The intensity ratio (I_D/I_G) of the D-band to the G-band of the CoO@rGO composite is about 1.37, indicating that there are many defects in the CoO@rGO 3D mesh composite. Compared with the Raman spectrum of the rGO flexible film (Figure S5), the CoO@rGO flexible composite film has a stronger D-band intensity, indicating more defects or disorder sites in the heterogeneous structure, which helps to provide more active regions for lithium storage. Figure 3d shows the FTIR spectra of CoO@rGO, which exhibits characteristic stretching frequencies at 823, 1039, 1385, 1762, and 3343 cm^{-1}, corresponding to C=C, Co-O-Co, S-O, C=O, and O-H functional groups, respectively. One of the peaks at 1385 cm^{-1} represents the S-O stretching vibration of the sulfate group, which can be attributed to the addition of ammonium sulfide etching in the Co-MOF@GO composite. The above analysis confirms the successful preparation of CoO@rGO 3D mesh composites.

As shown in Figure 4a,b, the adsorption–desorption experiments of N2 were performed on co-MOF@rGO before and after etching, and the pore size distribution was calculated by the BJH method. The type IV isotherm plot with a distinct characteristic hysteresis loop after etching (Figure 4b) indicates the dominant number of mesopores compared to the unetched sample. The obtained specific surface area is 22.4 m^2 g^{-1}, and the pore size distribution is in the range of 2–20 nm. It is noteworthy that the specific surface area of the etched composites increases compared to the pre-etching composites possessing a specific surface area of 11.7 m^2 g^{-1}. This is due to the coupling effect between the N,S doping introduced during the etching process and the hollow structure produced by the etching. The large specific surface area helps to improve the utilization of the active electrode material, while the abundant mesopores of the material facilitate the rapid diffusion of lithium ions and increase the lithium storage sites. X-ray photoemission spectroscopy (XPS) measurements show the surface chemistry of CoO@rGO flexible films. The XPS spectra in Figure 4c confirm the presence of C, CO, O, N, and S elements in the CoO@rGO flexible film. The low content of N and S elements can be attributed to the introduction of heteroatoms caused by the reduction of $(NH_4)_2S$ during the etching process. The introduction of N and S into the carbon structure can add more active sites, thus increasing the lithium storage capacity and providing high specific capacity and rate performance for Li-ion batteries. As shown in the C 1s spectrum (Figure 4d), the XPS C 1s spectrum of CoO@rGO flexible film can be divided into two main peaks, where 284.76 eV corresponds to the C=C/C–C bond and 286.32 eV corresponds to the C–O bond. The peak of C 1s is attributed to rGO. The stronger peak with a binding energy of 284.76 indicates the deoxygenation process accompanying rGO reduction in CoO@rGO flexible films, which is consistent with the previously reported results [32]. In the fitted high-resolution Co 2p spectra (Figure 4e), the two peaks located at 781.26 and 797.05 eV can be attributed to Co^{2+} with Co 2p$_{3/2}$ and Co2p$_{1/2}$, which indicates the presence of Co^{2+} in the CoO@rGO flexible film, consistent with its theoretical chemical state. The high-resolution O1spectrum (Figure 4f) can be decomposed into three peaks corresponding to Co–OH (533.68 eV), Co–O–C (531.83 eV), and Co–O (530.29 eV), respectively. One of them at the binding energy of 531.83 eV corresponds to the Co–O–C bond of the oxygen-containing functional group on the surface of rGO. This result indicates that CoO and rGO successfully hybridize through Co–O–C, which provides a strong binding strength at the mutual interface of CoO and rGO [37].

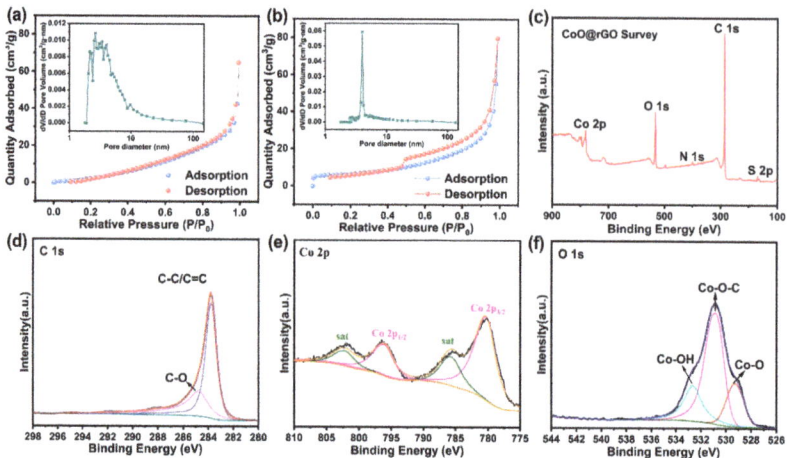

Figure 4. Nitrogen adsorption–desorption isotherms of (**a**) Co-MOF@rGO before etching and (**b**) Co-MOF@rGO after etching; (**c**) XPS survey spectrum and corresponding (**d**) C 1s, (**e**) Co 2p, and (**f**) O 1s XPS spectra of the CoO@rGO flexible film.

3.2. Electrochemical Performances

Our designed CoO@rGO flexible film has an excellent three-dimensional network that facilitates overcoming significant volume variations, promotes rapid lithium-ion transport, and provides reasonable rate capability and cycling performance for the assembled Li-ion batteries. More importantly, benefiting from its flexible self-supporting feature, the CoO@rGO flexible film can be quickly prepared and directly utilized as a stand-alone negative electrode for Li-ion batteries and avoids using conductive binders. We have assembled CoO@rGO flexible films into half cells and evaluated their electrochemical performance. Figure 5a shows the charge and discharge curves at a current density of 1 A g^{-1} with a voltage window of 0.01–3.0 V. It is clear that the first turn discharge and charge capacities of the CoO@rGO flexible membrane electrode are about 1020 mA h g^{-1} and 722 mA h g^{-1}, respectively, thus corresponding to an initial coulombic efficiency (ICE) of 72%. The irreversible capacity loss can be attributed to the formation of the SEI membrane, some undecomposed Li$_2$O, and the irreversible decomposition of the electrolyte [38]. The rate performance of the CoO@rGO flexible-film electrode was evaluated by gradually increasing the current density from 0.1 to 2.0 A g^{-1} (Figure 5b). It can be seen that the CoO@rGO flexible film exhibits the best rate of electrochemical performance compared to the comparison sample. Reversible capacities of 556, 437, 373, 336, 320, 307, and 251 mA h g^{-1} were obtained at 0.1, 0.2, 0.4, 0.6, 0.8, 1.0, and 2.0 A g^{-1}, respectively. Moreover, the specific capacity can be easily recovered to 509 mA h g^{-1} when the current density is reduced to 0.1 A g^{-1}, proving its excellent rate performance. In contrast, the comparison electrodes CoO and pure rGO films exhibit relatively low specific capacities and do not recover their initial specific capacities well when the current density is restored to the initial current.

To explain the excellent kinetic origin of CoO@rGO, CV measurements at different scan rates were used to evaluate the kinetic reactions (Figure 5c). In the CV curve, the scan voltage from negative to positive can be seen as the anodic oxidation process, which corresponds to the oxidation peak, and vice versa as the cathodic reduction process, which corresponds to the reduction peak. It can be seen that when the scan rate is at 0.1 mVs^{-1}, the cathodic and anodic peaks in the CV curves are symmetrical in shape, and the peak heights are basically the same, indicating that the electrochemical reactions occurring at the CoO@rGO flexible-film electrode/electrolyte interface have good reversibility. With the increase in scanning rate, all CV curves exhibit similar shapes and well-preserved redox peaks, with peak current (i) increasing with scan rate (v). The relationship between peak current and scan rate is i = av^b, where a and b are parameters, and b is worth calculating by plotting a graph and fitting a straight line [39–41]. Determining the value of b allows a qualitative analysis of the charge storage mechanism. When the value of slope (b) is close to 0.5, the storage behavior is dominated by diffusion-type reactions; when it is close to 1.0, it tends to capacitively controlled processes [42]. The b-values of the anodic and cathodic peaks of the CoO@rGO flexible film are 0.84 and 0.82 (Figure 5d), suggesting that it is dominated by capacitance-controlled behavior. Specifically, the ratio between diffusion-controlled and capacitive-controlled can be further quantified with a relationship given by the Equation [43–45]:

$$i = k_1 \, v + k_2 \, v^{0.5}.$$

With this equation, the ratio of pseudocapacitive charge storage can be obtained for different scan rates. Figure 5e shows the capacitive contribution (68%) to the entire scan area at 0.2 mV s^{-1}. As shown in Figure 5f, when the CoO@rGO flexible film was tested at scan rates of 0.2, 0.4, 0.6, 0.8, and 1 mV s^{-1}, the contribution of capacitive behavior was 68%, 72%, 77%, 83%, and 93%, respectively. It can be seen that the contribution of capacitance increases with increasing scan rate. The high pseudo-capacitance rate may be due to the synergistic effect of the hollow structure formed by Co-MOF after etching by ammonium sulfide and after compounding with reduced graphene oxide, which provides more active sites for Li intercalation/delamination as well as lithium storage.

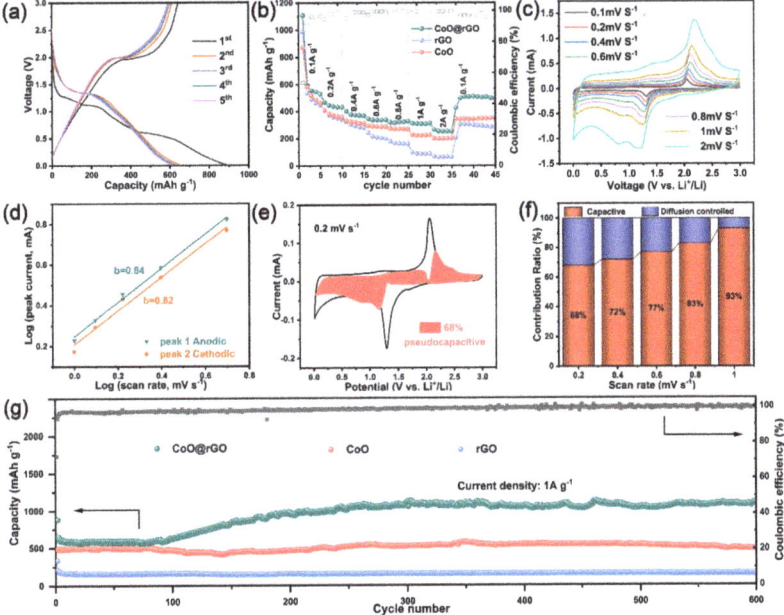

Figure 5. Electrochemical properties of the CoO@rGO flexible film for LIBs: (**a**) discharge/charge curves at a current density of 1 A g^{-1}; (**b**) rate properties; (**c**) CV curve at different sweep speeds; (**d**) fitting curves of the b-values; (**e**) the capacitive contribution to charge storage at a scan rate of 0.2 mV s^{-1}; (**f**) the capacitive and diffusion contributions at different scan rates; (**g**) cycling properties at 1.0 A g^{-1}.

Figure 5g shows the cycling performance of the CoO@rGO composite electrode compared to the comparison electrode CoO and pure rGO films at a current density of 1 A g^{-1}. It can be seen that the CoO@rGO composite electrode exhibits the most excellent cycling performance compared to the comparison electrode. The CoO@rGO composite electrode shows a trend of consistently increasing capacity over one hundred and fifty cycles dur-

Nanomaterials **2023**, *13*, 1986

ing the entire charge/discharge process with the voltage window of 0.01–3.0 V at 1 A g^{-1} current density, which is common for various nanostructured metal oxide electrodes whose capacity increases with cycling [46]. Activation of the electrode material during charge/discharge is one of the main reasons for this phenomenon, and the gradual increase in capacity may be due to the reversible growth of the polymer gel-like film caused by the degradation of the dynamically active electrolyte [47,48]. In addition, the hollow structure of CoO effectively restrains the volume expansion during cycling, while the rGO sheet layer provides an excellent buffering effect. After 600 cycles, the CoO@rGO composite electrode exhibits an impressive discharge capacity of 1103 mA h g^{-1}. In contrast, the discharge capacities of CoO and pure rGO electrodes are only 483 and 155 mA h g^{-1} (Table S1). This is significantly better than many other reported cobalt-based anode materials for lithium-ion batteries (Table S2). In addition, from three cycles onwards, for the CoO@rGO composite electrode, around 97% coulombic efficiency remains stable, showing its excellent reversible Li$^+$ insertion extraction performance. This fully reflects the important role of the hollow structure of the CoO and rGO three-dimensional conductive network for the excellent cycling performance of the CoO@rGO flexible film self-supported electrode.

4. Conclusions

In summary, we designed and synthesized a flexible self-supporting film electrode with hollow porous CoO nanocubes compounded with rGO. The CoO@rGO flexible-film composite electrode prepared using this method has many advantages: it can be prepared faster and more economically, on a large scale, with a simple process flow, which is beneficial for its industrial application in lithium-ion batteries, and the method is self-forming without adding conductive agents and binders, which can effectively improve the overall energy density of the electrode. The hollow and porous structure can effectively shorten the ion transport path and provide more active sites for lithium ions. In addition, the large-sized rGO sheet layer not only improves the overall conductivity of the material and promotes rapid charge transfer but also provides sufficient buffer space for the hollow Co-MOF nanocubes. Due to the synergistic effect of this hollow porous structure and the three-dimensional reduced graphene oxide network, the CoO@rGO flexible-film electrode shows extremely excellent electrochemical performance. We investigated the electrochemical performance of CoO@rGO flexible film as an anode material for Li-ion batteries, showing good cycling stability (1103 mA h g^{-1} after 600 cycles at 1.0 A g^{-1}). Even at a high current density of 5.0 A g^{-1}, a reversible capacity of 586 mA h g^{-1} was maintained, which is significantly better than many previously reported cobalt-based electrode materials. The CoO@rGO flexible-film composite electrode is expected to be an ideal next-generation anode candidate for industrial applications of lithium-ion batteries.

Supplementary Materials: The following supporting information can be downloaded at: https://www.mdpi.com/article/10.3390/nano13131986/s1, Figure S1: SEM images of Co-MOF nanocubes at different magnifications; Figure S2: Digital images of CoO@rGO films; Figure S3: SEM images of hollow porous CoO@rGO flexible film; Figure S4: EDS analysis of the hollow porous CoO@rGO flexible film; Figure S5: Raman spectra of rGo flexible film; Figure S6: HRTEM image of Co-MOF@rGO; Figure S7: Cycling performance at 5 A g^{-1} of CoO@rGO flexible film for LIBs; Table S1: Comparison of electrochemical performances for this work; Table S2: Test conditions and electrochemical performances of Co-based LIBs anodes. References [49–60] are cited in the supplementary materials.

Author Contributions: Methodology, J.Z., J.Y., Q.W. and Z.L.; Data curation, J.Z., J.Y. and Q.W.; Writing—original draft, J.Z. and J.Y.; Writing—review & editing, Z.L.; Supervision, J.-I.H. and Z.L.; Funding acquisition, Z.L. All authors have read and agreed to the published version of the manuscript.

Funding: This work was funded by the National Natural Science Foundation of China (Grant Nos. 21905152 and 52176076), the Taishan Scholar Project of Shandong Province of China (Grant Nos. tsqn202211160 and ts20190937), the Youth Innovation Team Project for Talent Introduction and Cultivation in Universities of Shandong Province, the China Postdoctoral Science Foundation (Grant Nos. 2022M713249), the Technology Innovation Program (20015786), and Development and

Demonstration for Improving Performance and Reliability of Core Components in Hydrogen Refueling Station) funded by the Ministry of Trade, Industry and Energy (MOTIE, Korea) and human resources development project of the Korea Institute of Energy Technology Evaluation and Planning (KETEP) for the support grant funded by the Korea government Ministry of Trade, Industry and Energy under the project titled: "Middle market enterprise specialized human resources development for residential and commercial fuel cell", numbered: 20224000000580.

Data Availability Statement: The data presented in this study is available on request from the corresponding authors.

Conflicts of Interest: The authors declare that they have no known competing financial interests or personal relationships that could have appeared to influence the work reported in this paper.

References

1. Tu, C.; Peng, A.; Zhang, Z.; Qi, X.; Zhang, D.; Wang, M.; Huang, Y.; Yang, Z. Surface-Seeding Secondary Growth for CoO@Co$_9$S$_8$ P-N Heterojunction Hollow Nanocube Encapsulated into Graphene as Superior Anode toward Lithium Ion Storage. *Chem. Eng. J.* **2021**, *425*, 130648. [CrossRef]
2. Nayak, P.K.; Yang, L.; Brehm, W.; Adelhelm, P. From Lithium-Ion to Sodium-Ion Batteries: Advantages, Challenges, and Surprises. *Angew. Chem. Int. Ed.* **2018**, *57*, 102–120. [CrossRef] [PubMed]
3. Li, M.; Lu, J.; Chen, Z.; Amine, K. 30 Years of Lithium-Ion Batteries. *Adv. Mater.* **2018**, *30*, 1800561. [CrossRef]
4. Guan, X.; Nai, J.; Zhang, Y.; Wang, P.; Yang, J.; Zheng, L.; Zhang, J.; Guo, L. CoO Hollow Cube/Reduced Graphene Oxide Composites with Enhanced Lithium Storage Capability. *Chem. Mater.* **2014**, *26*, 5958–5964. [CrossRef]
5. Li, H.; Zhou, H. Enhancing the Performances of Li-Ion Batteries by Carbon-Coating: Present and Future. *Chem. Commun.* **2012**, *48*, 1201–1217. [CrossRef] [PubMed]
6. Li, H.; Wang, Z.; Chen, L.; Huang, X. Research on Advanced Materials for Li-Ion Batteries. *Adv. Mater.* **2009**, *21*, 4593–4607. [CrossRef]
7. Chen, J.; Xia, X.; Tu, J.; Xiong, Q.; Yu, Y.-X.; Wang, X.; Gu, C. Co$_3$O$_4$ –C Core–Shell Nanowire Array as an Advanced Anode Material for Lithium Ion Batteries. *J. Mater. Chem.* **2012**, *22*, 15056–15061. [CrossRef]
8. Peng, C.; Chen, B.; Qin, Y.; Yang, S.; Li, C.; Zuo, Y.; Liu, S.; Yang, J. Facile Ultrasonic Synthesis of Coo Quantum Dot/Graphene Nanosheet Composites with High Lithium Storage Capacity. *ACS Nano* **2012**, *6*, 1074–1081. [CrossRef]
9. Buqa, H.; Goers, D.; Holzapfel, M.; Spahr, M.E.; Novák, P. High Rate Capability of Graphite Negative Electrodes for Lithium-Ion Batteries. *J. Electrochem. Soc.* **2005**, *152*, A474. [CrossRef]
10. Jeschull, F.; Brandell, D.; Edström, K.; Lacey, M.J. A Stable Graphite Negative Electrode for the Lithium–Sulfur Battery. *Chem. Commun.* **2015**, *51*, 17100–17103. [CrossRef]
11. Park, S.-K.; Choi, J.H.; Kang, Y.C. Unique Hollow NiO Nanooctahedrons Fabricated through the Kirkendall Effect as Anodes for Enhanced Lithium-Ion Storage. *Chem. Eng. J.* **2018**, *354*, 327–334. [CrossRef]
12. Li, L.; Seng, K.H.; Chen, Z.; Guo, Z.; Liu, H.K. Self-Assembly of Hierarchical Star-like Co$_3$O$_4$ Micro/Nanostructures and Their Application in Lithium Ion Batteries. *Nanoscale* **2013**, *5*, 1922–1928. [CrossRef] [PubMed]
13. Qin, Y.; Li, Q.; Xu, J.; Wang, X.; Zhao, G.; Liu, C.; Yan, X.; Long, Y.; Yan, S.; Li, S. CoO-Co Nanocomposite Anode with Enhanced Electrochemical Performance for Lithium-Ion Batteries. *Electrochim. Acta* **2017**, *224*, 90–95. [CrossRef]
14. Leng, X.; Ding, X.; Hu, J.; Wei, S.; Jiang, Z.; Lian, J.; Wang, G.; Jiang, Q.; Liu, J. In Situ Prepared Reduced Graphene Oxide/CoO Nanowires Mutually-Supporting Porous Structure with Enhanced Lithium Storage Performance. *Electrochim. Acta* **2016**, *190*, 276–284. [CrossRef]
15. Huang, Y.; Lin, Z.; Zheng, M.; Wang, T.; Yang, J.; Yuan, F.; Lu, X.; Liu, L.; Sun, D. Amorphous Fe$_2$O$_3$ Nanoshells Coated on Carbonized Bacterial Cellulose Nanofibers as a Flexible Anode for High-Performance Lithium Ion Batteries. *J. Power Sources* **2016**, *307*, 649–656. [CrossRef]
16. Zheng, M.; Tang, H.; Li, L.; Hu, Q.; Zhang, L.; Xue, H.; Pang, H. Hierarchically Nanostructured Transition Metal Oxides for Lithium-Ion Batteries. *Adv. Sci.* **2018**, *5*, 1700592. [CrossRef]
17. Li, X.; Dhanabalan, A.; Wang, C. Enhanced Electrochemical Performance of Porous NiO–Ni Nanocomposite Anode for Lithium Ion Batteries. *J. Power Sources* **2011**, *196*, 9625–9630. [CrossRef]
18. Su, L.; Zhou, Z.; Shen, P. Ni/C Hierarchical Nanostructures with Ni Nanoparticles Highly Dispersed in N-Containing Carbon Nanosheets: Origin of Li Storage Capacity. *J. Phys. Chem. C* **2012**, *116*, 23974–23980. [CrossRef]
19. Zhou, C.; Wu, C.; Liu, D.; Yan, M. Metal-Organic Framework Derived Hierarchical Co/C@ V$_2$O$_3$ Hollow Spheres as a Thin, Lightweight, and High-Efficiency Electromagnetic Wave Absorber. *Chem. –A Eur. J.* **2019**, *25*, 2234–2241. [CrossRef]
20. Zhang, S.; Shi, X.; Wen, X.; Chen, X.; Chu, P.K.; Tang, T.; Mijowska, E. Interconnected Nanoporous Carbon Structure Delivering Enhanced Mass Transport and Conductivity toward Exceptional Performance in Supercapacitor. *J. Power Sources* **2019**, *435*, 226811. [CrossRef]
21. Zhang, H.; Nai, J.; Yu, L.; Lou, X.W.D. Metal-Organic-Framework-Based Materials as Platforms for Renewable Energy and Environmental Applications. *Joule* **2017**, *1*, 77–107. [CrossRef]

22. Sun, P.-P.; Zhang, Y.-H.; Shi, H.; Shi, F.-N. Study on the Properties of Cu Powder Modified 3-D Co-MOF in Electrode Materials of Lithium Ion Batteries. *J. Solid State Chem.* **2022**, *307*, 122740. [CrossRef]
23. Xu, Y.; Wu, C.; Ao, L.; Jiang, K.; Shang, L.; Li, Y.; Hu, Z.; Chu, J. Three-Dimensional Porous Co_3O_4–CoO@GO Composite Combined with N-Doped Carbon for Superior Lithium Storage. *Nanotechnology* **2019**, *30*, 425404. [CrossRef] [PubMed]
24. Zhang, M.; Wang, Y.; Jia, M. Three-Dimensional Reduced Graphene Oxides Hydrogel Anchored with Ultrafine CoO Nanoparticles as Anode for Lithium Ion Batteries. *Electrochim. Acta* **2014**, *129*, 425–432. [CrossRef]
25. Liang, C.; Zhai, T.; Wang, W.; Chen, J.; Zhao, W.; Lu, X.; Tong, Y. Fe_3O_4/Reduced Graphene Oxide with Enhanced Electrochemical Performance towards Lithium Storage. *J. Mater. Chem. A* **2014**, *2*, 7214–7220. [CrossRef]
26. Yoo, E.; Kim, J.; Hosono, E.; Zhou, H.; Kudo, T.; Honma, I. Large Reversible Li Storage of Graphene Nanosheet Families for Use in Rechargeable Lithium Ion Batteries. *Nano Lett.* **2008**, *8*, 2277–2282. [CrossRef] [PubMed]
27. Wu, Z.-S.; Yang, S.; Sun, Y.; Parvez, K.; Feng, X.; Müllen, K. 3D Nitrogen-Doped Graphene Aerogel-Supported Fe_3O_4 Nanoparticles as Efficient Electrocatalysts for the Oxygen Reduction Reaction. *J. Am. Chem. Soc.* **2012**, *134*, 9082–9085. [CrossRef] [PubMed]
28. Shi, Y.H.; Wang, K.; Li, H.H.; Wang, H.F.; Li, X.Y.; Wu, X.L.; Zhang, J.P.; Xie, H.M.; Su, Z.M.; Wang, J.W.; et al. Fe_3O_4 Nanoflakes-RGO Composites: A High Rate Anode Material for Lithium-Ion Batteries. *Appl. Surf. Sci.* **2020**, *511*, 145465. [CrossRef]
29. Chen, T. Two-Dimensional MnO_2/Reduced Graphene Oxide Nanosheet as a High-Capacity and High-Rate Cathode for Lithium-Ion Batteries. *Int. J. Electrochem. Sci.* **2018**, *13*, 8575–8588. [CrossRef]
30. Wang, R.; Xu, C.; Sun, J.; Gao, L.; Lin, C. Flexible Free-Standing Hollow Fe_3O_4/Graphene Hybrid Films for Lithium-Ion Batteries. *J. Mater. Chem. A* **2013**, *1*, 1794–1800. [CrossRef]
31. Wang, F.; Ye, Y.; Wang, Z.; Lu, J.; Zhang, Q.; Zhou, X.; Xiong, Q.; Qiu, X.; Wei, T. MOF-Derived Co_3O_4@rGO Nanocomposites as Anodes for High-Performance Lithium-Ion Batteries. *Ionics* **2021**, *27*, 4197–4204. [CrossRef]
32. Cao, L.; Kang, Q.; Li, J.; Huang, J.; Cheng, Y. Assembly Control of CoO/Reduced Graphene Oxide Composites for Their Enhanced Lithium Storage Behavior. *Appl. Surf. Sci.* **2018**, *455*, 96–105. [CrossRef]
33. Zhang, W.; Hu, L.; Wu, F.; Li, J. Decreasing Co_3O_4 Particle Sizes by Ammonia-Etching and Catalytic Oxidation of Propane. *Catal. Lett.* **2017**, *147*, 407–415. [CrossRef]
34. Campos-Delgado, J.; Romo-Herrera, J.M.; Jia, X.; Cullen, D.A.; Muramatsu, H.; Kim, Y.A.; Hayashi, T.; Ren, Z.; Smith, D.J.; Okuno, Y.; et al. Bulk Production of a New Form of Sp2 Carbon: Crystalline Graphene Nanoribbons. *Nano Lett.* **2008**, *8*, 2773–2778. [CrossRef]
35. Zhang, M.; Uchaker, E.; Hu, S.; Zhang, Q.; Wang, T.; Cao, G.; Li, J. CoO–Carbon Nanofiber Networks Prepared by Electrospinning as Binder-Free Anode Materials for Lithium-Ion Batteries with Enhanced Properties. *Nanoscale* **2013**, *5*, 12342–12349. [CrossRef] [PubMed]
36. Li, G.-C.; Zhao, W. Zeolitic Imidazolate Frameworks Derived Co Nanoparticles Anchored on Graphene as Superior Anode Material for Lithium Ion Batteries. *J. Alloys Compd.* **2017**, *716*, 156–161. [CrossRef]
37. Wan, B.; Guo, J.; Lai, W.-H.; Wang, Y.-X.; Liu, M.; Liu, H.-K.; Wang, J.-Z.; Chou, S.-L.; Dou, S.-X. Layered Mesoporous CoO/Reduced Graphene Oxide with Strong Interfacial Coupling as a High-Performance Anode for Lithium-Ion Batteries. *J. Alloys Compd.* **2020**, *843*, 156050. [CrossRef]
38. Zhou, G.; Wang, D.-W.; Li, F.; Zhang, L.; Li, N.; Wu, Z.-S.; Wen, L.; Lu, G.Q.; Cheng, H.-M. Graphene-Wrapped Fe_3O_4 Anode Material with Improved Reversible Capacity and Cyclic Stability for Lithium Ion Batteries. *Chem. Mater.* **2010**, *22*, 5306–5313. [CrossRef]
39. Biswal, A.; Panda, P.K.; Acharya, A.N.; Mohapatra, S.; Swain, N.; Tripathy, B.C.; Jiang, Z.-T.; Minakshi Sundaram, M. Role of Additives in Electrochemical Deposition of Ternary Metal Oxide Microspheres for Supercapacitor Applications. *ACS Omega* **2020**, *5*, 3405–3417. [CrossRef] [PubMed]
40. Samal, R.; Dash, B.; Sarangi, C.K.; Sanjay, K.; Subbaiah, T.; Senanayake, G.; Minakshi, M. Influence of Synthesis Temperature on the Growth and Surface Morphology of Co_3O_4 Nanocubes for Supercapacitor Applications. *Nanomaterials* **2017**, *7*, 356. [CrossRef]
41. Li, Y.; Fu, Y.; Liu, W.; Song, Y.; Wang, L. Hollow Co-Co_3O_4@CNTs Derived from ZIF-67 for Lithium Ion Batteries. *J. Alloys Compd.* **2019**, *784*, 439–446. [CrossRef]
42. Augustyn, V.; Simon, P.; Dunn, B. Pseudocapacitive Oxide Materials for High-Rate Electrochemical Energy Storage. *Energy Environ. Sci.* **2014**, *7*, 1597–1614. [CrossRef]
43. Yu, Z.; Song, J.; Gordin, M.L.; Yi, R.; Tang, D.; Wang, D. Phosphorus-Graphene Nanosheet Hybrids as Lithium-Ion Anode with Exceptional High-Temperature Cycling Stability. *Adv. Sci.* **2015**, *2*, 1400020. [CrossRef] [PubMed]
44. Wickramaarachchi, K.; Minakshi, M. Status on Electrodeposited Manganese Dioxide and Biowaste Carbon for Hybrid Capacitors: The Case of High-Quality Oxide Composites, Mechanisms, and Prospects. *J. Energy Storage* **2022**, *56*, 106099. [CrossRef]
45. Wickramaarachchi, K.; Sundaram, M.M.; Henry, D. Surfactant-Mediated Electrodeposition of a Pseudocapacitive Manganese Dioxide a Twofer. *J. Energy Storage* **2022**, *55*, 105403. [CrossRef]
46. Wang, Z.; Luan, D.; Madhavi, S.; Hu, Y.; Lou, X.W.D. Assembling Carbon-Coated α-Fe_2O_3 Hollow Nanohorns on the CNT Backbone for Superior Lithium Storage Capability. *Energy Environ. Sci.* **2012**, *5*, 5252–5256. [CrossRef]
47. Wickramaarachchi, K.; Minakshi, M. Consequences of Electrodeposition Parameters on the Microstructure and Electrochemical Behavior of Electrolytic Manganese Dioxide (EMD) for Supercapacitor. *Ceram. Int.* **2022**, *48*, 19913–19924. [CrossRef]

48. Wickramaarachchi, K.; Minakshi, M.; Aravindh, S.A.; Dabare, R.; Gao, X.; Jiang, Z.-T.; Wong, K.W. Repurposing N-Doped Grape Marc for the Fabrication of Supercapacitors with Theoretical and Machine Learning Models. *Nanomaterials* **2022**, *12*, 1847. [CrossRef]
49. Li, F.; Zou, Q.-Q.; Xia, Y.-Y. CoO-Loaded Graphitable Carbon Hollow Spheres as Anode Materials for Lithium-Ion Battery. *J. Power Sources* **2008**, *177*, 546–552. [CrossRef]
50. Guan, H.; Wang, X.; Li, H.; Zhi, C.; Zhai, T.; Bando, Y.; Golberg, D. CoO Octahedral Nanocages for High-Performance Lithium Ion Batteries. *Chem. Commun.* **2012**, *48*, 4878–4880. [CrossRef]
51. Zhang, L.; Hu, P.; Zhao, X.; Tian, R.; Zou, R.; Xia, D. Controllable Synthesis of Core–Shell Co@CoO Nanocomposites with a Superior Performance as an Anode Material for Lithium-Ion Batteries. *J. Mater. Chem.* **2011**, *21*, 18279–18283. [CrossRef]
52. Wu, Z.-S.; Ren, W.; Wen, L.; Gao, L.; Zhao, J.; Chen, Z.; Zhou, G.; Li, F.; Cheng, H.-M. Graphene Anchored with Co₃O₄ Nanoparticles as Anode of Lithium Ion Batteries with Enhanced Reversible Capacity and Cyclic Performance. *ACS Nano* **2010**, *4*, 3187–3194. [CrossRef] [PubMed]
53. Yan, N.; Hu, L.; Li, Y.; Wang, Y.; Zhong, H.; Hu, X.; Kong, X.; Chen, Q. Co₃O₄ Nanocages for High-Performance Anode Material in Lithium-Ion Batteries. *J. Phys. Chem. C* **2012**, *116*, 7227–7235. [CrossRef]
54. Li, B.; Cao, H.; Shao, J.; Li, G.; Qu, M.; Yin, G. Co₃O₄@graphene Composites as Anode Materials for High-Performance Lithium Ion Batteries. *Inorg. Chem.* **2011**, *50*, 1628–1632. [CrossRef] [PubMed]
55. Huang, G.; Zhang, F.; Du, X.; Qin, Y.; Yin, D.; Wang, L. Metal Organic Frameworks Route to in Situ Insertion of Multiwalled Carbon Nanotubes in Co₃O₄ Polyhedra as Anode Materials for Lithium-Ion Batteries. *ACS Nano* **2015**, *9*, 1592–1599. [CrossRef] [PubMed]
56. Yan, C.; Chen, G.; Zhou, X.; Sun, J.; Lv, C. Template-Based Engineering of Carbon-Doped Co₃O₄ Hollow Nanofibers as Anode Materials for Lithium-Ion Batteries. *Adv. Funct. Mater.* **2016**, *26*, 1428–1436. [CrossRef]
57. Zhu, J.; Tu, W.; Pan, H.; Zhang, H.; Liu, B.; Cheng, Y.; Deng, Z.; Zhang, H. Self-Templating Synthesis of Hollow Co₃O₄ Nanoparticles Embedded in N,S-Dual-Doped Reduced Graphene Oxide for Lithium Ion Batteries. *ACS Nano* **2020**, *14*, 5780–5787. [CrossRef] [PubMed]
58. Wang, J.; Zhang, Q.; Li, X.; Xu, D.; Wang, Z.; Guo, H.; Zhang, K. Three-Dimensional Hierarchical Co₃O₄/CuO Nanowire Heterostructure Arrays on Nickel Foam for High-Performance Lithium Ion Batteries. *Nano Energy* **2014**, *6*, 19–26. [CrossRef]
59. Li, M.; Yin, Y.-X.; Li, C.; Zhang, F.; Wan, L.-J.; Xu, S.; Evans, D.G. Well-Dispersed Bi-Component-Active CoO/CoFe₂O₄ Nanocomposites with Tunable Performances as Anode Materials for Lithium-Ion Batteries. *Chem. Commun.* **2012**, *48*, 410–412. [CrossRef]
60. Wang, P.; Sun, Z.; Liu, H.; Gao, Z.-W.; Hu, J.; Yin, W.-J.; Ke, Q.; Lu Zhu, H. Strategic Synthesis of Sponge-like Structured SiOx@C@CoO Multifunctional Composites for High-Performance and Stable Lithium-Ion Batteries. *J. Mater. Chem. A* **2021**, *9*, 18440–18453. [CrossRef]

nanomaterials

Article

Facile Synthesis of Battery-Type CuMn₂O₄ Nanosheet Arrays on Ni Foam as an Efficient Binder-Free Electrode Material for High-Rate Supercapacitors

Chandu V. V. Muralee Gopi [1], R. Ramesh [2], Rajangam Vinodh [3], Salem Alzahmi [4,5,*] and Ihab M. Obaidat [5,6,*]

1   Department of Electrical Engineering, University of Sharjah, Sharjah P.O. Box 27272, United Arab Emirates
2   Department of Chemical Engineering, School of Mechanical, Chemical and Materials Engineering,
    Adama Science and Technology University, Adama P.O. Box 1888, Ethiopia
3   Green Hydrogen Lab (GH2Lab), Institute for Hydrogen Research (IHR), Université du Québec à
    Trois-Rivières (UQTR), 3351 Boulevard des Forges, Trois-Rivières, QC G9A 5H7, Canada
4   Department of Chemical & Petroleum Engineering, United Arab Emirates University,
    Al Ain P.O. Box 15551, United Arab Emirates
5   National Water and Energy Center, United Arab Emirates University,
    Al Ain P.O. Box 15551, United Arab Emirates
6   Department of Physics, United Arab Emirates University, Al Ain P.O. Box 15551, United Arab Emirates
*   Correspondence: s.alzahmi@uaeu.ac.ae (S.A.); iobaidat@uaeu.ac.ae (I.M.O.)

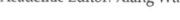

Citation: Gopi, C.V.V.M.; Ramesh, R.;
Vinodh, R.; Alzahmi, S.; Obaidat, I.M.
Facile Synthesis of Battery-Type
CuMn₂O₄ Nanosheet Arrays on Ni
Foam as an Efficient Binder-Free
Electrode Material for High-Rate
Supercapacitors. *Nanomaterials* 2023,
13, 1125. https://doi.org/10.3390/
nano13061125

Academic Editor: Xiang Wu

Received: 1 March 2023
Revised: 16 March 2023
Accepted: 20 March 2023
Published: 21 March 2023

Abstract: The development of battery-type electrode materials with hierarchical nanostructures has recently gained considerable attention in high-rate hybrid supercapacitors. For the first time, in the present study novel hierarchical CuMn₂O₄ nanosheet arrays (NSAs) nanostructures are developed using a one-step hydrothermal route on a nickel foam substrate and utilized as an enhanced battery-type electrode material for supercapacitors without the need of binders or conducting polymer additives. X-ray diffraction, scanning electron microscopy (SEM), and transmission electron microscopy (TEM) techniques are used to study the phase, structural, and morphological characteristics of the CuMn₂O₄ electrode. SEM and TEM studies show that CuMn₂O₄ exhibits a nanosheet array morphology. According to the electrochemical data, CuMn₂O₄ NSAs give a Faradic battery-type redox activity that differs from the behavior of carbon-related materials (such as activated carbon, reduced graphene oxide, graphene, etc.). The battery-type CuMn₂O₄ NSAs electrode showed an excellent specific capacity of 125.56 mA h g⁻¹ at 1 A g⁻¹ with a remarkable rate capability of 84.1%, superb cycling stability of 92.15% over 5000 cycles, good mechanical stability and flexibility, and low internal resistance at the interface of electrode and electrolyte. Due to their excellent electrochemical properties, high-performance CuMn₂O₄ NSAs-like structures are prospective battery-type electrodes for high-rate supercapacitors.

Keywords: CuMn₂O₄; nanosheet arrays; hydrothermal; battery-type; supercapacitors

1. Introduction

The exhaustion of fossil fuels, with related global climate issues and growing energy demand from society, has created an urgent call for the progress of high-rate electrochemical energy storage systems that can store vast amounts of energy (high specific energy) and have rapid charge–discharge (high specific power) [1]. Currently, lithium-ion batteries (LIB), supercapacitors (SCs), and sodium-ion (SIBs) storage systems are extensively used for the storage systems of electric vehicles, portable electronics, wearable electronics, etc. [2,3]. Supercapacitors have received the most attention from people among all types of energy storage devices, owing to their benefits including outstanding cycling stability, superior specific power, and rapid charge–discharge cycles [4]. Supercapacitors are classified into three varieties based on their energy storage capabilities: electrochemical double-layer capacitors (EDLCs), pseudocapacitors (PCs), and battery-type supercapacitors [5]. The EDLC materials accumulate the charges by the adsorption of electrolyte ions at the electrode and electrolyte interface, while the PCs and battery-type materials store charges by Faradaic electrochemical reactions. The morphology, structure, and conductivity of the electrode materials, the electrolyte, the complexity of the device's fabrication circumstances, and other aspects need to be considered while preparing supercapacitor devices. The electroactive material is the essential supercapacitor device component to determine electrochemical performance [6]. Commonly used electrode materials include conductive polymers, conductive carbon compounds, transition metal components, etc. [7]. However, the SCs could produce a tremendous specific power but low specific energy; in contrast, the LIBs could only give a low specific power [8,9]. In this case, creating a hybrid supercapacitor (HSC) with high specific power and high specific energy is particularly desirable [10]. These HSCs comprise a collective arrangement of a supercapacitor electrode material and a battery-type electrode material [11]. Hence, in recent years developing positive and negative electroactive materials for HSC has witnessed significant demand in electric vehicles, portable electronics, and wearable electronics applications.

Battery-type materials, such as NiO, Co_3O_4, $Ni(OH)_2$, NiS, $NiCo_2S_4$, etc., delivered superior energy storage capabilities compared to electric double layer capacitors (EDLC materials, such as carbon-related materials, such as reduced graphene oxide, graphene, etc.) and pseudocapacitor materials (such as MnO_2, RuO_2, etc.) [12–17]. Therefore, it is crucial to design and create highly effective battery-type electrodes to elevate the supercapacitor's performance, particularly its specific energy. Additionally, binary metal oxides ($CuCo_2O_4$, $NiCo_2O_4$, $NiMoO_4$, $CoMoO_4$, $FeMoO_4$, $CuNiO_2$, $NiMn_2O_4$, $MnCo_2O_4$, etc.) have drawn significant interest as battery-type electrodes for supercapacitors compared to the mono-metal oxides due to their enhanced electrical conductivity and additional oxidation states [18–21]. Manganese-based binary oxides have received the most attention among these oxides, because they offer numerous benefits including low toxicity, great abundance, various valence, and low cost [22]. Several research initiatives have been developed to produce binary metal oxides, including exploring novel electrode materials and constructing hierarchical nanostructures. For example, Krishnan et al. developed a $MnCo_2O_4$ nanoflakes battery-type material for a supercapacitor using the rapid microwave-assisted technique and obtained a specific capacity of 74.44 mA h g^{-1} at 0.5 A g^{-1} [23]. The $NiMn_2O_4$ synthesized by Krishna et al. delivered a specific capacity of 33.66 mA h g^{-1} at 0.5 A g^{-1} [24]. By a simple hydrothermal route, Wei et al. developed the nanostructured spinel $NiMn_2O_4$ electrode and obtained a specific capacity of 73.61 mA h g^{-1} at 1 A g^{-1} with superb cycling stability (96% over 1000 cycles) [25]. Recently, Cheng et al. synthesized hierarchical $CuMn_2O_4$ microspheres using a micro/nano $MnCO_3$ precursor and obtained a specific capacity of 73 mA h g^{-1} at 1 A g^{-1} [26]. On the other hand, Zhang et al. deposited the spinel $CuMn_2O_4$ on graphene nanosheets ($CuMn_2O_4$-RGO) using the sol-gel method and physical grinding and obtained a specific capacity of 95 mA h g^{-1} at 1 A g^{-1} [27]. However, it still remains a challenge to develop $CuMn_2O_4$ micro/nanostructures with outstanding energy storage performance for supercapacitors. Moreover, transition metal oxide performance is unsatisfactory in supercapacitor applications due to their limited electrical conductivity, intrinsic structural features, and lower capacitance value than the theoretical value. Therefore, exceptional efforts have been undertaken in this direction to create battery-type binary metal oxides with various morphologies for high-rate supercapacitors.

It is also necessary to consider the fabrication route to develop highly efficient electrode materials for hybrid supercapacitor applications. Several fabrication routes for producing metal oxides and sulfides include sol-gel, coprecipitation, chemical bath deposition, and hydrothermal and solvothermal reactions. Among these approaches, hydrothermal synthesis typically outperforms the others because of its low-temperature synthesis process, affordable equipment, and ability to modify composition and particle size by adjusting fabrication factors [16,28]. As a result, a simple hydrothermal approach was applied in this study to create a highly effective electrode material for supercapacitor applications.

Inspired by the latest findings of binary metal oxides, we created a $CuMn_2O_4$ nanosheet-like structure using a facile one-step hydrothermal approach and effectively utilized it as a battery-type electrode for supercapacitors. According to structural and morphological investigations, $CuMn_2O_4$ has a nanosheet array-like morphology that provides plenty of electroactive sites and promotes quick redox reactions. As a result, the battery-type $CuMn_2O_4$ NSA electrode displays an exceptional specific capacity (125.56 mA h g^{-1} at 1 A g^{-1}), excellent rate capability (84.1% even at 10 A g^{-1}), and extraordinary cycling performance (92.15% after 5000 cycles). Moreover, at various bending angles (flat, 45°, and 90°) the CV and GCD plots of the $CuMn_2O_4$ NSAs electrode are consistent in shape and no noticeable distortion occurs, suggesting outstanding mechanical stability and flexibility.

2. Experimental Method

2.1. Materials

The chemicals used in this study were acquired from Sigma-Aldrich, Seoul, South Korea, and utilized without additional purification; these are potassium hydroxide (KOH), copper nitrate hexahydrate ($Cu(NO_3)_2 \cdot 6H_2O$), ammonium fluoride (NH_4F), manganese nitrate hexahydrate ($Mn(NO_3)_2 \cdot 6H_2O$), hydrochloric acid (HCl), and urea (CH_4N_2O).

2.2. Fabrication of Battery-Type $CuMn_2O_4$ NSA Material on Ni Foam Surface

On nickel (Ni) foam substrate, binder-free $CuMn_2O_4$ NSA structures were deposited via a simple one-step hydrothermal technique. The 1×2 cm^2 Ni foam substrates were cleaned using an ultrasonic cleaner for 15 min each with 3 M HCl, acetone, ethanol, and deionized (DI) water before the electroactive material was deposited. Based on our previous work, various binary metal oxides were deposited on the Ni foam surface at the deposition temperature and time of 100 °C and 6 h. Hence, in the present study the deposition temperature and time of 100 °C and 6 h are used in the hydrothermal condition to deposit the $CuMn_2O_4$ NSA material on the Ni foam surface [20,29,30]. In 60 mL DI water, 0.05 M of $Cu(NO_3)_2 \cdot 6H_2O$, 0.1 M of $Mn(NO_3)_2 \cdot 6H_2O$, 0.24 M of CH_4N_2O and 0.12 M NH_4F were combined and stirred for 30 min to deposit the $CuMn_2O_4$ material on Ni foam. Pre-cleaned Ni foams and the $CuMn_2O_4$ reaction recipe were put into a 100 mL Teflon-lined autoclave for 6 h at 100 °C. After cooling, the electrodes were taken out of the autoclave, washed several times with deionized water and ethanol, and dried at 60 °C for a whole night. Ultimately, the electrode was heated for 2 h at 200 °C and given the designation $CuMn_2O_4$ NSAs. The active material loading on Ni foam is about 2.3 mg. The CuO material on Ni foam was also synthesized at the same fabrication method, except without the addition of $Mn(NO_3)_2 \cdot 6H_2O$ [31].

2.3. Material Characterization and Electrochemical Measurements

Transmission electron microscopy (TEM, CJ111), X-ray diffraction (D8 ADVANCE), and scanning electron microscopy (SEM, S4800, Hitachi, Pusan National University, Busan, South Korea) were used to analyze the morphology, crystalline structure, and phase purity of the as-prepared $CuMn_2O_4$ electrode. All electrochemical measurements were made using a Bio-Logic SP-150 electrochemical analyzer. In a three-electrode system configuration filled with 3 M KOH aqueous electrolyte, the electrochemical behaviors of the as-developed electrode were examined utilizing electrochemical impedance spectroscopy (EIS), galvanostatic charge–discharge (GCD), and cyclic voltammetry (CV). The as-prepared $CuMn_2O_4$ electrode was used as the working electrode, Ag/AgCl electrode was used as the reference electrode, and the platinum wire was used as the counter electrode. In a three-electrode arrangement, Equation (1) was used to compute the specific capacity (Q_{SC}, mA h g^{-1}) of the as-fabricated sample from the GCD plots [32].

$$Q_{SC} = \frac{I \times \Delta t}{m \times 3.6} \tag{1}$$

where I (A) is the current (A), Δt (s) is the discharge time, and m (g) is the electroactive material mass (g).

3. Results and Discussion

To validate the effective deposition of the $CuMn_2O_4$ active material on the Ni foam, an XRD analysis of the thin film was performed. Figure 1 shows the XRD spectrum of $CuMn_2O_4$, which displayed well-defined sharp peaks consistent with the previous literature reports [33,34], denoting the successful fabrication of the active material. Bright and intense Nickel diffraction peaks are visible in XRD due to the backdrop of Ni foam. The peaks obtained at 2θ positions of 30.6°, 35.6°, 43.4°, 57.5°, 63.2°, and 74.3° correspond to the (220), (311), (400), (511), (440), and (533) planes of crystalline $CuMn_2O_4$ (JCPDS no. 34-1400) [33,34]. The absence of additional peaks in the XRD pattern supports the $CuMn_2O_4$ sample's phase purity.

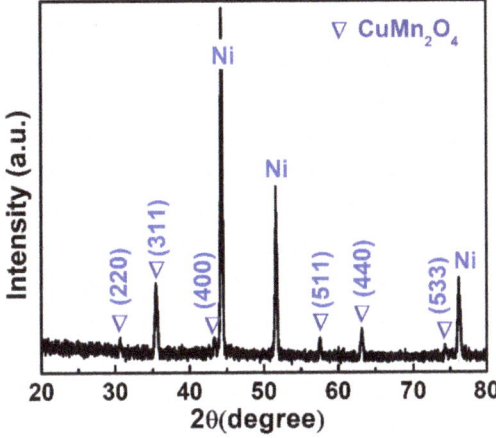

Figure 1. $CuMn_2O_4$ NSA's XRD pattern on the surface of Ni foam.

Figure 2a depicts the schematic representation for the fabrication of binder-free $CuMn_2O_4$ nanosheet arrays on the Ni foam using a simple one-step hydrothermal. Ni foam, as it is widely known, is a three-dimensional (3D) structure that is highly conductive and possesses an open porous structure with huge surface area stems. As a result, Ni foam could be utilized as an efficient conductive stage to increase the active material mass loading, and is also more suited for the sweep out of electrons produced from the electroactive material during redox processes. These fantastic benefits led us to select the Ni foam as a current collector and build the $CuMn_2O_4$ NSA sequentially using a one-step hydrothermal method. Urea and NH_4F were utilized as precipitants and complexing agents, respectively.

As shown in Figure 2b–d, the morphology and structure of the fabricated $CuMn_2O_4$ electrode were examined by SEM analysis. The SEM image with low magnification (Figure 2b) shows that the $CuMn_2O_4$ active material is completely coated on the Ni foam surface. The high-resolution SEM images in Figure 2c,d reveal that the microstructures are made up of interwoven ultrathin nanosheets with relative thicknesses ranging from ~7.5 to 30 nm. The ultrathin nanosheets are interconnected and generate a hierarchical nanosheet array (NSA) morphology. TEM and high-resolution-TEM characterization methods were used to evaluate the morphology and crystalline characteristics of the $CuMn_2O_4$ NSAs electrodes. Different magnification TEM pictures of $CuMn_2O_4$ NSAs are shown in Figure 2e–g, and they demonstrate the deposition of hierarchical $CuMn_2O_4$ NSAs over the Ni foam surface. The HR-TEM picture exhibits lattice fringes of the $CuMn_2O_4$ compound at a lattice spacing of 0.256 nm relating to the (311) plane, as shown in Figure 2g. The TEM results are consistent with the XRD analysis of the $CuMn_2O_4$ NSAs. The as-deposited NSAs are anticipated to improve the specific surface area, supply many electroactive sites, and enable the rapid diffusion of electrolyte ions, all of which will enhance the charge storage performance.

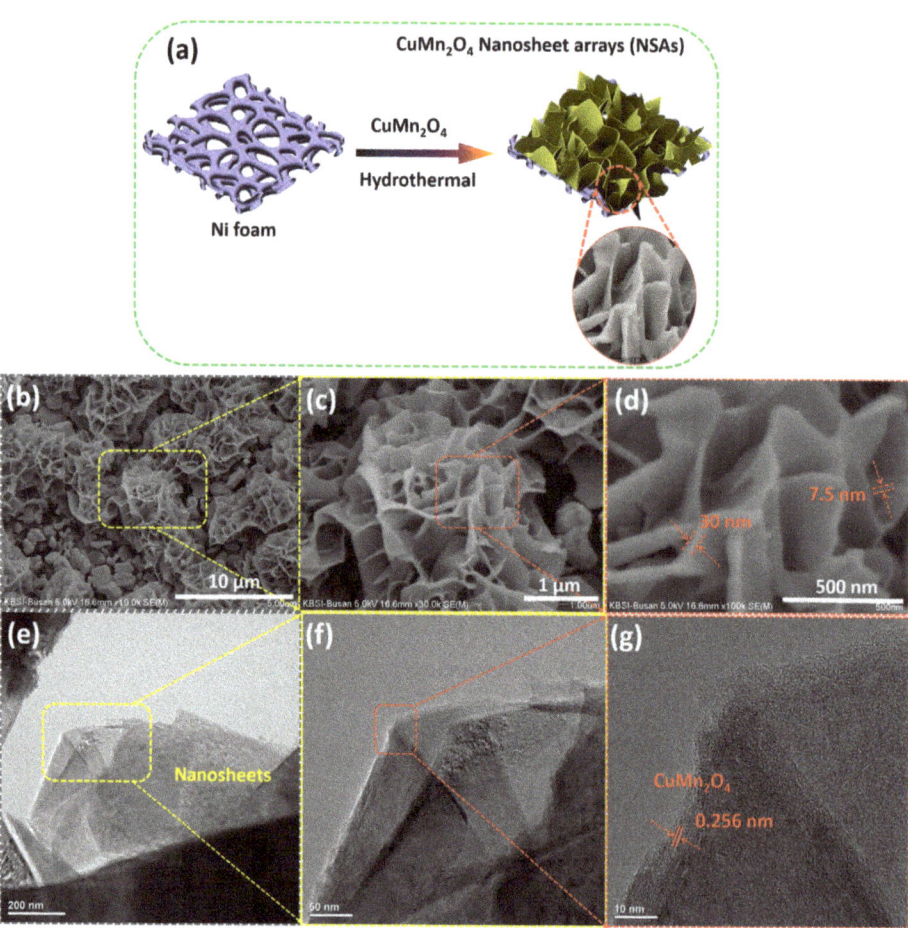

Figure 2. (**a**) Schematic diagram of the preparation of CuMn$_2$O$_4$ NSAs electrode. (**b–d**) SEM and high-resolution SEM images of CuMn$_2$O$_4$ NSAs electrode. (**e–g**) TEM and high-resolution TEM images of CuMn$_2$O$_4$ NSAs electrode.

The electrochemical battery-type supercapacitor performance of the CuMn$_2$O$_4$ NSAs electrode was investigated by CV, GCD, and EIS techniques in a three-electrode configuration using 3 M KOH as the electrolyte. As-prepared CuMn$_2$O$_4$ NSAs electrodes belong to the battery-type electrodes, as shown by the CV plots of the CuMn$_2$O$_4$ NSAs sample in Figure 3a, which show clear redox peaks obtained in the potential window range from 0 to 0.55 V (vs. Ag/AgCl) at various sweep rates of 2, 5, 10, 25, and 50 mV s^{-1}. The diffusion mechanism is mainly due to the redox reaction between the electrolyte ions and the active material. As a result, diffusion control mainly drives the active substance's contribution to specific capacity at low scan rates. Peak current levels of the pair of redox peaks were seen to increase as the scan rate increased, and the forms of the redox peaks were conserved even at a high sweep rate of 50 mV s^{-1}, implying a rapid redox reaction, elevated electro-conductivity, and excellent rate capacity of this electrode. Additionally, the oxidation and reduction peaks gradually moved to more positive and negative potentials when the sweep rate rose because of the electrochemical redox reaction's fast charge and discharge rates and lower material resistance [35]. According to the following equations, the reversible Faradaic redox reactions of Cu and Mn species provide the basis for the electrochemical reaction's mechanism:

$$CuMn_2O_4 + OH^- + H_2O \rightleftharpoons CuOOH + 2MnOOH + e^- \qquad (2)$$

$$CuOOH + OH^- \rightleftharpoons CuO_2 + H_2O + e^- \qquad (3)$$

$$MnOOH + OH^- \rightleftharpoons MnO_2 + H_2O + e^- \qquad (4)$$

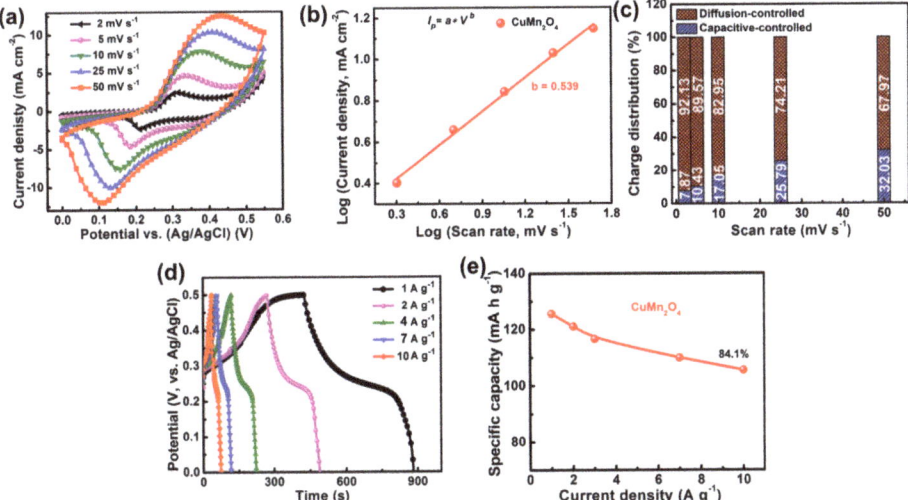

Figure 3. (**a**) CuMn₂O₄ NSAs electrode CV graphs at different sweep rates. (**b**) The b−value was calculated using a Log of cathodic peak current density vs. a Log of sweep rate. (**c**) The plot of the CuMn₂O₄ NSAs electrode's charge distribution vs. scan rates. (**d**) CuMn₂O₄ NSAs electrode GCD profiles at varied current densities. (**e**) Calculated specific capacity values vs. current density of the CuMn₂O₄ NSAs electrode.

Additionally, to determine the charge storage kinetics of the as-developed material, the *b* value was calculated using log (i) = blog(v) + log(a) where *v and I* denote the cathode sweep rate and peak current, respectively [36]. As depicted in Figure 3b, the *b* value of the CuMn₂O₄ NSAs electrode material is 0.539, which is very close to 0.5, indicating that its diffusion-control (battery-type) dominant kinetics behavior is entirely compatible with recent studies on battery-type material behavior [18,36].

Furthermore, to estimate the diffusive- and capacitive-controlled contribution in the CuMn₂O₄ NSAs electrode, we represented peak current (i_p) as the sum of a capacitive-regulated (k_1v) and diffusion-regulated (k_2v) process by the following Equations [37]:

$$i_p = k_1v + k_2v^{1/2} \qquad (5)$$

$$\frac{i_p}{v^{1/2}} = k_1v^{1/2} + k_2 \qquad (6)$$

where i_p, v, and k_1 and k_2 represent the peak current (A), sweep rate (V s⁻¹), and the constant parameters. The linearly fitted plot between $i/v^{1/2}$ vs. $v^{1/2}$ provided the k_1 and k_2 values. As illustrated in Figure 3c, the CuMn₂O₄ NSAs electrodes demonstrated a more dominating diffusion-controlled mechanism, with 92.13% overall capability at a low sweep rate of 2 mV s⁻¹. The intense battery-type mechanism was caused by the electrolyte's OH- ions having sufficient time to diffuse into the electrode material at low sweep rates. However, the CuMn₂O₄ NSAs electrode's diffusion-controlled contribution was reduced to 67.97% as the sweep rate rose from 2 to 50 mV s⁻¹, while the capacitive-controlled impact climbed to 32.03%. With increased scan rates, the reduced diffusion-controlled behavior was due to insufficient time for ion migration and intercalation; in contrast, the enhanced capacitive-regulated behavior was due to the rapid electrolyte ions transit that happens at the interface of electrode and electrolyte.

The GCD plots of the $CuMn_2O_4$ NSAs electrode are shown in Figure 3d, which depicts the current densities varying from 1 to 10 A g^{-1} with a potential window range of 0 to 0.5 V. The GCD plots' distinct plateau sections reveal their battery-like nature. The results of the GCD tests on the $CuMn_2O_4$ NSAs electrode agree well with those obtained from the CV. Due to the limited ion diffusion at high current density, the charging and discharging periods for the $CuMn_2O_4$ NSAs electrode steadily reduce as the current density increases. However, the ions and charges in the electrolyte will have enough time to diffuse and transfer when the current density is low [38]. Equation (1) was used to determine the specific capacities of the battery-type $CuMn_2O_4$ NSAs electrode at different current densities, and calculated values are depicted in Figure 3e. The calculated specific capacities of the $CuMn_2O_4$ NSAs electrode were 125.56, 121.11, 116.67, 109.92, and 105.56 mA h g^{-1} at 1, 2, 4, 7, and 10 A g^{-1}, respectively. As investigated in the electrochemical analyses, the $CuMn_2O_4$ NSA electrode delivered a high Q_{SC} value of 125.56 mA h g^{-1} at a current density of 1 A g^{-1}. Furthermore, the $CuMn_2O_4$ NSAs electrode material retained 84.1% of its initial specific capacity value even at 10 A g^{-1} current density, indicating the nanostructured $CuMn_2O_4$ electrode materials' high charge–discharge efficiency. The specific capacity value's steady reduction with increasing current density was mainly due to the loss of active materials on the Ni foam surface during the redox reaction [39]. The specific capacity values produced in this study are comparable to and even more significant than the other electroactive materials and other reported $CuMn_2O_4$ hierarchical structures for supercapacitors previously reported (Table 1).

Moreover, the $CuMn_2O_4$ NSAs electrode was further investigated using EIS analysis to assess the material's charge transfer kinetics and electrical conductivity. Figure 4 depicts the resulting Nyquist plot of the as-fabricated $CuMn_2O_4$ NSA electrode in the 0.01 Hz–100 kHz frequency range at the open-circuit potential. The obtained Nyquist plots are fit using the equivalent circuit shown in the inset of Figure 4. The Nyquist plot showed a small semicircle in the high-frequency and a straight line in the low-frequency regions, representing the charge transfer resistance (R_{ct}) and Warburg diffusion resistance, respectively. The intersection of the Nyquist plot with the X-axis in the high-frequency area represents the electrode material's equivalent series resistance (R_s). The fitting results show that low R_S ($\sim$0.58 Ω cm^2) and R_{ct} ($\sim$0.18 Ω cm^2) values on the $CuMn_2O_4$ NSAs electrode demonstrated excellent charge transfer kinetics and outstanding electronic conductivity. The vertical line in the low-frequency zone represents the lower Warburg diffusion resistance, representing the quick electrolyte diffusion and rapid ion charge transfer. The slope for the $CuMn_2O_4$ NSA electrode is near 45°, showing that the Faradic redox reaction for the $CuMn_2O_4$ NSA is mainly controlled by diffusion, indicating that the $CuMn_2O_4$ NSA corresponds to conventional battery-type electrode materials, as depicted in Figure 4.

Furthermore, the electroactive material's excellent cycling behavior is essential for supercapacitors' real-time application. The cycling behavior of the $CuMn_2O_4$ NSAs material evaluated at 4 A g^{-1} over 5000 cycles is shown in Figure 5a. The specific capacity values improved somewhat in the early cycles due to the complete activation of the $CuMn_2O_4$ NSAs material by the continual entry of electrolyte ions into their inner portions [40]. Over 5000 cycles, the specific capacity gradually decreased and it retained $\sim$92.15% of its primary specific capacity value, demonstrating exceptional cycling life. This was further verified by the SEM picture of the $CuMn_2O_4$ electrode taken after 5000 cycles (inset of Figure 5a). The shape of the nanosheet arrays was highly retained, and the electroactive material remained attached to the surface of the Ni foam, illustrating the remarkable cycling life of the $CuMn_2O_4$ NSAs electrode. Moreover, electrochemical measurements (CV, GCD, and EIS) were carried out both before and after the cycling assessment, and the related graphs are shown in Figure 5b–d. As shown in Figure 5b,c, before and after the cycling test, the CV and GCD curves preserved no change in the energy storage performance, revealing the excellent cycling stability of the electroactive material on the Ni foam surface. As shown in Figure 5d, the obtained R_S values before and after the cycling test were almost the same, denoting the stable electrochemical stability of the electroactive material. Moreover, EIS analysis reveals a minor change in the charge transfer resistance of the $CuMn_2O_4$ NSA electrode after the cycling test ($\sim$0.18 Ω cm^2 to $\sim$0.29 cm^2). Hence, the slightly decreased specific capacity value and a slight rise in R_{ct} value during the cycling test might be attributed to a slight solidification or dryness of the electrolyte [41]. Moreover, the tangent line in the low-frequency region of the Nyquist plot confirms the lower Warburg diffusion resistance, which facilitates the ionic diffusion process.

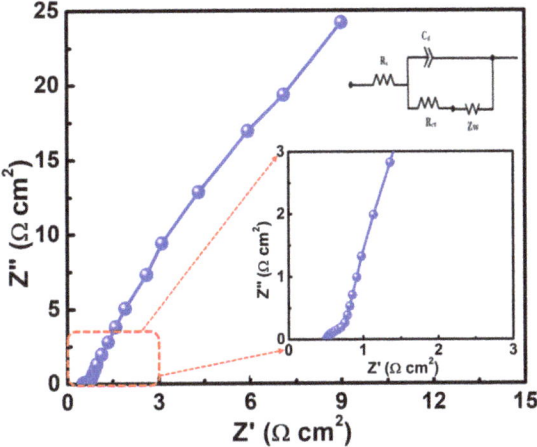

Figure 4. Nyquist plot of $CuMn_2O_4$ NSAs electrode (inset shows the enlarged Nyquist plot and equivalent circuit to fit the Nyquist plots).

The $CuMn_2O_4$ NSAs electrode delivered an excellent specific capacity, outstanding rate capability (84.1%), and good cycling life span. It is worth mentioning that the delivered higher *specific capacity* value from the hydrothermally fabricated battery-type $CuMn_2O_4$ NSA electrode exhibited improved electrochemical performance that was analogous to the performances of the previously reported battery-type binary metal oxide electrodes such as the hydrothermally prepared $NiCo_2O_4$ electrode (Q_{SC} = 34.02 mA h g^{-1} at 1 A g^{-1} and cycling life of 78.30% over 6000 cycles) [42], hydrothermally developed $CuCo_2O_4$ electrode (Q_{SC} = 84.22 mA h g^{-1} at 1 A g^{-1} and cycling life of 71.80% over 5000 cycles) [43], microwave-assisted $NiMn_2O_4$ electrode (Q_{SC} = 138.83 mA h g^{-1} at 1 A g^{-1} and cycling life of 85.80% over 6000 cycles) [44], hydrothermally synthesized $CuNiO_2$ electrode (Q_{SC} = 111.52 mA h g^{-1} at 2 A g^{-1} and cycling life of 89.13% over 3000 cycles) [45], hydrothermally prepared $CuCo_2O_4$ electrode (Q_{SC} = 86.37 mA h g^{-1} at 1 A g^{-1} and cycling life 93% over 6000 cycles) [46], and the hydrothermally developed $FeCo_2O_4$ electrode (Q_{SC} = 125.56 mA h g^{-1} at 1 A g^{-1} and cycling life 93.68% over 4000 cycles) [47], respectively. Table 1 compares and summarizes the total electrochemical behavior of different electrodes.

Moreover, CV and GCD assessments at different bending angles were used to examine the $CuMn_2O_4$ NSA electrode's flexibility and mechanical stability. The obtained curves are depicted in Figure 6a,b. The inset of Figure 6b shows the photographs of the $CuMn_2O_4$ NSA electrode bent at various angles. The $CuMn_2O_4$ NSA electrode is highly flexible and can be bent at 45° and 90° angles without destroying its physical structure, as is shown in the inset of Figure 6b. The $CuMn_2O_4$ electrode's CV and GCD profiles exhibit almost identical CV and GCD profiles at different bending angles (flat, 45°, and 90°), and show minimal performance change, demonstrating the material's exceptional mechanical stability and flexibility.

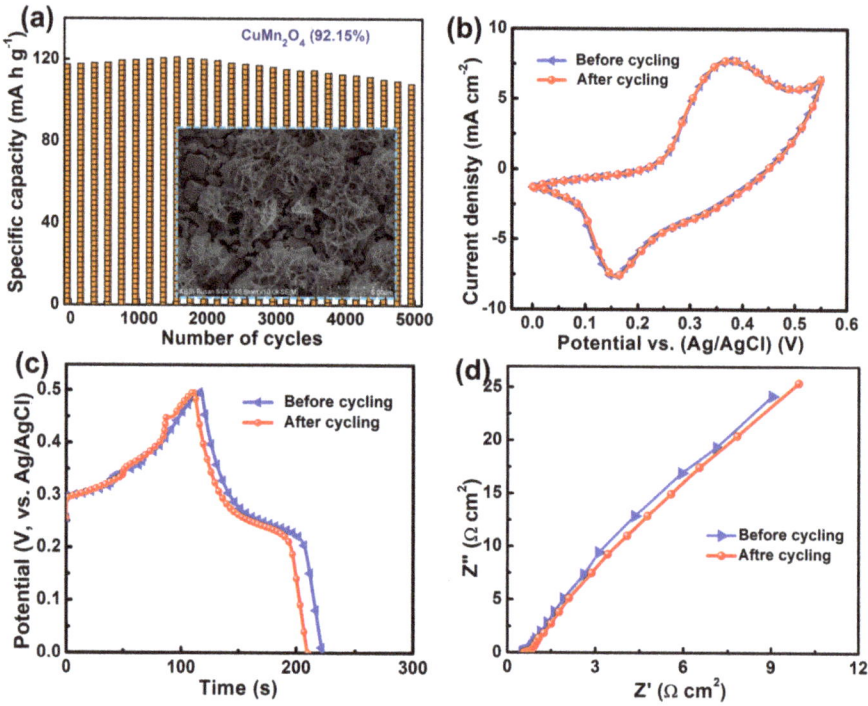

Figure 5. (**a**) Cycling stability (inset depicts the SEM image after the stability test). The CuMn$_2$O$_4$ NSA electrode (**b**) CV, (**c**) GCD, and (**d**) EIS graphs before and after cycle testing.

Table 1. The energy storage behavior of the current battery-type CuMn$_2$O$_4$ NSAs electrode and previously disclosed battery-type binary metal oxide electrodes and other reported CuMn$_2$O$_4$ hierarchical structures are compared.

Battery-Type Electrode	Preparation Route	Specific Capacity (mA h g^{-1})	Cycling Stability (Cycles)	Ref.
NiCo$_2$O$_4$ flower-like	Hydrothermal	34.02 at 1 A g^{-1}	78.30% (6000)	[42]
CuCo$_2$O$_4$ ultrathin nanosheets	Hydrothermal	84.22 at 1 A g^{-1}	71.80% (5000)	[43]
NiMn$_2$O$_4$ microspheres	Microwave-assisted	138.83 at 1 A g^{-1}	85.80% (6000)	[44]
CuNiO$_2$ dandelion flower-like	Hydrothermal	111.52 at 2 A g^{-1}	89.13% (3000)	[45]
CuCo$_2$O$_4$ microspheres	Hydrothermal	86.37 at 1 A g^{-1}	93.00% (6000)	[46]
FeCo$_2$O$_4$ chopsticks-like	Hydrothermal	113.32 at 1 A g^{-1}	93.68% (4000)	[47]
CuMn$_2$O$_4$ microspheres	Micro/nano MnCO$_3$ precursor	73 at 1 A g^{-1}	96% (3000)	[26]
Spinel CuMn$_2$O$_4$-RGO nanosheets	Sol-gel with physical griding	95 at 1 A g^{-1}	75.5% (1000)	[27]
CuMn$_2$O$_4$	Hydrothermal	125.56 at 1 A g^{-1}	92.15% (5000)	Present work

Figure 7 depicts the electrochemical behavior of the binder-free CuMn$_2$O$_4$ NSAs material on Ni-foam for the hybrid supercapacitors. The conductivity of the as−prepared materials could be improved by directly depositing ultra-thin CuMn$_2$O$_4$ NSAs−based hierarchical nanostructures on the Ni foam surface. Additionally, this would offer efficient transport routes and rapid ion diffusion channels that would be feasible for electrochemical processes. Therefore, as−prepared binder−free CuMn$_2$O$_4$ NSAs electrodes with improved electrochemical properties could be a battery-type electrode material for high−rate hybrid supercapacitor applications. Additionally, more advanced experimental research (combining CuMn$_2$O$_4$ nanostructures with other metal oxides or carbon materials), thorough analyses, and measurement methods are currently being investigated to improve the CuMn$_2$O$_4$ electroactive material's properties for electrochemical energy storage applications.

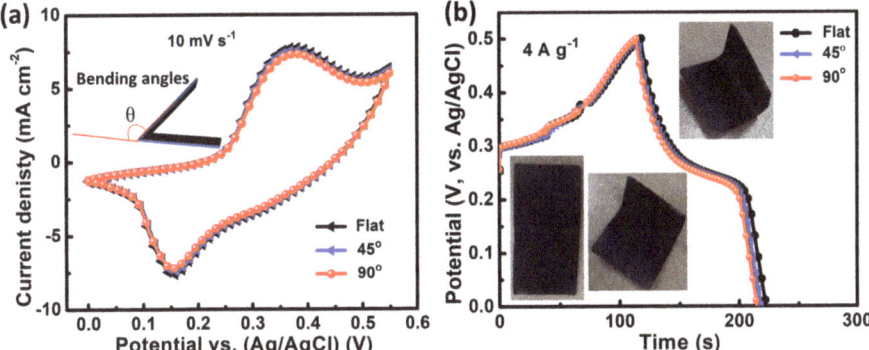

Figure 6. At different bending angles, the (**a**) CV and (**b**) GCD profiles of a CuMn$_2$O$_4$ NSA electrode. Photographs of CuMn$_2$O$_4$ NSA electrode bent in various angles showing the flexibility of the electrode (inset of Figure 6b).

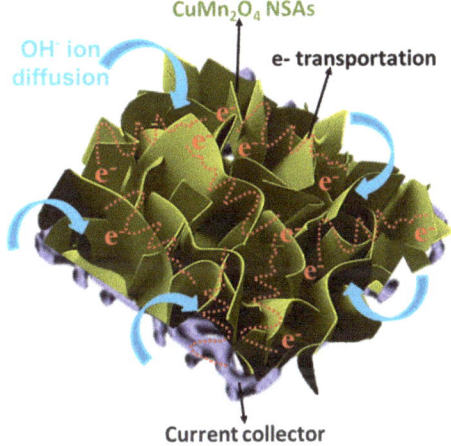

Figure 7. Diffusion of electrolyte ions is shown schematically in CuMn$_2$O$_4$ nanosheet array morphology.

4. Conclusions

A facile one-step hydrothermal process was employed to deposit novel CuMn$_2$O$_4$ NSAs structures on Ni foam. The structural and morphological studies of the as-fabricated CuMn$_2$O$_4$ NSAs electrode were studied using XRD, SEM, and TEM characterization techniques. The as-developed CuMn$_2$O$_4$ NSAs electrode displayed a Faradic battery-type redox activity, as evidenced by the potential plateaus from the CV and GCD techniques. As a battery-type electrode material, the CuMn$_2$O$_4$ NSAs electrode demonstrated exceptional electrochemical capabilities, including the maximum specific capacity (125.56 mA h g^{-1} at 1 A g^{-1}), rate capability (84.1% even at 10 A g^{-1}), and cycling stability (92.15% over 5000 cycles). Moreover, the as-prepared CuMn$_2$O$_4$ NSAs electrode material delivered good mechanical stability and flexibility at various bending angles (flat, 45°, and 90°). The as-prepared electrode's remarkable energy storage performance was due to the hierarchical interconnected nanosheet array architectures, which supplied numerous electroactive sites to facilitate the rapid Faradaic redox reactions. Hence, the excellent electrochemical properties of the CuMn$_2$O$_4$ NSAs electrode have significant potential for use in high-rate supercapacitors as a battery-type electrode material.

Author Contributions: Conceptualization, C.V.V.M.G., I.M.O. and S.A.; Supervision and validation, I.M.O. and S.A.; Investigation and writing, C.V.V.M.G., I.M.O. and S.A.; Investigation and visualization, R.R., C.V.V.M.G. and R.V.; Validation, C.V.V.M.G., I.M.O. All authors have read and agreed to the published version of the manuscript.

Funding: This work was financially supported by the UAEU-Strategic research program under grant no. 12R128.

Institutional Review Board Statement: Not applicable.

Informed Consent Statement: Not applicable.

Data Availability Statement: In this investigation, no fresh data were collected or examined. Sharing of data is not relevant to this subject.

Acknowledgments: The authors acknowledged the UAEU-Strategic research program under grant no. 12R128.

Conflicts of Interest: The authors declare no conflict of interest.

References

1. Wang, G.; Zhang, L.; Zhang, J. A review of electrode materials for electrochemical supercapacitors. *Chem. Soc. Rev.* **2012**, *41*, 797–828. [CrossRef] [PubMed]
2. Ma, Y.; Liu, P.; Xie, Q.; Zhang, G.; Zheng, H.; Cai, Y.; Li, Z.; Wang, L.; Zhu, Z.Z.; Mai, L.; et al. Double-shell Li-rich layered oxide hollow microspheres with sandwich-like carbon@spinel@layered@spinel@carbon shells as high-rate lithium ion battery cathode. *Nano Energy* **2019**, *59*, 184–196. [CrossRef]
3. Wang, L.; Xie, X.; Dinh, K.N.; Yan, Q.Y.; Ma, J.M. Synthesis, characterizations and utilization of oxygen-deficient metal oxides for lithium/sodium-ion batteries and supercapacitors. *Coord. Chem. Rev.* **2019**, *397*, 138–167. [CrossRef]
4. Zhao, W.; Xu, X.; Wu, N.; Zhao, X.; Gong, J. Dandelion-Like CuCo$_2$O$_4$@ NiMn LDH Core/Shell Nanoflowers for Excellent Battery-Type Supercapacitor. *Nanomaterials* **2023**, *13*, 730. [CrossRef] [PubMed]
5. Zang, J.H.; Zhang, G.X.; Zhou, T.; Sun, S.H. Recent developments of planar microsupercapacitors: Fabrication, properties and applications. *Adv. Funct. Mater.* **2020**, *30*, 1910000. [CrossRef]
6. Gong, Y.N.; Li, D.L.; Luo, C.Z.; Fu, Q.; Pan, C.X. Highly porous graphitic biomass carbon as advanced electrode materials for supercapacitors. *Green Chem.* **2017**, *19*, 4132–4140. [CrossRef]
7. Forouzandeh, P.; Kumaravel, V.; Pillai, S.C. Electrode Materials for Supercapacitors: A Review of Recent Advances. *Catalysts* **2020**, *10*, 969. [CrossRef]
8. Snook, G.A.; Kao, P.; Best, A.S. Conducting-polymer-based supercapacitor devices and electrodes. *J. Power Sources* **2011**, *196*, 1–12. [CrossRef]
9. Li, B.; Dai, F.; Xiao, Q.; Yang, L.i.; Shen, J.; Zhang, C.; Cai, M. Nitrogen-doped activated carbon for a high energy hybrid supercapacitor. *Energy Environ. Sci.* **2016**, *9*, 102–106. [CrossRef]
10. Shen, J.; Li, X.; Wan, L.; Liang, K.; Tay, B.K.; Kong, L.; Yan, X. An Asymmetric Supercapacitor with Both Ultra-High Gravimetric and Volumetric Energy Density Based on 3D Ni(OH)2/MnO2@Carbon Nanotube and Activated Polyaniline-Derived Carbon. *ACS Appl. Mater. Interfaces* **2017**, *9*, 668–676. [CrossRef]
11. Hsu, C.T.; Hu, C.C.; Wu, T.H.; Chen, J.C.; Rajkumar, M. How the Electrochemical Reversibility of a Battery-type Material Affects the Charge Balance and Performances of Asymmetric Supercapacitors. *Electrochim. Acta* **2014**, *146*, 759–768. [CrossRef]
12. Natarajan, S.; Ulaganathan, M.; Bajaj, H.C.; Aravindan, V. Transformation of spent Li-ion battery into high energy supercapacitors in asymmetric configuration. *Chem. Electron. Chem.* **2019**, *6*, 5283–5292. [CrossRef]
13. Divya, M.L.; Natarajan, S.; Lee, Y.S.; Aravindan, V. Biomass-derived carbon: A value added journey towards constructing high-energy supercapacitors in an asymmetric fashion. *ChemSusChem* **2019**, *12*, 4353–4382. [CrossRef]
14. Tang, Y.; Chen, S.; Mu, S.; Chen, T.; Qiao, Y.; Yu, S.; Gao, F. Synthesis of capsule-like porous hollow nano-nickel cobalt sulfides via cation exchange based on the kirkendall effect for high-performance supercapacitors. *ACS Appl. Mater. Interfaces* **2016**, *8*, 9721–9732. [CrossRef]
15. Nagaraju, G.; Sekhar, S.C.; Bharat, L.K.; Yu, J.S. Wearable Fabrics with Self-Branched Bimetallic Layered Double Hydroxide Coaxial Nanostructures for Hybrid Supercapacitors. *ACS Nano* **2017**, *11*, 10860–10874. [CrossRef]
16. Sekhar, S.C.; Nagaraju, G.; Yu, J.S. High-performance pouch-type hybrid supercapacitor based on hierarchical NiO-Co3O4-NiO composite nanoarchitectures as an advanced electrode material. *Nano Energy* **2018**, *48*, 81–92. [CrossRef]
17. Sun, J.; Tian, X.; Xu, C.; Chen, H. Porous CuCo2O4 microtubes as a promising battery-type electrode material for high-performance hybrid supercapacitors. *J. Materiomics* **2021**, *7*, 1358–1368. [CrossRef]

18. Gopi, C.V.V.M.; Sambasivam, S.; Raghavendra, K.V.G.; Vinodh, R.; Obaidat, I.M.; Kim, H.J. Facile synthesis of hierarchical flower-like NiMoO4-CoMoO4 nanosheet arrays on nickel foam as an efficient electrode for high rate hybrid supercapacitors. *J. Energy Storage* **2020**, *30*, 101550. [CrossRef]
19. Nam, H.W.; Gopi, C.V.V.M.; Sambasivam, S.; Vinodh, R.; Raghavendra, K.V.G.; Kim, H.J.; Obaidat, I.M.; Kim, S. Binder-free honeycomb-like FeMoO4 nanosheet arrays with dual properties of both battery-type and pseudocapacitive-type performances for supercapacitor applications. *J. Energy Storage* **2020**, *27*, 101055. [CrossRef]
20. Song, C.S.; Gopi, C.V.V.M.; Vinodh, R.; Sambasivam, S.; Kalla, R.M.N.; Obaidat, I.M.; Kim, H.J. Morphology-dependent binder-free CuNiO$_2$ electrode material with excellent electrochemical performances for supercapacitors. *J. Energy Storage* **2019**, *26*, 101037. [CrossRef]
21. Zhang, M.M.; Song, Z.X.; Liu, H.; Ma, T.J. Biomass-derived highly porous nitrogen-doped graphene orderly supported NiMn$_2$O$_4$ nanocrystals as efficient electrode materials for asymmetric supercapacitors. *Appl. Srf. Sci.* **2020**, *507*, 145065. [CrossRef]
22. Lia, L.; Jiang, G.X.; Ma, J.M. CuMn2O4/graphene nanosheets as excellent anode for lithium-ion battery. *Mater. Res. Bull* **2018**, *104*, 53–59. [CrossRef]
23. Krishnan, S.G.; Harilal, M.; Arshid, N.; Jagadish, P.; Khalid, M.; Li, L.P. Rapid microwave-assisted synthesis of MnCo2O4 nanoflakes as a cathode for battery-supercapacitor hybrid. *J. Energy Storage* **2021**, *44*, 103566. [CrossRef]
24. Krishna, B.N.V.; Bhagwan, J.; Yu, J.S. Sol-Gel Routed NiMn2O4 Nanofabric Electrode Materials for Supercapacitors. *J. Electrochem. Soc.* **2019**, *166*, A1950. [CrossRef]
25. Wei, H.; Wang, J.; Yu, L.; Zhang, Y.; Hou, D.; Li, T. Facile synthesis of NiMn2O4 nanosheet arrays grown on nickel foam as novel electrode materials for high-performance supercapacitors. *Ceram. Int.* **2016**, *42*, 14963–14969. [CrossRef]
26. Cheng, C.; Cheng, Y.; Lai, G. CuMn$_2$O$_4$ hierarchical microspheres as remarkable electrode of supercapacitors. *Mater. Lett.* **2022**, *317*, 132102. [CrossRef]
27. Zhang, C.; Xie, A.; Zhang, W.; Chang, J.; Liu, C.; Gu, L.; Duo, X.; Pan, F.; Luo, S. CuMn$_2$O$_4$ spinel anchored on graphene nanosheets as a novel electrode material for supercapacitor. *J. Energy Storage* **2021**, *34*, 102181. [CrossRef]
28. Chen, Y.; Qu, B.; Hu, L.; Xu, Z.; Li, Q.; Wang, T. High-performance supercapacitor and lithium-ion battery based on 3D hierarchical NH4F-induced nickel cobaltate nanosheet–nanowire cluster arrays as self-supported electrodes. *Nanoscale* **2013**, *5*, 9812–9820. [CrossRef]
29. Gopi, C.V.M.; Ramesh, R.; Kim, H.J. Designing nanosheet manganese cobaltate@manganese cobaltate nanosheet arrays as a battery-type electrode material towards high-performance supercapacitors. *J. Energy Storage* **2022**, *47*, 103603. [CrossRef]
30. Kim, H.J.; Kim, C.W.; Kim, S.Y.; Reddy, A.E.; Gopi, C.V.V.M. Facile synthesis of unique diamond-like structured CdMn2O4@CdMn2O4 composite material for high performance supercapacitors. *Materials Lett.* **2018**, *210*, 143–147. [CrossRef]
31. Gopi, C.V.V.M.; Vinodh, R.; Sambasivam, S.; Obaidat, I.M.; Kalla, R.M.N.; Kim, H.J. One-pot synthesis of copper oxide-cobalt oxide core-shell nanocactus-like heterostructures as binder-free electrode materials for high-rate hybrid supercapacitors. *Mater. Today Energy* **2019**, *14*, 100358. [CrossRef]
32. Sambasivam, S.; Gopi, C.V.M.; Arbi, H.M.; Kumar, Y.A.; Kim, H.J.; Zahmi, S.A.; Obaidat, I.M. Binder-free hierarchical core-shell-like CoMn$_2$O$_4$@MnS nanowire arrays on nickel foam as a battery-type electrode material for high-performance supercapacitors. *J. Energy Storage* **2021**, *36*, 102377. [CrossRef]
33. Yousef, R.; Al-Zoubi, A.; Sad-Din, N. A Study of Structural Properties of CuMn2O4 Synthesized by Solid State Method. *Adv. Phys. Theor. Appl.* **2018**, *71*, 24–30.
34. Zhu, B.; Qin, Y.; Du, J.; Zhang, F.; Lei, X. Ammonia Etching to Generate Oxygen Vacancies on CuMn2O4 for Highly Efficient Electrocatalytic Oxidation of 5-Hydroxymethylfurfural. *ACS Sustain. Chem. Eng.* **2021**, *9*, 11790–11797. [CrossRef]
35. Guo, Y.; Yu, L.; Wang, C.Y.; Lin, Z.; Lou, X.W.D. Hierarchical Tubular Structures Composed of Mn-Based Mixed Metal Oxide Nanoflakes with Enhanced Electrochemical Properties. *Adv. Funct. Mater.* **2015**, *25*, 5184–5189. [CrossRef]
36. Guan, B.Y.; Kushima, A.; Yu, L.; Li, S.; Li, J.; Lou, X.W.D. Coordination Polymers Derived General Synthesis of Multishelled Mixed Metal-Oxide Particles for Hybrid Supercapacitors. *Adv. Mater.* **2017**, *29*, 1605902. [CrossRef]
37. Fu, W.B.; Zhao, Y.Y.; Mei, J.F.; Wang, F.J.; Han, W.H.; Wang, F.C.; Xie, E.Q. Honeycomb-like Ni3S2 nanosheet arrays for high-performance hybrid supercapacitors. *Electrochim. Acta* **2018**, *283*, 737–743. [CrossRef]
38. Gao, M.J.; Le, K.; Xu, D.M.; Wang, Z.; Wang, F.L.; Liu, W.; Yu, H.J.; Liu, J.R.; Chen, C.Z. Controlled sulfidation towards achieving core-shell 1D-NiMoO4 @ 2D-NiMoS4 architecture for high-performance asymmetric supercapacitor. *J. Alloys Compd.* **2019**, *804*, 27–34. [CrossRef]
39. Wang, G.R.; Li, Y.B.; Xu, L.; Jin, Z.L.; Wang, Y.B. Facile synthesis of difunctional NiV LDH@ZIF-67 p-n junction: Serve as prominent photocatalyst for hydrogen evolution and supercapacitor electrode as well. *Renew. Energ.* **2020**, *162*, 535–549. [CrossRef]
40. Sambasivam, S.; Raghavendra, K.V.G.; Yedluri, A.K.; Arbi, H.M.; Narayanaswamy, V.; Gopi, C.V.V.M.; Choi, B.-C.; Kim, H.-J.; Alzahmi, S.; Obaidat, I.M. Facile Fabrication of MnCo$_2$O$_4$/NiO Flower-Like Nanostructure Composites with Improved Energy Storage Capacity for High-Performance Supercapacitors. *Nanomaterials* **2021**, *11*, 1424. [CrossRef]
41. Bhagwan, J.; Nagaraju, G.; Ramulu, B.; Sekhar, S.C.; Yu, J.S. Rapid synthesis of hexagonal NiCo$_2$O$_4$ nanostructures for high-performance asymmetric supercapacitors. *Electrochim. Acta* **2019**, *299*, 509–517. [CrossRef]
42. Jiang, W.; Hu, F.; Yan, Q.; Wu, X. Investigation on electrochemical behaviors of NiCo2O4 battery-type supercapacitor electrodes: The role of an aqueous electrolyte. *Inorg. Chem. Front.* **2017**, *4*, 1642–1648. [CrossRef]

43. Sun, J.; Du, X.; Wu, R.; Zhang, Y.; Xu, C.; Chen, H. Bundle-like CuCo2O4 microstructures assembled with ultrathin nanosheets as battery-type electrode materials for high-performance hybrid supercapacitors. *ACS Appl. Energy Mater.* **2020**, *3*, 8026–8037. [CrossRef]

44. Sun, Y.; Zhang, J.J.; Sun, X.N.; Huang, N.B. High-performance spinel NiMn2O4 microshphres self-assembled with nanosheets by microwave-assisted synthesis for supercapacitors. *Crystengcomm* **2020**, *22*, 1645–1652. [CrossRef]

45. Gopi, C.V.V.M.; Joo, H.H.; Vinodha, R.; Kim, H.J.; Sambasivam, S.; Obaidat, I.M. Facile synthesis of flexible and binder-free dandelion flower-like CuNiO2 nanostructures as advanced electrode material for high-performance supercapacitors. *J. Energy Storage* **2019**, *26*, 100914.

46. Li, G.F.; Liu, S.Q.; Pan, Y.; Zhou, T.Y.; Ding, J.D.; Sun, Y.M.; Wang, Y.Q. Self-templated formation of CuCo2O4 triple-shelled hollow microspheres for all-solid-state asymmetric supercapacitor. *J. Alloys Compd.* **2019**, *787*, 694–699. [CrossRef]

47. Huang, T.; Qiu, Z.; Hu, Z.; Zhang, Z. Porous chopsticks-like FeCo$_2$O$_4$ by the hydrothermal method for high-performance asymmetric supercapacitors. *J. Energy Storage* **2022**, *46*, 103898. [CrossRef]

nanomaterials

MDPI

Article

Chemical Vapor Transport Synthesis of Fibrous Red Phosphorus Crystal as Anodes for Lithium-Ion Batteries

Lei Liu [1,†], Xing Gao [2,†], Xuemei Cui [3], Bofeng Wang [1], Fangzheng Hu [1], Tianheng Yuan [1], Jianhua Li [4], Lei Zu [1,*], Huiqin Lian [1] and Xiuguo Cui [1,*]

1   College of New Materials and Chemical Engineering, Beijing Institute of Petrochemical Technology, Beijing 102617, China
2   School of Materials Science and Engineering, Beijing University of Chemical Technology, Beijing 100029, China
3   Department of Mechanical and Materials Engineering, College of Engineering and Applied Science, University of Cincinnati, 2600 Clifton Ave, Cincinnati, OH 45221, USA
4   Kailuan (Group) Limited Liability Corporation, Tangshan 064012, China
*   Correspondence: zulei@bipt.edu.cn (L.Z.); cuixiuguo@bipt.edu.cn (X.C.)
†   These authors contributed equally to this work.

Abstract: Red phosphorus (RP) is considered to be the most promising anode material for lithium-Ion batteries (LIBs) due to its high theoretical specific capacity and suitable voltage platform. However, its poor electrical conductivity (10^{-12} S/m) and the large volume changes that accompany the cycling process severely limit its practical application. Herein, we have prepared fibrous red phosphorus (FP) that possesses better electrical conductivity (10^{-4} S/m) and a special structure by chemical vapor transport (CVT) to improve electrochemical performance as an anode material for LIBs. Compounding it with graphite (C) by a simple ball milling method, the composite material (FP-C) shows a high reversible specific capacity of 1621 mAh/g, excellent high-rate performance and long cycle life with a capacity of 742.4 mAh/g after 700 cycles at a high current density of 2 A/g, and coulombic efficiencies reaching almost 100% for each cycle.

Keywords: fibrous phosphorus; lithium-Ion battery; anode

check for updates

Citation: Liu, L.; Gao, X.; Cui, X.; Wang, B.; Hu, F.; Yuan, T.; Li, J.; Zu, L.; Lian, H.; Cui, X. Chemical Vapor Transport Synthesis of Fibrous Red Phosphorus Crystal as Anodes for Lithium-Ion Batteries. *Nanomaterials* 2023, *13*, 1060. https://doi.org/10.3390/nano13061060

Academic Editor: Carlos Miguel Costa

Received: 28 February 2023
Revised: 13 March 2023
Accepted: 13 March 2023
Published: 15 March 2023

1. Introduction

With the growing urgency of the global energy crisis and environmental pollution, the development and application of clean energy must be vigorously promoted. Great progress has been made in sustainable energy technologies based on wind or solar energy. Therefore, it is desirable to develop efficient, safe and inexpensive energy storage technologies. Rechargeable (secondary) batteries are typically used in small to medium scale energy storage. Among these, LIBs have been utilized as the predominant power source of portable electronic devices due to their relatively high energy density, long life, lack of memory effects and environmental friendliness [1–3].

Recently, phosphorus (P) has turned out to be the most promising anode material candidate for LIBs with a high theoretical specific capacity (2596 mAh/g, based on the alloying process of P-LiP-Li$_2$P-Li$_3$P) [4]. The element phosphorus exists in three main forms: white phosphorus (WP), RP and black phosphorus (BP). Figure S1 shows the structures of different allotropes of phosphorus. WP is very unstable and flammable due to the weak bonding energy of tetrahedral P$_4$, which is very dangerous and makes WP unsuitable for use in LIBs [5]. BP is also widely recognized for its high electrical conductivity (−100 S/m, close to hard carbon) and its application in LIBs [6], but the difficulty of preparation and the high price of BP severely limit the commercialization of BP. Compared to WP and BP, RP is a more prospective commercial anode material that is not only cheaper but also environmentally friendly. However, there are two main challenges limiting the practical application of RP: (1) RP has an extremely low electronic conductivity of 10^{-14} S/m and is almost non-conductive. (2) RP suffers from a huge volume expansion of about 300% during cycling [5,7–9].

These problems can lead to material fracture and pulverization of the red phosphorus material during cycling, as well as rapid capacity decay. To address these problems, many researchers have demonstrated the need to build microstructures and reduce the size of the active material to achieve fast kinetics. Park et al. [10] prepared phosphorus–carbon composites by high-energy ball milling, which significantly improved the electrical conductivity of P and enabled its application in LIBs. Zhou et al. [11] successfully prepared hollow phosphorus nanospheres with porous shells and controlled diameters by a solvothermal method. The hollow nanosphere structures of these porous shells effectively adapt to volume changes and avoid the pulverization of active material, exhibiting excellent long-cycle performance in LIBs.

Wang et al. [12] first compounded porous carbon with RP by the evaporation–condensation method, resulting in a stable battery cycle of 55 turns. The evaporation–condensation method is considered to be an effective strategy to address the large volume expansion caused by conductivity and lithiation. However, there are two important issues in the evaporative condensation approach; firstly, the low P mass loading (~30 wt%) in the carbon-based framework severely limits the energy density. Secondly [13,14], the residual WP can lead to safety issues of flammability and high toxicity [15,16]. To address the low P mass loading and safety issues, Zhang et al. [17] systematically investigated the interactions between P_4 and various functional groups in carbon materials (C_{surf}, C_{edge}, C−N−5, C−N−1, C−N$_g$, C−S−5, C−S−6, C−S−6, C=O, C−O−C, C−OH and COOH) and provided a structural design strategy for the carbon framework to build up the edge sp^2 carbon atoms. These edge carbon atoms, in turn, provide high adsorption energy for P_4 molecules through the strong P-C bonds. The RP-PC anodes they prepared exhibit extremely high P mass loading (close to the theoretical limit for loading), strong and stable P-C bonds, and significantly improved electron and Li^+ transfer, resulting in excellent high-rate, long-cycle stability.

Inspired by this previous work, we prepared FP materials by the CVT method in order to further improve the rate capability of P to Li storage. FP materials possess many electrochemical advantages: (1) the electrical conductivity of FP is 10^{-4} S/m (0.2 V), which is 8 orders of magnitude higher than that of commercial red phosphorus (RP) (10^{-12} S/m), and FP possesses better electrical conductivity compared to RP [18]. (2) FP is a 1-dimensional (1D) material, but it also possesses a 2-dimensional (2D) layered structure, and this 2D structure can effectively reduce the Li^+ diffusion length and also enrich its electrochemical active center [19]. (3) FP can be reduced in size to the nanoscale by liquid phase exfoliation, and its small-enough size can effectively reduce the volume change of the material during cycling without structural damage [20]. The fibrous red phosphorus–graphite (FP-C) composite, prepared by FP with graphite by ball milling exhibited excellent cycling stability and high-rate performance, maintaining a high reversible specific capacity of 742.4 mAh/g after 700 cycles at high current densities.

2. Materials and Methods

2.1. Preparation of FP

(1) Synthesis of the FP: 6 g of commercial amorphous red phosphorus and 0.6 g of iodine were loaded into a quartz tube while maintaining a vacuum inside the tube. The tube was placed in a muffle furnace. The muffle furnace increased the temperature to 500 °C at a rate of 2.5 °C per minute and maintained it for 24 h, followed by natural cooling to room temperature. The quartz tube was broken, and the lump of FP was removed and ground into a powder, which was subsequently cleaned and dried using ethanol and acetone.

(2) Control experiment: 6 g of commercial amorphous red phosphorus was loaded into a quartz tube while maintaining a vacuum inside the tube. The subsequent heat and treatment process is as described above.

(3) Preparation of crystalline red phosphorus nanoribbons (FP NR): The cleaned FP powder was added to 250 mL of NMP and sonicated (600 W) for 12 h to disperse the FP in the NMP.

2.2. Preparation of FP-C and RP-C

Synthesis of the phosphorus–graphite composites: The phosphorus–carbon composites were prepared by a simple ball milling process. FP or RP was mixed with graphite powder (mass ratio 7:3), 1 g of the mixture was removed and added to zirconia (25 mL) and zirconia ball milling beads were placed (mass ratio 90:1). The above operations were carried out in a glove box. The jar was placed in a planetary wheel ball mill with a speed of 500 r/min and ball milling was carried out without interruption for 12 h.

2.3. Measurement of Material Characteristics

Analysis of the crystal phase of the material was carried out by X-ray diffraction (XRD) using a Panalytical Empyrean diffractometer. Raman spectra of the analyzed materials were collected using a Renishwa Raman instrument using a 532 nm laser wavelength. Analysis of the chemical composition of the material was determined using X-ray photoelectron spectroscopy (XPS), carried out on a Thermo ESCALAB 250XI using a monochromatic Al-Ka source (1486.6 eV). Observation of the morphology and structure of the prepared materials was carried out using a scanning electron microscope (SEM) and transmission electron microscopy (TEM). SEM images were collected using a ZEISS Gemini 300 device and TEM images were gathered using a JEOL JEM-F200 device operating at 200 kV.

2.4. Electrochemical Characterization of Materials

A CR2032 button cell, consisting of a phosphorus electrode, diaphragm and lithium metal, was assembled in an argon glove box with a glove box water oxygen content of less than 0.01 PPm. Preparation of the phosphorus based electrode: A homogeneous slurry was made by mixing phosphorus–carbon composite, acetylene black and polyvinylidene fluoride (PVDF) binder in a mass ratio of 7:2:1 into a N-Methyl pyrrolidinone (NMP) solution. The slurry was coated onto copper foil and baked in a vacuum drying oven at 120 °C for 12 h. The mass loading of electrodes was around 1.9 mg. A common electrolyte was used, 1 M/L LiPF$_6$ at EC/DEC/EMC = 1:1:1, v/v, 10% FEC. Celgard 2400 film was used as the separator. Constant current charge/discharge tests were carried out using a LANDHE cell test system with a voltage window ranging from 0.01 to 2.5 V (for Li electrodes). Cyclic voltammetric curve testing was carried out using a CH1660B electrochemical workstation with an electrochemical window ranging from 0.01 to 3.0 V. Electrochemical impedance spectra were tested on a CH1660B electrochemical workstation in the frequency range from 10^5 to 0.01 Hz. All of these electrochemical performances were tested under ambient temperature.

3. Results and Discussion

3.1. Characterization of FP

We prepared bulk FP by the low-temperature CVT method. As can be seen in Figure S2b, the addition of iodine (I$_2$) causes the RP to nucleate and crystallize in the low-temperature region of the quartz tube to form FP blocks that attach to the inner wall of the quartz tube. This suggests that I$_2$ plays a transport and catalytic role throughout CVT. RP as a solid source and I$_2$ as a sublimed mineralizer produce gaseous intermediate iodophosphorus compounds (PI$_x$) in the sublimation zone when the temperature reaches the sublimation temperature, which are transported towards the deposition zone by the PI$_x$ partial pressure gradient [21]. The gaseous intermediate is converted to deposit products (FP) and releases mineralizer molecules. The released mineralizer molecules increase the partial pressure of the mineralizer in the deposition zone, thus they can transport back to the sublimation zone for further reactions until all the RP is converted to FP [22]. The bulks of FP (Figure S3) were ground and cleaned by ultrasonication with alcohol to remove I$_2$ and dried by filtration to obtain FP powder. As shown in Figure S4, from optical microscope images of RP and FP powders, it is evident that RP is a deep red color, while FP appears orange-red. It can be seen that RP is a particle of approximately 40 μm in size, in contrast to FP which is a strip of fibers. There is a clear morphological difference between RP and FP, and RP can induce a transition from granular RP to one-dimensional fibrous FP after catalysis by I_2. The FP was observed by SEM, as can be seen in Figure 1a, and the bulk FP after growth by CVT consisted of 50 μm long micron rods arranged neatly and showing a dispersive shape. A partial magnification of one end of one of the micrometer rods shows that the FP has a lamellar structure (Figure 1b). After ultrasonic exfoliation, the length and diameter of the FP powder were significantly shortened, both consisting of rods of a few tens of microns (Figure 1c). The morphology and structure of the FP nanoribbons were observed by TEM (Figure 1d), and it was found that the FP nanoribbons showed a lamellar shape. In the high-resolution TEM (HR-TEM) image (Figure 1e), the lattice stripe spacing of 0.58 nm corresponded to the (001) plane in the vertical view. In addition, the surface of the FP nanoribbons is locally oxidized, which is consistent with the P-O bonding results appearing by XPS. The selected area electron diffraction (SAED) pattern of a single nanoribbon in Figure 1f shows that single-crystal FP nanoribbons were formed along the [1$\bar{4}$1] axis.

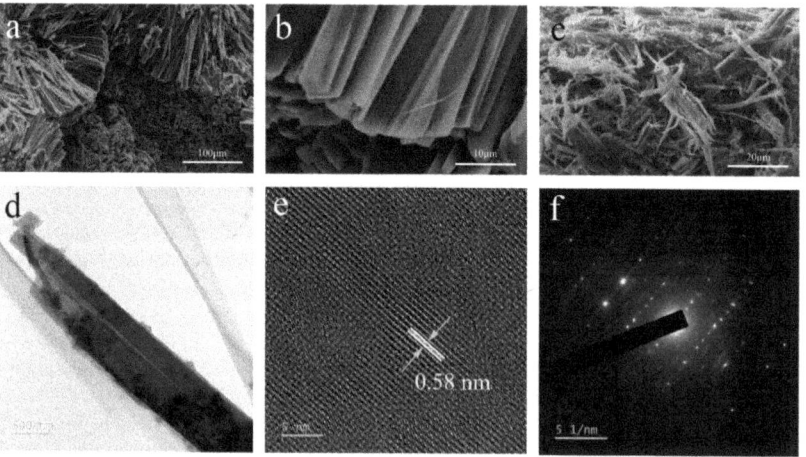

Figure 1. (**a**,**b**) SEM images of FP bulk; (**c**) SEM image of FP powder; (**d**) TEM image of FP NR; (**e**) HR-TEM image of the FP NR; (**f**) SAED pattern of a single of FP NR.

FP was further compared with RP by XRD, as shown in Figure 2a. The XRD pattern of RP is a broad-peaked amorphous phase in the range of 12–18° and 26–36°, corresponding to the characteristic peaks of commercial amorphous RP. The XRD peak of FP corresponds highly to the XRD diffraction peak of triclinic P. Also based on the study of Du et al. [23] for FP, it shows that the sample we prepared was 1D FP. In their study of FP, Liu et al. [19] found that FP also has a 2D band structure with a nanoribbon thickness of approximately 7.8 nm, which indicates that FP has not only a 1D structure but also a 2D band structure. As can be seen from the structural diagram of FP (Figure 2a–c), FP consists of P-cage chains of P_8 and P_9 interconnected to form tubular structures, and the layers of these tubular structures are parallel to each other, with two tubular structures connected by two dumbbell-shaped p-atom structures. It should be noted that if the layers of the tubular structure are perpendicular to each other they are Hittorf's phosphorus (HP) or violet phosphorus (Figure S1), which also results in a high degree of similarity between the XRD and Raman spectrum of FP and HP [18]. Moreover, based on Winchester et al.'s summary [24], the preparation of FP did not match with type II RP and matched best with type IV RP. Figure 2b contrasts the Raman spectra of FP and RP in the frequency range of 200–500 cm^{-1}. It can be seen that RP has no obvious characteristic peaks, and only broad peaks appear in the range of 345–365 cm^{-1}, which mainly comes from the long-range disordered structure of polymeric phosphides, while the Raman spectra of FP show complex vibrational patterns, mainly due to the low symmetry of trigonal crystals and the diversity of atomic coordination. The peak at 368 cm^{-1} is the stretching vibration of the P$_8$ phosphorus cage and the peak at 353 cm^{-1} is the stretching vibration peak of the P$_9$ phosphorus cage [25]. XPS was used to perform compositional analysis of the FP powders. As shown in Figure 2c, the presence of element I$_2$ was not detected in the XPS survey spectrum, suggesting that the addition of I$_2$ only facilitated the transition from RP to FP and did not adulterate the FP. In addition, the 2p$^{1/2}$ and 2p$^{3/2}$ peaks at 130.0 eV and 131.4 eV, respectively, correspond to the P-P bond [18]. The peak at 134.7 eV corresponds to the P-O bond (Figure 2d), which indicates that FP undergoes oxidation in air, which is identical to the oxidative nature of BP [26,27].

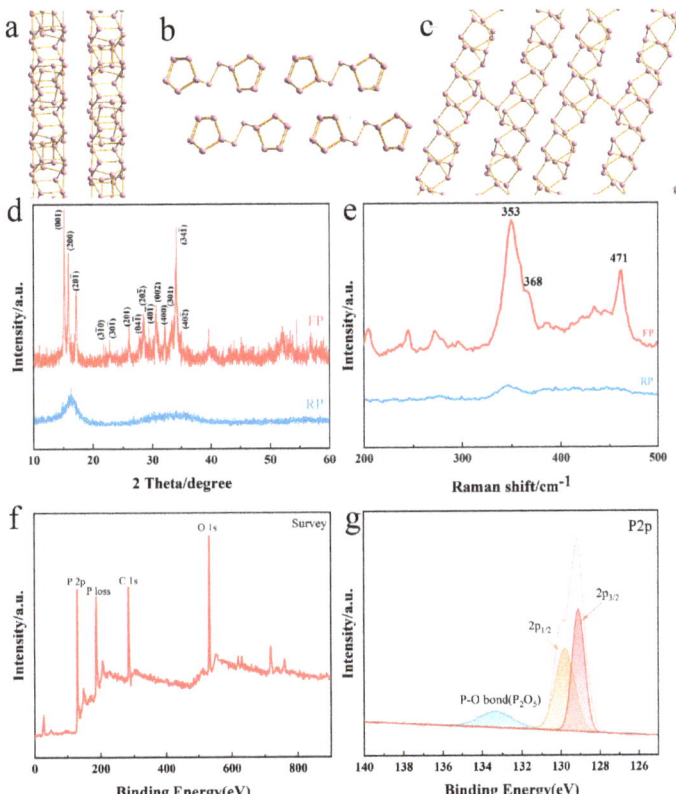

Figure 2. Schematic structural view of FP along the a (**a**), b (**b**) and c (**c**); (**d**) XRD curves and (**e**) Raman spectra of RP and FP; (**f**) XPS survey spectrum of FP; (**g**) P2p spectrum of FP.

3.2. Electrochemical Properties of FP-C and RP-C

FP-C and RP-C composites were prepared by ball milling FP and RP with graphite and characterized by XRD and Raman (Figure S5), and the electrochemical properties of the composites were studied. The lithiation and de-lithiation processes of FP-C and RP-C composites under half-cells were investigated by cyclic voltammetry, tested at a scan rate of 0.2 mV/s with a voltage window of 0.01–3 V. As shown in Figure 3b,c, a broad peak appeared at about 1 V during the first cathodic scan of the CV curve of FP-C, which is attributed to the formation of the SEI film of the solid electrolyte [28], and the subsequent broad peak is the lithiation process of the P-based material. Two peaks of approximately 1.1 V and 1.6 V appeared in the anode, corresponding to a de-lithiation process of the P-based material [29,30]. In the subsequent cycles, the cathode peak shifted to 0.6 V and the peak area increased, followed by a high overlap of the CV curves, indicating the high reversibility and stability of the FP-C negative electrode. In contrast, the anodic peak of the RP-C negative electrode decayed severely and the peak area decreased, which suggests severe polarization and low reversible stability of the RP-C negative electrode. It is noteworthy that the redox peak potentials of FP and RP are almost identical, which illustrates that the electrochemical properties of FP and RP as anode materials in LIBs are highly similar, but the electrochemical properties owing to their different structures show great differences.

Figure 3d compares the cycling stability performance of FP and RP. FP-C shows excellent cycling stability and still holds a high specific capacity of 1621 mAh/g after 80 cycles, whereas the capacity of RP-C only remains at 494 mAh/g. The capacity of the RP-C electrode drops rapidly before 40 cycles, which is attributed to a rupture of the SEI film due to the huge volume expansion of RP during lithiation. The thicker SEI film creates such electrochemical segregation that the Li^+ in the formed Li-P alloy cannot return to the Li metal to form dead Li^+, while the larger volume expansion leads to pulverization of the active material, and this part of the P is no longer involved in the normal lithiation/delithiation process. The FP, however, may have a special 1D structure and nanoscale scale such that it does not continue to expand after a certain range of expansion [20], which results in the FP maintaining a very high reversible specific capacity after 80 cycles. It was observed that FP-C maintained a stable and high coulombic efficiency throughout the cycle, while the coulombic efficiency of RP-C fluctuated remarkably, which was attributed to the stable structure of FP-C (Figure 3a). The structural stability of the composite facilitates the stabilization of the SEI passivation layer during cycling (Figure S6), maintaining nearly 100% coulombic efficiency and excellent cyclability. Figure S6 shows that the FP remains a relatively intact structure after 20 cycles at a current density of 0.5 A/g. Figure 3c,f show the constant current charge/discharge curves for FP-C and RP-C at a current density of 0.2 mA/g. The hysteresis of the charge-discharge curve of RP-C increased significantly after cycling, rising to 1158 mV after 70 cycles, whereas the overpotential of FP-C was more stable and much lower than that of RP-C, at 338 mV after 70 cycles, a phenomenon that is highly consistent with the results of the CV curve.

The high-rate capability is also an important indicator to evaluate the performance of the battery. Figure 3g displays that FP-C exhibits a higher-rate capability than RP-C in the current density range of 0.28–2 A/g. Specifically, the average reversible capacities of FP-C at current densities of 0.28, 0.8, 1.2, 1.6, 2, 0.5 and 0.28 A/g were 1726, 1444, 1312, 1203, 1060, 1316 and 1638 mAh/g, respectively, while the average reversible capacities of RP-C were only 1116, 839, 661, 591, 449, 609 and 599 mAh/g, respectively. The contrast between the two is remarkable, as the reversible discharge capacity of RP-C is only 45% that of FP-C at a high current density of 2 A/g. After high multiplication cycles, when the current density returned to 0.28 mA/g, FP-C still had a specific capacity of 1401 mAh/g compared to 567 mAh/g for RP-C, indicating that FP-C exhibits excellent high-rate performance and fast response kinetics. It is noteworthy that the initial discharged capacity of FP-C is higher than that of RP-C in both the cycling test of low current density and the high-rate capability test. This may be due to the fact that the layered structure of FP allows Li^+ to be embedded in the layers before the alloy reaction, similar to the insertion of BP into the layers before the alloy reaction [31], so that Li^+ can alloy with more P atoms, while the 1D structure of FP allows Li^+ to pass through faster [32]. Figure 3h shows that the reversible discharge capacity of FP-C remains at 742.4 mAh/g after 700 cycles at a high current density of 2 A/g, demonstrating the excellent high rate cycling stability performance of FP-C. Figure 3i and Table S1 show how this work compares to previous work, with the FP-C electrode exhibiting excellent electrochemical performance [5,6,12,17,32–42].

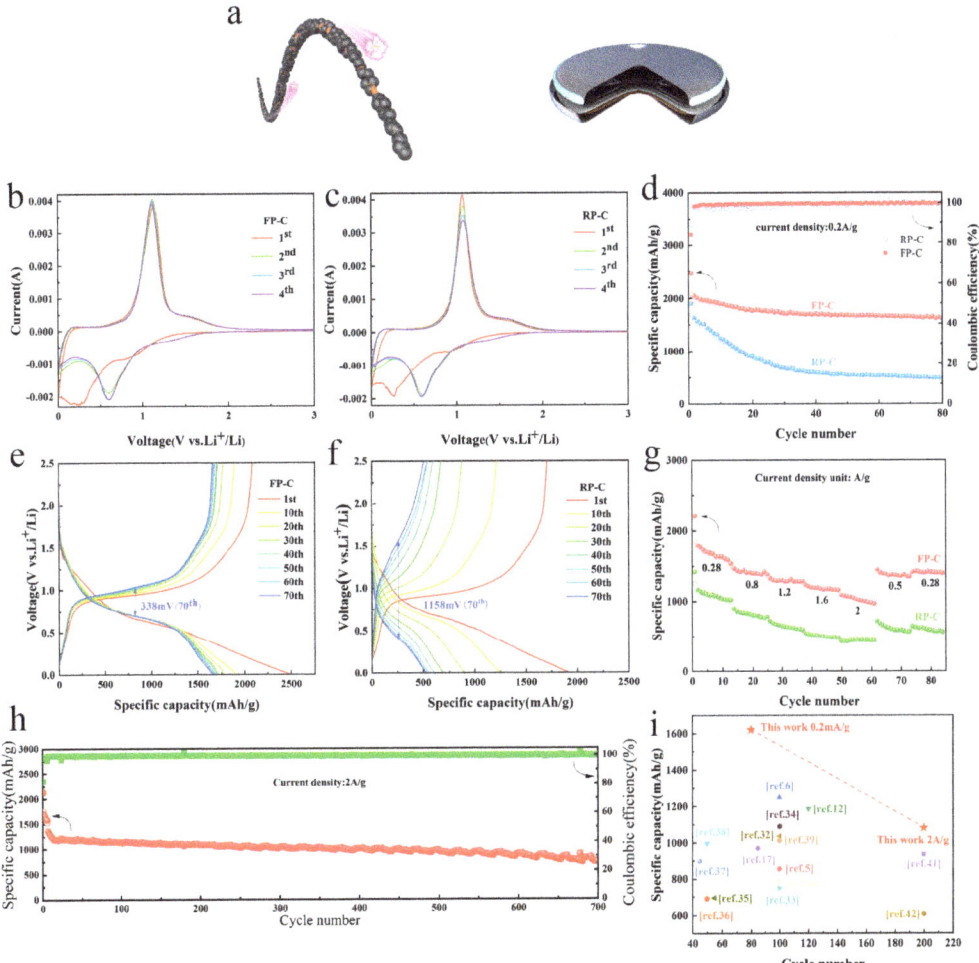

Figure 3. Electrochemical performance of RP-C and FP-C electrodes for LIBs. (**a**) Schematic illustration of the FP-C and button cell. (**b,c**) CV of the FP-C and RP-C anodes with a scan rate of 0.2 mV/s between 0.01 and 3.0 V vs. Li$^+$/Li. Cycling stability of FP-C and RP-C anodes for 80 cycles at 0.2 A/g (**d**) and the corresponding voltage profiles of FP-C (**e**) and RP-C (**f**). (**g**) Rate performance of FP-C and RP-C anodes at the varied rate from 0.28 A/g to 2 A/g. (**h**) Capacity and coulombic efficiency during cycling of the FP-C anode at 2 A/g after 5 initial cycles at a low current density of 0.2 A/g. Overall capacity and cycling stability compared with other reported RP- and BP-based anode materials (**i**). The data and current density of each material are summarized in Table S1 (Supporting Information).

We prepared BP-C composites using the same method and investigated their cyclic stability and high-rate capability. As can be seen from Figure S7, the remaining reversible capacity of the BP-C negative electrode after 80 cycles at a current density of 0.2 A/g is 1555.7 mAh/g, a lower rate than that of the FP-C (1621 mAh/g). However, BP-C performed well in terms of high-rate capability with an average reversible capacity of 1250.13 mAh/g at a high current density of 2 A/g, compared to 1060 mAh/g for FP-C, suggesting that FP has better cycling stability compared to BP, and BP has better high-rate performance. The higher conductivity (100 S/m) of BP also contributes to the better high-rate performance of

BP, which can be attributed to the special 1D structure of FP that allows FP-C to exhibit better cycling stability. Although the conductivity of FP is 8 orders of magnitude better than that of RP, the conductivity of FP is still lower than that of BP, indicating that FP is more suitable for energy storage applications than BP, while BP is more suitable for power battery applications.

The kinetic factors were further investigated on the electrochemical performance by testing the CV of FP-C and RP-C at a series of scan rates of 0.1, 0.2, 0.4, 0.6, 0.8 and 1.0 mV/s, as shown in Figure 4a,b. The peak potentials of the CV curves for both FP-C and RP-C electrodes rise with increasing scan rate and cover the potential range of the lower scan rate, which also indicates the capacitive behavior of both FP-C and RP-C electrodes. According to equations 1 and 2 [36], there is a power law relationship between the peak current of the CV curve and the scan rate, and the electrochemical reactions can be divided into diffusion-Controlled interactions and capacitive processes.

$$i = a\nu^b \tag{1}$$

$$\log(i) = b\log(\nu) + \log(a) \tag{2}$$

where i is the peak current (mA), v is the scan rate (mV/s) and a and b are adjustable parameters.

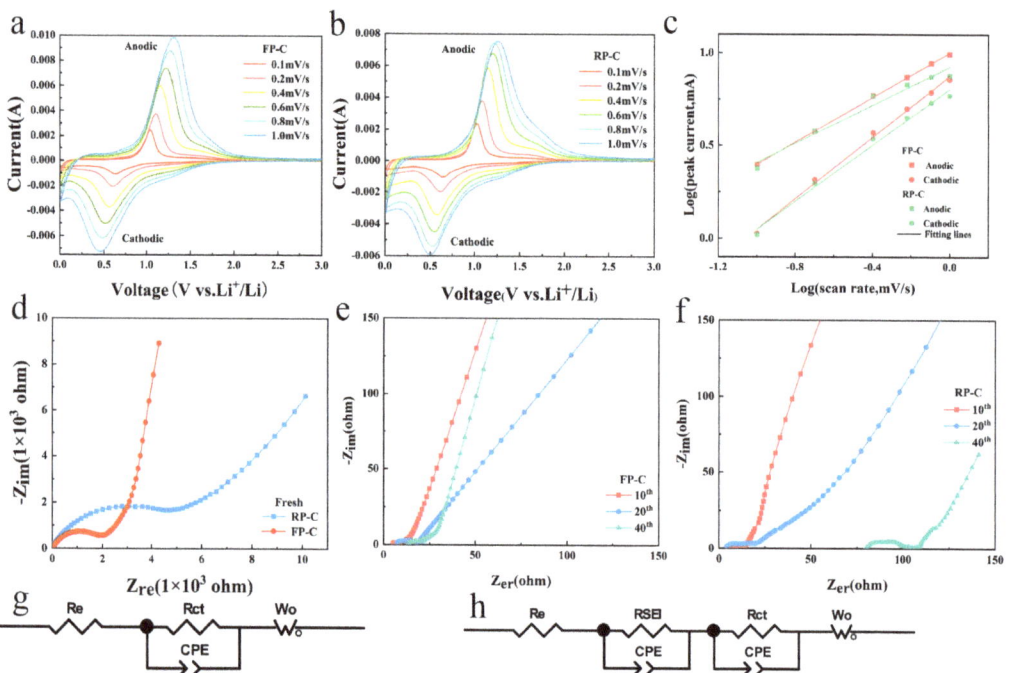

Figure 4. Electrochemical reaction kinetics of FP-C and RP-C electrodes for LIBs. CV curves of FP-C (**a**) and RP-C (**b**) at various scan rates from 0.1 to 1.0 mV/s for LIBs. (**c**) Log (peak current)-log (scan rate) curves for the observed cathodic and anodic peaks in (**a,b**). EIS Nyquist plots of fresh FP-C and RP-C electrodes (**d**). EIS Nyquist plots of FP-C (**e**) and RP-C (**f**) electrodes were tested after 10, 20, and 40 cycles at 0.2 A/g. Equivalent circuit (**g**) corresponding to (**d**) EIS spectra. (**h**) is the equivalent circuit for the EIS spectra of (**e,f**).

As shown in Figure 4c, the b values for the FP-C anode and cathode peaks were calculated to be 0.604 and 0.827, respectively, indicating that both diffusion-Controlled alloying reactions and the capacitive processes dominate the electrochemical reactions of FP-C. Similarly, the b values of 0.516 and 0.757 for the anode and cathode of RP-C, respectively, suggest that the capacity of both FP-C and RP-C are diffusion-Controlled alloy reactions.

Electrochemical impedance spectroscopy (EIS) was also used to interpret the detailed electrochemical reaction kinetics. Figure 4d–f show the EIS spectra of FP-C and RP-C electrodes after testing at 0, 10, 20, and 40 cycles turns at a current density of 0.2 A/g. All Nyquist plots consist of a semicircle in the high-frequency region and a line in the low-frequency region, representing the charge transfer process and Li^+ diffusion in the solid state, respectively. The equivalent circuits corresponding to the impedance data include the electrolyte and electrode ohmic resistance (R_e), the SEI film resistance (R_{SEI}), the charge-transfer resistance (R_{ct}) and the parallel constant phase element (CPE), as well as the Warburg impedance (W_o). The fresh RP-C electrode had an R_{ct} of ~4750 Ω, while the FP-C electrode had an R_{ct} of ~2035 Ω, showing the higher electronic conductivity of the FP-C. As shown in the graphs of turn 10 versus turn 20 of the cycle, the R_{ct} of both the RP-C and FP-C electrode decrease to a small stable value, indicating that a stable interface is formed between the electrolyte and the electrode (Figure 4e,f), while the R_{ct} of the RP-C is slightly larger than that of the FP-C. However, at the 40th cycle, the FP-C and RP-C show a more distinct difference, with the RP-C electrode having a higher internal cell resistance and being heavily polarized, corresponding to the cycle performance test in Figure 3d, which coincides with the time of maximum cell capacity decay. In the case of the FP-C electrode, the tendency for the battery capacity to decay decreases in 20 cycles and becomes more stable inside the battery at 40 cycles.

4. Conclusions

In conclusion, we have prepared FP by the low-temperature CVT method. FP significantly increases its electrochemical performance as a P anode electrode due to its special structure and high electrical conductivity. While FP was prepared with C by the ball milling method to produce FP-C composites, the composites exhibited much better electrochemical performance with excellent cycle stability and multiplicative properties than RP-C. The design strategy of FP-C offers great guidance implications for achieving mass-produced and high-performance P-based negative electrodes.

Supplementary Materials: The following supporting information can be downloaded at: https://www.mdpi.com/article/10.3390/nano13061060/s1. Table S1: Comparison of the electrochemical performance of our work with that of previous phosphorus-based LIBs. Figure S1. Schematic structures of different allotropes of phosphorus. Figure S2: Optical photographs of the quartz tube reactor after the reaction (a) without I2, (b) with I2. Figure S3: Optical photographs of FP bulk lumps. Figure S4: Optical photographs of filamentous red phosphorus. Figure S5: FP-C and RP-C composites' (a) XRD diagram (b) Raman spectra. Figure S6. SEM images of the FP-C electrodes in LIBs after 20 cycles at a current density of 0.5 A/g. Figure S7: (a) Cycling stability of BP-C anodes for 80 cycles at 0.2 A/g. (b) Rate performance of FP-C and RP-C anodes at the varied rate from 0.28 A/g to 2 A/g.

Author Contributions: Conceptualization, X.C. (Xuemei Cui); formal analysis, F.H. and T.Y.; investigation, B.W.; resources, H.L.; writing—original draft preparation, L.L.; writing—review and editing, X.G.; visualization, J.L.; supervision, L.Z.; funding acquisition, X.C. (Xiuguo Cui) All authors have read and agreed to the published version of the manuscript.

Funding: This research was funded by [National Natural Science Foundation of China Projects] grant number [51573021].

Data Availability Statement: The data presented in this study are available on request on request form the corresponding author.

Conflicts of Interest: The authors declare that they have no known competing financial interest or personal relationships that could have appeared to influence the work reported in this paper.

References

1. Marom, R.; Amalraj, S.F.; Leifer, N.; Jacob, D.; Aurbach, D. A review of advanced and practical lithium battery materials. *J. Mater. Chem.* **2011**, *21*, 9938–9954. [CrossRef]
2. Etacheri, V.; Marom, R.; Elazari, R.; Salitra, G.; Aurbach, D. Challenges in the development of advanced Li-ion batteries: A review. *Energy Environ. Sci.* **2011**, *4*, 3243–3262. [CrossRef]
3. Yu, H.-C.; Ling, C.; Bhattacharya, J.; Thomas, J.C.; Thornton, K.; Van der Ven, A. Designing the next generation high capacity battery electrodes. *Energy Environ. Sci.* **2014**, *7*, 1760–1768. [CrossRef]
4. Ramireddy, T.; Xing, T.; Rahman, M.M.; Chen, Y.; Dutercq, Q.; Gunzelmann, D.; Glushenkov, A.M. Phosphorus-carbon nanocomposite anodes for lithium-ion and sodium-ion batteries. *J. Mater. Chem. A* **2015**, *3*, 5572–5584. [CrossRef]
5. Sun, J.; Zheng, G.; Lee, H.-W.; Liu, N.; Wang, H.; Yao, H.; Yang, W.; Cui, Y. Formation of Stable Phosphorus-Carbon Bond for Enhanced Performance in Black Phosphorus Nanoparticle-Graphite Composite Battery Anodes. *Nano Lett.* **2014**, *14*, 4573–4580. [CrossRef]
6. Jin, H.; Xin, S.; Chuang, C.; Li, W.; Wang, H.; Zhu, J.; Xie, H.; Zhang, T.; Wan, Y.; Qi, Z.; et al. Black phosphorus composites with engineered interfaces for high-rate high-capacity lithium storage. *Science* **2020**, *370*, 192–197. [CrossRef]
7. Sun, J.; Lee, H.-W.; Pasta, M.; Yuan, H.; Zheng, G.; Sun, Y.; Li, Y.; Cui, Y. A phosphorene-graphene hybrid material as a high-capacity anode for sodium-ion batteries. *Nat. Nanotechnol.* **2015**, *10*, 980–985. [CrossRef]
8. Zhu, Y.; Wen, Y.; Fan, X.; Gao, T.; Han, F.; Luo, C.; Liou, S.-C.; Wang, C. Red Phosphorus–Single-Walled Carbon Nanotube Composite as a Superior Anode for Sodium Ion Batteries. *ACS Nano* **2015**, *9*, 3254–3264. [CrossRef]
9. Yu, Z.; Song, J.; Gordin, M.L.; Yi, R.; Tang, D.; Wang, D. Phosphorus-Graphene Nanosheet Hybrids as Lithium-Ion Anode with Exceptional High-Temperature Cycling Stability. *Adv. Sci.* **2015**, *2*, 1400020. [CrossRef]
10. Park, C.-M.; Sohn, H.-J. Black Phosphorus and its Composite for Lithium Rechargeable Batteries. *Adv. Mater.* **2007**, *19*, 2465–2468. [CrossRef]
11. Zhou, J.; Liu, X.; Cai, W.; Zhu, Y.; Liang, J.; Zhang, K.; Lan, Y.; Jiang, Z.; Wang, G.; Qian, Y. Wet-Chemical Synthesis of Hollow Red-Phosphorus Nanospheres with Porous Shells as Anodes for High-Performance Lithium-Ion and Sodium-Ion Batteries. *Adv. Mater.* **2017**, *29*, 1700214. [CrossRef]
12. Wang, L.; He, X.; Li, J.; Sun, W.; Gao, J.; Guo, J.; Jiang, C. Nano-Structured Phosphorus Composite as High-Capacity Anode Materials for Lithium Batteries. *Angew. Chem.* **2012**, *124*, 9168–9171. [CrossRef]
13. Hao, G.; Jiao, R.; Deng, Z.; Liu, Y.; Lan, D.; He, W.; Lang, Z.; Cui, J. Red phosphorus infiltrated into porous C/SiOx derived from rice husks to improve its initial Coulomb efficiency in lithium-ion batteries. *Colloids Surf. A Physicochem. Eng. Asp.* **2023**, *665*, 131180. [CrossRef]
14. Vorfolomeeva, A.A.; Stolyarova, S.G.; Asanov, I.P.; Shlyakhova, E.V.; Plyusnin, P.E.; Maksimovskiy, E.A.; Gerasimov, E.Y.; Chuvilin, A.L.; Okotrub, A.V.; Bulusheva, L.G. Single-Walled Carbon Nanotubes with Red Phosphorus in Lithium-Ion Batteries: Effect of Surface and Encapsulated Phosphorus. *Nanomaterials* **2022**, *13*, 153. [CrossRef]
15. Zhou, J.; Shi, Q.; Ullah, S.; Yang, X.; Bachmatiuk, A.; Yang, R.; Rummeli, M.H. Phosphorus-Based Composites as Anode Materials for Advanced Alkali Metal Ion Batteries. *Adv. Funct. Mater.* **2020**, *30*, 2004648. [CrossRef]
16. Zhou, J.; Jiang, Z.; Niu, S.; Zhu, S.; Zhou, J.; Zhu, Y.; Liang, J.; Han, D.; Xu, K.; Zhu, L. Self-standing hierarchical P/CNTs@ rGO with unprecedented capacity and stability for lithium and sodium storage. *Chem* **2018**, *4*, 372–385. [CrossRef]
17. Zhang, S.; Liu, C.; Wang, H.; Wang, H.; Sun, J.; Zhang, Y.; Han, X.; Cao, Y.; Liu, S.; Sun, J. A Covalent P–C Bond Stabilizes Red Phosphorus in an Engineered Carbon Host for High-Performance Lithium-Ion Battery Anodes. *ACS Nano* **2021**, *15*, 3365–3375. [CrossRef]
18. Chen, Z.; Zhu, Y.; Wang, Q.; Liu, W.; Cui, Y.; Tao, X.; Zhang, D. Fibrous phosphorus: A promising candidate as anode for lithium-ion batteries. *Electrochim. Acta* **2019**, *295*, 230–236. [CrossRef]
19. Liu, Q.; Zhang, X.; Wang, J.; Zhang, Y.; Bian, S.; Cheng, Z.; Kang, N.; Huang, H.; Gu, S.; Wang, Y.; et al. Crystalline Red Phosphorus Nanoribbons: Large-Scale Synthesis and Electrochemical Nitrogen Fixation. *Angew. Chem.* **2020**, *132*, 14489–14493. [CrossRef]
20. Chan, C.K.; Peng, H.; Liu, G.; McIlwrath, K.; Zhang, X.F.; Huggins, R.A.; Cui, Y. High-performance lithium battery anodes using silicon nanowires. *Nat. Nanotechnol.* **2007**, *3*, 31–35. [CrossRef]
21. Wang, D.; Luo, F.; Lu, M.; Xie, X.; Huang, L.; Huang, W. Chemical Vapor Transport Reactions for Synthesizing Layered Materials and Their 2D Counterparts. *Small* **2019**, *15*, e1804404. [CrossRef] [PubMed]
22. Sun, Z.; Zhang, B.; Zhao, Y.; Khurram, M.; Yan, Q. Synthesis, Exfoliation, and Transport Properties of Quasi-1D van der Waals Fibrous Red Phosphorus. *Chem. Mater.* **2021**, *33*, 6240–6248. [CrossRef]
23. Du, L.; Zhao, Y.; Wu, L.; Hu, X.; Yao, L.; Wang, Y.; Bai, X.; Dai, Y.; Qiao, J.; Uddin, G.; et al. Giant anisotropic photonics in the 1D van der Waals semiconductor fibrous red phosphorus. *Nat. Commun.* **2021**, *12*, 4822. [CrossRef] [PubMed]
24. Winchester, R.A.L.; Whitby, M.; Shaffer, M.S.P. Synthesis of Pure Phosphorus Nanostructures. *Angew. Chem.* **2009**, *121*, 3670–3675. [CrossRef]
25. Fasol, G.; Cardona, M.; Hönle, W.; Von Schnering, H. Lattice dynamics of Hittorf's phosphorus and identification of structural groups and defects in amorphous red phosphorus. *Solid State Commun.* **1984**, *52*, 307–310. [CrossRef]
26. Wood, J.D.; Wells, S.A.; Jariwala, D.; Chen, K.-S.; Cho, E.; Sangwan, V.K.; Liu, X.; Lauhon, L.J.; Marks, T.J.; Hersam, M.C. Effective Passivation of Exfoliated Black Phosphorus Transistors against Ambient Degradation. *Nano Lett.* **2014**, *14*, 6964–6970. [CrossRef]

27. Zhao, Y.; Sun, Z.; Zhang, B.; Yan, Q. Unveiling the degradation chemistry of fibrous red phosphorus under ambient conditions. *ACS Appl. Mater. Interfaces* **2022**, *14*, 9925–9932. [CrossRef]
28. Li, X.; Zhang, S.; Du, J.; Liu, L.; Mao, C.; Sun, J.; Chen, A. Strong interaction between phosphorus and wrinkle carbon sphere promote the performance of phosphorus anode material for lithium-ion batteries. *Nano Res.* **2023**. [CrossRef]
29. Marino, C.; Boulet, L.; Gaveau, P.; Fraisse, B.; Monconduit, L. Nanoconfined phosphorus in mesoporous carbon as an electrode for Li-ion batteries: Performance and mechanism. *J. Mater. Chem.* **2012**, *22*, 22713–22720. [CrossRef]
30. Kim, D.-Y.; Ahn, H.-J.; Kim, J.-S.; Kim, I.-P.; Kweon, J.-H.; Nam, T.-H.; Kim, K.-W.; Ahn, J.-H.; Hong, S.-H. The Electrochemical Properties of Nano-Sized Cobalt Powder as an Anode Material for Lithium Batteries. *Electron. Mater. Lett.* **2009**, *5*, 183–186. [CrossRef]
31. Hembram, K.P.S.S.; Jung, H.; Yeo, B.C.; Pai, S.J.; Lee, H.J.; Lee, K.-R.; Han, S.S. A comparative first-principles study of the lithiation, sodiation, and magnesiation of black phosphorus for Li-, Na-, and Mg-ion batteries. *Phys. Chem. Chem. Phys.* **2016**, *18*, 21391–21397. [CrossRef]
32. Li, W.; Yang, Z.; Li, M.; Jiang, Y.; Wei, X.; Zhong, X.; Gu, L.; Yu, Y. Amorphous Red Phosphorus Embedded in Highly Ordered Mesoporous Carbon with Superior Lithium and Sodium Storage Capacity. *Nano Lett.* **2016**, *16*, 1546–1553. [CrossRef]
33. Kong, W.; Wen, Z.; Zhou, Z.; Wang, G.; Yin, J.; Cui, L.; Sun, W. A self-healing high-performance phosphorus composite anode enabled by *in situ* preformed intermediate lithium sulfides. *J. Mater. Chem. A* **2019**, *7*, 27048–27056. [CrossRef]
34. Li, M.; Li, W.; Hu, Y.; Yakovenko, A.A.; Ren, Y.; Luo, J.; Holden, W.M.; Shakouri, M.; Xiao, Q.; Gao, X. New insights into the high-performance black phosphorus anode for lithium-ion batteries. *Adv. Mater.* **2021**, *33*, 2101259. [CrossRef]
35. Zhang, Y.; Wang, L.; Xu, H.; Cao, J.; Chen, D.; Han, W. 3D Chemical Cross-Linking Structure of Black Phosphorus@CNTs Hybrid as a Promising Anode Material for Lithium Ion Batteries. *Adv. Funct. Mater.* **2020**, *30*, 1909372. [CrossRef]
36. Liu, B.; Zhang, Q.; Li, L.; Jin, Z.; Wang, C.; Zhang, L.; Su, Z.-M. Encapsulating Red Phosphorus in Ultralarge Pore Volume Hierarchical Porous Carbon Nanospheres for Lithium/Sodium-Ion Half/Full Batteries. *ACS Nano* **2019**, *13*, 13513–13523. [CrossRef]
37. Wang, L.; Guo, H.; Wang, W.; Teng, K.; Xu, Z.; Chen, C.; Li, C.; Yang, C.; Hu, C. Preparation of sandwich-like phosphorus/reduced graphene oxide composites as anode materials for lithium-ion batteries. *Electrochim. Acta* **2016**, *211*, 499–506. [CrossRef]
38. Han, X.; Zhang, Z.; Han, M.; Cui, Y.; Sun, J. Fabrication of red phosphorus anode for fast-charging lithium-ion batteries based on TiN/TiP2-enhanced interfacial kinetics. *Energy Storage Mater.* **2019**, *26*, 147–156. [CrossRef]
39. Yuan, D.; Cheng, J.; Qu, G.; Li, X.; Ni, W.; Wang, B.; Liu, H. Amorphous red phosphorous embedded in carbon nano-tubes scaffold as promising anode materials for lithium-ion batteries. *J. Power Sources* **2016**, *301*, 131–137. [CrossRef]
40. Li, J.; Jin, H.; Yuan, Y.; Lu, H.; Su, C.; Fan, D.; Li, Y.; Wang, J.; Lu, J.; Wang, S. Encapsulating phosphorus inside carbon nanotubes via a solution approach for advanced lithium ion host. *Nano Energy* **2019**, *58*, 23–29. [CrossRef]
41. Liu, H.; Zou, Y.; Tao, L.; Ma, Z.; Liu, D.; Zhou, P.; Liu, H.; Wang, S. Sandwiched Thin-Film Anode of Chemically Bonded Black Phosphorus/Graphene Hybrid for Lithium-Ion Battery. *Small* **2017**, *13*, 1700758. [CrossRef] [PubMed]
42. Yan, C.; Zhao, H.; Li, J.; Jin, H.; Liu, L.; Wu, W.; Wang, J.; Lei, Y.; Wang, S. Mild-Temperature Solution-Assisted Encapsulation of Phosphorus into ZIF-8 Derived Porous Carbon as Lithium-Ion Battery Anode. *Small* **2020**, *16*, 1907141. [CrossRef] [PubMed]

Article

Boosting the Initial Coulomb Efficiency of Sisal Fiber-Derived Carbon Anode for Sodium Ion Batteries by Microstructure Controlling

Yuan Luo [1], Yaya Xu [1], Xuenuan Li [1], Kaiyou Zhang [1], Qi Pang [2] and Aimiao Qin [1,2,*]

[1] Key Lab New Processing Technology for Nonferrous Metal and Materials, Ministry of Education, College of Material Science and Engineering, Guilin University of Technology, Guilin 541004, China
[2] Guangxi Key Laboratory of Electrochemical Energy Materials, School of Chemistry and Chemical Engineering, Guangxi University, Nanning 530004, China
* Correspondence: 2005032@glut.edu.cn

Abstract: As anode material for sodium ion batteries (SIBs), biomass-derived hard carbon has attracted a great deal of attention from researchers because of its renewable nature and low cost. However, its application is greatly limited due to its low initial Coulomb efficiency (ICE). In this work, we employed a simple two-step method to prepare three different structures of hard carbon materials from sisal fibers and explored the structural effects on the ICE. It was determined that the obtained carbon material, with hollow and tubular structure (TSFC), exhibits the best electrochemical performance, with a high ICE of 76.7%, possessing a large layer spacing, a moderate specific surface area, and a hierarchical porous structure. In order to better understand the sodium storage behavior in this special structural material, exhaustive testing was performed. Combining the experimental and theoretical results, an "adsorption-intercalation" model for the sodium storage mechanism of the TSFC is proposed.

Keywords: bio-derived hard carbon; Coulomb efficiency; Na-ion batteries; storage mechanism

Citation: Luo, Y.; Xu, Y.; Li, X.; Zhang, K.; Pang, Q.; Qin, A. Boosting the Initial Coulomb Efficiency of Sisal Fiber-Derived Carbon Anode for Sodium Ion Batteries by Microstructure Controlling. *Nanomaterials* **2023**, *13*, 881. https://doi.org/10.3390/nano13050881

Academic Editor: Carlos Miguel Costa

Received: 5 February 2023
Revised: 19 February 2023
Accepted: 22 February 2023
Published: 26 February 2023

1. Introduction

Although lithium-ion batteries (LIBs), with a long cycle life and high energy density, are considered one of the most promising and successful energy storage systems today, they also face a number of problems, such as high cost and lack of resources. As a result, a new technology has emerged in the field of energy storage-sodium ion batteries (SIBs), which have electrochemical properties similar to LIBs [1–3]. Researchers have focused on the following types of anode materials: titanium-based oxides, sodium alloys, binary transition metal oxides (such as $NiMoO_4$ [4] and $MgMoO_4$ [5]), and carbon materials. Among the carbon-based materials that have been widely studied are graphite, graphene, soft carbon, and hard carbon. It is well known that graphite is the most common anode material in lithium-ion batteries, but its performance in sodium-ion batteries is not as expected because the Na^+ radius (0.102 nm) is larger than the Li^+ radius (0.076 nm) [6], Na^+ cannot be stably embedded in the graphite structure, and only a very small number of binary sodium-graphite embedding compounds can be embedded in graphite [7]. Since the insertion and removal of large size sodium ions leads to a slowing down of the kinetic process, the structure undergoes irreversible phase changes, thus accelerating the degradation of the electrochemical properties, which implies a greater degree of limitation on the structure of the material [8–10].

Biomass hard carbon has attracted great attention due to its low cost, renewable nature, and green properties. Many biomass hard carbon materials, such as biowaste [11], eggshell [12], mango seed husk [13], etc., have been proven to have excellent storage performance when used in energy storage systems. However, most of them exhibit low initial

Coulombic efficiency (ICE) for SIBs, i.e., 53.1%, 64.0%, and 69.0% of ICE for hard carbon from kelp [14], rice husk [15], and tea [16], respectively. Therefore, many researchers have discussed in depth the issues of defects, structural design, surface area, and conductivity to effectively improve the Coulomb efficiency. The properties of biomass char are mainly dependent on the nature of the raw material and thermal transformation. In order to make these biochar products more suitable for use as electrochemical energy storage materials, appropriate modifications are required to enhance the specific surface area, the pore volume, or the formation of functional groups. Usually, there are two methods for enhancement: physical or chemical. Physical activation usually involves a two-step process. Biomass materials are first pyrolyzed to generate biochar (400 °C to 850 °C) and then activated using gases such as CO_2, air, or their mixtures [17]. These methods can reasonably design the structure of carbon material and improve the electrochemical properties. For example, Yan et al. [18] reported char balls obtained from nitrogen-rich oatmeal by hydrothermal and subsequent charring processes, exhibiting a smooth surface with an average diameter of about 2 μm; the results show that NCSs treated at 500 °C exhibit a high maximum charge capacity of 336 mAh g^{-1} after 50 cycles at a current density of 50 mA g^{-1}. Duan et al. [19] prepared N-doped carbon microspheres by the pyrolysis of chitin from seafood waste (crab and shrimp shells), which consisted of nanofiber entanglements forming an interlinked nanofiber framework structure, and the highest deliverable energy density reached up to 58.7 Wh/kg. Jin et al. [20] prepared a series of porous hollow charcoal spheres using various spores (stone pine grass, Ganoderma lucidum, and multi-spike stone pine spores) as charcoal precursors and self-templates using high-temperature charring and activation treatments; the obtained electrodes showed remarkable electrical double-layer storage performances, such as high specific capacitance (308 F g^{-1} in organic electrolytes), ultrafast rate capability (retaining 263 F g^{-1} at a very high current density of 20 A g^{-1}), and good cycling stability (93.8% retention after 10,000 charge-discharge cycles). Dong et al. [21] prepared a nitrogen-doped foamy charcoal plate by the charring and activation of teak peel and the prepared charcoal plate, with a macroporous network consisting of hollow tubes with diameters of 20–50 μm; the obtained electrodes showed a high specific capacitance of up to 338 F g^{-1} at 1 A g^{-1} and good rate capability with a capacitance retention of 59% at 20 A g^{-1}.

In our group's early efforts, we have conducted great work on the application of biomass carbon materials in LIBs. Among many studies, it was found that the hard carbon derived from cellulose-rich sisal fibers, with the advantages of low cost, green, and sustainable bioresource, exhibits excellent electrochemical performance when used as an anode for LIBs. The main focus of our previous work was to modulate the structure of sisal fibers using chemical activation and calcination methods, and to obtain sisal fiber carbon with various morphologies and structures [22–24] (its chemical activation reagents were KOH and HCl). Generally, the ICE of hard carbon can be effectively improved by controlling the defect concentration and specific surface area [25], but developing appropriate structures to ensure structural stability and high Coulombic efficiency, without affecting the electrolyte–electrolyte interface behavior, remains a challenge. In this work, in order to carry out an in-depth study on the relationship between the biomass hard carbon structures and their energy storage properties, for the first time, we chose sisal fibers as the study material for SIBs. We prepared three structure sisal fiber carbons: tubular sisal fiber carbon (TSFC), sheet sisal fiber carbon (SSFC), and spherical sisal fiber carbon (GSFC), using hydrothermal and calcination methods, and systematically investigated the different effects on ICE with the change of sisal fiber carbon structure. It was found that the special structure of TSFC can helps to improve electrochemical performance, especially greatly improving the ICE (76.7%), as well as the cycling stability. In addition, the diffusion kinetics and storage behavior of sodium ions in TSFC were further investigated.

2. Materials and Methods

2.1. Materials

All the chemical reagents used in the experiments were of analytical grade and were used without further purification. All aqueous solutions were prepared with deionized water. The sisal fiber used was from Guangxi Sisal Group, KOH reagent (Purity 85%, Guangdong Guanghua Chemical Factory Co., Ltd., Guangdong, China), and HCl reagent (AR 36–38%, Xilong Chemical Co., Shanghai, China).

2.2. Preparation of Tubular Sisal Fiber Carbon

The tubular sisal fiber carbon was prepared based on the methods used in our previous work [22], and the preparation process is as follows:

First, the sisal fiber was cleaned by washing, and 5 g of the clean, dry sisal fiber was placed into 2.5 mol L^{-1} KOH solution for hydrothermal reaction (160 °C, 14 h). When the reaction was finished and the autoclave was naturally cooled to room temperature, the precursor was collected after filtering and washing it to neutral. Then, it was dried in a blast drying oven (60 °C) for 24 h. Finally, the dried precursor was put into a crucible and calcined in a tube furnace at nitrogen atmosphere and kept at 900 °C for 1 h. The obtained black sample was collected in a crucible and named TSFC.

2.3. Preparation of Sheet Sisal Fiber Carbon

The sheet sisal fiber carbon was prepared based on the methods used in our previous work [23], and the preparation process is as follows:

First, the sisal fiber was cleaned by washing, and 5 g of the clean, dry sisal fiber was placed into 2.5 mol L^{-1} KOH solution for hydrothermal reaction (160 °C, 14 h). When the reaction finished and the autoclave was naturally cooled to room temperature, the precursor was collected after filtering and washing it to neutral. Then, it was dried in a blast drying oven (60 °C) for 24 h. A total of 3 g of the dried sample was weighed into a beaker, heated (80 °C), and stirred for 2 h (with 100 mL of 2.5 mol L^{-1} KOH solution), followed by drying. Finally, the dried precursor was put into a crucible and calcined in a tube furnace (nitrogen atmosphere, 900 °C, 1 h). The obtained black sample was collected in a crucible and named SSFC.

2.4. Preparation of Sphere Sisal Fiber Carbon

The sphere sisal fiber carbon was prepared based on the methods used in our previous work [24], the preparation process is as follows:

First, sisal fiber was cleaned by washing, and 5 g of the clean, dry sisal fiber was placed into a 2 mol L^{-1} HCl solution for hydrothermal reaction (180 °C, 12 h). When the reaction finished and the autoclave was naturally cooled to room temperature, the precursor was collected after filtering and washing it to neutral. Then, it was dried in a blast drying oven (60 °C) for 24 h. Finally, the dried precursor was put into a crucible and calcined in a tube furnace (nitrogen atmosphere, 900 °C, 1 h). The obtained black sample was collected in a crucible and named GSFC. In order to investigate the optimal conditions, we adjusted the concentration of HCl (0.1 M, 1 M, 4 M), which we named GSFC-0.1, GSFC-1, and GSFC-4, respectively.

2.5. Material Characterization

The morphology of the samples was observed by SEM (S4800) and TEM (JEM-2100F). The structure of the hard carbon material was characterized by XRD (X'Pert PRO, PANalytical B.V., Cu Kα = 1.54056 Å, 2Θ = 10–80) and Raman (Thermo Fisher Scientific DXR, Waltham, MA, USA), with the 532 nm excitation wavelength. The micropore volume and average pore size of the materials were characterized by nitrogen adsorption and desorption experiments (Quantachrome, Autosorb, Boynton Beach, FL, USA) at 77 k. The electrical conductivity of the material was measured by the four-probe method using model RTS-2A (Guangzhou Four-Probe Technology, Guangzhou, China) equipment.

2.6. Electrochemical Measurements

The prepared active material was homogeneously mixed with conductive carbon black and polyvinylidene fluoride (PVDF) in the mass ratio of 8:1:1; then, N-methyl-2-pyrrolidone (NMP) was used as the solvent, and finally, the mixture was made into a paste of appropriate concentration. The electrolyte was $1.0 \, mol \, L^{-1}$ $NaClO_4$ in ethylene carbonate (EC) and dimethyl carbonate (DEC), with a volume ratio of 1:1 and 5 vol% fluoroethylene carbonate. The electrochemical properties were tested using the CHI-760D electrochemical workstation. The voltage window for cyclic voltammetry (CV) measurement is 0.01–3 V and the electrochemical impedance spectra (EIS) were in the frequency range of 0.01–100,000 Hz. The Neware GCD testing system was utilized for galvanostatic charging/discharging profiles.

3. Results and Discussion

3.1. Morphology and Structure of Carbon Materials

The morphology and structure of the samples were represented by SEM and TEM. The SEM images of TSFC, SSFC, and GSFC at different magnifications are shown in Figure 1. From Figure 1a,b, it can be seen that the microstructure of TSFC consists of a hollow tube, with a wall thickness of ~1.4 μm and an inner tube diameter of ~4.5 μm. The special hollow tubular structure with a bumpy striped surface can promote the transfer of electrons. As shown in Figure 1c,d, the SSFC displays porous nanosheet structures, with a sheet thickness of about 10 nm and a pore size of 1–2 μm, and these interconnected nanosheets form a porous channel structure. Figure 1e,f presents the solid spherical structure of GSFC of about ~3 μm in size, with a smooth surface. In order to investigate the pretreatment condition effects on the morphology and structure of the materials, we take GSFC as an example to observe the evolution process of the spherical structure under different concentrations of HCl (0.1 M, 1 M, 2 M, and 4 M). The SEM results are shown in Figure S1. It can be seen that the morphology and structure of the material change significantly with the increase in the concentration of HCl: (i) in small concentrations of 0.1 M, small sphere particles grow on the surface of fibrous tubular carbon; (ii) when the concentration increases to 1 M, the elongated tubular fibers gradually disappear and become irregular spheres; (iii) when the concentration reaches 2 M, regular and more uniform spheres are obtained; and (iv) when the concentration continues to increase to 4 M, a large number of spheres are bonded together and the uniformity of the morphology decreases. We can assume that this structural change is due to the breakage of a large number of ester and ether bonds in the material during the degradation process [26]. Previously, the Austrian chemist Skrabal also mentioned that the decomposition of a compound is affected by a certain concentration of H^+ or OH^- [26]. When cellulose hydrolyzes in an acidic environment, it produces a conjugate acid, which cleaves the β-1,4 glycosidic bond and interacts with H_2O to produce H^+, yielding more glucose [26]. Glucose is cleaved to form a large number of carboxylic acids, aldehydes, and furans, which are dehydrated and condensed to form aromatic compounds, which in turn form spherical carbon materials [26,27]. On the contrary, in an alkaline environment [26], cellulose will form a large number of organic acids during hydrolysis to neutralize the alkaline substances of the reaction system, which will cause a large amount of H^+ to be consumed, resulting in inadequate cellulose dissolution and stripping the cellulose away. The formation of a lamellar porous structure is mainly due to the reaction of KOH and C at high temperatures to produce H_2, CO_2, and CO, which promote the formation of porous carbon materials [28].

As depicted in Figure 2a–f, the TEM diagram clearly exhibits the microstructure of the materials. Figure 2b,d,f shows the high resolution transmission microscopy (HRTEM) images of TSFC, SSFC, and GSFC, respectively. We can clearly see that the crystal plane spacing of all three structures is larger than the graphite layer spacing (graphite layer spacing is 0.335 nm), and the crystal plane spacing of TSFC (0.37 nm) is significantly larger than the other two morphologies. It has also been demonstrated in the literature that Na ions cannot be inserted into graphite with a layer spacing smaller than 0.335 nm, but they can be easily inserted into graphite with a layer spacing of 0.37 nm [29], so we can expect that the TSFC anode will exhibit superior sodium storage performance compared to the other two structures.

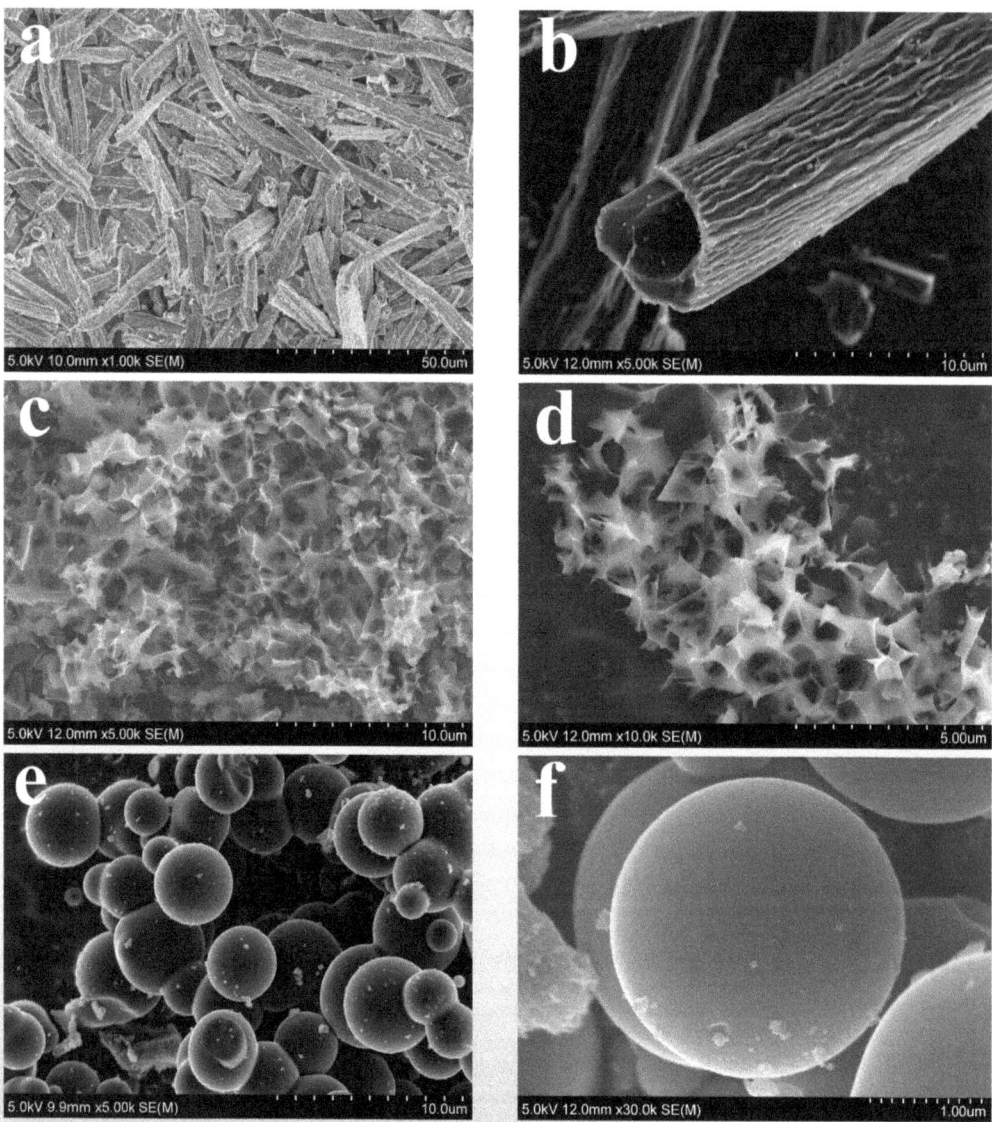

Figure 1. The different magnification SEM images of (**a**,**b**) TSFC, (**c**,**d**) SSFC, (**e**,**f**) GSFC.

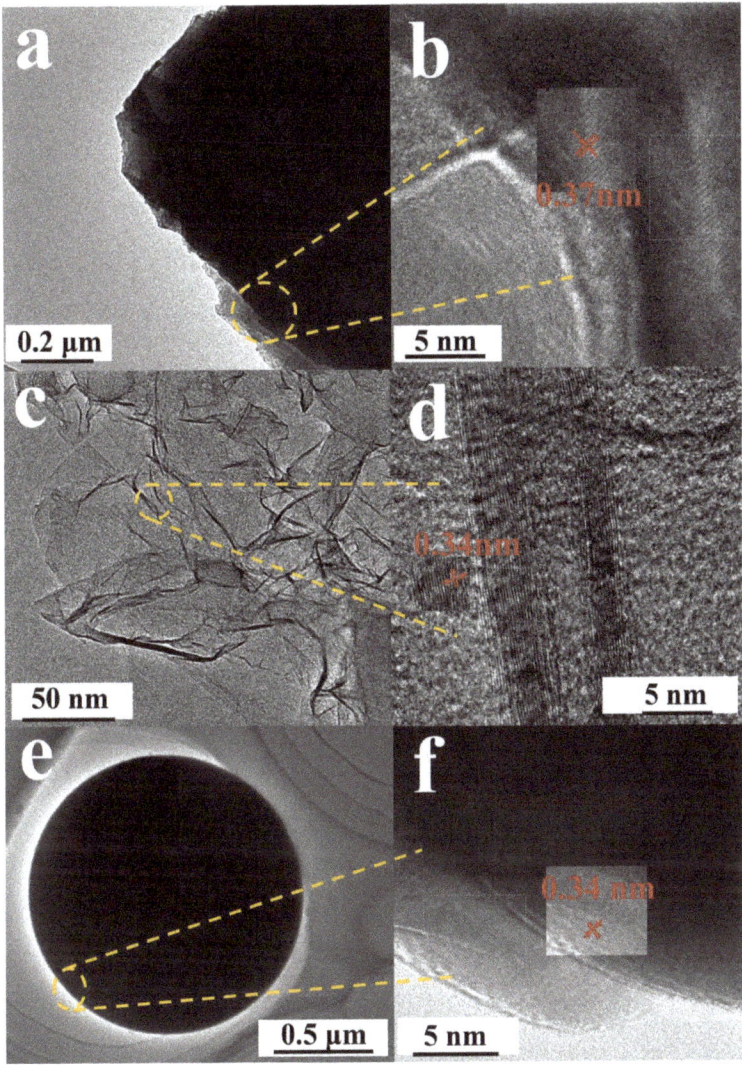

Figure 2. TEM images of (**a**,**b**) TSFC, (**c**,**d**) SSFC, (**e**,**f**) GSFC.

As shown in Figure 3a,b, the microstructure of samples was analyzed by XRD and Raman spectroscopy. The XRD patterns of TSFC, SSFC, and GSFC display two main broad peaks at about ~23° and 43° corresponding to (002) and (101) lattice planes, respectively [30], demonstrating the amorphous state of the samples. A shift in the (002) peak toward a small angle can be observed for the TSFC material, indicating that the interlayer spacing of the structure also changes to some extent. This is consistent with the TEM results, indicating that the layer spacing of biogenic hard carbon changes with the change of morphology [31]. Additionally, the disordered and graphitized structures of samples have been assessed by Raman spectroscopy. The Raman spectra show two separate characteristic bands of the D-band peak at ~1330 cm^{-1} and the G-band peak at 1580 cm^{-1}, corresponding to the D-band, with sp3 defects, and the G-band, with ordered graphite sp2 features [32]. The value of I_D/I_G is often used to characterize the defects in carbon materials [33]. It can be seen that the I_D/I_G value of TSFC sample (0.852) is higher than that of the SSFC (0.847) and

GSFC (0.835) samples, indicating that TSFC has more defects and may provide more active sites for sodium storage [34]. The porosity of the samples was further investigated by N_2 adsorption-desorption tests. The N_2 adsorption-desorption isotherms of TSFC, SSFC, and GSFC are shown in Figure 3c, where we can observe that the three structures exhibit the type IV isotherms. For TSFC and SSFC, their isotherms show a sharp rise at relative pressures of less than 0.1, followed by bending into a platform accompanying the H4 reversible hysteresis loop, implying the presence of well-developed microporous and mesoporous structures. The average pore size and pore volume were calculated by the Barrett–Joyner–Halenda (BJH) method and the single point method. The pore size distribution of the three structures in Figure 3d indicates the presence of well-developed mesoporous structures in the three hard carbon structure materials [32]. The porous structure details are given in Table S1, and compared with other two samples, TSFC has the most developed porous structure, with an average pore diameter of about 2.79 nm and a medium specific surface area of 426.02 m^2 g^{-1}, as well as a moderate open pore volume of 0.049 cm^3 g^{-1}. However, it is worth noting that the increase in microporous volume leads not only to a low density of the hard carbon material, but also to a decrease in intercalation capacity due to the reduction of graphite-like nanodomains in the hard carbon structure [35]. Therefore, combining all the above data references, the hollow tubular structure is favorable for soaking the electrolyte and will promote the ion transport from electrolyte to electrode.

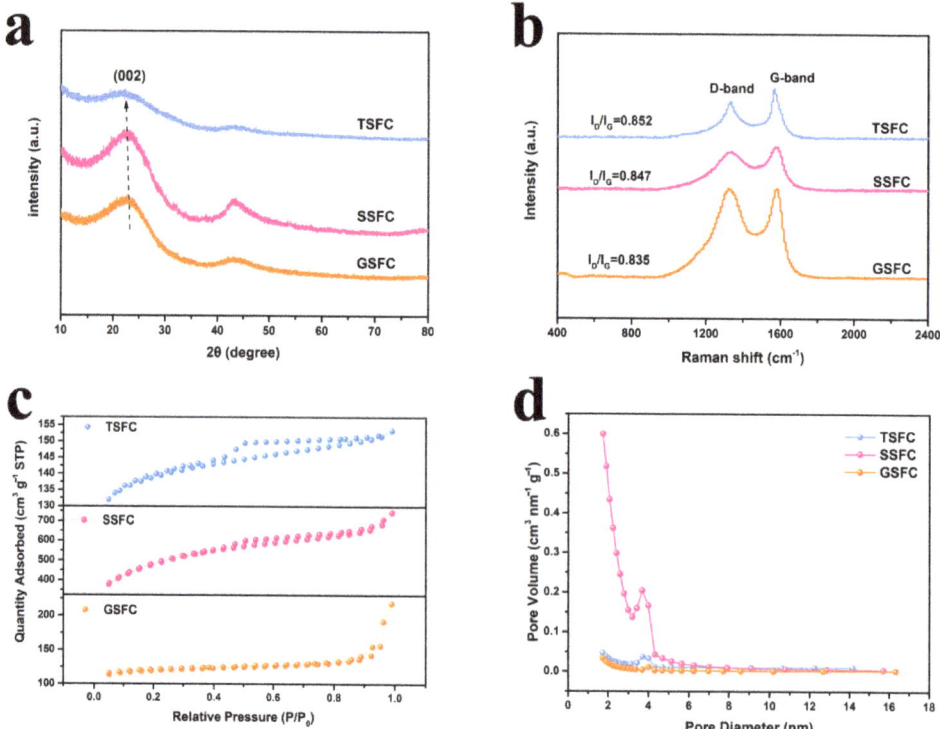

Figure 3. Structural characterization of TSFC, SSFC, and GSFC: (**a**) XRD patterns; (**b**) Raman spectrum; (**c**) N_2 adsorption/desorption isotherms; (**d**) pore size distributions.

3.2. Electrochemical Performances

To further investigate the effect of structure on the sodium ion storage properties, we performed cyclic voltammetry CV tests for the first three turns at a scan rate of 0.5 mV s^{-1}. As shown in Figure 4a–c, it can be observed that the CV curves of the first three turns of the three electrode materials are very similar, with a pair of redox peaks at around 0.01 V/0.16 V, corresponding to the insertion and extraction process of Na$^+$. In the initial cathodic scan, the three electrodes show irreversible reduction peaks, with different intensities. This may be related to the irreversible reactive binding of Na$^+$ on the surface functional groups or other defective sites and the formation of SEI films in the first cycle [36]. Compared to SSFC and GSFC, TSFC shows a large sharp peak below 0.1 V, demonstrating an enhanced sodium storage [37]. All the CV curves for the TSFC (Figure 4a) have an excellent overlap during the subsequent cycles, indicating that the electrode material has good reversible capacity and cycle stability. The EIS plots and four-probe resistance tests for the three electrodes are shown in Figure 4d,e, respectively. The EIS plots are composed of a semicircle at high frequency and a slope line at low frequency, which correspond to the interaction of Na$^+$ with the SEI layer/charge transfer between the electrolyte and the active material and the diffusion of Na$^+$ in the active material, respectively [38]. It is clear that the TSFC anode exhibits a lower Rct (32.8 Ω) than do SSFC and GSFC, which can be attributed to its large aperture nanochannels, increasing surface wettability and making the electrolyte flow easy throughout the electrode [39]. To further demonstrate the electrical conductivity of the material, we tested the positive and negative currents of the electrode sheet using a four-probe resistivity tester (Figure 4e). The test results show that the TSFC electrode material possesses the lowest conductivity (2.49 Ω·cm), indicating that the structure plays an important role in the resistance of the material and helps to improve the sodium ion transport, which corroborates with the EIS results. Figure 4f shows the discharge curves of TSFC, SSFC, and GSFC at a current density of 0.1 A g^{-1}. All discharge curves show two areas: the sloping portion (3.0 V~0.1 V) and the plateau portion (<0.1 V). The initial charge/discharge capacities of SSFC, GSFC, and TSFC are 131.11/461.6, 170.8/330.8, and 265.2/345.9 mAh g^{-1}, respectively. The ICE of the three structures also varies from 28.4% to 76.7%, depending on their structures. It is obvious that TSFC has a better first charge/discharge efficiency than the other two samples. The reason is that a proper specific surface area will reduce the occurrence of side reactions, thus reducing the irreversible capacity and thus exhibiting a high initial Coulomb efficiency [40]. The discharge curves of TSFC, SSFC, and GSFC exhibit sloping capacities of 216 mAh g^{-1}, 340 mAh g^{-1}, and 265 mAh g^{-1}, and plateau capacities of 132 mAh g^{-1}, 120 mAh g^{-1}, and 66 mAh g^{-1}, respectively. Obviously, the capacities of the three anodes are mainly derived from the adsorption of Na$^+$ (sloping part). The plateau contributions of TSFC, SSFC, and GSFC are 38%, 26%, and 20% of the total capacity, respectively. There is no doubt that the plateau part of TSFC contributes more capacity than that of the other two, which is mainly attributed to the larger Na$^+$ intercalation layer provided by the layer spacing of TSFC. Cyclic stability over 100 cycles was recorded for the samples at 100 mA g^{-1}, as shown in Figure 4g. Compared to SSFC and GSFC, TSFC performed relatively well, with an initial specific capacity of 265.2 mAh g^{-1} and an ICE of 76.7%, while the ICE of SSFC and GSFC were only 28.4% and 51.6%. This indicates that the specific surface area has a greater effect on ICE, which has been confirmed in the literature [41]. Table S2 provides a comparison of the electrochemical properties of various previously reported biomass-derived carbon materials with the materials prepared in this work. It is clear that the TSFC anode material possesses a higher ICE. In addition, the TSFC also has a higher ICE than that of some previously reported transition metal materials (such as NiMoO$_4$ [4] and MgMoO$_4$ [5]). For the stability of the electrode material structure, multiplicative performance tests were performed on three electrodes (as shown in Figure 4h). At specific current densities of 0.02, 0.05, 0.1, 0.5, 1.0, and 2.0 A g^{-1}, the specific capacities of TSFC are 281.2, 257.7, 235.0, 103.6, 70.14, and 24.6 mAh g^{-1}, respectively. When the current density returns to 0.02 A g^{-1}, a capacity of 255.6 mAh g^{-1} is obtained, demonstrating a good rate performance and the capacity retention rate remained at 90.8%. On the contrary, an inferior rate performance of the SSFC and GSFC is observed under the same condition. The long-cycle performance of the TSFC electrode is also evaluated, as shown in Figure 4i. The capacity of TSFC could be maintained at 110 mAh g^{-1} after 400 cycles at 50 mA g^{-1}, and the charge/discharge efficiency was always maintained at about 100%, indicating its excellent durability and potential.

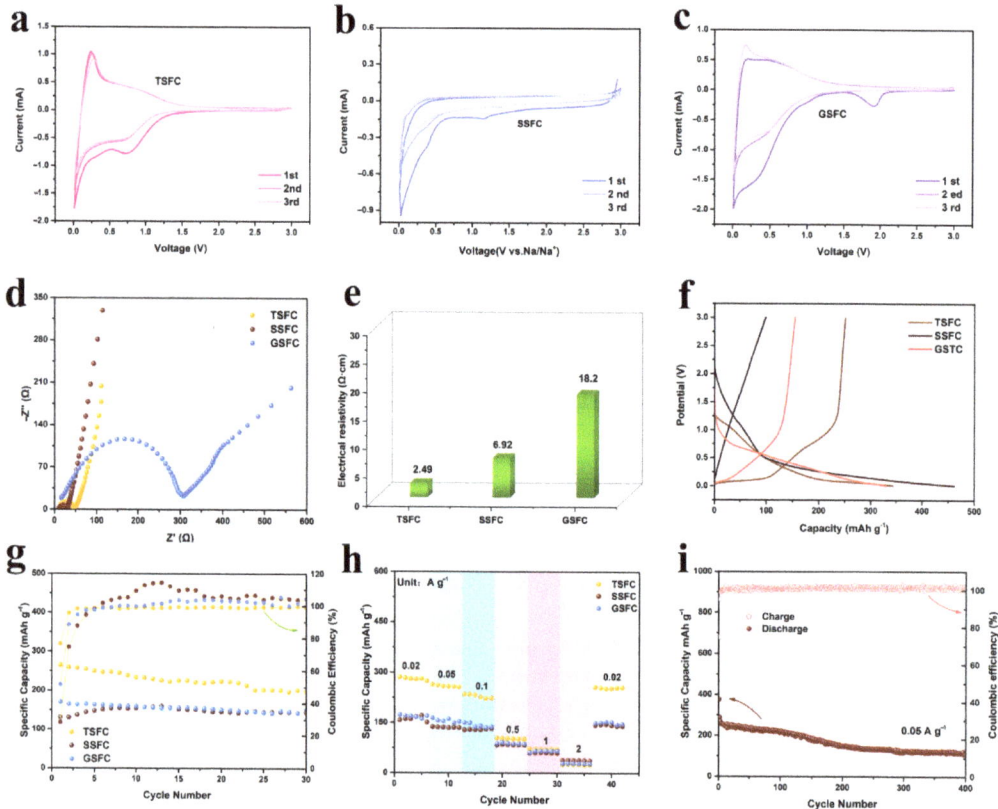

Figure 4. (**a–c**) CV curves of the first three circles, (**d**) EIS spectra, and (**e**) the electronic conductivity of the TSFC, SSFC, and GSFC anodes. (**f**) The initial galvanostatic discharge-charge curves at 0.1 A g^{-1}; (**g**) comparison of the three anodes cycling performance at 0.1 A g^{-1}; (**h**) rate performance of the three anodes; (**i**) prolonged cycling performance of the TSFC anode at 0.05 mA g^{-1}.

In order to explain the excellent sodium storage performance of TSFC, the CV curve at different scanning rates of 0.2, 0.4, 0.6, 0.8, and 1 mV s^{-1} (vs. Na/Na$^+$) was measured, as shown in Figure 5a, in which the CV curves exhibit a certain deviation from the rectangle, demonstrating the combination of two different charge storage mechanisms of faradaic and non-faradaic reactions [42]. As the scan rate increases, the shape of the CV curve remains constant, without serious distortion, indicating its high reversibility and excellent multiplicative performance [42]. Generally, the relationship between peak current (*i*) and scan rate (*v*) obeys Equation (1) [43]:

$$log(i) = log(a) + blog(v) \tag{1}$$

Here, *b* represents the slope of log(*v*) and log(*i*), where *b* values were in the range of 0.5 to 1, indicating a pseudo-capacitance contribution, in addition to diffusion-controlled intercalation behavior at these potentials. The calculation result of *b* values can be seen in Figure 5b. The b-value of TSFC is 0.75, indicating that it has a hybrid mechanism of pseudo-capacitance contribution and diffusion-controlled intercalation behavior. Furthermore, the percentage contribution of the capacitive process in the electrochemical reaction process can be calculated according to Equation (2) [43]:

$$i(v) = k_1 v + k_2 v^{\frac{1}{2}} \tag{2}$$

where $k_1 v$ is the surface capacitive contribution, and $k_2 v^{1/2}$ is the diffusion contribution. The calculated results are shown in Figure 5c,d, where it can be observed that the pseudocapacitance contribution takes up an increasing percentage as the scan speed increases. This phenomenon is probably due to the special hierarchical porous structure of TSFC, which induces the shortened transport pathway, faster delocalization, and the embedding of sodium ions. The pseudocapacitance tests were also analyzed for SSFC and GSFC under the same conditions, and the detailed data are shown in Figures S2 and S3. It can be seen that the b-value of the SSFC and GSFC electrodes is 0.57 and 0.82, respectively. Thus, a hybrid mechanism exists in the three electrode materials. The pseudocapacitance percentage increases as the scanning speeds increase, which are 32%, 39%, 45%, 48%, and 51% for TSFC; 46%, 53%, 57%, 63%, and 65% for SSFC; and 44%, 52%, 58%, 61%, and 64% for GSFC. However, the contribution of the diffusion-controlled interpolation behavior of TSFC is higher than that of the other two at different sweep speeds, and this result is exactly consistent with the above analysis results in Figure 4f.

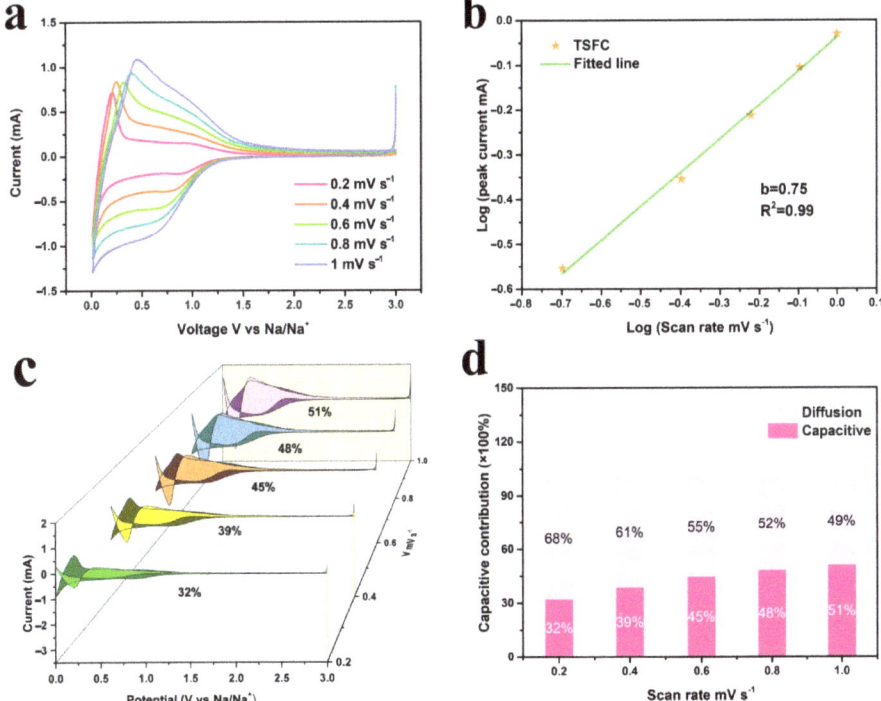

Figure 5. (**a**) Voltammetry curves of TSFC at different scan rates between 0.2 and 1 mV s^{-1}; (**b**) the plots of log(i) versus log(v) of TSFC; (**c**) pseudocapacitive contribution waterfall diagram of TSFC; (**d**) pseudocapacitive contribution of TSFC at different scan rates between 0.2 and 1 mV s^{-1}.

Figure 6a–f show the GITT reaction curves of TSFC, SSFC, and GSFC; the Na$^+$ diffusion coefficients of the three electrode materials fluctuate between 10^{-2} cm^2 s^{-1} and 10^{-6} cm^2 s^{-1}, but the Na$^+$ diffusion coefficient of TSFC has small fluctuation at the beginning of discharge, indicating that an enhanced sodium diffusion occurred during sodium adsorption [44]. As the voltage slowly decreased, the diffusion of sodium ions gradually smoothed out. When the voltage dropped down below 0.1 V, the sodium ion diffusion coefficient decayed rapidly. We believe that the diffusion of D_{Na}^+ is caused by the change in the sodium ion storage mechanism from the adsorption to insertion type [45]. Interestingly, the D_{Na}^+ of TSFC tends to rise at the end of the discharge process. This phenomenon is common in the high-temperature treated hard carbon materials, suggesting that TSFC can exhibit similar structural properties after electrochemical reactions [45]. In addition, when using TSFC as an electrode material, the charging and discharging duration of TSFC is up to 73 h, while the charging and discharging duration of SSFC and GSFC is only 39 h and 68 h, respectively, which is clearly longer than that of the other two samples. This result is also consistent with the results of the above comprehensive analysis.

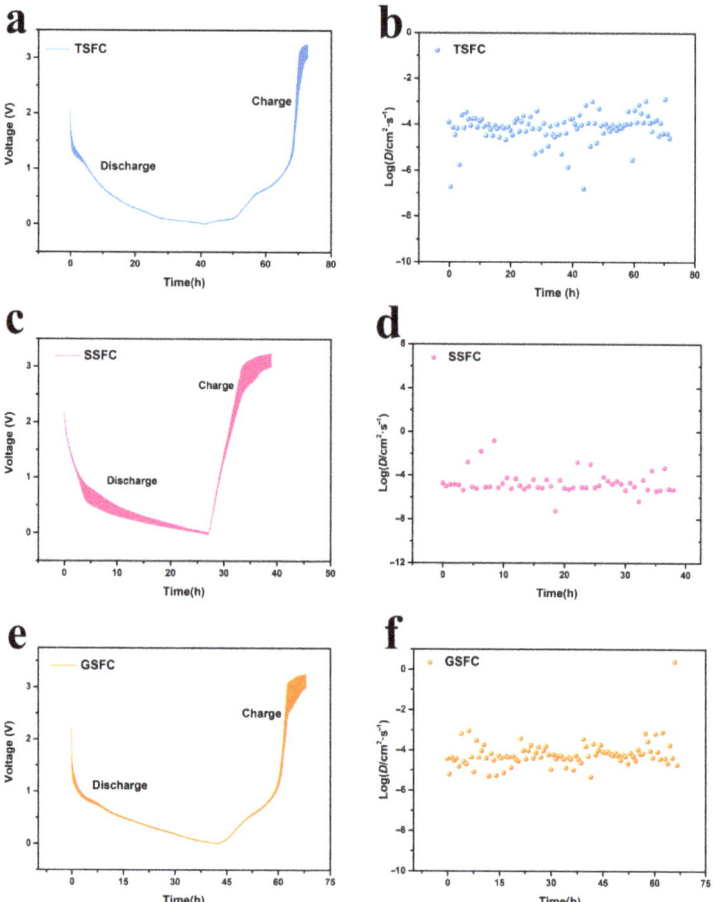

Figure 6. (a–f) Shows the GITT reaction curves of the composite of TSFC, SSFC, and GSFC.

To investigate the sodium storage mechanism of the TSFC anode materials in detail, we recorded the capacity variation of the TSFC discharge curve from the 1st cycle to the 400th cycle at a current density of 0.05 mA g^{-1}; the direct effect on the low voltage plateau capacity and high voltage sloping capacity was also explored in detail. As shown in Figure 7, the sloping capacity of TSFC increases from 62% in the 1st cycle to 82.5% in the 400th. The curves show that the contribution of ramp capacity predominates during high voltage discharge, indicating that in TSFC, the special hierarchical porous structure provides more active sites for the physical/chemical adsorption of Na$^+$. As the number of cycles increases, the number of electrochemical reactions increases accordingly from the reactions that initially occur on the electrode surface, slowly deepening to the interior, while the micropores open at this time with the successive sodiation/desodiation cycle reactions, at which time the electrolyte enters the micropores more easily, leading to an increase in sloping capacity [46]. Moreover, the combined contribution of intercalation and micropore filling to the plateau capacity is believed to play a role in the overall sodium storage process when the voltage is gradually reduced to 0.1 V. According to the literature, a layer spacing of 0.36–0.40 nm is required for the viability of sodium ion insertion/extraction from amorphous carbon in the low-voltage plateau region [47]; the layer spacing of TSFC is 0.37 nm, which is right in this range. On the other hand, hard carbon materials possess a complex structure where porosity and layer spacing are the main objective conditions for sodium storage. The presence of inflection points at the end of the potential of the discharge curve also becomes the basis for determining the predominance of interlayer intercalation and microporous filling [35]. As for the low-voltage plateau region of TSFC, there was no electrochemical inflection point when the discharge gradually reached 0 V. Therefore, based on the above analysis we reasonably believe that the sodium storage mechanism of TSFC is "adsorption-intercalation/filling," and the model is shown in Figure 7.

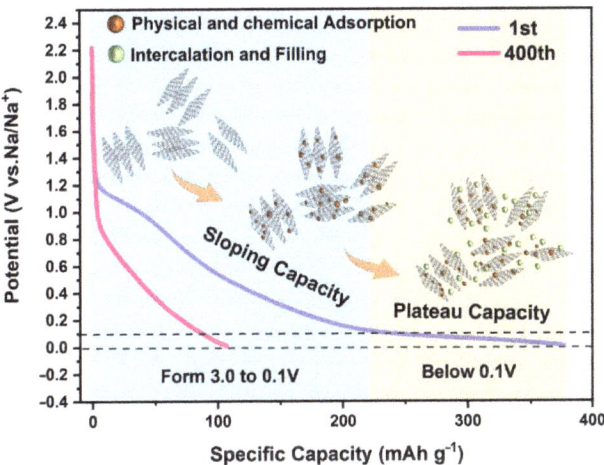

Figure 7. Sodium–ion storage mechanism of TSFC.

Based on the above analysis, it can be concluded that the TSFC anode exhibits excellent ICE and multiplicative performance, which may be attributed to the following characteristics: (i) the specific surface area can effectively reduce the generation of side reactions and further improve the Coulomb efficiency; (ii) the larger layer spacing can provide efficient ion channels and accelerate the Na$^+$ transport; and (iii) the specific hierarchical porous structure facilitates the rapid diffusion of electrolyte ions into the interior of the electrode material, providing more active surface sites and thus increasing the slope capacity contribution.

4. Conclusions

In conclusion, three different structure hard carbon materials (TSFC, SSFC, and GSFC) were prepared via a low-cost and simple method to treat sisal fibers through varying the pretreatment conditions under acid or alkali systems, and a detailed comparative study on the morphology and the electrochemical properties of these structures was conducted. The effect of biomass hard carbon structures on the initial Coulomb efficiency was carefully explored. It was found that the tubular structure of sisal fiber carbon (TSFC) shows a medium specific surface area, larger layer spacing, and a special hierarchical porous structure, which favors efficient ion transport channels, ion migration, and storage. The reversible specific capacity of TSFC can maintain 110 mAh g^{-1} at a current density of 0.05 A g^{-1} for 400 cycles, and the charge/discharge efficiency was maintained at about 100%. TSFC also exhibits a great increase in ICE (76.7%), along with cycling stability, compared with SSFC and GSFC. To investigate the effect of structural changes on its sliding and plateau capacities, the discharge curves of the TSFC with different numbers of cycles were analyzed, and the "adsorption-intercalation/filling" model for the sodium storage mechanism of TSFC was proposed; that is, at low voltage, the plateau capacity is related to the intercalation/filling of sodium ions in materials, and the slope capacity at high voltage is related to the adsorption behavior of the sodium ions. In summary, this study provides a new approach for obtaining high ICE for sodium ion battery anodes.

Supplementary Materials: The following supporting information can be downloaded at: https://www.mdpi.com/article/10.3390/nano13050881/s1, Figure S1: SEM images of (**a**) GSFC-0.1, (**b**) GSFC-1, (**c**) GSFC, and (**d**) GSFC-4; Figure S2: (a,e) Voltammetry curves of SSFC at different scan rates between 0.2 and 1 mV s^{-1}; (b,f) the plots of log(*i*) versus log(*v*) of SSFC; (c,g) pseudocapacitive contribution waterfall diagram of SSFC; (d,h) pseudocapacitive contribution of SSFC at different scan rates between 0.2 and 1 mV s^{-1}; Figure S3: (a,e) Voltammetry curves of GSFC at different scan rates between 0.2 and 1 mV s^{-1}; (b,f) the plots of log(*i*) versus log(*v*) of GSFC; (c,g) pseudocapacitive contribution waterfall diagram of GSFC; (d,h) pseudocapacitive contribution of GSFC at different scan rates between 0.2 and 1 mV s^{-1}; Table S1: N_2 adsorption/desorption isotherms parameters for the samples; Table S2: The comparison of the electrochemical performance (Coulomb efficiency) of TSFC with other biomass hard carbons for SIB. References [48–53] are cited in the supplementary materials.

Author Contributions: Conceptualization, methodology, writing—original draft preparation, investigation, and resources, Y.L.; data curation and visualization, Y.X. and X.L.; validation, Q.P.; formal analysis, K.Z.; writing—review and editing, supervision, project administration, and funding acquisition, A.Q. All authors have read and agreed to the published version of the manuscript.

Funding: This work was supported by the Foundation of Guangxi Key Laboratory of Optical and Electronic Materials Devices (20AA17), the Innovation Project of Guangxi Graduate Education (YCBZ2021062), and the Natural Science Foundation of Guangxi Province (2018JJA160029), National Natural Science Foundation of China (51564009).

Institutional Review Board Statement: Not applicable.

Informed Consent Statement: Not applicable.

Data Availability Statement: Data available within the article or its Supplementary Materials.

Conflicts of Interest: The authors declare that they have no known competing financial interests or personal relationships that could have appeared to influence the work reported in this paper.

References

1. Jing, W.; Lei, Y.; Ren, Q.; Fan, L.; Zhang, F.; Shi, Z. Facile hydrothermal treatment route of reed straw-derived hard carbon for high-performance sodium ion battery. *Electrochim. Acta* **2018**, *219*, 188–196.
2. Yang, B.; Wang, J.; Zhu, Y.; Ji, K.; Xia, Y. Engineering hard carbon with high initial coulomb efficiency for practical sodium-ion batteries. *J. Power Sources* **2021**, *492*, 229656. [CrossRef]
3. Palomares, V.; Serras, P.; Villaluenga, I.; Hueso, K.B.; Carretero-González, J.; Rojo, T. Na-ion batteries recent advances and present challenges to become low cost energy storage systems. *Energy Environ. Sci.* **2012**, *5*, 5884–5901. [CrossRef]
4. Zhao, J.; Hu, Z.; Chen, S.; Zhang, W.; Liu, X. Electrospinning synthesis of amorphous $NiMoO_4$/graphene dendritic nanofibers as excellent anodes for sodium ion batteries. *Nanotechnology* **2020**, *31*, 505401. [CrossRef]
5. Zhang, L.; He, W.; Ling, M.; Shen, K.; Liu, X. Self-standing $MgMoO_4$/Reduced Graphene Oxide Nanosheet Arrays for Lithium and Sodium Ion Storage. *ELectrochim. Acta* **2017**, *252*, 322–330. [CrossRef]
6. Xiao, L.; Cao, Y.; Xiao, J.; Wang, W.; Liu, J. High capacity, reversible alloying reactions in SnSb/C nanocomposites for Na-ion battery applications. *Chem. Commun.* **2012**, *48*, 3321–3323. [CrossRef]
7. Qian, J.; Wu, X.; Cao, Y.; Ai, X.; Yang, H. High capacity and rate capability of amorphous phosphorus for sodium ion batteries. *Angew. Chem. Int. Ed.* **2013**, *125*, 4731–4734. [CrossRef]

8. Xiao, J.; Mei, D.; Li, X.; Xu, W.; Wang, D.; Graff, G.; Bennett, W.; Nie, Z.; Saraf, L.; Aksay, I.; et al. Hierarchically porous graphene as a lithium-air battery electrode. *Nano Lett.* **2011**, *11*, 5071–5078. [CrossRef]

9. Xiong, H.; Slater, M.D.; Balasubramanian, M.; Johnson, C.S.; Rajh, T. Amorphous TiO_2 nanotube anode for rechargeable sodium ion batteries. *J. Phys. Chem. Lett.* **2011**, *2*, 2560–2565. [CrossRef]

10. Wang, J.; Yang, J.; Yin, W.; Hirano, S.I. Carbon-coated graphene/antimony composite with a sandwich-like structure for enhanced sodium storage. *J. Mater. Chem. A* **2017**, *5*, 20623–20630. [CrossRef]

11. Wickramaarachchi, K.; Minakshi, M. Status on electrodeposited manganese dioxide and biowaste carbon for hybrid capacitors: The case of high-quality oxide composites, mechanisms, and prospects. *J. Energy Storage* **2022**, *56*, 106099. [CrossRef]

12. Minakshi, M.; Visbal, H.; Mitchell, D.; Fichtner, M. Bio-waste chicken eggshells to store energy. *Dalton Trans.* **2018**, *47*, 16828–16834. [CrossRef]

13. Wickramaarachchi, K.; Minakshi, M.; Aravindh, S.; Dabare, R.; Gao, X.; Jiang, Z.; Wong, K. Repurposing N-Doped Grape Marc for the Fabrication of Supercapacitors with Theoretical and Machine Learning Models. *Nanomaterials* **2022**, *12*, 1847. [CrossRef]

14. Wang, P.; Zhu, X.; Wang, Q.; Xu, X.; Zhou, X.; Bao, J. Kelp-derived hard carbons as advanced anode materials for sodium-ion batteries. *J. Mater. Chem. A* **2017**, *5*, 5761–5769. [CrossRef]

15. Rybarczyk, M.K.; Li, Y.M.; Qiao, M.; Hu, Y.S.; Titirici, M.M.; Lieder, M. Hard carbon derived from rice husk as low cost negative electrodes in Na-ion batteries. *J. Energy Chem.* **2019**, *29*, 17–22. [CrossRef]

16. Pei, L.; Cao, H.; Yang, L.; Liu, P.; Zhao, M.; Xu, B.; Guo, J. Hard carbon derived from waste tea biomass as high-performance anode material for sodium-ion batteries. *Ionics* **2020**, *26*, 5535–5542. [CrossRef]

17. Gao, Z.; Zhang, Y.; Song, N.; Li, X. Biomass-derived renewable carbon materials for electrochemical energy storage. *Mater. Res. Lett.* **2017**, *5*, 69–88. [CrossRef]

18. Yan, D.; Yu, C.; Zhang, X.; Lu, T.; Hu, B.; Li, H.; Pan, L. Nitrogen-doped CMs derived from oatmeal as high capacity and superior long life anode material for sodium ion battery. *Electrochim. Acta* **2016**, *191*, 385–391. [CrossRef]

19. Duan, B.; Gao, X.; Yao, X.; Fang, Y.; Huang, L.; Zhou, J.; Zhang, L. Unique elastic N-doped carbon nanofibrous microspheres with hierarchical porosity derived from renewable chitin for high rate supercapacitors. *Nano Energy* **2016**, *27*, 482–491. [CrossRef]

20. Jin, Y.; Tian, K.; Wei, L.; Zhang, X.; Guo, X. Hierarchical porous microspheres of activated carbon with a high surface area from spores for electrochemical double-layer capacitors. *J. Mater. Chem. A* **2016**, *4*, 15968–15979. [CrossRef]

21. Dong, Y.; Wang, W.; Quan, H.; Huang, Z.; Chen, D.; Guo, L. Nitrogen-doped foam-like carbon plate consisting of carbon tubes as high performance electrode materials for supercapacitors. *ChemElectroChem* **2016**, *3*, 814–821. [CrossRef]

22. Liu, Y.; Qin, A.; Chen, S.; Liao, L.; Zhang, K.; Mo, Z. Hybrid Nanostructures of MoS_2/Sisal fiber tubular carbon as anode material for lithium ion batteries. *Int. J. Electrochem. Sci.* **2018**, *13*, 2054–2068. [CrossRef]

23. Wei, L.; Zhuang, S.; Qin, A.; Liao, L.; Chen, S.; Zhang, K. 2D Hybrid nanostructures of $MoSe_2\perp$Sisal fiber activated carbon for enhanced Li storage performance. *Mater. Express* **2020**, *10*, 964–973. [CrossRef]

24. Wang, D.; Zhang, K.; Liao, L.; Chen, S.; Qin, A. Synthesis of nitrogen and sulfur Co-doped sisal fiber carbon and its electrochemical performance in lithium-ion battery. *Int. J. Electrochem. Sci.* **2019**, *14*, 102–113. [CrossRef]

25. Xu, Z.Q.; Chen, J.C.; Wu, M.Q.; Chen, C.; Song, Y.C.; Wang, Y.S. Effects of different atmosphere on electrochemical performance of hard carbon electrode in sodium ion battery. *Electron. Mater. Lett.* **2019**, *15*, 428–436. [CrossRef]

26. Lu, H.; Chen, X.; Jia, Y.; Chen, H.; Wang, Y.; Ai, X.; Yang, H.; Cao, Y. Engineering Al_2O_3 atomic layer deposition: Enhanced hard carbon-electrolyte interface towards practical sodium ion batteries. *Nano Energy* **2019**, *64*, 103903. [CrossRef]

27. He, H.; Sun, D.; Tang, Y.; Wang, H.; Shao, M. Understanding and improving the initial coulombic efficiency of high-capacity anode materials for practical sodium ion batteries. *Energy Storage Mater.* **2019**, *23*, 233–251. [CrossRef]

28. Bobleter, O. Hydrothermal degradation of polymers derived from plants. *Prog. Polym. Sci.* **1994**, *19*, 797–841. [CrossRef]

29. Cao, Y.; Xiao, L.; Sushko, M.L.; Wang, W.; Schwenzer, B.; Xiao, J.; Nie, Z.; Saraf, L.; Yang, Z.; Liu, J. Sodium ion insertion in hollow carbon nanowires for battery applications. *Nano Lett.* **2012**, *12*, 3783–3787. [CrossRef]

30. Ding, J.; Wang, H.; Li, Z.; Cui, K.; Karpuzov, D.; Tan, X.; Kohandehghan, A.; Mitlin, D. Peanut shell hybrid sodium ion capacitor with extreme energy-power rivals lithium ion capacitors. *Energy Environ. Sci.* **2015**, *8*, 941–955. [CrossRef]

31. Simone, V.; Boulineau, A.; De Geyer, A.; Rouchon, D.; Simonin, L.; Martinet, S. Hard carbon derived from cellulose as anode for sodium ion batteries: Dependence of electrochemical properties on structure. *J. Energy Chem.* **2016**, *25*, 761–768. [CrossRef]

32. Wang, P.T.; Wang, H.N.; Liang, C.; Yu, K.F. Two-dimensional porous flake biomass carbon with large layer spacing as an anode material for sodium ion batteries. *Diam. Relat. Mater.* **2023**, *131*, 109601. [CrossRef]

33. Wei, C.H.; Dang, W.L.; Li, M.J.; Ma, X.; Li, M.Q.; Zhang, Y. Hard-soft carbon nanocomposite prepared by pyrolyzing biomass and coal waste as sodium-ion batteries anode material. *Mater. Lett.* **2023**, *330*, 133368. [CrossRef]

34. Lee, G.H.; Moon, S.H.; Kim, M.C.; Kim, S.J.; Choi, S.; Kim, E.S.; Han, S.B.; Park, K.W. Molybdenum carbide embedded in carbon nanofiber as a 3D flexible anode with superior stability and high-rate performance for Li-ion batteries. *Ceram. Int.* **2018**, *44*, 7972–7977. [CrossRef]

35. Chen, X.Y.; Tian, J.Y.; Li, P.; Fang, Y.L.; Fang, Y.J.; Liang, X.M.; Feng, J.W.; Dong, J.; Ai, X.P.; Yang, H.X. Overall understanding of sodium storage behaviors in hard carbons by an "Adsorption-Intercalation/Filling" hybrid mechanism. *Adv. Energy Mater.* **2022**, *12*, 2200886. [CrossRef]

36. Lyu, T.; Wang, R.; Liang, L.; Chen, J.; Shen, P.K. Hierarchical porous oviform carbon capsules with double-layer shells derived from mushroom spores for efficient sodium ion storage. *J. Electroanal. Chem.* **2020**, *8*, 114310. [CrossRef]

37. Wang, P.F.; Zhu, K.; Ye, K.; Gong, Z.; Liu, R.; Cheng, K.; Wang, G.L.; Yan, J.; Cao, D.X. Three-dimensional biomass derived hard carbon with reconstructed surface as a free-standing anode for sodium-ion batteries. *J. Colloid Interface Sci.* **2020**, *561*, 203–210. [CrossRef]

38. Li, X.; Zeng, X.; Ren, T.; Zhao, J.; Zhu, Z.; Sun, S.; Zhang, Y. The transport properties of sodium-ion in the low potential platform region of oatmeal-derived hard carbon for sodium-ion batteries. *J. Alloys Compd.* **2019**, *787*, 229–238. [CrossRef]

39. Li, D.; Chen, H.; Liu, G.; Wei, M.; Ding, L.-X.; Wang, S.; Wang, H. Porous nitrogen doped carbon sphere as high performance anode of sodium-ion battery. *Carbon* **2015**, *94*, 888–894. [CrossRef]

40. Wang, K.; Jin, Y.; Sun, S.; Huang, Y.; Peng, J.; Luo, J.; Zhang, Q.; Qiu, Y.; Fang, C.; Han, J. Low-cost and high-performance hard carbon anode materials for sodium-ion batteries. *ACS Omega* **2017**, *2*, 1687–1695. [CrossRef]

41. Yang, X.; Zheng, X.C.; Yan, Z.H.; Huang, Z.Y.; Zhou, H.H.; Kuang, Y.F.; Li, H.X. Construction and preparation of nitrogen-doped porous carbon material based on waste biomass for lithium-ion batteries. *Int. J. Hydrogen Energy* **2021**, *46*, 17267–17281. [CrossRef]

42. Li, S.; Chen, J.; Xiong, J.; Gong, X.; Ciou, J.; Lee, P. Encapsulation of MnS Nanocrystals into N, S-Co-doped Carbon as Anode Material for Full Cell Sodium-Ion Capacitors. *Nano-Micro Lett.* **2020**, *12*, 34. [CrossRef]

43. Yang, L.; Lei, Y.Y.; Liang, X.M.; Qu, L.; Xu, K.; Hua, Y.N.; Feng, J.W. SnO$_2$ nanoparticles composited with biomass N-doped carbon microspheres as low cost, environmentally friendly and high-performance anode material for sodium-ion and lithium-ion batteries. *J. Power Sources* **2022**, *547*, 232032. [CrossRef]

44. Guo, S.; Chen, Y.M.; Tong, L.P.; Cao, Y.; Jiao, H.; Long, Z.; Qiu, X.Q. Biomass hard carbon of high initial coulombic efficiency for sodium-ion batteries: Preparation and application. *Electrochim. Acta* **2022**, *410*, 140017. [CrossRef]

45. Zhang, H.; Ming, H.; Zhang, W.; Cao, G.; Yang, Y. Coupled carbonization strategy toward advanced hard carbon for high-energy sodium-ion battery. *ACS Appl. Mater. Interfaces* **2017**, *9*, 23766–23774. [CrossRef]

46. Campbell, B.; Ionescu, R.; Favors, Z.; Ozkan, C.S.; Ozkan, M. Bio-derived; binderless, hierarchically porous carbon anodes for Li-ion batteries. *Sci. Rep.* **2015**, *5*, 14575. [CrossRef]

47. Sun, N.; Qiu, J.S.; Xu, B. Understanding of sodium storage mechanism in hard carbons: Ongoing development under debate. *Adv. Energy Mater.* **2022**, *12*, 200715. [CrossRef]

48. Yu, K.; Wang, X.; Yang, H.; Bai, Y.; Wu, C. Insight to defects regulation on sugarcane waste-derived hard carbon anode for sodium-ion batteries. *J. Energy Chem.* **2021**, *55*, 499–508. [CrossRef]

49. Yu, K.; Zhao, H.; Wang, X.; Zhang, M.; Dong, R.; Li, Y.; Bai, Y.; Xu, H.; Wu, C. Hyperaccumulation route to Ca-rich hard carbon materials with cation self-incorporation and interlayer spacing optimization for high-performance sodium-ion batteries. *ACS Appl. Mater. Interfaces* **2020**, *12*, 10544–10553. [CrossRef]

50. Liu, H.; Liu, H.; Di, S.; Zhai, B.; Li, L.; Wang, S. Advantageous tubular structure of biomass-derived carbon for high-performance sodium storage. *ACS Appl. Energy Mater.* **2021**, *4*, 4955–4965. [CrossRef]

51. Cheng, H.; Tang, Z.; Luo, X.; Zheng, Z. Spartina alterniflora-derived porous carbon using as anode material for sodium-ion battery. *Sci. Total Environ.* **2021**, *777*, 146120. [CrossRef]

52. Shaji, N.; Ho, C.; Nanthagopal, M.; Santhoshkumar, P.; Sim, G.; Lee, C. Biowaste-derived heteroatoms-doped carbon for sustainable sodium-ion storage. *J. Alloys Compd.* **2021**, *872*, 159670. [CrossRef]

53. Jia, Y.; Chen, X.; Lu, H.; Zhong, F.; Feng, X.; Chen, W.; Ai, X.; Yang, H.; Cao, Y. Hard carbon anode derived from camellia seed shell with superior cycling performance for sodium-ion batteries. *J. Phys. D Appl. Phys.* **2020**, *53*, 414002. [CrossRef]

Review

A Review on the Application of Cobalt-Based Nanomaterials in Supercapacitors

Lin Yang, Qinghan Zhu, Ke Yang, Xinkai Xu, Jingchun Huang, Hongfeng Chen * and Haiwang Wang *

A Key Laboratory of Dielectric and Electrolyte Functional Material Hebei Province, Northeastern University at Qinhuangdao, Qinhuangdao 066004, China
* Correspondence: chf1405104586@126.com (H.C.); whwdbdx@126.com (H.W.)

Abstract: Among many electrode materials, cobalt-based nanomaterials are widely used in supercapacitors because of their high natural abundance, good electrical conductivity, and high specific capacitance. However, there are still some difficulties to overcome, including poor structural stability and low power density. This paper summarizes the research progress of cobalt-based nanomaterials (cobalt oxide, cobalt hydroxide, cobalt-containing ternary metal oxides, etc.) as electrode materials for supercapacitors in recent years and discusses the preparation methods and properties of the materials. Notably, the focus of this paper is on the strategies to improve the electrochemical properties of these materials. We show that the performance of cobalt-based nanomaterials can be improved by designing their morphologies and, among the many morphologies, the mesoporous structure plays a major role. This is because mesoporous structures can mitigate volume changes and improve the performance of pseudo capacitance. This review is dedicated to the study of several cobalt-based nanomaterials in supercapacitors, and we hope that future scholars will make new breakthroughs in morphology design.

Keywords: supercapacitor; cobalt-containing nanomaterials; morphological design

Citation: Yang, L.; Zhu, Q.; Yang, K.; Xu, X.; Huang, J.; Chen, H.; Wang, H. A Review on the Application of Cobalt-Based Nanomaterials in Supercapacitors. *Nanomaterials* **2022**, *12*, 4065. https://doi.org/10.3390/nano12224065

Academic Editor: Sergio Brutti

Received: 26 October 2022
Accepted: 16 November 2022
Published: 18 November 2022

Publisher's Note: MDPI stays neutral with regard to jurisdictional claims in published maps and institutional affiliations.

1. Introduction

1.1. Background

Using non-renewable resources such as fossil fuels will cause severe environmental pollution, and their prices are rising yearly due to their dwindling reserves. Therefore, it is urgent to develop sustainable green energy, among which wind and solar energy have been used on a large scale [1]. To better store and transport electricity from sustainable energy sources, energy storage technology has been developed significantly. Rechargeable batteries and supercapacitors (SCs) have been the major chemical energy storage devices.

At present, rechargeable lithium-ion batteries with good safety performance, high voltage and high energy density are widely used. However, with the rising demand for lithium-ion batteries, lithium resources are facing an extremely tight situation. Thus, sodium, an alkali metal, has attracted increasing attention in recent years due to its abundant content and low cost. However, poor cycle performance is still the most significant problem hindering the development of sodium-ion batteries. Compared to rechargeable batteries, SCs have faster charging and discharging processes (SCs: 1–10 s and batteries: 0.5–5 h), higher power density (SCs: 500–10,000 W kg^{-1} and batteries < 1000 W kg^{-1}), longer lifetime (SCs > 500,000 h and batteries: 500–1000 h) and safer operation [2–5]. However, SCs have a disadvantage in terms of low energy density (SCs: 1–10 W h kg^{-1} and batteries: 10–100 W h kg^{-1}) [2,6–9]. To get over the barrier of low energy density, one of the most common approaches is to develop high-performance electrode materials for SCs.

1.2. Transition Group Metals Electrode Materials

Transition group metal materials have been widely used as electrode materials for SCs in recent years, and include oxides/hydroxides [10–13], sulfides [14–17], phosphides [18], and other categories. Among these materials, RuO_2, the most representative one, was considered the most desirable pseudocapacitive material for its theoretical specific capacitance (1300–2200 F g^{-1}) [19]. However, insufficient resources and the environmental toxicity of RuO_2 has unfortunately limited its further development [20]. This has led the relevant research on RuO_2 to its compound materials and other transition group metals to reduce the cost. Among them, cobalt-based materials are promising electrode materials for SCs because of their natural abundance, good cycle stability, abundant electroactive sites, high specific capacitance, and high electronic conductivity. In recent years, various cobalt-based materials, such as Co_3O_4, $Co(OH)_2$, cobalt-based ternary metal oxides, and sulfides, have been widely studied and many advances have been made.

1.3. Contents of This Review

Scholars have done much research on cobalt-based nano-material electrodes. However, their broad application is limited due to low electrochemical potential window, poor structural stability, unsatisfactory cycle stability and low power density. Generally speaking, the morphology, chemical composition and crystal defects of cobalt-based electrode materials have a great influence on the electrochemical performance of energy storage devices. Researchers have explored this issue, including doping other elements, introducing oxygen vacancies, and controlling synthesis conditions to construct different spatial structures of materials to improve the performance of the above electrode materials.

As far as we know, most of the existing reviews classify cobalt-based nanomaterials into a specific class of materials for a brief overview, while few reviews summarize their applications in SC electrodes alone. To promote future breakthroughs in this field, we provide a more comprehensive description of the application of cobalt-based nanomaterials in supercapacitors. Starting from nano-structured cobalt-based materials (cobalt tetroxide, cobalt hydroxide, cobalt-containing ternary metal oxides) and their composites, the application of cobalt-based materials in supercapacitor electrodes is introduced. First, the working principle and classification of SCs are introduced. Second, the applications of cobalt-based nano-compounds in SCs are studied, including the structure and electrochemical properties of cobalt-based nano-materials, the synthesis methods of electrode materials, the construction of different nano-structures and composites with other materials. In addition, the influence of morphology on the properties of cobalt-based nanomaterial electrodes is emphasized. Finally, we look forward to the development and challenges of SCs and cobalt-based materials.

2. Cobalt-Based Nanomaterials for SC Applications

With the popularity of mobile electronic devices, electric vehicles, and new energy vehicles, energy storage systems have become an integral part of modern society. Among them, SCs have become electrochemical containers, and have attracted significant attention because of their safe operation, good cycle performance, fast charging capacity and high-power density.

As shown in Figure 1a, a SC mainly consists of a pair of parallel plate electrodes, an electrolyte solution, electrode materials and an ion-permeable separator [21]. The separator can separate the two electrodes effectively to prevent mutual contact and short circuit [22]. The energy storage mechanism of SCs include (1) reversible ion adsorption and desorption processes between active materials and electrolytes, and (2) reversible faradaic redox reactions during charging and discharging. Furthermore, according to the charge storage mechanism of SCs, they can be divided into three categories: electronic double-layer capacitors (EDLCs), pseudo-capacitors (PCs) and battery-type capacitors. The specific mechanisms of these three types of capacitors are explained below.

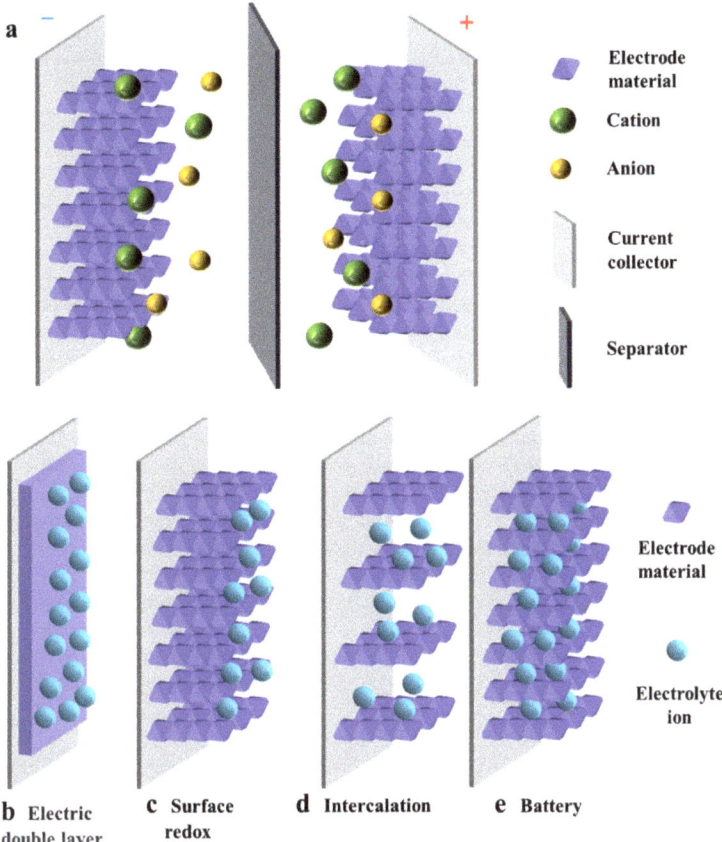

Figure 1. (**a**) Schematic diagram of the structure of an SC and energy storage mechanism of (**b**) an electric double layer capacitor, (**c**) a surface redox capacitor, (**d**) intercalation capacitor and (**e**) a battery-type capacitor.

EDLC is controlled by reversible adsorption/desorption of electrolyte ions at the electrode/electrolyte interface (Figure 1b), a process involving only the physical adsorption of ions but not any chemical reaction [23]. During the charging process, electrons migrate from the negative electrode to the positive electrode, accumulating positive and negative charges at the two electrodes. Then, the anions in the electrolyte solution move toward the positive electrode and the cations move toward the negative electrode. During the discharge process, the reverse procedure takes place. Since the potential drop is primarily limited to a small range (0.1–10 nm), EDLC has a higher energy density than the conventional capacitor, and its capacitance is related to the interface area of the electrodes. Therefore, common electrode materials mainly include porous carbon-based electrode materials with high specific surface area [24–26]. However, due to the absence of Faraday redox reactions in the energy storage process, the charging mechanism confines the capacitance to a lower range, exhibiting a higher power density but lower energy density and specific capacitance.

Based on the Faraday redox reaction, the pseudo-capacitance gives the SCs higher charge storage capacity. Similar to the charging and discharging processes occurring in batteries, the energy storage process in such SCs is a fast reversible Faraday reaction at or near the surface of the active material, but without causing phase changes in the electrode material [22,27]. PCs can be divided into two types: PCs controlled by surface redox

Nanomaterials **2022**, *12*, 4065

reactions (Figure 1c) and PCs controlled by intercalation layers (Figure 1d). For the former PCs, during the redox pseudo-capacitance process, electron transfer occurs when ions in the electrolyte solution are attracted to or near the electrode surface. For the latter PCs, electron transfer occurs when ions are transferred into the gap or interlayer of the electrode and layered electrodes expose a larger area in an electrolyte solution. However, electrode materials are prone to shrinkage and expansion during charging and discharging due to the redox reaction at the electrode, leading to poor cycling performance [28]. Both capacitance and energy density of PCs are much larger than those of EDLCs. This is mainly attributed to the unique charge storage mechanism of the Faraday redox reaction rather than the fully reversible physical charge/discharge processes.

Battery-type SCs (their structures are shown in Figure 1e) are distinguished from PCs by their distinctive feature of exhibiting phase change behavior during charging and discharging [29–31]. The charge storage mechanism in battery materials involves the reaction with OH^- in alkaline medium, which is controlled by the diffusion of electrolyte ions [31]. Battery-like materials usually have high charge storage capacity. However, the slow phase change of the material during charging and discharging reduces its kinetic performance, making its multiplicative performance low. In contrast, battery-type materials with unique nanostructures have a high specific surface area, creating great active sites for redox reactions and providing a shorter distance for the diffusion of electrolyte ions. Moreover, the rapid phase transition of battery-like materials during charge storage is mitigated by designing their nanostructures.

Transition metal oxides are widely studied as SC electrode materials because they possess higher energy density than carbon materials due to the Faraday electrolysis reaction involved in the electrochemical process. Among them, cobalt nanomaterial is a typical transition metal SC material. In recent years, research on SC electrode materials of Co_3O_4, $Co(OH)_2$, $MnCo_2O_4$, $NiCo_2O_4$, $ZnCo_2O_4$ and their derivatives have been widely reported.

2.1. Cobalt Oxide

In recent years, transition metal oxides have attracted more and more attention as electrode materials with ultra-high electrochemical activity for SCs [32–38]. Among various transition metal oxides, Co_3O_4 electrode materials and related composites have been widely studied because of their high specific capacitance, low price, and environmental friendliness. In addition, the Co_3O_4 electrode material, with special microstructure and morphology, has excellent electrochemical capacitance behavior.

At present, several processes are used to prepare Co_3O_4, the common ones being hydrothermal [39,40], electrochemical deposition [41], thermal decomposition [42], and sol-gel methods [43]. The hydrothermal method is a process in which the dissolution and recrystallization of insoluble substances occurs in a closed reactor at high temperature and pressure. Experimental parameters, such as temperature, time and molar ratio of additives, have been found to have a significant effect on the morphology of the product [44]. Electrochemical deposition is another important method to prepare electrode materials. During the deposition process, electrical energy can provide a strong driving force for the redox reaction, thus ensuring the uniform growth of electrode materials on conductive substrates, such as stainless steel, nickel foam, and carbon cloth [45–47]. Meanwhile, the conductive substrate is used as the working electrode, and deposition conditions such as scan rate, number of cycles, electrolyte concentration and pH are used as control parameters to achieve high surface area and uniform deposition. On the other hand, the thermal decomposition method usually relies on the conversion of certain substances at high temperatures to achieve the modification of electrode materials. This avoids complex multiple synthesis steps and minimizes the use of solvents, making it simple and environmentally friendly. As for the sol-gel method, the process can be described as follows: precursors such as metal alcohol salts or inorganic compounds are hydrolyzed under certain conditions to form a stable and transparent sol system, then are agglomerated into a gel, and finally dried and sintered to form a solid. The advantages of this method are low reaction temperature, easy control of the reaction, and high homogeneity of the sample down to the molecular or atomic level. The shape and size of the nanoparticles are usually controlled by adjusting the ratio of raw materials and the initial pH of the solution.

2.1.1. Co_3O_4

As mentioned above, Co_3O_4 as a transition metal oxide, has a theoretical specific capacitance of 3560 F g^{-1}, good reversibility, and excellent electrochemical properties [48]. Therefore, it is one of the most attractive electrode materials for SCs. However, the capacitive degradation of Co_3O_4 at high current densities results in its poor reversibility [49,50]. This phenomenon leads to the actual obtained Co_3O_4 specific capacitance being much lower than the theoretical value, so the application of Co_3O_4 in SCs is severely limited. It has been reported that the electrochemical performance of Co_3O_4 can be greatly improved by regulating the micromorphology of Co_3O_4.

In recent years, various morphologies of Co_3O_4 have been synthesized by different methods (shown in Figure 2), such as Co_3O_4 nanofibers [51], layered Co_3O_4 [52], Co_3O_4 nanoparticles [53], Co_3O_4 nanorod arrays [54], core-shell Co_3O_4 [55], porous Co_3O_4 nanowires [56], and hollow coral-shaped Co_3O_4 [57]. Several Co_3O_4 electrode materials with typical morphologies are briefly described below, including their preparation processes, unique spatial structures, and their principles. For the convenience of readers, the electrical property data of these materials is listed separately in Table 1.

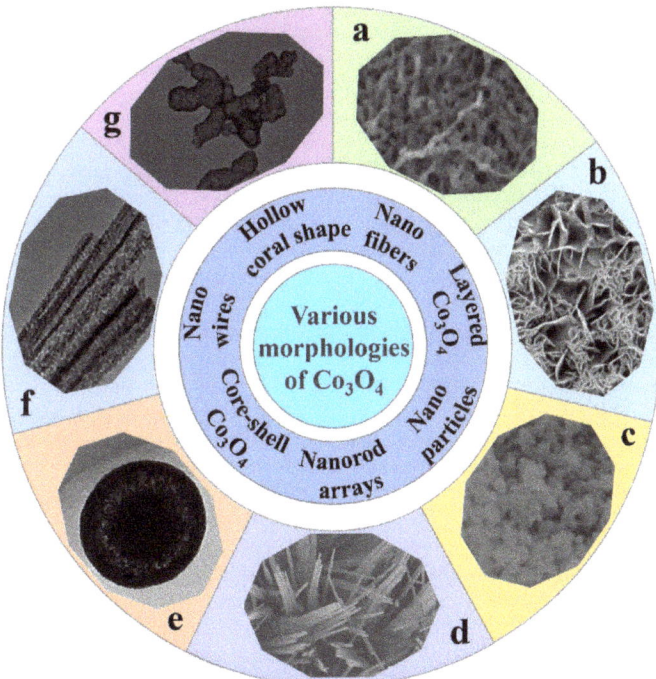

Figure 2. Various morphologies of Co_3O_4. (**a**) Nano fibers; reprinted with permission from ref. [51]. (**b**) Layered Co_3O_4; reprinted with permission from ref. [52]. (**c**) Nano particles; reprinted with permission from ref. [53]. (**d**) Nanorod arrays; reprinted with permission from ref. [54]. (**e**) Core-shell Co_3O_4; reprinted with permission from ref. [55]. (**f**) Nano wires; reprinted with permission from ref. [56]. (**g**) Hollow coral shape; reprinted with permission from ref. [57].

Table 1. Electrochemical properties of each microscopic morphology.

Morphology	Specific Capacitance (Current Density)	Cycling Performance (Cycles, Current Density)	Year	Ref.
nanofibers	407 F g^{-1} (5 mV s^{-1})	94% (1000, 1 A g^{-1})	2014	[51]
layered Co$_3$O$_4$	352 F g^{-1} (2 A g^{-1})	129% (2500, 2 A g^{-1})	2012	[52]
nanoparticles	362.8 F g^{-1} (0.2 A g^{-1})	73.5% (1000, 1 A g^{-1})	2014	[53]
nanorod arrays	154.9 C g^{-1} (1 A g^{-1})	88% (1000, 1 A g^{-1})	2019	[54]
core-shell Co$_3$O$_4$	837.7 F g^{-1} (1 A g^{-1})	87.0% (2000, 5 A g^{-1})	2018	[55]
porous nanowires	2815.7 F g^{-1} (1 A g^{-1})	88.8% (1100, 1 A g^{-1})	2018	[56]
hollow coral shape	626.5 F g^{-1} (5 mV s^{-1})	≈100% (5000, 10 A g^{-1})	2019	[57]

Manish Kumar et al. prepared Co$_3$O$_4$ nanofibers (shown in Figure 2a) by electrospinning technology [51]. Due to the large specific surface area and unique porous network morphology of this structure, the electrolyte solution can better contact with the electrode material. This is conducive to the transport of ions and electrons at the electrode-electrolyte interface, thus accelerating the redox progress. Duan et al. synthesized layered porous Co$_3$O$_4$ films by a hydrothermal method [52]. As shown in Figure 2b, the prepared Co$_3$O$_4$ films display a two-layer structure in which the lower structure consists of an array of Co$_3$O$_4$ monolayer hollow spheres and the upper structure consists of porous mesh-like Co$_3$O$_4$ nanosheets. The high porosity and large specific surface area provide a short path for ion/electron transfer, and the close contact between the active material and the electrolyte leads to high electrochemical activity, which enhances the pseudocapacitive performance. In addition, the graded porous structure can also moderate the volume changes caused by redox reactions, thus improving cycling performance. Deng et al. synthesized cobalt oxides (Co$_3$O$_4$ and Co$_3$O$_4$/CoO) by burning a mixture of Co(NO$_3$)$_2$·6H$_2$O and citric acid (Figure 2c) [53]. They experimentally confirmed that the morphology of the electrode materials could be influenced by adjusting the citric acid/Co(NO$_3$)$_2$·6H$_2$O molar ratio. Based on this, they produced electrode materials with the best performance. As shown in Figure 2d, unique Co$_3$O$_4$ nanorod arrays were synthesized through a simple chemical bath deposition and annealing process by Chen et al. [54]. Due to their high specific surface area and novel structure, the specific capacitance of Co$_3$O$_4$ nanorod arrays is high. It was found that Co$_3$O$_4$ nanorod arrays have good cycling stability, conductivity, and ion diffusion behavior. Liu et al. prepared Co$_3$O$_4$ mesoporous nanospheres with a homogeneous core-shell by the solvothermal and rapid calcination methods (Figure 2e) [55]. The accumulation density of sub-nanoparticles and the thickness of Co$_3$O$_4$ shell layer can be controlled by changing the annealing time. Both the tunable mesoporous and core-shell structures can facilitate the ion and electron transport efficiently while adapting to the volume change of the oxide electrode during cycling. Xu et al. successfully prepared one-dimensional porous Co$_3$O$_4$ nanowires by thermal decomposition of coordination polymers with nitrilotriacetic acid as a chelating agent using a solvothermal method (Figure 2f) [56]. The porous structure of Co$_3$O$_4$ nanowires consists of many nanoparticles. The special structure maximizes the exposure of the active material to the alkaline electrolyte, resulting in high specific capacity and good cycling stability. Wang et al. obtained hollow coral-shaped Co$_3$O$_4$ nanostructures by calcining cobalt oxalate precursors in the air (Figure 2g) [57]. The hollow structure allows it to withstand volume changes during the reaction process and thus exhibits excellent cycling performance.

Starting from improving the contact area between electrode and electrolyte, Lu et al. prepared layered Co$_3$O$_4$ electrode material by combining 2-methylimidazole cobalt salt and electro-spun nanofibers [58]. Its unique three-dimensional (3D) network and nano porous structure reduced the ion diffusion distance and increased the contact area between electrode and electrolyte, thus improving its electrochemical performance. The synthesized

Co_3O_4 electrode can provide a high specific capacitance of 970 F g^{-1} at a current density of 1 A g^{-1}, an energy density of 54.6 W h kg^{-1} at a power density of 360.6 W kg^{-1}, and a capacitance retention rate of 77.5% after 5000 cycles at 6 A g^{-1}.

The above study showed that the electrochemical performance of Co_3O_4 can be significantly improved by adjusting its morphology. By designing a unique structure, the contact area can be increased, and the close contact between the active material and the electrolyte can lead to high electrochemical activity, which enhances the pseudocapacitive performance. In addition, the graded porous structure can also moderate the volume changes caused by redox reactions, thus improving the cycling performance.

2.1.2. Co_3O_4 Composites

To further improve the performance of Co_3O_4, and meet the needs of various applications, one of the main means is to prepare Co_3O_4 composites by anchoring Co_3O_4 on a carbon-based material with high electrical conductivity. Among many carbon-based materials, graphene with large specific area, unique mechanical, and excellent electrochemical properties is considered to be an ideal carrier for loading Co_3O_4 nanostructures [59]. Therefore, graphene-based Co_3O_4 composites have become a research hotspot in recent years. For example, Tan et al. made self-supporting and non-adhesive Co_3O_4 nano sheet arrays/graphene/Ni hybrid foams by in-situ synthesis of graphene and Co_3O_4 nanosheets on nickel foam [60]. The SEM image shows that the porous structure supported by the composite remain good. At the same time, the substrate is completely covered by Co_3O_4 nanosheets and there is no agglomeration. This self-supporting and adhesive free characteristic avoids the disadvantage of the high resistance of traditional graphene-based Co_3O_4 composites due to the contact between hybrid particles, additives, adhesives, and collectors. The cycle performance of Co_3O_4 nano sheet/graphene/Ni hybrid electrode has been studied. It was found that after 5000 cycles at a current density of 10 mA cm^{-2}, it had 112.2% of the initial capacitance. This indicates that the ability of this unique Co_3O_4 nano sheet/graphene/Ni hybrid electrode can meet the requirements of good capacity and long cycle life at high current density.

Younis et al. synthesized Co_3O_4 nanosheets by one-step electrochemical deposition on carbon foam followed by annealing [41]. The electrochemical properties of the Co_3O_4 nanosheets were improved due to the good electrical conductivity of the composite carbon foam. In addition, a dense mesoporous structure could be observed in the SEM images, which may be one of the main reasons for the improved electrochemical properties. Electrochemical tests showed that the prepared Co_3O_4 nanosheets had ideal capacitive properties with a maximum specific capacitance of 106 F g^{-1} in 1 M NaOH solution at a scan rate of 0.5 V s^{-1}. In this report, the prepared ultrathin nanosheets were simple in process, low in cost, and suitable for industrial applications, which have high reference value.

Introducing oxygen vacancies into transition metal oxides can change their geometric and electronic structures, improve their intrinsic conductivity and electrochemical activity, and improve their properties [61–64]. For example, Xiang and others prepared Co_3O_4 nano sheet electrode materials with different oxygen vacancy content by different reduction methods [65]. They showed that Co_3O_4 electrode with high oxygen vacancy content has better electrochemical performance. At the current density of 2 A g^{-1}, the capacity retention percentage can reach 95% after 3000 cycles, while the capacitance retention rate of the original Co_3O_4 nanosheet electrode was only 90% under the same conditions. This indicates that the introduction of oxygen vacancy can improve the conductivity, increase the capacitance, and significantly improve the electrochemical performance.

Yang et al. used the one-step laser irradiation method for the first time to synthesize ultrafine Co_3O_4 nanoparticles/graphene composites with rich oxygen vacancies by laser-induced reduction and fragmentation [66]. Compared with the traditional method, the one-step laser irradiation method is simple, does not need to add reducing agents and additives, and solves the pollution problem of organic additives. At 10 A g^{-1} current density, the capacitance retention of the composites after 2000 cycles could reach 99.3%,

Nanomaterials **2022**, *12*, 4065

while the capacitance retention of porous Co_3O_4 nanorods electrodes was only about 84.7%, indicating that Co_3O_4 nanoparticles/graphene composites have excellent cycle stability.

2.2. Co(OH)$_2$

Similar to transition metal oxides, transition metal hydroxides have excellent pseudocapacitive properties [67]. Among them, $Co(OH)_2$ has become one of the promising materials in SCs due to its high theoretical capacitance (3460 F g^{-1}) and low cost. With electrode materials, reversible redox reactions take place during charge and discharge. The specific process is that $Co(OH)_2$ stores charge by participating in the O-H bond breaking and recombination reaction in the electrolyte. The redox reaction can be expressed as [68]:

$$Co(OH)_2 + OH^- \rightarrow CoOOH + H_2O + e^- \tag{1}$$

The oxidation product CoOOH can further undergo a deprotonation reaction and carry out the second redox reaction [22]:

$$CoOOH + OH^- \rightarrow CoO_2 + H_2O + e^- \tag{2}$$

Although the theoretical capacitance of $Co(OH)_2$ is very high, it is difficult to meet the requirements of fast electron transport rate at high power density because it is a P-type semiconductor. An effective way to alleviate the above problems is to construct conductive matrix hybrid nanostructures of $Co(OH)_2$. For example, Pan et al. synthesized $Co(OH)_2$/Ni nano-lake array with porous structure by hydrothermal and electrodeposition methods [69]; its microstructure is shown in Figure 3. As a conductive substrate, nickel foam forms a porous conductive network, which can shorten the diffusion path of ions and electrons, and improve the charge efficiency, thus effectively improving the electrochemical performance of SC. When the charge and discharge rate changes from 1 A g^{-1} to 40 A g^{-1}, the capacitance retention rate reaches 87.6%, while that of pure $Co(OH)_2$ nano-lake array is only 76.4% under the same conditions.

Li et al. prepared a 3D independent $Co(OH)_2$/Ni heterostructure electrode by depositing sea urchin-like $Co(OH)_2$ microspheres on nickel foam using a one-step hydrothermal method [70]. According to the analysis of its electrochemical performance, the capacitance could reach 1916 F g^{-1} at 10 mA cm^{-2}, and 79.3% of the original capacitance was maintained after 5000 charge and discharge cycles at 80 mA cm^{-2} current density. The reason for this decrease in capacitance is that some sea urchin-like $Co(OH)_2$ microspheres become inconspicuous rod-like and stacked plate-like CoOOH due to changes in composition and structure during charging and discharging.

To improve the density of SCs while maintaining their flexibility, Zhao and his colleagues deposited $Co(OH)_2$ on nickel oxide/hydroxide coated nano porous nickel (np-NiO_xH_y@Ni) by electrochemical deposition [71]. Then they successfully synthesized a $Co(OH)_2$/np-NiO_xH_y@Ni hybrid electrode with a hierarchical porous structure and excellent flexibility. The layered porous structure improves the surface area and effectively promotes ion diffusion. At the same time, the coordination between $Co(OH)_2$ and NiO_xH_y electroactive materials significantly improves the electrochemical reaction activity of electrode materials. The capacitance of $Co(OH)_2$/np-NiO_xH_y@Ni electrode was 1421.1 F cm^{-3} at 0.5 A cm^{-3} current density, and 81.6% of the original capacitance remained after 8000 cycles at 2 A cm^{-3} current density.

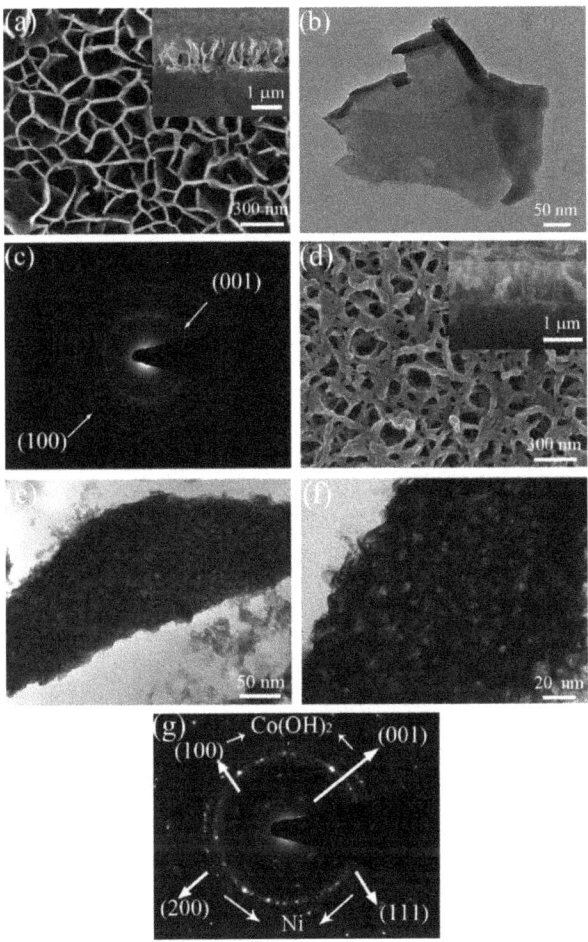

Figure 3. (**a**) SEM image, (**b**) TEM image and (**c**) SAED pattern of a Co(OH)$_2$ nanoflake array. (**d**) SEM image, (**e,f**) TEM images and (**g**) SAED pattern of a Co(OH)$_2$/Ni composite nanoflake array grown on nickel foam. Reproduced with permission from G.X. Pan, Porous Co(OH)$_2$/Ni composite nanoflake array for high performance supercapacitors; published by Elsevier, 2012 [69].

2.3. Cobalt-Containing Ternary Metal Oxide

Cobalt-containing ternary metal oxides are typical spinel structures, and the cells of spinel consist of eight small cubic cells, which are four A-type cells and four B-type cells interconnected (Figure 4). Each A-type or B-type unit has four O^{2-} for a total of 32. M ions are in the center of the A-type unit (tetrahedral gap) and half of the vertices of the eight small cubic units for a total of eight. Cobalt ions occupy each of four octahedral gaps, for a total of 16. The cell general formula of cobalt-based spinel is M$_8$Co$_{16}$O$_{32}$, and the chemical formula is summarized as MCo$_2$O$_4$. Furthermore, in general, the alkaline electrolytes of different Co-based spinel MCo$_2$O$_4$ (M = Co, Ni, Fe, and Mn) undergo approximately the same reversible electrochemical redox reactions with the discharge products of M ions as hydroxyl oxides MOOH [72–74]. The resulting MOOH (M = Co, Fe, and Mn) further discharges and produces the corresponding CoO$_2$ [72], FeO$_4$$^{2-}$ [75] and MnO$_2$. Because of the presence of Cu(I)/Cu(II) pairs, the discharge products of Zn^{2+} and Cu^{2+} are Zn(II) [76] and Cu(I)/Cu(II) [77] hydroxides.

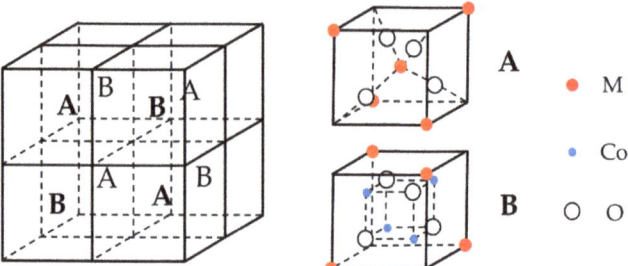

Figure 4. Schematic crystal structure of MCo_2O_4.

Metal oxides with multiple metal cations generally have higher conductivity and capacitive activity than single metal oxides [78]. Among them, ternary transition metal oxides provide more active sites for redox reaction and improve electronic conductivity because they have two different cations [79]. Compared with binary transition metal oxides such as Co_3O_4, the electrochemical properties of ternary transition metal oxides ($MnCo_2O_4$, $NiCo_2O_4$, $ZnCo_2O_4$, etc.) are significantly improved under the influence of the synergistic effect generated by the coupling of two transition metals [36].

2.3.1. $MnCo_2O_4$

$MnCo_2O_4$ is a typical compound with a spinel structure. It can show two lattice structures: (a) normal spinel [80,81], (b) anti spinel [82]. Due to the diversity of crystal structure, the variation of charges (Mn and Co) occupied in octahedron and tetrahedron makes it have excellent redox stability [83]. Manganese transmits more electrons and has higher capacity, while cobalt has higher oxidation potential. Many experiments have proved that $MnCo_2O_4$ improves the electrochemical performance of single Co_3O_4 and shows better conductivity, structural stability, and cycle performance [84–86]. The reaction principle of $MnCo_2O_4$ is as follows:

$$MnCo_2O_4 + OH^- + H_2O \rightarrow MnOOH + 2CoOOH + e^- \tag{3}$$

$$MnOOH + OH^- \rightarrow MnO_2 + H_2O + e^- \tag{4}$$

$$CoOOH + OH^- \rightarrow CoO_2 + H_2O + e^- \tag{5}$$

$MnCo_2O_4$ reacts under alkaline conditions to form MnOOH and CoOOH, and the resulting MnOOH and CoOOH continue to react with OH^- to form MnO_2 and CoO_2, while releasing electrons.

Based on the above studies, $MnCo_2O_4$ is considered an ideal candidate material for SCs, so it has been widely studied. Various forms of $MnCo_2O_4$ materials have been prepared, such as flower shaped hollow microspheres [87], core-shell structures [88], nano cages [89], nano needles [90], ellipsoids [91], and sea urchins [92]. For example, Dong et al. synthesized $MnCo_2O_4$ with a hierarchical nanocage structure using a bimetallic zeolite imidazolate framework as the precursor and template [89]. The preparation process and morphological characterization are shown in Figure 5. Through the analysis of its micro morphology, it can be found that many interconnected nanoparticles form a highly porous nanocage structure. This unique nanocage structure exposes a large area of surface and mesoporous structure, which promotes the diffusion of ions and ensures its excellent electrochemical performance in SCs. By testing the electrochemical performance of $MnCo_2O_4$ electrode, it was found that it can show 95% capacitance retention after 4500 cycles at 1 A g^{-1}, which proves its superior cycle stability. Che et al. synthesized flower-shaped $MnCo_2O_4$ hollow microspheres with a nano flower structure by the template free method of mixing and heating the solvent to 180 °C [87], and then calcining at 350 °C for two hours. An SEM microscopic image is shown in the Figure 6. The larger surface area and porous structure provide more active sites, promote the transfer of ions and electrons, accelerate the reaction rate, and greatly enhance its electrochemical storage performance. The capacity retention rate of the electrode was 93.6% after 2000 consecutive cycles at a high current density of $1\ A\ g^{-1}$.

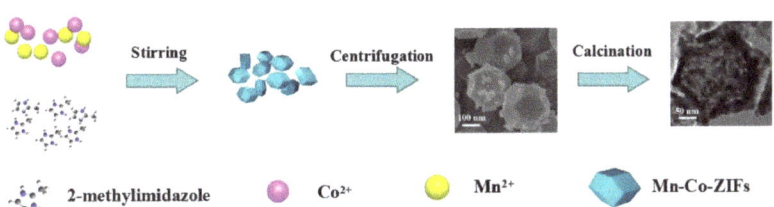

Figure 5. Preparation process of a nanocage $MnCo_2O_4$ electrode. Reproduced with permission from Yanying Dong, Facile synthesis of hierarchical nanocage $MnCo_2O_4$ for high performance supercapacitor; published by Elsevier, 2016 [89].

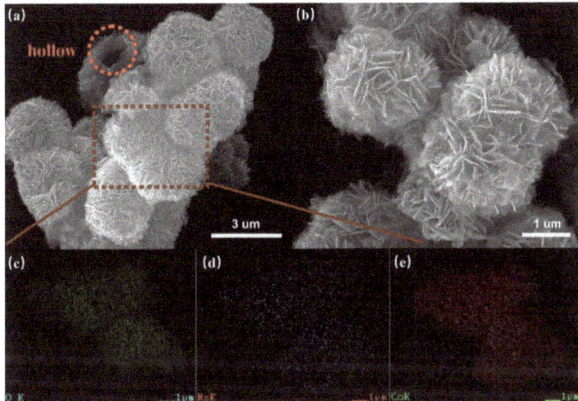

Figure 6. SEM images (**a,b**) of the calcined $MnCo_2O_4$ products and the corresponding elements mapping (**c–e**) taken from the square area marked in Figure 6a. Reproduced with permission from Hongwei Che, Template-free synthesis of novel flower-like $MnCo_2O_4$ hollow microspheres for application in supercapacitors; published by Elsevier, 2016 [87].

Although the electrochemical performance of an $MnCo_2O_4$ electrode is significantly improved compared with a single Co_3O_4 electrode, its development is limited by its poor cycle stability in long-term use. To solve this problem, one of the effective methods is to compound it with some carbonaceous material with light electric power or other pseudo-capacitive oxide or hydroxide to improve its cycle stability. For example, Wang et al. synthesized a 3D porous structure based on $MnCo_2O_4$ modified graphene [93]. The specific capacitance reached $503\ F\ g^{-1}$ at a current density of $1\ A\ g^{-1}$. After 5000 charge-discharge cycles (current density of $10\ A\ g^{-1}$), 97.4% of the specific capacitance was retained.

Zhao et al. synthesized an $MnCo_2O_4@Ni(OH)_2$ multicomponent composite by a stepwise hydrothermal method [88]. The synthesis process is shown in Figure 7. First, layered double hydroxides of cobalt and manganese were generated with hexamethylenetetramine as a structure guide. With the directional attachment process as the driving force, the nanoparticles finally grew into $MnCo_2O_4$ nanostructures. Then, using nickel chloride and hexamethylenetetramine as the lead solution, an ultra-thin $Ni(OH)_2$ nano sheet was fixed on the nano alloy by hydrothermal method to produce a layered $MnCo_2O_4@Ni(OH)_2$ core-shell structure. The discharge time of $MnCo_2O_4@Ni(OH)_2$ was about four times that of $MnCo_2O_4$.

Figure 7. Schematic illustration of the general electrode design process. Reproduced with permission from Limin Wu, Preparation of $MnCo_2O_4$@$Ni(OH)_2$ Core-Shell Flowers for Asymmetric Supercapacitor Materials with Ultrahigh Specific Capacitance; published by Wiley-VCH Verlag, 2016 [88].

The specific capacitance of activated carbon electrode can reach 328 F g^{-1} at 0.2 A g^{-1}, and the maximum energy density of asymmetric SC (ASC) can reach 48 W h kg^{-1} when the mean power density is 1.4 kW kg^{-1}, which is significantly higher than that of most commercial batteries. In addition, the capacitance retention of the hybrid electrode is about 90% after 2500 cycles at a current density of 6 A g^{-1}, and the structure of the nano alloy remains good. The above results show that the electrochemical performance of $MnCo_2O_4$ is significantly improved and its cycle stability is higher by compounding $MnCo_2O_4$ with $Ni(OH)_2$.

As mentioned above, compared with a single $MnCo_2O_4$ material, $MnCo_2O_4$ compounded with other materials has higher cycle stability and greater prospects. Although the electrochemistry of the material can be improved to some extent by changing the morphology and structure or compounding with other materials, the low conductivity of $MnCo_2O_4$ has hindered its wide application as an energy storage device. At the same time, how to accurately control the micro morphology of the composite still needs further exploration.

2.3.2. $NiCo_2O_4$

As a typical cobalt-containing ternary metal oxide, $NiCo_2O_4$ is also a transition metal oxide with a spinel structure. It has the advantages of high electrochemical activity, good conductivity, high theoretical capacitance, low cost, and simple synthesis. Therefore, $NiCo_2O_4$ is also one of the most attractive electrode materials in SCs [94–98]. In its structure, nickel ions occupy octahedral sites, and cobalt ions diffuse in octahedral and tetrahedral sites [99]. The electronic conductivity and electrochemical activity of $NiCo_2O_4$ are significantly higher than those of nickel oxide and cobalt oxide alone due to the synergistic effect of Ni with Co.

At present, various nanostructures of $NiCo_2O_4$ have been prepared, such as nanowires [100], nanosheets [101], nanoflowers [102], and nanorods [103]. Among them, hollow nano materials have a large surface area, large gap and short effective transmission distance of electrolyte ions [104]. They provide more electroactive sites for rapid ion insertion of the whole electrode material and show excellent electrochemical performance. Xu et al. synthesized hollow $NiCo_2O_4$ nanospheres with a layered structure [104]. When using them as electrodes, the specific capacitance at 1 A g^{-1} is 1229 F g^{-1}, which is higher than that of $NiCo_2S_4$ hollow spheres (1036 F g^{-1} at 1 A g^{-1}) [105], $NiCo_2O_4$ hollow spheres (1141 F g^{-1} at 1 A g^{-1}) [106], hollow $NiCo_2O_4$ sub microspheres (678 F g^{-1} at 1 A g^{-1}) [107], urchin-like $NiCo_2O_4$ hollow microspheres (942.2 F g^{-1} at 0.5 A g^{-1}) [108], and mesoporous $NiCo_2O_4$ hollow microspheres (987 F g^{-1} at 1 A g^{-1}) [109]. After 3000 cycles at 50 mV s^{-1}, the total specific capacitance retention of hollow $NiCo_2O_4$ nanosphere electrode is 86.3%, while the total specific capacitance retention of $NiCo_2O_4$ microsphere electrode is 83.7%.

Although hollow microspheres can effectively improve surface area, the single-structure $NiCo_2O_4$ electrode material still has the disadvantages of low conductivity, limited kinetics, and poor electrochemical performance [110–114]. To improve its electrochemical performance, constructing $NiCo_2O_4$ layered nanostructure composites has become an important means [115–119]. For example, Zhou and others synthesized 3D porous graphene/$NiCo_2O_4$ hybrid films with copper oxide as a template [120]. Its unique 3D porous structure can store many electrolytes and provide rich active centers, thus improving the electrochemical performance. At 1 A g^{-1}, the specific capacitance can reach 708.36 F g^{-1}. After 6000 cycles at 10 A g^{-1}, the initial capacitance of 94.3% is maintained. Li et al. prepared flower-like hollow C@$MnCo_2O_4$ with high specific surface area. At a discharge current density of 1 A g^{-1}, the discharge capacitance reached 728.4 F g^{-1}, and after 1000 cycles at 8 A g^{-1} the initial capacitance retention of the composite was 95.9% [121]. Zhao et al. synthesized ultra-thin $NiCo_2O_4$/NiO nanosheets grown on silicon nitride [122]. After 2000 cycles at 20 mA cm^{-2} current density, the specific capacitance retention was 90.9%, and the energy density was 60 W h kg^{-1} when the power density was 1.66 kW kg^{-1}. Cheng et al. prepared a 3D layered $NiCo_2O_4$@$NiMoO_4$ nuclear shell nanowires/nanowire sheet array on nickel foam, with a capacitance retention rate of 85.2% after 3000 cycles at a current density of 20 mA cm^{-2} [123]. After a long cycle, the volume resistance of the ASC device increased slightly from the initial 0.40 Ω to 0.42 Ω. The above shows that the prepared composites have good cycle stability. Lee et al. synthesized $MnCo_2O_4$-$NiCo_2O_4$ composite with layered nanostructure by one-step chemical bath deposition method [124]. When used as an electrode, the specific capacitance reached 1152 F g^{-1} at 1 A g^{-1}. After 3000 cycles at 6 A g^{-1}, the specific capacitance retention of the composite was 95.38%, while $NiCo_2O_4$ is 86.14% and $MnCo_2O_4$ was 61.65%, indicating that the composite of the two materials significantly increased the cycle stability of the material.

The electrochemical properties of the above $NiCo_2O_4$ composite have been significantly improved. However, due to lattice mismatch between $NiCo_2O_4$ and other components, this leads to poor structural stability, lower specific capacitance and cycle life. Therefore, Wang et al. compounded $NiCo_2O_4$ and $NiCo_2O_4$ with the same lattice type to prepare 3D delamination $NiCo_2O_4$@$NiCo_2O_4$ [94]. The preparation process of the core-shell nano cone array is shown in Figure 8. First, $NiCo_2O_4$ is grown vertically on nickel foam by hydrothermal method. After annealing, neat $NiCo_2O_4$ nano-cone arrays is formed first. Then, the $NiCo_2O_4$ nanosheet is coated on the $NiCo_2O_4$ surface formed in the previous step. Finally, layered core-shell $NiCo_2O_4$@$NiCo_2O_4$ nanostructures are fabricated on nickel foams after subsequent annealing.

After 21,000 cycles at 4 A g^{-1}, the capacitance retention rate of the electrode reached 85.3%, and the structure did not change significantly during charge and discharge. When used in SCs, $NiCo_2O_4$@$NiCo_2O_4$ core-shell nanostructure had a capacitance of 2045.2 F g^{-1} at a current density of 1 A g^{-1}, which is better than the single component of $NiCo_2O_4$ nanosheet (346.4 F g^{-1}) and $NiCo_2O_4$ nano cone (1381.8 F g^{-1}).

2.3.3. ZnCo$_2$O$_4$

Similar to $MnCo_2O_4$ and $NiCo_2O_4$ mentioned above, $ZnCo_2O_4$ has the advantages of high theoretical capacitance, high conductivity, environmental friendliness, and low cost, and is considered as a potential SC material [125]. At present, $ZnCo_2O_4$ materials with various nanostructures, such as nanowires [126], nanosheets [127,128] nanoparticles [129], and nanospheres [130,131], have been prepared. For example, Wang and colleagues synthesized $ZnCo_2O_4$ nanowire electrode materials grown on nickel foam [132]. First, the precursor $ZnCo_2O_4$ nanowire arrays were grown on nickel foam by a hydrothermal reaction and then calcined in air. Finally, $ZnCo_2O_4$ nanowire arrays supported by nickel foam were obtained. The synthesized $ZnCo_2O_4$ nanowires have a porous structure, which makes the material have large specific surface area and can promote the diffusion of reactants. The prepared $ZnCo_2O_4$ nanowire/nickel foam electrode had a specific capacitance of 1625 F g^{-1} at a current density of 5 A g^{-1}, and 94% of the original capacitance was maintained after 5000 cycles at 20 A g^{-1}.

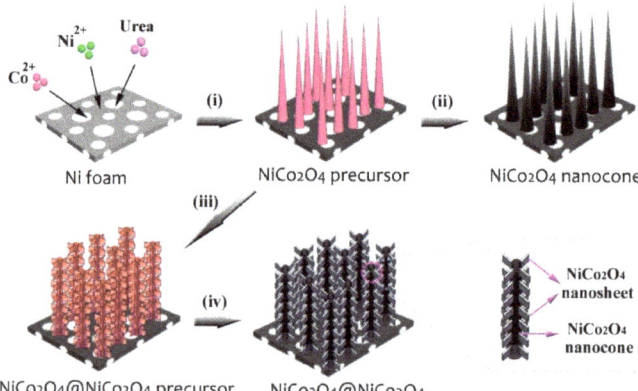

Figure 8. Schematic of the fabrication process for 3D $NiCo_2O_4$@$NiCo_2O_4$ hierarchical core-shell NCAs on Ni foam. Reproduced with permission from Xiuhua Wang, Three-Dimensional $NiCo_2O_4$@$NiCo_2O_4$ Core-Shell Nanocones Arrays for High-Performance Supercapacitors; published by Elsevier, 2018 [94].

Xu et al. prepared a $ZnCo_2O_4$ nanostructure with a porous structure and found that the conversion between nanosheets and nanowires was obtained by regulating hydrothermal temperature [133]. When the current density was 1 A g^{-1}, $ZnCo_2O_4$ had a specific capacitance of 776.2 F g^{-1}, and the energy density was 84.48 W h kg^{-1} when the mean power density was 0.4 kW kg^{-1}. It had 84.3% capacity retention after 1500 cycles (3 A g^{-1}). Venkatachalam et al. prepared hexagonal-like $ZnCo_2O_4$ nanomaterials by a simple hydrothermal method [134]. The prepared electrode materials had a specific capacitance of 845.7 F g^{-1} at a current density of 1 A g^{-1}, and retained 95.3% of the original capacitance after 5000 cycles at 5 A g^{-1}. Shang et al. synthesized 3D layered peony flower-like $ZnCo_2O_4$ electrode nanomaterials by a simple solvothermal method and annealing without additives [135]. The microstructure is shown in Figure 9. The assembled ASC $ZnCo_2O_4$//active carbon had an energy density of 29.76 W h kg^{-1} at a power density of 398.53 W kg^{-1}. In addition, the peony-shaped $ZnCo_2O_4$ electrode material had a specific capacitance of 440 F g^{-1} at a current density of 1 A g^{-1}, and the capacitance was maintained at 155.6% after 3000 cycles (2 A g^{-1}).

Although the above nano $ZnCo_2O_4$ materials have specific applications in SCs, the insufficient utilization efficiency and poor conductivity of the materials limit their electrochemical properties to a certain extent and there are difficulties in them meeting the needs of practical applications. To solve this problem, one of the commonly used methods is to introduce oxygen vacancies. The existence of an oxygen vacancy can significantly improve the conductivity of $ZnCo_2O_4$, adjust the electronic structure, increase the active sites, and promote the electrochemical performance of SCs. For example, Xiang and his colleagues prepared two-dimensional (2D) $ZnCo_2O_4$ nanosheets rich in oxygen vacancies [136]. The nanoscale thickness and large surface area effectively improved the utilization of the electrode while promoting electron transfer. A specific capacitance of 2111 F g^{-1} was attained at a current density of 1 A g^{-1}, while the specific capacitance of the original $ZnCo_2O_4$ nano sheet at the same current density was only 1121 F g^{-1}. When the power density was 160 W kg^{-1}, the energy density of ASC constructed by $ZnCo_2O_4$ nanosheet (with oxygen vacancy)//activated carbon is 34.6 W h kg^{-1}, and 93% of the original capacitance was maintained after 3000 cycles at 2 A g^{-1}.

Combining $ZnCo_2O_4$ nanostructures with conductive metal or carbon materials to construct composites is one of the methods to alleviate the above problems. For example, Wu et al. synthesized a series of $ZnCo_2O_4$@$Ni(OH)_2$ nanostructures grown on nickel foam by a two-step hydrothermal method; the preparation process is shown in Figure 10 [137].

First, the $ZnCo_2O_4$ nanowires were uniformly covered on the nickel foam by a hydrothermal method and then $Ni(OH)_2$ nanosheets were grown on the $ZnCo_2O_4$ nanowire after a second hydrothermal reaction. $ZnCo_2O_4$ nanowires were used as the substrate and $Ni(OH)_2$ nanosheets were used as the upper layer. The strong binding force between them reduced the contact resistance and promote the transfer of electrons to enhance the electrochemical reaction activity of the material. The synthesized hybrid structure was used to fabricate capacitors with an energy density of 57.3 W kg^{-1} at 4675.3 W h kg^{-1}, and an initial capacitance of 48.6 C g^{-1} at 1 A g^{-1}, which retained 90.5% after 10,000 cycles at the same current density.

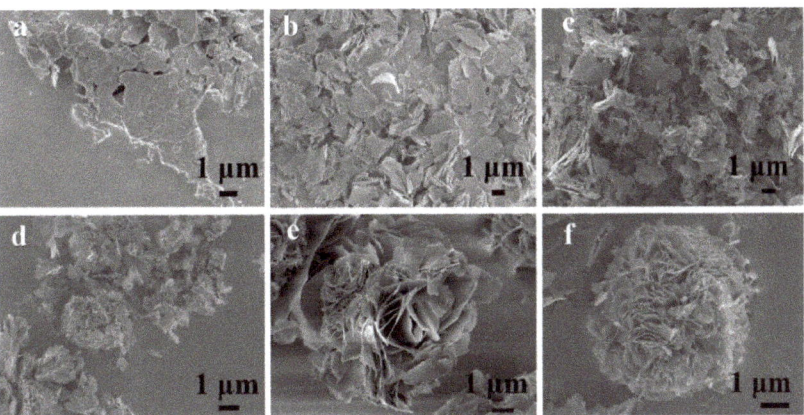

Figure 9. SEM images of ZnCo-preprepared with different reaction times: (**a**) 4 h, (**b**) 6 h, (**c**) 12 h, (**d**) 18 h, (**e**) 21 h, and (**f**) 24 h. Reproduced with permission from Liangyu Shang, Self-assembled hierarchical peony-like $ZnCo_2O_4$ for high-performance asymmetric supercapacitors; published by Elsevier, 2017 [135].

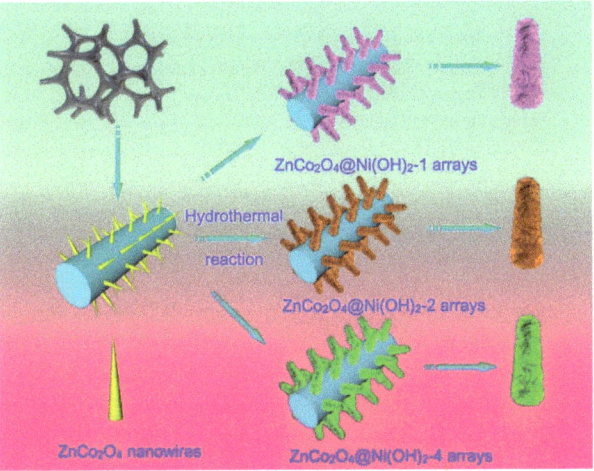

Figure 10. Synthesis schematic of synthesized $ZnCo_2O_4@Ni(OH)_2$ samples. Reproduced with permission from Meizhen Dai, Ni Foam Substrates Modified with a $ZnCo_2O_4$ Nanowire-Coated $Ni(OH)_2$ Nanosheet Electrode for Hybrid Capacitors and Electrocatalysts; published by ACS Publications, 2021 [137].

Xie et al. synthesized a $ZnCo_2O_4@ZnWO_4$ nanowire array with a core-shell structure on nickel foam, and the synergistic effect between $ZnCo_2O_4$ nanowire and $ZnWO_4$ sheet effectively improved the electrochemical performance of hybrid electrode [138]. The synthesis process is like that of $ZnCo_2O_4@Ni(OH)_2$. As shown in Figure 11, $ZnCo_2O_4$ nanowires are first grown on nickel foam, and then $ZnWO_4$ nanosheets arrays are produced by a simple hydrothermal method using $ZnCo_2O_4$ nanowires as skeletons. The constructed $ZnCo_2O_4@ZnWO_4$//active carbon ASC had an energy density of 24 W h kg^{-1} at a power density of 400 W kg^{-1}. The original capacitance retention was 98.5% after 5000 cycles at a current density of 100 mA cm^{-2}.

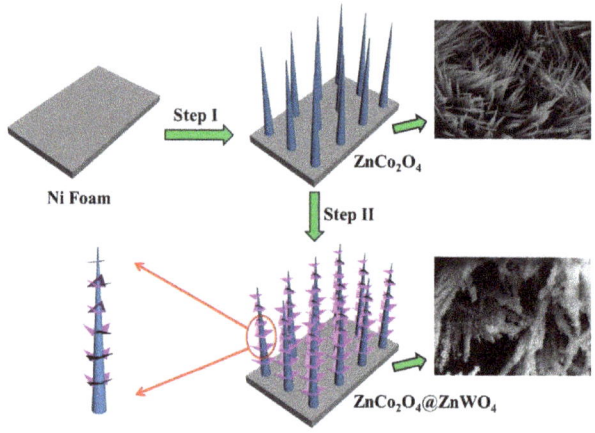

Figure 11. Schematic illustrating the fabrication process of the $ZnCo_2O_4@ZnWO_4$ core-shell nanowire arrays on a nickel foam substrate. Reproduced with permission from Li Xie, Core-shell structured $ZnCo_2O_4@ZnWO_4$ nanowire arrays on nickel foam for advanced asymmetric supercapacitors; published by Elsevier, 2018 [138].

Table 2 summarizes the structure, specific capacity and cycling performance of $MnCo_2O_4$, $NiCo_2O_4$ and $ZnCo_2O_4$. The electrochemical performance of the electrodes was significantly improved after designing unique morphologies for the materials. By constructing effective structures, such as spherical, rod-like, and hollow structures, the contact area can be increased, resulting in close contact between the active material and the electrolyte, which leads to high electrochemical activity and enhanced pseudocapacitive performance. In addition, the porous structure can alleviate the volume change caused by the redox reaction, thus improving the cycle performance, so the materials in the table are often designed as porous structures. Among the various unique morphologies, 2D microstructures are an important category because such structures increase the contact area between the electrolyte and the electrode material. For example, Younis et al. designed various micro morphologies including nanowires, nanocables, nano-micro biscuits, and micro-walls [139]. Among them, nano-micro biscuits with distinct 2D structural features exhibited the best electrochemical performance. Xiang et al. designed $ZnCo_2O_4$ nanosheets with nanoscale thickness and large surface area, which could improve the electron transfer efficiency and electrode utilization [136]. In addition, Zhang et al. prepared $NiCo_2O_4$ nanosheets with a more ordered crystal structure, high specific surface area and diffusion channels [140]. Liu et al. prepared $MnCo_2O_4$ with a nanoflower-like morphology and porous structure [141]. Because of its unique nanostructure, the prepared electrode had high capacity and good rate performance. In conclusion, 2D nanostructures usually have a large surface area and dense porous structure. This structure is beneficial to increase the contact area between electrolyte and electrode, thus improving the electron transfer efficiency.

Table 2. Summary of materials, structures, and electrochemical properties of cobalt-containing ternary metal oxides.

Materials	Structure	Specific Capacitance (Current Density)	Cycling Performance (Cycles, Current Density)	Year	Refs.
$MnCo_2O_4$	polyhedral nanostructure	1763 F g^{-1} (1 A g^{-1})	95% (4500, 1 A g^{-1})	2017	[89]
	flower-like hollow microspheres	235.7 F g^{-1} (1 A g^{-1})	93.6% (2000, 1 A g^{-1})	2016	[87]
	3D porous structure	503 F g^{-1} (1 A g^{-1})	97.4% (5000, 10 A g^{-1})	2019	[93]
	belt-based core-shell nanoflowers	2154 F g^{-1} (5 A g^{-1})	90% (2500, 6 A g^{-1})	2016	[88]
$NiCo_2O_4$	hollow nanospheres with layered structure	1229 F g^{-1} (1 A g^{-1})	86.3% (3000, 50 mV s^{-1})	2018	[104]
	hollow spheres	1036 F g^{-1} (1 A g^{-1})	78.6% (10,000, 5 A g^{-1})	2015	[105]
	hollow sub microspheres	678 F g^{-1} (1 A g^{-1})	87% (3500, 10 A g^{-1})	2013	[107]
	urchin-like hollow microspheres	942.2 F g^{-1} (0.5 A g^{-1})	90% (1000, 2.5 mA cm^{-2})	2017	[108]
	mesoporous hollow microspheres	987 F g^{-1} (1 A g^{-1})	≈100% (5000, 5 A g^{-1})	2015	[109]
	3D porous graphene/NiCo$_2$O$_4$ hybrid films	708.36 F g^{-1} (1 A g^{-1})	94.3% (6000, 10 A g^{-1})	2020	[120]
	flower-like hollow	728.4 F g^{-1} (1 A g^{-1})	95.9% (1000, 8 A g^{-1})	2014	[121]
	ultra-thin nanosheets	1801 F g^{-1} (1 mA cm^{-2})	90.9% (2000, 20 mA cm^{-2})	2016	[122]
	3D layered nuclear shell nanowires/nanowires sheet array	—	85.2% (3000, 20 mA cm^{-2})	2015	[123]
	layered nanostructure	1152 F g^{-1} (1 A g^{-1})	95.38% (3000, 6 A g^{-1})	2018	[124]
	layered core-shell nanostructures	2045.2 F g^{-1} (1 A g^{-1})	85.3% (21000, 4 A g^{-1})	2018	[94]
$ZnCo_2O_4$	nanowire	1625 F g^{-1} (5 A g^{-1})	94% (5000, 20 A g^{-1})	2014	[132]
	porous structure	776.2 F g^{-1} (1 A g^{-1})	84.3% (1500, 3 A g^{-1})	2017	[133]
	hexagonal like nano materials	845.7 F g^{-1} (1 A g^{-1})	95.3% (5000, 5 A g^{-1})	2017	[134]
	3D layered peony flower like material	440 F g^{-1} (1 A g^{-1})	155.6% (3000, 2 A g^{-1})	2017	[135]
	2D nanosheets	2111 F g^{-1} (1 A g^{-1})	93% (3000, 2 A g^{-1})	2021	[136]
	nanowires	48.6 C g^{-1} (1 A g^{-1})	90.5% (10,000, 1 A g^{-1})	2021	[137]
	nanowire array with core-shell structure	13.4 F cm^{-2} (4 mA cm^{-2})	98.5% (5000, 100 mA cm^{-2})	2018	[138]

2.4. Cobalt-Containing Ternary Metal Oxide Derivatives

As mentioned above, cobalt-containing ternary metal oxides have great potential in the application of SCs. To further improve their electrochemical performance, researchers have focused on the derivatives of these metal oxides. Transition metal sulfides have high electronic conductivity, two orders of magnitude higher than the corresponding oxides, because the valence states of the transition metals in the sulfides closely resemble those of the metals [142–144]. At the same time, because sulfur is less electronegative than oxygen, it can produce a more flexible structure instead of oxygen. This can effectively avoid the structural disintegration of transition metal sulfide-based electrodes due to interlayer elongation, which facilitates the transport of electrons in the internal structure [145]. In addition, combining two or more sulfides can improve the electrical properties of transition metal sulfides, resulting in a richer redox reaction [146–148] because bimetallic sulfides possess more prosperous diverse states, smaller optical band gaps, and better chemical stability than single-metal sulfides [144,148]. Compared with single metal oxide, transition metal sulfides such as Co-Mo-S, NiCo$_2$S$_4$ have higher capacitance, multivalent redox reactions and higher conductivity [149], so they have great potential.

2.4.1. Co-Mo-S

Co-Mo-S matrix composites have great potential as SC electrode materials because of their advantages of reversible redox reaction band gap, high conductivity, and low electronegativity [149–159]. For example, Balamurugan et al. used ion exchange reaction technology to synthesize a porous nano foam support structure composed of ultra-thin Co-Mo-S nanosheets [160]. When Co-Mo-S nanosheets are used as the electrode of the SC, they can provide an ultra-high specific capacitance of 2343 F g^{-1} at a current density of 1 mA cm^{-2}, and the capacitance remains 96.6% after 20,000 cycles. In addition, the energy density and power density of Co-Mo-S/nitrogen doped graphene nanosheets assembled in ASC are 89.6 W h kg^{-1} and 20.07 W kg^{-1}. The capacitance retention rate can reach 86.8% after 50,000 cycles. The unique electrochemical properties of Co-Mo-S nanosheets are attributed to the ultra-high contact area with 3D nickel foam and electrolyte.

Xu et al. prepared amorphous $CoMoS_4$ by a simple precipitation method and used it as an SC material for the first time [161]. Changing the current density from 1 A g^{-1} to 3 A g^{-1}, the galvanostatic charge/discharge curves are shown in Figure 12 when the potential is from 0 V to 0.6 V. The specific capacitance was calculated according to these curves. The results show that it had a specific capacitance of 661 F g^{-1} at a current density of 1 A g^{-1}. Simultaneously, the constructed $CoMoS_4$//reduced graphene oxide hybrid SC had a particular capacity of 77 F g^{-1} at a current density of 0.5 A g^{-1}, and its energy density was 27.2 W h kg^{-1} at a power density of 400 W kg^{-1}. In addition, after 10,000 cycles at 80 mV s^{-1}, the original capacitance was maintained at about 86%.

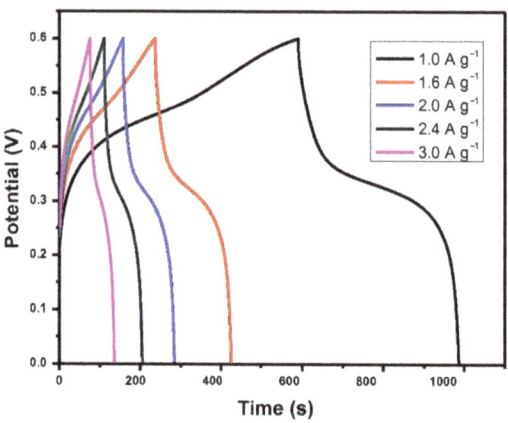

Figure 12. Galvanostatic charge/discharge curves of $CoMoS_4$. Reproduced with permission from Xiaoyang Xu, Amorphous $CoMoS_4$ for a valuable energy storage material candidate; published by Elsevier, 2016 [161].

Recently, Sun et al. synthesized Co-Mo-S nanosheet networks by a simple two-step hydrothermal method [162]. The ASC assembled with the product as the cathode had an energy density of 72.25 W h kg^{-1} at 2700 W kg^{-1}. After 9000 cycles at 2 A g^{-1}, the capacitance retention rate reached 83.4%.

Although Co-Mo-S has excellent potential in SCs, its relatively poor rate capability and cycle stability limit its application. Overcoming these disadvantages and improve its electrochemical properties has become a key problem of Co-Mo-S capacitor materials. A practical method is to achieve excellent cycle capacity and rate performance by construction of the electrode material structure. Ma et al. designed and constructed hollow core-shell $CoMoS_4$@Ni-Co-S nanotubes on carbon cloth for the first time by a hydrothermal method and electrodeposition process [163]. The preparation process of Co-S nanotubes is shown in Figure 13. First, Co(OH)F nanowire arrays are synthesized by hydrothermal reaction under high temperature and high pressure with carbon cloth as a current collector. Then, Co(OH)F nanowires and $(NH_4)_2MoS_4$ precursor solution ae transformed into $CoMoS_4$ nanotubes. Finally, 3D layered $CoMoS_4$@Ni-Co-S nanotube hybrid arrays are synthesized by electrochemical deposition method. Among them, Ni-Co-S nanosheets are closely arranged around $CoMoS_4$ hollow nanotubes, which is conducive to the exposure of electrochemical active sites and keeps the structure stable to a certain extent during charge and discharge. At the same time, the core-shell structure facilitates the close contact of the electrode/electrolyte and avoids the aggregation of Ni-Co-S. The novel $CoMoS_4$@Ni-Co-S electrode had an excellent specific capacitance of 2208.5 F g^{-1} at 1 A g^{-1} and good cycle life (91.3% capacitance retention over 5000 cycles at 3 A g^{-1}). In addition, the assembled $CoMoS_4$@Ni-Co-S//activated carbon ASC had an energy density of 49.1 W h kg^{-1} at 800 W kg^{-1} and a capacity retention rate of 90.3% after 10,000 cycles.

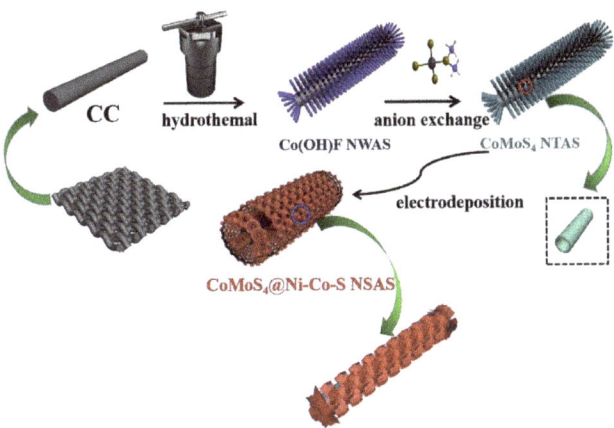

Figure 13. Schematic illustration of the fabrication of a hierarchical core-shell hollow CoMoS$_4$@Ni-Co-S nanotubes electrode. Reproduced with permission from Fei Ma, Hierarchical core-shell hollow CoMoS$_4$@Ni-Co-S nanotubes hybrid arrays as advanced electrode material for supercapacitors; published by Elsevier, 2019 [163].

2.4.2. NiCo$_2$S$_4$

As mentioned earlier, NiCo$_2$O$_4$ and its composites have great potential in SCs. The conductivity of NiCo$_2$S$_4$ is 100 times that of NiCo$_2$O$_4$, and NiCo$_2$S$_4$ shows higher electrochemical activity and capacitance than other cobalt nickel compounds because of its inherent redox reaction center. However, NiCo$_2$S$_4$-based electrodes suffer from defects such as easy oxidation in alkaline electrolytes and poor long-term cycling stability [164]. Therefore, effective space structures need to be designed to improve their drawbacks. So far, 3D NiCo$_2$S$_4$ nanostructures such as nanoflowers, core-shell and dendrites have been synthesized.

For example, Shi et al. synthesized layered sea urchin-like hollow NiCo$_2$S$_4$ by a template-free solvothermal method [165]. The capacitance reached 1398 F g^{-1} at 1 A g^{-1}, and the specific capacity retention rate reached 74.4% after 5000 cycles at 10 A g^{-1}. Zhang et al. synthesized nano NiCo$_2$S$_4$ with 3D honeycomb structure by a hydrothermal method and vulcanization method [166]. When the current density was 1 mA cm^{-2}, its maximum specific capacity exceeded 14 mA h cm^{-2}. After 1000 cycles at a current density of 10 mA cm^{-2}, the specific capacity remained at 96.96%.

These structures have been widely used in electrode materials. However, their poor electronic conductivity and potential risk of structural collapse and damage during long-term use limit the application of NiCo$_2$S$_4$ materials. One of the main methods to solve this problem is to build 3D hierarchical structure materials and increase the contact area with electrolyte.

Li et al. successfully synthesized layered dendritic NiCo$_2$S$_4$@NiCo$_2$S by a three-step continuous hydrothermal method, and the layered microstructure of the highly porous structure facilitated ion transport during charge and discharge, resulting in a significant improvement in electrochemical performance. When the current density was 240 mA cm^{-2}, the electrode discharge specific capacity of the dendritic structure reached 4.43 mA h cm^{-2}. When the current density was increased from 40 mA cm^{-2} to 240 mA cm^{-2}, its rate capability reached 70.1% [167]. Tang et al. synthesized ultra-high load (10.33 mA cm^{-2}) 3D layered NiCo$_2$S$_4$/Ni$_3$S$_2$ nanosheets with an energy density of 4.69 W h m^{-2} (power density of 10.33 W m^{-2}), and a stability of 91.4% after 8000 cycles at 20.66 mA cm^{-2} [168].

Zhang et al. synthesized NiCo$_2$S$_4$ spheres with granular nuclei by a simple two-step hydrothermal reaction [169]. A NiCo$_2$(OH)$_6$/C precursor was prepared using a carbon pellet cluster as a template. Granular NiCo$_2$S$_4$ was synthesized by reacting with sodium sulfide, and then the $NiCo_2S_4$ precursor was grown on the periphery of the granular $NiCo_2S_4$ to form a unique structure. The specific surface area of the prepared $NiCo_2S_4$ ball was 26.61 m^2 g^{-1}, which is about twice that of the particle $NiCo_2S_4$ (11.41 m^2 g^{-1}). This higher specific surface area increased the electroactive sites that can transfer charge and shortens the transmission path, which is conducive to improving the electrochemical activity of the material. When the current density was 1 A g^{-1}, the specific capacitance of the granular $NiCo_2S_4$ spherical electrode reached 1156 F g^{-1}, which was 71% higher than that of the $NiCo_2S_4$ electrode. In addition, after 1000 charge-discharge cycles (5 A g^{-1}), the $NiCo_2S_4$ sphere electrode with granular nuclear showed 82% capacitance retention, and the cycle stability was significantly better than that of the granular $NiCo_2S_4$ electrode.

Wu et al. prepared a hierarchical nanostructured $NiCo_2S_4$ nanoflower@$NiCo_2S_4$ nanosheet material by a hydrothermal method (Figure 14) [170]. Using this composite as the electrode in the SC, it had a specific capacity of 338.1 mA h g^{-1} at 2 mV s^{-1}, which is about three times higher than that of a single $NiCo_2S_4$ nanosheet. In addition, 90% of the original capacity was maintained after 4000 reaction cycles at a current density of 20 A g^{-1}. The synthesized $NiCo_2S_4$ nanoflowers@$NiCo_2S_4$ nanosheets//$NiCo_2S_4$ nanoflowers@$NiCo_2S_4$ nanosheets symmetrical SC device had an energy density of 18.05 W h kg^{-1} at a power density of 750 W kg^{-1}. The capacitance retention rate of the symmetrical SC device was 89% after 4000 cycles (10 A g^{-1}). The multilayer 3D structure can explain this improvement in electrochemical performance. The upper nanoflowers are composed of many rough nanotubes, which increase the surface volume ratio and the contact range of the electrolyte. This unique structure can provide more electrochemical active sites, promote ion adsorption, and reduce the volume expansion in the charge and discharge process. Furthermore, the lower layer nanosheet arrays on the nickel foam can avoid damage and increase the stability of the electrochemical reaction.

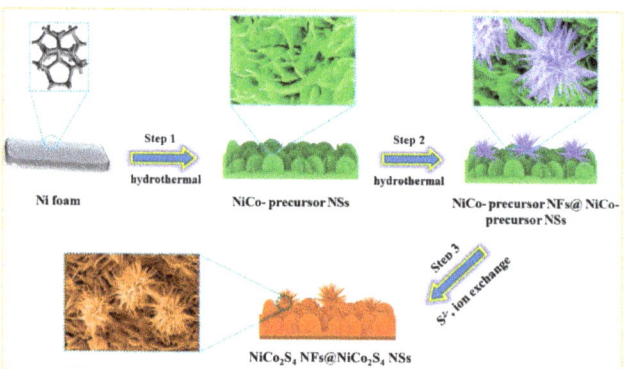

Figure 14. Schematic diagram showing the fabrication of $NiCo_2S_4$ nanoflowers @$NiCo_2S_4$ nanosheets. Reproduced with permission from Wenling Wu, Hierarchical structure of Self-Supported $NiCo_2S_4$ Nanoflowers@$NiCo_2S_4$ nanosheets as high rate-capability and cycling-stability electrodes for advanced supercapacitor; published by Elsevier, 2021 [170].

Densely arranged and structurally stable nanosheets can act as a charge transport interconnectors with nickel foam, further improving the charge transport rate. This unique synergistic effect between nanoflower and nanosheet structure effectively increases the structural stability and electrochemical active sites of the material. The effect also promotes charge transfer and ion transport, which is conducive to accelerating the electrochemical reaction rate and improving the energy storage effect of the material.

As mentioned above, the construction of 3D multilayer hierarchical structures can improve the electrochemical properties of materials and in-use stability. At the same time, the construction of nanostructured composites by doping other impurity atoms is the foremost solution to the problem of low electronic conductivity and poor stability of

$NiCo_2S_4$ materials. Among them, carbon material has superior conductivity [170], which can promote charge transfer. Due to the strong coupling between the carbon substrate and metal-based oxide, the composite of carbon material and $NiCo_2S_4$ can effectively increase the electrochemical activity of electrode material [171]. For example, Pezzotti and co-workers synthesized a kelp-like $NiCo_2S_4$-C-MoS_2 composite by hydrothermal and solvothermal methods [172]. It had a specific capacitance of 1601 F g^{-1} at a current density of 0.5 A g^{-1} and 75% of the initial specific capacity after 2000 cycles at a current density of 2 A g^{-1}. Shim et al. synthesized a hollow C-$NiCo_2S_4$ nano-lake sheet structure with a one-step solvent method [173]. The specific capacitance reached 1722 F g^{-1} at a current density of 1 A g^{-1}, and 95.60% capacity retention after 5000 cycles at a current density of 10 A g^{-1}.

In addition, since the electronegativity and atomic radius of P and S atoms are similar, introducing P atoms results in lattice distortion, providing more active sites. Therefore, the introduction of the P atom is also a way to improve the electrochemical activity of materials. Based on the above, Liu et al. introduced P and C elements into a $NiCo_2S_4$ electrode material by a one-step solvothermal method and phosphating process [174]. As the electrode material of SCs, it had a specific capacity of 1026 C g^{-1} at a current density of 1 A g^{-1}, and an original capacity retention rate of 89% after 20,000 cycles at 10 A g^{-1}. In comparison, $NiCo_2S_4$ only reached 65% of the original capacity under the same conditions. The ASC had an energy density of 131.40 W h kg^{-1} at a power density of 1355.37 W kg^{-1}, and 96.3% of the original capacity was maintained after 10,000 cycles at a current density of 2 A g^{-1}.

Dai et al. prepared relatively stable $ZnCo_2O_4$@Ni [171]. The specific capacity of Ni-Co-S composite electrode material was 1396.9 C g^{-1} at a current density of 1 A g^{-1}, while $ZnCo_2O_4$ nanorods and Ni-Co-S showed a specific capacity of 1025.5 C g^{-1} and 1026 C g^{-1}, respectively, under the same conditions. At the same time, the device showed a capacity retention rate of 85.5% after 1000 charge-discharge cycles at 4 A g^{-1}. Bai et al. prepared 2D Co_3O_4@Ni(OH)$_2$ [175]. The SC synthesized by this method had a specific capacitance of 98.4 F g^{-1} in the potential range of 0–1.7 V at 5 mA cm^{-2} and an energy density of 40.0 W h kg^{-1} at a power density of 349.6 W g^{-1}. In addition, the original specific volume retention rate was 90.5% after 5000 cycles (1.61 A g^{-1}). This proved that the composite with core-shell structure can retain the advantages of each component, and the synergistic effect between them can be used to improve the electrochemical properties of the material. Based on the above, Zhang et al. synthesized layered core-shell polyporrole nanotubes@$NiCo_2S_4$ materials by coating $NiCo_2S_4$ nanosheets on conductive polypyrrole nanotubes [144]; the formation process is shown in Figure 15. The material had a specific capacitance of 911 F g^{-1} at a current density of 1 A g^{-1} and maintained a capacitance of 592 F g^{-1} at a current density of 20 A g^{-1}. After 4000 cycles at a current density of 5 A g^{-1}, the original capacitance was 93.2%.

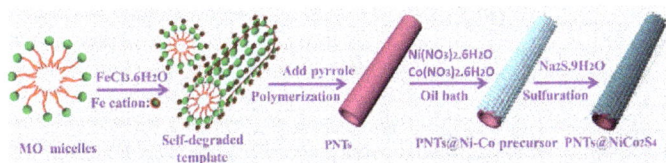

Figure 15. Schematic illustration of polyporrole nanotubes@$NiCo_2S_4$ core-shell formation. Reproduced with permission from Jun Zhang, Hierarchical polypyrrole nanotubes@$NiCo_2S_4$ nanosheets core-shell composites with improved electrochemical performance as supercapacitors; published by Elsevier, 2017 [144].

2.5. Other Cobalt-Containing Materials

Among other cobalt-containing materials, Co_3O_4@$NiMoO_4$ has been most studied because $NiMoO_4$ has good conductivity, which can improve the energy storage capacity of Co_3O_4.

Zhang et al. used hydrothermal and annealing methods to synthesize flower-like hybridized arrays on nickel foam [176]. Using Co_3O_4 nanowire arrays as scaffolds, $NiMoO_4$ nanosheets were grown on the surface to form a new type of 3D layered battery electrode Co_3O_4@$NiMoO_4$. The specific capacity of the hybrid array of the prepared Co_3O_4@$NiMoO_4$ was 636.8 C g^{-1} at 5 mA cm^{-2}. Moreover, the retention rate was 84.1% at 20 mA cm^{-2} after 2000 cycles and showed excellent electrochemical performance. The prepared hybrid capacitor (Co_3O_4@$NiMoO_4$ as the positive electrode and activated carbon as the negative electrode) reached a high energy density of 58.5 W h kg^{-1} at 389 W kg^{-1}.

Yang et al. adopted a similar method using mesoporous Co_3O_4 nanowires directly grown on the nickel foam as the skeleton to support the $NiMoO_4$ nanosheet coating, and obtained Co_3O_4@$NiMoO_4$ [177]. The high specific capacitance of the synthesized Co_3O_4@$NiMoO_4$ was 3.61 F cm^{-2} at a current density of 2 mA cm^{-2}. After 9000 cycles, about 101.3% of the initial capacity was still retained. Such a unique structure can significantly improve the permeability of electrolyte ions in the material.

Li et al. designed and synthesized nanowire/nanosheet arrays directly grown on carbon cloth by a two-step hydrothermal method [178]. Growing uniformly on carbon cloth collectors, the crystalline Co_3O_4 nanowires were used as backbone supports and provided reliable electrical connections for $NiMoO_4$ nanosheet coatings with mesoporous structures. This enabled $NiMoO_4$ to be fully utilized by creating faster electron/ion conductivity and electroactive sites. When the current density was 3 mA cm^{-2}, the specific capacitance of the prepared 3D hybrid nanocomposites was 3.61 F cm^{-2}, and when the current density increased from 3 mA cm^{-2} to 15 mA cm^{-2}, the capacitance retention was 82%. The combined effect of the 3D nanostructure and the pseudo capacitance of the electrode materials resulted in superior electrochemical performance.

Cai et al. fabricated a 3D structure Co_3O_4@$NiMoO_4$ using a similar method as above [179]. A shown in Figure 16, the prepared material showed a significantly enhanced surface capacitance of 5.69 F cm^{-2} when the current density was 30 mA cm^{-2}, which was five times that of the original Co_3O_4 electrode (1.10 F cm^{-2}). With a power density of 5000 W kg^{-1}, the energy density of the hybrid electrode was 56.9 W h kg^{-1}.

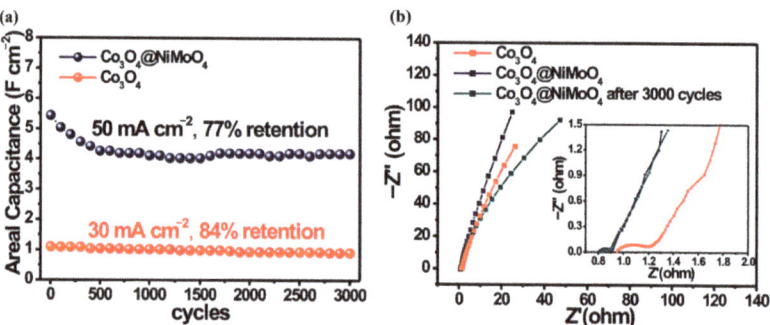

Figure 16. (a) Long-term cycling stability of the Co_3O_4 and Co_3O_4@$NiMoO_4$ hybrid electrodes. (b) Impedance Nyquist plots of the Co_3O_4 electrode and the Co_3O_4@$NiMoO_4$ hybrid electrode before and after 3000 cycles. Reproduced with permission from Daoping Cai, Three-Dimensional Co_3O_4@$NiMoO_4$ Core/Shell Nanowire Arrays on Ni Foam for Electrochemical Energy Storage; published by ACS Publications, 2014 [179].

Dong et al. first prepared a layered tubular yolk-shell composite by electrospinning and hydrothermal methods, and then calcination to prepare a Co_3O_4@$NiMoO_4$ composite [180]. As shown in Figure 17, the Co_3O_4@$NiMoO_4$ composite was made into an electrode with a specific enhanced capacitance of 913.25 F g^{-1} at a high current density of 10 A g^{-1}, and a capacitance retention of 88% due to its unique structure and chemical composition. When the current density changes from 0.5 A g^{-1} to 20 A g^{-1}, it had remarkable cycle stability.

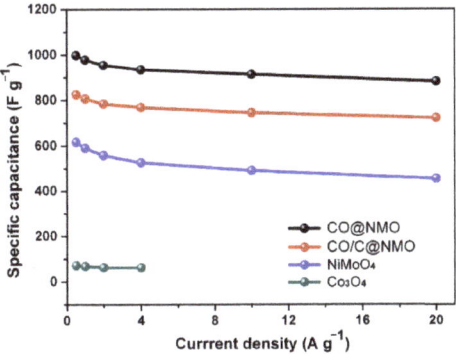

Figure 17. Specific capacitance of CO@NMO, CO/C@NMO, $NiMoO_4$ andCo_3O_4 electrodes. Reproduced with permission from Ping Yang, Synthesis of hierarchical tube-like yolk-shell Co_3O_4@$NiMoO_4$ for enhanced supercapacitor performance; published by Elsevier, 2018 [180].

Hong et al. prepared a uniform 2D Co_3O_4 structure by a simple chemical etching assisted method followed by thermal annealing, and then synthesized Co_3O_4@$NiMoO_4$ by a simple hydrothermal method [181]. The specific capacitance of the 3D hybrid nanostructures was1526 F g^{-1} at the current density of 3 mA cm^{-2}, and the capacitance retention was 72% when the current density increased from 3 mA cm^{-2} to 30 mA cm^{-2}. On this basis, a Co_3O_4@$NiMoO_4$ ASC was designed, and the maximum energy density of activated carbon was 37.8 W h kg^{-1} when the power density was 482 W kg^{-1}.

The above describes another cobalt-containing material, Co_3O_4@$NiMoO_4$. Among them, $NiMoO_4$ can improve the electrochemical performance of Co_3O_4. The electrochemical performance and stability of the two materials can be greatly improved by rational design of their microstructure, which has great potential.

3. Summary and Outlook

In conclusion, this paper reviews the application of cobalt-based nanomaterials in supercapacitors and presents the contributions of many scholars in this field in recent years. These scholars have tried many approaches to improve the electrode materials and enhance the supercapacitor performance. The properties of cobalt-based materials and the issues related to supercapacitors are also discussed.

In this paper, we first introduce the classification and working principle of SCs. According to the charge storage mechanism of SCs, they can be classified into three categories: EDLCs, PCs and battery-type capacitors. EDLCs store charge through a physical adsorption process controlled by reversible adsorption/desorption of electrolyte ions at the electrode/electrolyte interface without any chemical reaction involved. In contrast, PCs and battery-type capacitors benefit from Faraday redox reactions and have a unique charge storage mechanism with much larger capacitance and power density than EDLCs. Among the many electrode materials for these SCs, common cobalt-based materials include cobalt oxide, cobalt hydroxide, and cobalt-containing ternary metal oxides. Among them, the theoretical specific capacitance of Co_3O_4 (3560 F g^{-1}) is slightly higher than that of $Co(OH)_2$ (3460 F g^{-1}), and electrodes made from Co_3O_4 usually exhibit better cycling performance than that of $Co(OH)_2$. Compared to these two substances, the ternary metal oxides ($MnCo_2O_4$, $NiCo_2O_4$ and $ZnCo_2O_4$) show significantly higher performance due to the synergistic effect of the two transition metals coupled together.

To further enhance the performance of the above cobalt-based materials, the main methods are: (1) designing the morphology of the electrode materials; (2) introducing other elements, such as S, P, and Mn, among others; (3) compounding with other materials, and (4) improving the preparation process. First, designing unique morphologies is an effective and commonly used means to enhance the electrochemical performance of electrode materials. Microstructures such as nanoparticles, nanowires, nanotubes, nanosheets, and nanospheres are mainly used in the many studies reported in this paper. Among these morphologies, mesoporous structures play a major role. On the one hand, a mesoporous structure can significantly increase the surface area and shorten the diffusion length for electron and ion transport, thus accelerating the redox process and improving pseudocapacitance performance. On the other hand, it can moderate the volume change during the charging/discharging process, thus improving the cycling capability. Second, the introduction of other elements can further improve the performance of cobalt-based nanomaterials. As mentioned above, transition metal sulfides have significantly higher electrical conductivity and redox ability than corresponding metal oxides. Meanwhile, compounding cobalt-based nanomaterials with other materials can combine the advantages of both materials and improve the performance of electrodes. As mentioned above, many scholars have compounded cobalt-based nanomaterials with carbon-based materials, which are very commonly used today. Among the many carbon-based materials, graphene, which has a large specific surface area and excellent mechanical and electrochemical properties, is an ideal carrier. As a result, many graphene-cobalt-based nanomaterial composites have emerged in recent years. Finally, the electrode performance can also be enhanced by improving the current process. Among the many studies presented in this paper, hydrothermal methods have been widely used, which can easily alter the morphology and structure of nanomaterials. In addition, processes such as electrochemical deposition, electrostatic spinning, and sol-gel methods are also widely used due to their advantages in preparing nanostructures.

Some researchers have investigated the effects of some external factors (e.g., ultraviolet radiation, annealing temperature, deposition potential, etc.) on the performance of SCs. Ultraviolet irradiation increases the crystallinity of raw materials, and the electrochemical performance of supercapacitors made from ultraviolet-irradiated electrode materials was significantly improved [182]. During the annealing process, the grains agglomerate to form large particles, resulting in a uniform and dense porous microstructure [183]. This porous microstructure facilitates electrolytic ion insertion and electron transfer at the electrode/electrolyte interface, resulting in effective charge storage. As for the deposition potential, it has been shown that lower deposition potential leads to lower mass transfer rate and lower electrochemical performance.

In recent years, the field of energy storage devices has been developing rapidly, and sodium-ion batteries, potassium-ion batteries, and various kinds of SCs are being widely and deeply researched, among which miniature SCs (MSCs) are gradually attracting the attention of many researchers. MSCs are miniaturized SCs that have a similar composition to conventional SCs, but with significant structural differences. Conventional SCs have a vertical sandwich structure with inherent limitations including short-circuiting within a narrow distance between two electrodes, increased ion transport resistance, and high mass loading of active materials at an appropriately long distance [184,185]. By contrast, MSCs have a planar structure with a narrow insulating gap between the two electrodes, which avoids the use of a separator. This increases the mass loading of the active material, resulting in high power and energy density, low ion transport resistance, and short electrolyte ion diffusion distance [186]. Due to their small size and excellent electrochemical properties, MSCs could soon be widely used in various applications. Therefore, it is important to study the application of cobalt-based nanomaterials in MSCs.

It should also be noted that studies have shown that a deficit in cobalt supply could occur as early as 2030 [187]. This means that the advantage of the low cost of cobalt-based materials compared to RuO_2 will gradually decrease. The solutions to this problem are as follows: (1) finding alternative materials, such as Ni, Mn, Zn, and other transition group metals with good performance; (2) hybridizing cobalt-based materials with conductive materials with good performance to reduce the content of cobalt in monoliths while ensuring performance (there have been studies on doping polyaniline, polypyrrole, carbon nanotubes, and graphene, among other substances, into cobalt-based materials), and (3) further developing more efficient, convenient, and low-cost SCs recycling technology.

Author Contributions: Writing—original draft preparation and data curation, L.Y. and H.C.; writing—review and editing, H.W., Q.Z., K.Y., X.X. and J.H.; supervision, H.W.; project administration, H.W.; funding acquisition, H.W. All authors have read and agreed to the published version of the manuscript.

Funding: This work was supported by the Natural Science Foundation of China under Grant 21604007.

Institutional Review Board Statement: Not applicable.

Informed Consent Statement: Not applicable.

Data Availability Statement: Not applicable.

Conflicts of Interest: The authors declare no conflict of interest.

References

1. Liu, R.; Zhou, A.; Zhang, X.; Mu, J.; Che, H.; Wang, Y.; Wang, T.-T.; Zhang, Z.; Kou, Z. Fundamentals, advances and challenges of transition metal compounds-based supercapacitors. *Chem. Eng. J.* **2021**, *412*, 128611. [CrossRef]
2. Salunkhe, R.R.; Kaneti, Y.V.; Kim, J.; Kim, J.H.; Yamauchi, Y. Nanoarchitectures for Metal–Organic Framework-Derived Nanoporous Carbons toward Supercapacitor Applications. *Acc. Chem. Res.* **2016**, *49*, 2796–2806. [CrossRef] [PubMed]
3. Raza, W.; Ali, F.; Raza, N.; Luo, Y.; Kim, K.-H.; Yang, J.; Kumar, S.; Mehmood, A.; Kwon, E.E. Recent advancements in supercapacitor technology. *Nano Energy* **2018**, *52*, 441–473. [CrossRef]
4. Hashemi, M.; Rahmanifar, M.S.; El-Kady, M.F.; Noori, A.; Mousavi, M.F.; Kaner, R.B. The use of an electrocatalytic redox electrolyte for pushing the energy density boundary of a flexible polyaniline electrode to a new limit. *Nano Energy* **2018**, *44*, 489–498. [CrossRef]
5. Wang, S.; Yang, Y.; Dong, Y.; Zhang, Z.; Tang, Z. Recent progress in Ti-based nanocomposite anodes for lithium ion batteries. *J. Adv. Ceram.* **2019**, *8*, 1–18. [CrossRef]
6. Miller, J.R.; Simon, P. Electrochemical Capacitors for Energy Management. *Science* **2008**, *321*, 651–652. [CrossRef]
7. Kumar, S.; Saeed, G.; Zhu, L.; Hui, K.N.; Kim, N.H.; Lee, J.H. 0D to 3D carbon-based networks combined with pseudocapacitive electrode material for high energy density supercapacitor: A review. *Chem. Eng. J.* **2021**, *403*, 126352. [CrossRef]
8. Gao, X.R.; Wang, P.K.; Pan, Z.H.; Claverie, J.P.; Wang, J. Recent Progress in Two-Dimensional Layered Double Hydroxides and Their Derivatives for Supercapacitors. *ChemSusChem* **2020**, *13*, 1226–1254. [CrossRef]
9. Zhao, C.; Zheng, W. A Review on Aqueous Electrochemical Supercapacitors. *Front. Energy Res.* **2015**, *3*, 12. [CrossRef]
10. Jiang, J.; Li, Y.Y.; Liu, J.P.; Huang, X.T.; Yuan, C.Z.; Lou, X.W. Recent Advances in Metal Oxide-based Electrode Architecture Design for Electrochemical Energy Storage. *Adv. Mater.* **2012**, *24*, 5166–5180. [CrossRef]
11. Zhang, Y.; Li, L.; Su, H.; Huang, W.; Dong, X. Binary metal oxide: Advanced energy storage materials in supercapacitors. *J. Mater. Chem. A* **2015**, *3*, 43–59. [CrossRef]
12. Tan, H.T.; Sun, W.P.; Wang, L.B.; Yan, Q.Y. 2D Transition Metal Oxides/Hydroxides for Energy-Storage Applications. *ChemNanoMat* **2016**, *2*, 562–577. [CrossRef]
13. Liu, M.S.; Su, B.; Tang, Y.; Jiang, X.C.; Yu, A.B. Recent Advances in Nanostructured Vanadium Oxides and Composites for Energy Conversion. *Adv. Energy Mater.* **2017**, *7*, 1700885. [CrossRef]
14. Xu, J.; Zhang, J.J.; Zhang, W.J.; Lee, C.S. Interlayer Nanoarchitectonics of Two-Dimensional Transition-Metal Dichalcogenides Nanosheets for Energy Storage and Conversion Applications. *Adv. Energy Mater.* **2017**, *7*, 1700571. [CrossRef]
15. Zhang, Y.; Zhou, Q.; Zhu, J.X.; Yan, Q.Y.; Dou, S.X.; Sun, W.P. Nanostructured Metal Chalcogenides for Energy Storage and Electrocatalysis. *Adv. Funct. Mater.* **2017**, *27*, 1702317. [CrossRef]
16. Gao, Y.P.; Huang, K.J. $NiCo_2S_4$ Materials for Supercapacitor Applications. *Chem. Asian J.* **2017**, *12*, 1969–1984. [CrossRef]
17. Yu, X.Y.; Yu, L.; Lou, X.W. Hollow Nanostructures of Molybdenum Sulfides for Electrochemical Energy Storage and Conversion. *Small Methods* **2017**, *1*, 1600020. [CrossRef]
18. Li, X.; Elshahawy, A.M.; Guan, C.; Wang, J. Metal Phosphides and Phosphates-based Electrodes for Electrochemical Supercapacitors. *Small* **2017**, *13*, 1701530. [CrossRef]
19. Hu, C.-C.; Chang, K.-H.; Lin, M.-C.; Wu, Y.-T. Design and Tailoring of the Nanotubular Arrayed Architecture of Hydrous RuO_2 for Next Generation Supercapacitors. *Nano Lett.* **2006**, *6*, 2690–2695. [CrossRef]

20. Vadivel, S.; Naveen, A.N.; Theerthagiri, J.; Madhavan, J.; Santhoshini Priya, T.; Balasubramanian, N. Solvothermal synthesis of BiPO$_4$ nanorods/MWCNT (1D-1D) composite for photocatalyst and supercapacitor applications. *Ceram. Int.* **2016**, *42*, 14196–14205. [CrossRef]

21. Zhao, J.; Gong, J.W.; Wang, G.L.; Zhu, K.; Ye, K.; Yan, J.; Cao, D.X. A self-healing hydrogel electrolyte for flexible solid-state supercapacitors. *Chem. Eng. J.* **2020**, *401*, 125456. [CrossRef]

22. Wang, G.; Zhang, L.; Zhang, J. A review of electrode materials for electrochemical supercapacitors. *Chem. Soc. Rev.* **2012**, *41*, 797–828. [CrossRef] [PubMed]

23. Yang, X.; Cheng, C.; Wang, Y.; Qiu, L.; Li, D. Liquid-mediated dense integration of graphene materials for compact capacitive energy storage. *Science* **2013**, *341*, 534–537. [CrossRef] [PubMed]

24. Ma, F.; Ding, S.; Ren, H.; Liu, Y. Sakura-based activated carbon preparation and its performance in supercapacitor applications. *RSC Adv.* **2019**, *9*, 2474–2483. [CrossRef] [PubMed]

25. Ren, J.; Bai, W.Y.; Guan, G.Z.; Zhang, Y.; Peng, H.S. Flexible and Weaveable Capacitor Wire Based on a Carbon Nanocomposite Fiber. *Adv. Mater.* **2013**, *25*, 5965–5970. [CrossRef]

26. Guan, Y.; Mu, J.; Che, H.; Zhang, X.; Zhang, Z. Preparation of hierarchical porous carbon with high capacitance. *World J. Eng.* **2018**, *15*, 323–329. [CrossRef]

27. Kang, J.L.; Zhang, S.F.; Zhang, Z.J. Three-Dimensional Binder-Free Nanoarchitectures for Advanced Pseudocapacitors. *Adv. Mater.* **2017**, *29*, 1700515. [CrossRef]

28. Zhu, Z.-Z.; Wang, G.-C.; Sun, M.-Q.; Li, X.-W.; Li, C.-Z. Fabrication and electrochemical characterization of polyaniline nanorods modified with sulfonated carbon nanotubes for supercapacitor applications. *Electrochim. Acta* **2011**, *56*, 1366–1372. [CrossRef]

29. Simon, P.; Gogotsi, Y.; Dunn, B. Where Do Batteries End and Supercapacitors Begin? *Science* **2014**, *343*, 1210–1211. [CrossRef]

30. Brousse, T.; Bélanger, D.; Long, J.W. To Be or Not To Be Pseudocapacitive? *J. Electrochem. Soc.* **2015**, *162*, A5185–A5189. [CrossRef]

31. Wang, Y.; Song, Y.; Xia, Y. Electrochemical capacitors: Mechanism, materials, systems, characterization and applications. *Chem. Soc. Rev.* **2016**, *45*, 5925–5950. [CrossRef] [PubMed]

32. Song, X.S.; Li, X.F.; Bai, Z.M.; Yan, B.; Li, D.J. Controlling the Growth of Ni$_3$S$_2$ Anode with Tunable Sodium Storage. *Adv. Mater. Interfaces* **2018**, *5*, 8. [CrossRef]

33. Ouyang, Y.; Huang, R.; Xia, X.; Ye, H.; Jiao, X.; Wang, L.; Lei, W.; Hao, Q. Hierarchical structure electrodes of NiO ultrathin nanosheets anchored to NiCo$_2$O$_4$ on carbon cloth with excellent cycle stability for asymmetric supercapacitors. *Chem. Eng. J.* **2019**, *355*, 416–427. [CrossRef]

34. Wang, Y.; Zhang, W.; Guo, X.; Liu, Y.; Zheng, Y.; Zhang, M.; Li, R.; Peng, Z.; Zhang, Y.; Zhang, T. One-step microwave-hydrothermal preparation of NiS/rGO hybrid for high-performance symmetric solid-state supercapacitor. *Appl. Surf. Sci.* **2020**, *514*, 146080. [CrossRef]

35. Zhang, X.; Wang, J.; Ji, X.; Sui, Y.; Wei, F.; Qi, J.; Meng, Q.; Ren, Y.; He, Y. Nickel/cobalt bimetallic metal-organic frameworks ultrathin nanosheets with enhanced performance for supercapacitors. *J. Alloys Compd.* **2020**, *825*, 154069. [CrossRef]

36. Li, Y.; Han, X.; Yi, T.; He, Y.; Li, X. Review and prospect of NiCo$_2$O$_4$-based composite materials for supercapacitor electrodes. *J. Energy Chem.* **2019**, *31*, 54–78. [CrossRef]

37. Zhang, Y.F.; Ma, M.Z.; Yang, J.; Su, H.Q.; Huang, W.; Dong, X.C. Selective synthesis of hierarchical mesoporous spinel NiCo2O4 for high-performance supercapacitors. *Nanoscale* **2014**, *6*, 4303–4308. [CrossRef]

38. Fan, Y.; Liu, S.; Han, X.; Xiang, R.; Gong, Y.; Wang, T.; Jing, Y.; Maruyama, S.; Zhang, Q.; Zhao, Y. Ni–Co-Based Nanowire Arrays with Hierarchical Core–Shell Structure Electrodes for High-Performance Supercapacitors. *ACS Appl. Energy Mater.* **2020**, *3*, 7580–7587. [CrossRef]

39. Jadhav, S.; Kalubarme, R.S.; Suzuki, N.; Terashima, C.; Kale, B.; Gosavi, S.W.; Ashokkumar, M.; Fujishima, A. Probing electrochemical charge storage of 3D porous hierarchical cobalt oxide decorated rGO in ultra-high-performance supercapacitor. *Surf. Coat. Technol.* **2021**, *419*, 127287. [CrossRef]

40. Shaheen, A.; Hussain, S.; Qiao, G.J.; Mahmoud, M.H.; Fouad, H.; Akhtar, M.S. Nanosheets Assembled Co$_3$O$_4$ Nanoflowers for Supercapacitor Applications. *J. Nanoelectron. Optoelectron.* **2021**, *16*, 1357–1362. [CrossRef]

41. Xu, Z.; Younis, A.; Chu, D.; Ao, Z.; Xu, H.; Li, S. Electrodeposition of Mesoporous Co$_3$O$_4$ Nanosheets on Carbon Foam for High Performance Supercapacitors. *J. Nanomater.* **2014**, *2014*, 902730. [CrossRef]

42. Gong, H.; Bie, S.G.; Zhang, J.; Ke, X.B.; Wang, X.X.; Liang, J.Q.; Wu, N.; Zhang, Q.C.; Luo, C.X.; Jia, Y.M. In Situ Construction of ZIF-67-Derived Hybrid Tricobalt Tetraoxide@Carbon for Supercapacitor. *Nanomaterials* **2022**, *12*, 1571. [CrossRef]

43. Khalid, N.R.; Batool, A.; Ali, F.; Nabi, G.; Tahir, M.B.; Rafique, M. Electrochemical study of Mo-doped Co$_3$O$_4$ nanostructures synthesized by sol-gel method. *J. Mater. Sci.-Mater. Electron.* **2021**, *32*, 3512–3521. [CrossRef]

44. Hoa, N.D.; Van Tong, P.; Hung, C.M.; Van Duy, N.; Van Hieu, N. Urea mediated synthesis of Ni(OH)$_2$ nanowires and their conversion into NiO nanostructure for hydrogen gas-sensing application. *Int. J. Hydrogen Energy* **2018**, *43*, 9446–9453. [CrossRef]

45. Zhang, W.; Li, H.; Firby, C.J.; Al-Hussein, M.; Elezzabi, A.Y. Oxygen-Vacancy-Tunable Electrochemical Properties of Electrodeposited Molybdenum Oxide Films. *ACS Appl. Mater. Interfaces* **2019**, *11*, 20378–20385. [CrossRef] [PubMed]

46. Serrapede, M.; Rafique, A.; Fontana, M.; Zine, A.; Rivolo, P.; Bianco, S.; Chetibi, L.; Tresso, E.; Lamberti, A. Fiber-shaped asymmetric supercapacitor exploiting rGO/Fe2O3 aerogel and electrodeposited MnOx nanosheets on carbon fibers. *Carbon* **2019**, *144*, 91–100. [CrossRef]

47. Kulandaivalu, S.; Mohd Azahari, M.N.; Azman, N.H.N.; Sulaiman, Y. Ultrahigh specific energy of layer by layer polypyrrole/graphene oxide/multi-walled carbon nanotube | polypyrrole/manganese oxide composite for supercapacitor. *J. Energy Storage* **2020**, *28*, 101219. [CrossRef]

48. Hu, X.R.; Wei, L.S.; Chen, R.; Wu, Q.S.; Li, J.F. Reviews and Prospectives of Co₃O₄-Based Nanomaterials for Supercapacitor Application. *ChemistrySelect* **2020**, *5*, 5268–5288. [CrossRef]

49. Yuan, C.; Yang, L.; Hou, L.; Shen, L.; Zhang, X.; Lou, X.W. Growth of ultrathin mesoporous Co₃O₄ nanosheet arrays on Ni foam for high-performance electrochemical capacitors. *Energy Environ. Sci.* **2012**, *5*, 7883–7887. [CrossRef]

50. Pawar, S.A.; Patil, D.S.; Shin, J.C. Transition of hexagonal to square sheets of Co₃O₄ in a triple heterostructure of Co₃O₄/MnO₂/GO for high performance supercapacitor electrode. *Curr. Appl. Phys.* **2019**, *19*, 794–803. [CrossRef]

51. Kumar, M.; Subramania, A.; Balakrishnan, K. Preparation of electrospun Co₃O₄ nanofibers as electrode material for high performance asymmetric supercapacitors. *Electrochim. Acta* **2014**, *149*, 152–158. [CrossRef]

52. Duan, B.R.; Cao, Q. Hierarchically porous Co₃O₄ film prepared by hydrothermal synthesis method based on colloidal crystal template for supercapacitor application. *Electrochim. Acta* **2012**, *64*, 154–161. [CrossRef]

53. Deng, J.; Kang, L.; Bai, G.; Li, Y.; Li, P.; Liu, X.; Yang, Y.; Gao, F.; Liang, W. Solution combustion synthesis of cobalt oxides (Co₃O₄ and Co₃O₄/CoO) nanoparticles as supercapacitor electrode materials. *Electrochim. Acta* **2014**, *132*, 127–135. [CrossRef]

54. Chen, M.; Ge, Q.; Qi, M.; Liang, X.; Wang, F.; Chen, Q. Cobalt oxides nanorods arrays as advanced electrode for high performance supercapacitor. *Surf. Coat. Technol.* **2019**, *360*, 73–77. [CrossRef]

55. Liu, Z.Z.; Zhou, W.W.; Wang, S.S.; Du, W.; Zhang, H.L.; Ding, C.Y.; Du, Y.; Zhu, L.J. Facile synthesis of homogeneous core-shell Co₃O₄ mesoporous nanospheres as high performance electrode materials for supercapacitor. *J. Alloys Compd.* **2019**, *774*, 137–144. [CrossRef]

56. Xu, Y.; Ding, Q.; Li, L.; Xie, Z.; Jiang, G. Facile fabrication of porous Co₃O₄ nanowires for high performance supercapacitors. *New J. Chem.* **2018**, *42*, 20069–20073. [CrossRef]

57. Wang, X.; Zhang, N.; Chen, X.; Liu, J.; Lu, F.; Chen, L.; Shao, G. Facile precursor conversion synthesis of hollow coral-shaped Co₃O₄ nanostructures for high-performance supercapacitors. *Colloids Surf. A Physicochem. Eng. Asp.* **2019**, *570*, 63–72. [CrossRef]

58. Lu, Y.; Liu, Y.; Mo, J.; Deng, B.; Wang, J.; Zhu, Y.; Xiao, X.; Xu, G. Construction of hierarchical structure of Co₃O₄ electrode based on electrospinning technique for supercapacitor. *J. Alloys Compd.* **2021**, *853*, 157271. [CrossRef]

59. Ding, K.; Yang, P.; Hou, P.; Song, X.; Wei, T.; Cao, Y.; Cheng, X. Ultrathin and Highly Crystalline Co₃O₄ Nanosheets In Situ Grown on Graphene toward Enhanced Supercapacitor Performance. *Adv. Mater. Interfaces* **2017**, *4*, 1600884. [CrossRef]

60. Tan, H.Y.; Yu, B.Z.; Cao, L.L.; Cheng, T.; Zheng, X.L.; Li, X.H.; Li, W.L.; Ren, Z.Y. Layer-dependent growth of two-dimensional Co₃O₄ nanostructure arrays on graphene for high performance supercapacitors. *J. Alloys Compd.* **2017**, *696*, 1180–1188. [CrossRef]

61. Lu, X.; Zeng, Y.; Yu, M.; Zhai, T.; Liang, C.; Xie, S.; Balogun, M.S.; Tong, Y. Oxygen-deficient hematite nanorods as high-performance and novel negative electrodes for flexible asymmetric supercapacitors. *Adv Mater* **2014**, *26*, 3148–3155. [PubMed]

62. Zhai, T.; Xie, S.; Yu, M.; Fang, P.; Liang, C.; Lu, X.; Tong, Y. Oxygen vacancies enhancing capacitive properties of MnO₂ nanorods for wearable asymmetric supercapacitors. *Nano Energy* **2014**, *8*, 255–263. [CrossRef]

63. Wang, G.; Yang, Y.; Han, D.; Li, Y. Oxygen defective metal oxides for energy conversion and storage. *Nano Today* **2017**, *13*, 23–39. [CrossRef]

64. Kim, H.S.; Cook, J.B.; Lin, H.; Ko, J.S.; Tolbert, S.H.; Ozolins, V.; Dunn, B. Oxygen vacancies enhance pseudocapacitive charge storage properties of MoO₃₋ₓ. *Nat. Mater.* **2017**, *16*, 454–460. [CrossRef]

65. Xiang, K.; Xu, Z.; Qu, T.; Tian, Z.; Zhang, Y.; Wang, Y.; Xie, M.; Guo, X.; Ding, W.; Guo, X. Two dimensional oxygen-vacancy-rich Co₃O₄ nanosheets with excellent supercapacitor performances. *Chem. Commun.* **2017**, *53*, 12410–12413. [CrossRef]

66. Yang, S.; Liu, Y.; Hao, Y.; Yang, X.; Goddard, W.A., 3rd; Zhang, X.L.; Cao, B. Oxygen-Vacancy Abundant Ultrafine Co₃O₄/Graphene Composites for High-Rate Supercapacitor Electrodes. *Adv. Sci.* **2018**, *5*, 1700659. [CrossRef] [PubMed]

67. Haoxiang, W.; Zhang, W.; Chen, H.; Zheng, W. Towards unlocking high-performance of supercapacitors: From layered transition-metal hydroxide electrode to redox electrolyte. *Sci. China Technol. Sci.* **2015**, *58*, 1779–1798. [CrossRef]

68. Arico, A.S.; Bruce, P.; Scrosati, B.; Tarascon, J.-M.; van Schalkwijk, W. Nanostructured materials for advanced energy conversion and storage devices. *Nat. Mater.* **2005**, *4*, 366–377. [CrossRef]

69. Pan, G.X.; Xia, X.; Cao, F.; Tang, P.S.; Chen, H.F. Porous Co(OH)₂/Ni composite nanoflake array for high performance supercapacitors. *Electrochim. Acta* **2012**, *63*, 335–340. [CrossRef]

70. Li, D.; Zhu, S.; Gao, X.; Jiang, X.; Liu, Y.; Meng, F. Anchoring sea-urchin-like Co(OH)₂ microspheres on nickel foam as three-dimensional free-standing electrode for high-performance supercapacitors. *Ionics* **2021**, *27*, 789–799. [CrossRef]

71. Zhao, F.; Zheng, D.; Liu, Y.; Pan, F.; Deng, Q.; Qin, C.; Li, Y.; Wang, Z. Flexible Co(OH)₂/NiOₓHᵧ@Ni hybrid electrodes for high energy density supercapacitors. *Chem. Eng. J.* **2021**, *415*, 128871. [CrossRef]

72. Xuan, L.; Chen, L.; Yang, Q.; Chen, W.; Hou, X.; Jiang, Y.; Zhang, Q.; Yuan, Y. Engineering 2D multi-layer graphene-like Co₃O₄ thin sheets with vertically aligned nanosheets as basic building units for advanced pseudocapacitor materials. *J. Mater. Chem. A* **2015**, *3*, 17525–17533. [CrossRef]

73. Yan, D.; Wang, W.; Luo, X.; Chen, C.; Zeng, Y.; Zhu, Z. NiCo₂O₄ with oxygen vacancies as better performance electrode material for supercapacitor. *Chem. Eng. J.* **2018**, *334*, 864–872. [CrossRef]

74. Pendashteh, A.; Palma, J.; Anderson, M.; Marcilla, R. Nanostructured porous wires of iron cobaltite: Novel positive electrode for high-performance hybrid energy storage devices. *J. Mater. Chem. A* **2015**, *3*, 16849–16859. [CrossRef]

75. Mohamed, S.G.; Attia, S.Y.; Hassan, H.H. Spinel-structured FeCo$_2$O$_4$ mesoporous nanosheets as efficient electrode for supercapacitor applications. *Microporous Mesoporous Mater.* **2017**, *251*, 26–33. [CrossRef]

76. Xiao, X.C.; Wang, G.F.; Zhang, M.M.; Wang, Z.Z.; Zhao, R.J.; Wang, Y.D. Electrochemical performance of mesoporous ZnCo$_2$O$_4$ nanosheets as an electrode material for supercapacitor. *Ionics* **2018**, *24*, 2435–2443. [CrossRef]

77. Kaverlavani, S.K.; Moosavifard, S.E.; Bakouei, A. Designing graphene-wrapped nanoporous CuCo$_2$O$_4$ hollow spheres electrodes for high-performance asymmetric supercapacitors. *J. Mater. Chem. A* **2017**, *5*, 14301–14309. [CrossRef]

78. Wang, T.; Chen, H.C.; Yu, F.; Zhao, X.S.; Wang, H. Boosting the cycling stability of transition metal compounds-based supercapacitors. *Energy Storage Mater.* **2019**, *16*, 545–573. [CrossRef]

79. Zhang, K.; Zeng, W.; Zhang, G.; Hou, S.; Wang, F.; Wang, T.; Duan, H. Hierarchical CuCo$_2$O$_4$ nanowire@NiCo$_2$O$_4$ nanosheet core/shell arrays for high-performance supercapacitors. *RSC Adv.* **2015**, *5*, 69636–69641. [CrossRef]

80. Liang, Y.; Wang, H.; Zhou, J.; Li, Y.; Wang, J.; Regier, T.; Dai, H. Covalent hybrid of spinel manganese-cobalt oxide and graphene as advanced oxygen reduction electrocatalysts. *J. Am. Chem. Soc.* **2012**, *134*, 3517–3523. [CrossRef]

81. Wang, H.; Yang, Y.; Liang, Y.; Zheng, G.; Li, Y.; Cui, Y.; Dai, H. Rechargeable Li–O2 batteries with a covalently coupled MnCo$_2$O$_4$–graphene hybrid as an oxygen cathode catalyst. *Energy Environ. Sci.* **2012**, *5*, 7931–7935. [CrossRef]

82. Kim, K.J.; Heo, J.W. Electronic structure and optical properties of inverse-spinel MnCo$_2$O$_4$ thin films. *J. Korean Phys. Soc.* **2012**, *60*, 1376–1380. [CrossRef]

83. Krittayavathananon, A.; Pettong, T.; Kidkhunthod, P.; Sawangphruk, M. Insight into the charge storage mechanism and capacity retention fading of MnCo$_2$O$_4$ used as supercapacitor electrodes. *Electrochim. Acta* **2017**, *258*, 1008–1015. [CrossRef]

84. Li, J.; Xiong, S.; Li, X.; Qian, Y. Spinel Mn$_{1.5}$Co$_{1.5}$O$_4$ core–shell microspheres as Li-ion battery anode materials with a long cycle life and high capacity. *J. Mater. Chem.* **2012**, *22*, 23254–23259. [CrossRef]

85. Li, L.; Zhang, Y.Q.; Liu, X.Y.; Shi, S.J.; Zhao, X.Y.; Zhang, H.; Ge, X.; Cai, G.F.; Gu, C.D.; Wang, X.L.; et al. One-dimension MnCo$_2$O$_4$ nanowire arrays for electrochemical energy storage. *Electrochim. Acta* **2014**, *116*, 467–474. [CrossRef]

86. Yi, T.-F.; Sari, H.M.K.; Li, X.; Wang, F.; Zhu, Y.-R.; Hu, J.; Zhang, J.; Li, X. A review of niobium oxides based nanocomposites for lithium-ion batteries, sodium-ion batteries and supercapacitors. *Nano Energy* **2021**, *85*, 105955. [CrossRef]

87. Che, H.; Liu, A.; Mu, J.; Wu, C.; Zhang, X. Template-free synthesis of novel flower-like MnCo$_2$O$_4$ hollow microspheres for application in supercapacitors. *Ceram. Int.* **2016**, *42*, 2416–2424. [CrossRef]

88. Zhao, Y.; Hu, L.; Zhao, S.; Wu, L. Preparation of MnCo$_2$O$_4$@Ni(OH)$_2$ Core-Shell Flowers for Asymmetric Supercapacitor Materials with Ultrahigh Specific Capacitance. *Adv. Funct. Mater.* **2016**, *26*, 4085–4093. [CrossRef]

89. Dong, Y.; Wang, Y.; Xu, Y.; Chen, C.; Wang, Y.; Jiao, L.; Yuan, H. Facile synthesis of hierarchical nanocage MnCo$_2$O$_4$ for high performance supercapacitor. *Electrochim. Acta* **2017**, *225*, 39–46. [CrossRef]

90. Anjana, P.M.; Sarath Kumar, S.R.; Rakhi, R.B. MnCo$_2$O$_4$ nanoneedles self-organized microstructures for supercapacitors. *Mater. Today Commun.* **2021**, *28*, 102720. [CrossRef]

91. Liao, F.; Han, X.; Zhang, Y.; Han, X.; Xu, C.; Chen, H. Hydrothermal synthesis of mesoporous MnCo$_2$O$_4$/CoCo$_2$O$_4$ ellipsoid-like microstructures for high-performance electrochemical supercapacitors. *Ceram. Int.* **2019**, *45*, 7244–7252. [CrossRef]

92. Tan, S.F.; Ji, Y.J.; Chen, F.; Ouyang, W.M. Three-dimensional sea urchin-like MnCo$_2$O$_4$ nanoarchitectures on Ni foam towards high-performance asymmetric supercapacitors. *Front. Mater. Sci.* **2021**, *15*, 611–620. [CrossRef]

93. Wang, H.; Shen, C.; Liu, J.; Zhang, W.; Yao, S. Three-dimensional MnCo$_2$O$_4$/graphene composites for supercapacitor with promising electrochemical properties. *J. Alloys Compd.* **2019**, *792*, 122–129. [CrossRef]

94. Wang, X.; Fang, Y.; Shi, B.; Huang, F.; Rong, F.; Que, R. Three-dimensional NiCo$_2$O$_4$@NiCo$_2$O$_4$ core–shell nanocones arrays for high-performance supercapacitors. *Chem. Eng. J.* **2018**, *344*, 311–319. [CrossRef]

95. Wu, S.; Hui, K.S.; Hui, K.N.; Kim, K.H. Ultrathin porous NiO nanoflake arrays on nickel foam as an advanced electrode for high performance asymmetric supercapacitors. *J. Mater. Chem. A* **2016**, *4*, 9113–9123. [CrossRef]

96. Ma, Z.; Shao, G.; Fan, Y.; Feng, M.; Shen, D.; Wang, H. Fabrication of High-Performance All-Solid-State Asymmetric Supercapacitors Based on Stable α-MnO$_2$@NiCo$_2$O$_4$ Core–Shell Heterostructure and 3D-Nanocage N-Doped Porous Carbon. *ACS Sustain. Chem. Eng.* **2017**, *5*, 4856–4868. [CrossRef]

97. Li, Y.; Tang, F.; Wang, R.; Wang, C.; Liu, J. Novel Dual-Ion Hybrid Supercapacitor Based on a NiCo$_2$O$_4$ Nanowire Cathode and MoO$_2$-C Nanofilm Anode. *ACS Appl. Mater Interfaces* **2016**, *8*, 30232–30238. [CrossRef]

98. Li, L.; Peng, S.; Cheah, Y.; Teh, P.; Wang, J.; Wee, G.; Ko, Y.; Wong, C.; Srinivasan, M. Electrospun porous NiCo$_2$O$_4$ nanotubes as advanced electrodes for electrochemical capacitors. *Chemistry* **2013**, *19*, 5892–5898. [CrossRef]

99. Wang, C.; Zhou, E.; He, W.; Deng, X.; Huang, J.; Ding, M.; Wei, X.; Liu, X.; Xu, X.J. NiCo$_2$O$_4$-Based Supercapacitor Nanomaterials. *Nanomaterials* **2017**, *7*, 41. [CrossRef]

100. Shen, L.; Che, Q.; Li, H.; Zhang, X. Mesoporous NiCo$_2$O$_4$ Nanowire Arrays Grown on Carbon Textiles as Binder-Free Flexible Electrodes for Energy Storage. *Adv. Funct. Mater.* **2014**, *24*, 2630–2637. [CrossRef]

101. Venkatesh, K.; Karuppiah, C.; Palani, R.; Periyasamy, G.; Ramaraj, S.K.; Yang, C.C. 2D/2D nanostructures based on NiCo$_2$O$_4$/graphene composite for high-performance battery-type supercapacitor. *Mater. Lett.* **2022**, *323*, 132609. [CrossRef]

102. Xu, K.; Yang, J.; Li, S.; Liu, Q.; Hu, J. Facile synthesis of hierarchical mesoporous NiCo$_2$O$_4$ nanoflowers with large specific surface area for high-performance supercapacitors. *Mater. Lett.* **2017**, *187*, 129–132. [CrossRef]

103. Sethi, M.; Bhat, D.K. Facile solvothermal synthesis and high supercapacitor performance of NiCo$_2$O$_4$ nanorods. *J. Alloys Compd.* **2019**, *781*, 1013–1020. [CrossRef]

104. Xu, K.; Yang, J.; Hu, J. Synthesis of hollow NiCo$_2$O$_4$ nanospheres with large specific surface area for asymmetric supercapacitors. *J. Colloid Interface Sci.* **2018**, *511*, 456–462. [CrossRef] [PubMed]

105. Shen, L.; Yu, L.; Wu, H.B.; Yu, X.Y.; Zhang, X.; Lou, X.W. Formation of nickel cobalt sulfide ball-in-ball hollow spheres with enhanced electrochemical pseudocapacitive properties. *Nat. Commun.* **2015**, *6*, 6694. [CrossRef] [PubMed]

106. Shen, L.F.; Yu, L.; Yu, X.Y.; Zhang, X.G.; Lou, X.W. Self-Templated Formation of Uniform NiCo$_2$O$_4$ Hollow Spheres with Complex Interior Structures for Lithium-Ion Batteries and Supercapacitors. *Angew. Chem.-Int. Ed.* **2015**, *54*, 1868–1872. [CrossRef]

107. Yuan, C.; Li, J.; Hou, L.; Lin, J.; Pang, G.; Zhang, L.; Lian, L.; Zhang, X. Template-engaged synthesis of uniform mesoporous hollow NiCo$_2$O$_4$ sub-microspheres towards high-performance electrochemical capacitors. *RSC Adv.* **2013**, *3*, 18573–18578. [CrossRef]

108. Ji, C.; Liu, F.; Xu, L.; Yang, S. Urchin-Like NiCo$_2$O$_4$ hollow microspheres and FeSe$_2$ micro-snowflakes for flexible solid-state asymmetric supercapacitors. *J. Mater. Chem. A* **2017**, *5*, 5568–5576. [CrossRef]

109. Zhu, Y.; Wang, J.; Wu, Z.; Jing, M.; Hou, H.; Jia, X.; Ji, X. An electrochemical exploration of hollow NiCo$_2$O$_4$ submicrospheres and its capacitive performances. *J. Power Source* **2015**, *287*, 307–315. [CrossRef]

110. Liu, X.; Shi, S.; Xiong, Q.; Li, L.; Zhang, Y.; Tang, H.; Gu, C.; Wang, X.; Tu, J. Hierarchical NiCo$_2$O$_4$@NiCo$_2$O$_4$ core/shell nanoflake arrays as high-performance supercapacitor materials. *ACS Appl. Mater. Interfaces* **2013**, *5*, 8790–8795. [CrossRef]

111. Liu, B.; Kong, D.; Huang, Z.X.; Mo, R.; Wang, Y.; Han, Z.; Cheng, C.; Yang, H.Y. Three-dimensional hierarchical NiCo$_2$O$_4$ nanowire@Ni$_3$S$_2$ nanosheet core/shell arrays for flexible asymmetric supercapacitors. *Nanoscale* **2016**, *8*, 10686–10694. [CrossRef]

112. Kong, D.; Ren, W.; Cheng, C.; Wang, Y.; Huang, Z.; Yang, H.Y. Three-Dimensional NiCo$_2$O$_4$@Polypyrrole Coaxial Nanowire Arrays on Carbon Textiles for High-Performance Flexible Asymmetric Solid-State Supercapacitor. *ACS Appl. Mater. Interfaces* **2015**, *7*, 21334–21346. [CrossRef]

113. Ma, L.; Shen, X.; Zhou, H.; Ji, Z.; Chen, K.; Zhu, G. High performance supercapacitor electrode materials based on porous NiCo$_2$O$_4$ hexagonal nanoplates/reduced graphene oxide composites. *Chem. Eng. J.* **2015**, *262*, 980–988. [CrossRef]

114. Cheng, J.; Lu, Y.; Qiu, K.; Yan, H.; Xu, J.; Han, L.; Liu, X.; Luo, J.; Kim, J.K.; Luo, Y. Hierarchical Core/Shell NiCo$_2$O$_4$@NiCo$_2$O$_4$ Nanocactus Arrays with Dual-functionalities for High Performance Supercapacitors and Li-ion Batteries. *Sci. Rep.* **2015**, *5*, 12099. [CrossRef] [PubMed]

115. Niu, H.; Yang, X.; Jiang, H.; Zhou, D.; Li, X.; Zhang, T.; Liu, J.Y.; Wang, Q.; Qu, F.Y. Hierarchical core-shell heterostructure of porous carbon nanofiber@ZnCo$_2$O$_4$ nanoneedle arrays: Advanced binder-free electrodes for all-solid-state supercapacitors. *J. Mater. Chem. A* **2015**, *3*, 24082–24094. [CrossRef]

116. Li, L.; Li, R.M.; Gai, S.L.; Gao, P.; He, F.; Zhang, M.L.; Chen, Y.J.; Yang, P.P. Hierarchical porous CNTs@NCS@MnO$_2$ composites: Rational design and high asymmetric supercapacitor performance. *J. Mater. Chem. A* **2015**, *3*, 15642–15649. [CrossRef]

117. Xiong, W.; Hu, X.; Wu, X.; Zeng, Y.; Wang, B.; He, G.H.; Zhu, Z.H. A flexible fiber-shaped supercapacitor utilizing hierarchical NiCo$_2$O$_4$@polypyrrole core-shell nanowires on hemp-derived carbon. *J. Mater. Chem. A* **2015**, *3*, 17209–17216. [CrossRef]

118. Yang, Q.; Lu, Z.Y.; Li, T.; Sun, X.M.; Liu, J.F. Hierarchical construction of core-shell metal oxide nanoarrays with ultrahigh areal capacitance. *Nano Energy* **2014**, *7*, 170–178. [CrossRef]

119. Lim, E.; Jo, C.; Kim, H.; Kim, M.-H.; Mun, Y.; Chun, J.; Ye, Y.; Hwang, J.; Ha, K.-S.; Roh, K.C.; et al. Facile Synthesis of Nb2O5@Carbon Core–Shell Nanocrystals with Controlled Crystalline Structure for High-Power Anodes in Hybrid Supercapacitors. *ACS Nano* **2015**, *9*, 7497–7505. [CrossRef]

120. Zhou, Y.; Huang, Z.; Liao, H.; Li, J.; Wang, H.; Wang, Y. 3D porous graphene/NiCo$_2$O$_4$ hybrid film as an advanced electrode for supercapacitors. *Appl. Surf. Sci.* **2020**, *534*, 147598. [CrossRef]

121. Li, L.; He, F.; Gai, S.; Zhang, S.; Gao, P.; Zhang, M.; Chen, Y.; Yang, P. Hollow structured and flower-like C@MnCo$_2$O$_4$ composite for high electrochemical performance in a supercapacitor. *CrystEngComm* **2014**, *16*, 9873–9881. [CrossRef]

122. Zhao, J.; Li, Z.; Zhang, M.; Meng, A.; Li, Q. Direct Growth of Ultrathin NiCo$_2$O$_4$/NiO Nanosheets on SiC Nanowires as a Free-Standing Advanced Electrode for High-Performance Asymmetric Supercapacitors. *ACS Sustain. Chem. Eng.* **2016**, *4*, 3598–3608. [CrossRef]

123. Cheng, D.; Yang, Y.; Xie, J.; Fang, C.; Zhang, G.; Xiong, J. Hierarchical NiCo$_2$O$_4$@NiMoO$_4$ core–shell hybrid nanowire/nanosheet arrays for high-performance pseudocapacitors. *J. Mater. Chem. A* **2015**, *3*, 14348–14357. [CrossRef]

124. Lee, H.-M.; Gopi, C.V.V.M.; Rana, P.J.S.; Vinodh, R.; Kim, S.; Padma, R.; Kim, H.-J. Hierarchical nanostructured MnCo$_2$O$_4$–NiCo$_2$O$_4$ composites as innovative electrodes for supercapacitor applications. *New J. Chem.* **2018**, *42*, 17190–17194. [CrossRef]

125. Li, X.; Zhang, M.; Wu, L.; Fu, Q.; Gao, H. Annealing temperature dependent ZnCo$_2$O$_4$ nanosheet arrays supported on Ni foam for high-performance asymmetric supercapacitor. *J. Alloys Compd.* **2019**, *773*, 367–375. [CrossRef]

126. Guan, B.; Guo, D.; Hu, L.; Zhang, G.; Fu, T.; Ren, W.; Li, J.; Li, Q. Facile synthesis of ZnCo$_2$O$_4$ nanowire cluster arrays on Ni foam for high-performance asymmetric supercapacitors. *J. Mater. Chem. A* **2014**, *2*, 16116–16123. [CrossRef]

127. Moon, I.K.; Chun, K.Y. Ultra-high pseudocapacitance of mesoporous ZnCo$_2$O$_4$ nanosheets on reduced graphene oxide utilizing a neutral aqueous electrolyte. *RSC Adv.* **2015**, *5*, 807–811. [CrossRef]

128. Vijayakumar, S.; Nagamuthu, S.; Lee, S.H.; Ryu, K.S. Porous thin layered nanosheets assembled ZnCo$_2$O$_4$ grown on Ni-foam as an efficient electrode material for hybrid supercapacitor applications. *Int. J. Hydrogen Energy* **2017**, *42*, 3122–3129. [CrossRef]

129. Mallem, S.P.R.; Koduru, M.; Chandrasekhar, K.; Vattikuti, S.V.P.; Manne, R.; Reddy, V.R.; Lee, J.H. Potato Chip-Like 0D Interconnected ZnCo$_2$O$_4$ Nanoparticles for High-Performance Supercapacitors. *Crystals* **2021**, *11*, 469. [CrossRef]

130. Chen, H.; Wang, J.; Han, X.; Liao, F.; Zhang, Y.; Gao, L.; Xu, C. Facile synthesis of mesoporous $ZnCo_2O_4$ hierarchical microspheres and their excellent supercapacitor performance. *Ceram. Int.* **2019**, *45*, 8577–8584. [CrossRef]
131. Chen, H.; Wang, J.; Han, X.; Liao, F.; Zhang, Y.; Han, X.; Xu, C. Simple growth of mesoporous zinc cobaltite urchin-like microstructures towards high-performance electrochemical capacitors. *Ceram. Int.* **2019**, *45*, 4059–4066. [CrossRef]
132. Wang, S.; Pu, J.; Tong, Y.; Cheng, Y.; Gao, Y.; Wang, Z. $ZnCo_2O_4$ nanowire arrays grown on nickel foam for high-performance pseudocapacitors. *J. Mater. Chem. A* **2014**, *2*, 5434–5440. [CrossRef]
133. Xu, L.; Zhao, Y.; Lian, J.; Xu, Y.; Bao, J.; Qiu, J.; Xu, L.; Xu, H.; Hua, M.; Li, H. Morphology controlled preparation of $ZnCo_2O_4$ nanostructures for asymmetric supercapacitor with ultrahigh energy density. *Energy* **2017**, *123*, 296–304. [CrossRef]
134. Venkatachalam, V.; Alsalme, A.; Alswieleh, A.; Jayavel, R. Double hydroxide mediated synthesis of nanostructured $ZnCo_2O_4$ as high performance electrode material for supercapacitor applications. *Chem. Eng. J.* **2017**, *321*, 474–483. [CrossRef]
135. Shang, Y.; Xie, T.; Gai, Y.; Su, L.; Gong, L.; Lv, H.; Dong, F. Self-assembled hierarchical peony-like $ZnCo_2O_4$ for high-performance asymmetric supercapacitors. *Electrochim. Acta* **2017**, *253*, 281–290. [CrossRef]
136. Xiang, K.; Wu, D.; Fan, Y.; You, W.; Zhang, D.; Luo, J.-L.; Fu, X.-Z. Enhancing bifunctional electrodes of oxygen vacancy abundant $ZnCo_2O_4$ nanosheets for supercapacitor and oxygen evolution. *Chem. Eng. J.* **2021**, *425*, 130583. [CrossRef]
137. Dai, M.; Liu, H.; Zhao, D.; Zhu, X.; Umar, A.; Algarni, H.; Wu, X. Ni Foam Substrates Modified with a $ZnCo_2O_4$ Nanowire-Coated $Ni(OH)_2$ Nanosheet Electrode for Hybrid Capacitors and Electrocatalysts. *ACS Appl. Nano Mater.* **2021**, *4*, 5461–5468. [CrossRef]
138. Xie, L.; Liu, Y.; Bai, H.; Li, C.; Mao, B.; Sun, L.; Shi, W. Core-shell structured $ZnCo_2O_4@ZnWO_4$ nanowire arrays on nickel foam for advanced asymmetric supercapacitors. *J. Colloid Interface Sci.* **2018**, *531*, 64–73. [CrossRef]
139. Younis, A.; Chu, D.; Li, S. Ethanol-directed morphological evolution of hierarchical CeOx architectures as advanced electrochemical capacitors. *J. Mater. Chem. A* **2015**, *3*, 13970–13977. [CrossRef]
140. Zhang, X.; Yang, F.; Chen, H.; Wang, K.; Chen, J.; Wang, Y.; Song, S. In Situ Growth of 2D Ultrathin $NiCo_2O_4$ Nanosheet Arrays on Ni Foam for High Performance and Flexible Solid-State Supercapacitors. *Small* **2020**, *16*, e2004188. [CrossRef]
141. Liu, Y.; Du, X.; Li, Y.; Bao, E.; Ren, X.; Chen, H.; Tian, X.; Xu, C. Nanosheet-assembled porous $MnCo_2O_{4.5}$ microflowers as electrode material for hybrid supercapacitors and lithium-ion batteries. *J. Colloid Interface Sci.* **2022**, *627*, 815–826. [CrossRef] [PubMed]
142. Liu, Y.; Xiang, C.; Chu, H.; Qiu, S.; McLeod, J.; She, Z.; Xu, F.; Sun, L.; Zou, Y. Binary Co–Ni oxide nanoparticle-loaded hierarchical graphitic porous carbon for high-performance supercapacitors. *J. Mater. Sci. Technol.* **2020**, *37*, 135–142. [CrossRef]
143. Gao, M.-R.; Xu, Y.-F.; Jiang, J.; Yu, S.-H. Nanostructured metal chalcogenides: Synthesis, modification, and applications in energy conversion and storage devices. *Chem. Soc. Rev.* **2013**, *42*, 2986–3017. [CrossRef]
144. Zhang, J.; Guan, H.; Liu, Y.; Zhao, Y.; Zhang, B. Hierarchical polypyrrole nanotubes@$NiCo_2S_4$ nanosheets core-shell composites with improved electrochemical performance as supercapacitors. *Electrochim. Acta* **2017**, *258*, 182–191. [CrossRef]
145. Zhu, Y.; Wu, Z.; Jing, M.; Yang, X.; Song, W.; Ji, X. Mesoporous $NiCo_2S_4$ nanoparticles as high-performance electrode materials for supercapacitors. *J. Power Source* **2015**, *273*, 584–590. [CrossRef]
146. Mao, X.; Wang, Y.; Xiang, C.; Zhan, D.; Zhang, H.; Yan, E.; Xu, F.; Hu, X.; Zhang, J.; Sun, L.; et al. Core-shell structured $CuCo_2S_4@CoMoO_4$ nanorods for advanced electrode materials. *J. Alloys Compd.* **2020**, *844*, 156133. [CrossRef]
147. Zhu, J.; Tang, S.; Wu, J.; Shi, X.; Zhu, B.; Meng, X. Wearable High-Performance Supercapacitors Based on Silver-Sputtered Textiles with $FeCo_2S_4$-$NiCo_2S_4$ Composite Nanotube-Built Multitripod Architectures as Advanced Flexible Electrodes. *Adv. Energy Mater.* **2016**, *7*, 1601234. [CrossRef]
148. Chen, S.; Yang, Y.; Zhan, Z.; Xie, J.; Xiong, J. Designed construction of hierarchical $NiCo_2S_4$@polypyrrole core–shell nanosheet arrays as electrode materials for high-performance hybrid supercapacitors. *RSC Adv.* **2017**, *7*, 18447–18455. [CrossRef]
149. Yang, J.; Liu, W.; Niu, H.; Cheng, K.; Ye, K.; Zhu, K.; Wang, G.; Cao, D.; Yan, J. Ultrahigh energy density battery-type asymmetric supercapacitors: $NiMoO_4$ nanorod-decorated graphene and graphene/Fe_2O_3 quantum dots. *Nano Res.* **2018**, *11*, 4744–4758. [CrossRef]
150. Xu, X.; Liu, Q.; Liang, L.; Gu, H.; Zhao, Y.; Xing, X.; Zhang, X.; Hu, Y. Well-designed nanosheet-constructed porous $CoMoS_4$ arrays for ultrahigh-performance supercapacitors. *Ceram. Int.* **2020**, *46*, 4878–4888. [CrossRef]
151. Zhang, Y.; Sun, W.; Rui, X.; Li, B.; Tan, H.T.; Guo, G.; Madhavi, S.; Zong, Y.; Yan, Q. One-Pot Synthesis of Tunable Crystalline Ni_3S_4 @Amorphous MoS_2 Core/Shell Nanospheres for High-Performance Supercapacitors. *Small* **2015**, *11*, 3694–3702. [CrossRef]
152. Zhang, W.-J.; Huang, K.-J. A review of recent progress in molybdenum disulfide-based supercapacitors and batteries. *Inorg. Chem. Front.* **2017**, *4*, 1602–1620. [CrossRef]
153. Wang, T.; Chen, S.; Pang, H.; Xue, H.; Yu, Y. MoS_2-Based Nanocomposites for Electrochemical Energy Storage. *Adv. Sci.* **2017**, *4*, 1600289. [CrossRef]
154. Bao, J.; Zeng, X.-F.; Huang, X.-J.; Chen, R.-K.; Wang, J.-X.; Zhang, L.-L.; Chen, J.-F. Three-dimensional MoS_2/rGO nanocomposites with homogeneous network structure for supercapacitor electrodes. *J. Mater. Sci.* **2019**, *54*, 14845–14858. [CrossRef]
155. Yuan, C.; Wu, H.B.; Xie, Y.; Lou, X.W. Mixed transition-metal oxides: Design, synthesis, and energy-related applications. *Angew. Chem. Int. Ed.* **2014**, *53*, 1488–1504. [CrossRef]
156. Zhu, H.; Zhang, J.; Yanzhang, R.; Du, M.; Wang, Q.; Gao, G.; Wu, J.; Wu, G.; Zhang, M.; Liu, B.; et al. When cubic cobalt sulfide meets layered molybdenum disulfide: A core-shell system toward synergetic electrocatalytic water splitting. *Adv. Mater.* **2015**, *27*, 4752–4759. [CrossRef]

157. Duay, J.; Gillette, E.; Hu, J.; Lee, S.B. Controlled electrochemical deposition and transformation of hetero-nanoarchitectured electrodes for energy storage. *Phys. Chem. Chem. Phys.* **2013**, *15*, 7976–7993. [CrossRef]
158. Dai, Y.-H.; Kong, L.-B.; Yan, K.; Shi, M.; Zhang, T.; Luo, Y.-C.; Kang, L. Simple synthesis of a CoMoS₄ based nanostructure and its application for high-performance supercapacitors. *RSC Adv.* **2016**, *6*, 7633–7642. [CrossRef]
159. Umeshbabu, E.; Rajeshkhanna, G.; Rao, G.R. Urchin and sheaf-like NiCo₂O₄ nanostructures: Synthesis and electrochemical energy storage application. *Int. J. Hydrogen Energy* **2014**, *39*, 15627–15638. [CrossRef]
160. Balamurugan, J.; Li, C.; Peera, S.G.; Kim, N.H.; Lee, J.H. High-energy asymmetric supercapacitors based on free-standing hierarchical Co-Mo-S nanosheets with enhanced cycling stability. *Nanoscale* **2017**, *9*, 13747–13759. [CrossRef]
161. Xu, X.; Song, Y.; Xue, R.; Zhou, J.; Gao, J.; Xing, F. Amorphous CoMoS₄ for a valuable energy storage material candidate. *Chem. Eng. J.* **2016**, *301*, 266–275. [CrossRef]
162. Sun, Y.; Wang, X.; Zhang, W.-c.; Wu, X. Mesoporous Co–Mo–S nanosheet networks as cathode materials for flexible electrochemical capacitors. *CrystEngComm* **2021**, *23*, 7671–7678. [CrossRef]
163. Ma, F.; Dai, X.; Jin, J.; Tie, N.; Dai, Y. Hierarchical core-shell hollow CoMoS₄@Ni–Co–S nanotubes hybrid arrays as advanced electrode material for supercapacitors. *Electrochim. Acta* **2020**, *331*, 135459. [CrossRef]
164. Kang, L.; Zhang, M.; Zhang, J.; Liu, S.; Zhang, N.; Yao, W.; Ye, Y.; Luo, C.; Gong, Z.; Wang, C.; et al. Dual-defect surface engineering of bimetallic sulfide nanotubes towards flexible asymmetric solid-state supercapacitors. *J. Mater. Chem. A* **2020**, *8*, 24053–24064. [CrossRef]
165. Shi, Z.; Shen, X.; Zhang, Z.; Wang, X.; Gao, N.; Xu, Z.; Chen, X.; Liu, X. Hierarchically urchin-like hollow NiCo₂S₄ prepared by a facile template-free method for high-performance supercapacitors. *J Colloid Interface Sci* **2021**, *604*, 292–300. [CrossRef]
166. Zhang, Z.; Huang, X.; Li, H.; Zhao, Y.; Ma, T. 3-D honeycomb NiCo₂S₄ with high electrochemical performance used for supercapacitor electrodes. *Appl. Surf. Sci.* **2017**, *400*, 238–244. [CrossRef]
167. Li, W.; Zhang, B.; Lin, R.; Ho-Kimura, S.; He, G.; Zhou, X.; Hu, J.; Parkin, I.P. A Dendritic Nickel Cobalt Sulfide Nanostructure for Alkaline Battery Electrodes. *Adv. Funct. Mater.* **2018**, *28*, 1705937. [CrossRef]
168. Tang, T.; Cui, S.; Chen, W.; Hou, H.; Mi, L. Bio-Inspired nano-Engineering of an ultrahigh loading 3D hierarchical Ni@NiCo₂S₄/Ni₃S₂ electrode for high energy density supercapacitors. *Nanoscale* **2019**, *11*, 1728–1736. [CrossRef]
169. Zhang, Y.M.; Sui, Y.W.; Qi, J.Q.; Hou, P.H.; Wei, F.X.; He, Y.Z.; Meng, Q.K.; Sun, Z. Facile synthesis of NiCo₂S₄ spheres with granular core used as supercapacitor electrode materials. *J. Mater. Sci. Mater. Electron.* **2016**, *28*, 5686–5695. [CrossRef]
170. Wu, H.B.; Zhang, G.; Yu, L.; Lou, X.W.D. One-dimensional metal oxide-Carbon hybrid nanostructures for electrochemical energy storage. *Nanoscale Horiz.* **2016**, *1*, 27–40. [CrossRef]
171. Dai, M.; Zhao, D.; Liu, H.; Zhu, X.; Wu, X.; Wang, B. Nanohybridization of Ni–Co–S Nanosheets with ZnCo₂O₄ Nanowires as Supercapacitor Electrodes with Long Cycling Stabilities. *ACS Appl. Energy Mater.* **2021**, *4*, 2637–2643. [CrossRef]
172. Wang, D.; Zhu, W.; Yuan, Y.; Du, G.; Zhu, J.; Zhu, X.; Pezzotti, G. Kelp-like structured NiCo₂S₄-C-MoS₂ composite electrodes for high performance supercapacitor. *J. Alloys Compd.* **2018**, *735*, 1505–1513. [CrossRef]
173. Mohamed, S.G.; Hussain, I.; Shim, J.J. One-step synthesis of hollow C-NiCo₂S₄ nanostructures for high-performance supercapacitor electrodes. *Nanoscale* **2018**, *10*, 6620–6628. [CrossRef] [PubMed]
174. Liu, C.; Wu, X.; Wang, B. Performance modulation of energy storage devices: A case of Ni-Co-S electrode materials. *Chem. Eng. J.* **2020**, *392*, 123651. [CrossRef]
175. Bai, X.; Liu, Q.; Liu, J.; Zhang, H.; Li, Z.; Jing, X.; Liu, P.; Wang, J.; Li, R. Hierarchical Co₃O₄@Ni(OH)₂ core-shell nanosheet arrays for isolated all-solid state supercapacitor electrodes with superior electrochemical performance. *Chem. Eng. J.* **2017**, *315*, 35–45. [CrossRef]
176. Zhang, Y.; Yang, Y.; Mao, L.; Cheng, D.; Zhan, Z.; Xiong, J. Growth of three-dimensional hierarchical Co₃O₄@NiMoO₄ core-shell nanoflowers on Ni foam as electrode materials for hybrid supercapacitors. *Mater. Lett.* **2016**, *182*, 298–301. [CrossRef]
177. Yang, F.; Xu, K.; Hu, J. Hierarchical multicomponent electrode with NiMoO₄ nanosheets coated on Co₃O₄ nanowire arrays for enhanced electrochemical properties. *J. Alloys Compd.* **2019**, *781*, 1127–1131. [CrossRef]
178. Li, Y.; Wang, H.; Jian, J.; Fan, Y.; Yu, L.; Cheng, G.; Zhou, J.; Sun, M. Design of three dimensional hybrid Co₃O₄@NiMoO₄ core/shell arrays grown on carbon cloth as high-performance supercapacitors. *RSC Adv.* **2016**, *6*, 13957–13963. [CrossRef]
179. Cai, D.; Wang, D.; Liu, B.; Wang, L.; Liu, Y.; Li, H.; Wang, Y.; Li, Q.; Wang, T. Three-dimensional Co₃O₄@NiMoO₄ core/shell nanowire arrays on Ni foam for electrochemical energy storage. *ACS Appl. Mater. Interfaces* **2014**, *6*, 5050–5055. [CrossRef]
180. Dong, T.; Li, M.; Wang, P.; Yang, P. Synthesis of hierarchical tube-like yolk-shell Co₃O₄@NiMoO₄ for enhanced supercapacitor performance. *Int. J. Hydrogen Energy* **2018**, *43*, 14569–14577. [CrossRef]
181. Hong, W.; Wang, J.; Gong, P.; Sun, J.; Niu, L.; Yang, Z.; Wang, Z.; Yang, S. Rational construction of three dimensional hybrid Co₃O₄@NiMoO₄ nanosheets array for energy storage application. *J. Power Source* **2014**, *270*, 516–525. [CrossRef]
182. Xu, Z.; Younis, A.; Xu, H.; Li, S.; Chu, D. Improved super-capacitive performance of carbon foam supported CeOₓ nanoflowers by selective doping and UV irradiation. *RSC Adv.* **2014**, *4*, 35067–35071. [CrossRef]
183. Sadale, S.B.; Patil, S.B.; Teli, A.M.; Masegi, H.; Noda, K. Effect of deposition potential and annealing on performance of electrodeposited copper oxide thin films for supercapacitor application. *Solid State Sci.* **2022**, *123*, 106780. [CrossRef]
184. Beidaghi, M.; Gogotsi, Y. Capacitive energy storage in micro-scale devices: Recent advances in design and fabrication of micro-supercapacitors. *Energy Environ. Sci.* **2014**, *7*, 867–884. [CrossRef]

185. Liu, N.S.; Gao, Y.H. Recent Progress in Micro-Supercapacitors with In-Plane Interdigital Electrode Architecture. *Small* **2017**, *13*, 1701989. [CrossRef]
186. Qi, D.P.; Liu, Y.; Liu, Z.Y.; Zhang, L.; Chen, X.D. Design of Architectures and Materials in In-Plane Micro-supercapacitors: Current Status and Future Challenges. *Adv. Mater.* **2017**, *29*, 1602802. [CrossRef]
187. Patricia, A.D.; Darina, B.; Claudiu, P.; Nikolaos, A. *Cobalt: Demand-Supply Balances in the Transition to Electric Mobility*; Publications Office of the European Union: Luxembourg, 2018.

Article

Influence of the Composition and Testing Modes on the Electrochemical Performance of Li-Rich Cathode Materials

Lidia Pechen [1,*], Elena Makhonina [1], Anna Medvedeva [1], Yury Politov [1], Aleksander Rumyantsev [2], Yury Koshtyal [2], Alexander Goloveshkin [3] and Igor Eremenko [1]

1. Kurnakov Institute of General and Inorganic Chemistry of the Russian Academy of Sciences, 31 Leninsky pr., 119991 Moscow, Russia
2. Ioffe Institute of the Russian Academy of Sciences, 26 Politekhnicheskaya ul., 194021 St. Petersburg, Russia
3. A.N. Nesmeyanov Institute of Organoelement Compounds of the Russian Academy of Sciences, 28 Vavilova ul., 119334 Moscow, Russia
* Correspondence: lidia.s.maslennikova@gmail.com; Tel.: +7-4957756585 (ext. 2-41)

Abstract: Li-rich oxides are promising cathode materials for Li-ion batteries. In this work, a number of different compositions of Li-rich materials and various electrochemical testing modes were investigated. The structure, chemical composition, and morphology of the materials synthesized were studied by XRD with Rietveld refinement, ICP-OES, and SEM. The particle size distributions were determined by a laser analyzer. The galvanostatic intermittent titration technique and galvanostatic cycling with different potential limits at various current densities were used to study the materials. The electrochemical study showed that gradual increase in the upper voltage limit (formation cycles) was needed to improve further cycling of the cathode materials under study. A comparison of the data obtained in different voltage ranges showed that a lower cut-off potential of 2.5 V (2.5–4.7 V range) was required for a good cyclability with a high discharge capacity. An increase in the low cut-off potential to 3.0 V (3.0–4.8 V voltage range) did not improve the electrochemical performance of the oxides and, on the contrary, considerably decreased the discharge capacity and increased the capacity fade. The LMR35 cathode material ($Li_{1.149}Ni_{0.184}Mn_{0.482}Co_{0.184}O_2$) demonstrated the best functional properties among all the compositions studied.

Keywords: Li-rich cathode material; lithium-ion battery; voltage and capacity fade; testing mode

Citation: Pechen, L.; Makhonina, E.; Medvedeva, A.; Politov, Y.; Rumyantsev, A.; Koshtyal, Y.; Goloveshkin, A.; Eremenko, I. Influence of the Composition and Testing Modes on the Electrochemical Performance of Li-Rich Cathode Materials. *Nanomaterials* **2022**, *12*, 4054. https://doi.org/10.3390/nano12224054

Academic Editors: Wen Lei, Gaoran Li, Dan Luo and Shiqiang (Rob) Hui

Received: 20 October 2022
Accepted: 16 November 2022
Published: 17 November 2022

Publisher's Note: MDPI stays neutral with regard to jurisdictional claims in published maps and institutional affiliations.

1. Introduction

Today, there is continued interest in the development of energy storage devices, especially supercapacitors and lithium-ion batteries (LIBs) [1–3]. Li-rich cathode materials for LIBs of the general formula $Li_{1+x}M_{1-x}O_2$ (M = Ni, Mn, Co, etc.) are superior to layered NCM-like cathode materials due to much greater specific capacities and energies. The high capacity of Li-rich materials is provided by the oxygen redox process in addition to the redox reactions of transition metals (TM) when cycling to a high voltage (above 4.5 V) [4–9]. The first charge–discharge cycle of these materials to a high voltage generates a high capacity, however, with a large irreversibility, and leads to a structural transformation, whose nature is the subject of many recent discussions [4,8,10]. In the literature, the structural transformation is attributed to partial losses of oxygen and lithium mainly from the particle surface [3,11,12]. An increase in the discharge capacity is correlated with the formation of an electrochemically active manganese-containing phase [13,14]. During further cycling, TM ions migrate to the lithium sites, which gradually leads to the transformation of the layered structure to the spinel-like one [15,16].

The structure of Li-rich materials is considered in the literature as both a two-phase composite (nanocomposite) consisting of the trigonal (sp. g. $R\bar{3}m$) and the monoclinic (sp. g. $C2/m$) phases [17–20] and a solid solution based on the monoclinic [21,22] or the trigonal phase [23]. The formation of intergrowth structures and mirror twins is also discussed in the literature [23,24]. The structure complexity, a large first cycle irreversibility, and the structural transformation leading to a voltage decay and capacity fade do not allow one to apply Li-rich cathode materials in practice [10,25]. For the second decade, scientists have actively investigated the mechanisms of degradation of Li-rich cathode material [26–28]; however, a number of processes taking place during electrochemical cycling remain unclear.

It should be noted that many controversial studies concerning both the structure and functional properties of the Li-rich materials were published, which is likely due to the influence of many factors such as the synthesis method and conditions, morphology, tap density, etc. [29–31]. Moreover, the cell assembly, including the cell components, and the cycling mode, and other factors affect the functional properties of the cathode materials.

To improve the electrochemical properties of the Li-rich cathode material, it is necessary to understand the nature of the structural transformations and to find the methods to suppress unfavorable processes. The variation of the main components (Li, Mn, Ni, Co) is one of the methods to influence on the structure and properties of the Li-rich materials. Herein, we studied an effect of the different Li/(Mn + Ni + Co) ratios from 1.2 to 1.65 on the morphology, structure, and electrochemical properties of the materials such as cathodes for LIBs. We also studied the influence of the different testing modes, including so called formation cycles, in which the upper voltage limit increased gradually from cycle to cycle at different current densities. It was found that the slow formation cycles were required for the better electrochemical performance of the Li-rich oxides. All the materials were also tested in three different voltage ranges to estimate contributions of the low and high voltages to the material degradation. It was found that the cycling to the low voltage limit of 2.5 V was necessary to obtain the high capacity and stable cyclability of the Li-rich cathode materials.

2. Materials and Methods

The cathode materials were synthesized by coprecipitation of the transition metal (nickel, manganese, cobalt) carbonates from the corresponding nitrate salts. Potassium or sodium carbonate was used as a precipitator. The synthesis procedure was described in detail in our previous works [32,33]. The carbonate precursor was thoroughly mixed with lithium hydroxide monohydrate in ethanol and annealed at 480 °C for 6 h and 900 °C for 12 h. The targeted compositions of the cathode materials with sample designations and metal ratios are listed in Table 1.

Table 1. Designations of the samples and their targeted compositions.

Samples	Targeted Composition	Li/(Mn + Ni + Co)
LMR20	$Li_{1.091}Ni_{0.242}Mn_{0.424}Co_{0.242}O_2$	1.20
LMR35	$Li_{1.149}Ni_{0.184}Mn_{0.482}Co_{0.184}O_2$	1.35
LMR50	$Li_{1.200}Ni_{0.134}Mn_{0.534}Co_{0.134}O_2$	1.50
LMR65	$Li_{1.245}Ni_{0.088}Mn_{0.579}Co_{0.088}O_2$	1.65

The cathode materials obtained were characterized by XRD with the structure parameter refinement by the Rietveld method. The X-ray diffraction studies were carried out with a Bruker D8 Advance (Bruker AXS, Germany, Cu Kα, (Ni filter), λ = 0.15418 nm, 40 kW/40 mA, LynxEye 1D detector) diffractometer at room temperature in the 2θ range of 10°–90° with a step of 0.02°. The data collection was performed using the BrukerDIFFRAC-plus software package (Bruker AXS, Karlsruhe, Germany); the analysis was carried out with the EVA and TOPAS programs. The morphology of the materials was studied by SEM (NVision-40 (Carl Zeiss, Oberkochen, Germany)) with EDX microanalysis. The size particle distribution analysis was performed with the use of a laser particle sizer Analysette 22 MicroTec Plus (Idar-Oberstein, Germany). The compositions of the synthesized compounds were determined by ICP-OES (Thermo Scientific iCAP XP, Waltham, MA, USA).

All the compounds synthesized were tested in CR2032 coin-type cells as the active material (92 wt%). Polyvinylidene difluoride Solef 513 (Solvay, 3 wt%) and carbon black Super C65 (Timcal, 5 wt%) were used as a binder and electroconductive additive, respectively. The lithium foil was used as an anode, and two layers of Celgard 2325 were used as a separator. The used electrolyte (TC-E918, Tinci, Guangzhou, China) contained $LiPF_6$ salt dissolved in a mixture of ethylene, diethyl, ethylmethyl and propylene carbonates.

The electrochemical studies were performed using a Neware CT-4008W-5V10mA battery tester in the galvanostatic cycling mode in the voltage ranges of 2.5–4.7, 2.5–4.3, and 3.0–4.8 V at the current densities of 20 and 80 mA/g. The results of all the electrochemical tests were averaged over 4–6 cells. The formation cycles (by 2 cycles in the ranges of 2.5–4.3, 2.5–4.5, 2.5–4.6, and 2.5–4.7 V, successively) were conducted at the current density of 20 or 80 mA/g before the further cycling tests. The cycling tests at the current density of 80 mA/g without formation cycles were performed in the range of 2.5–4.7 V. The rate capability was examined in the range of 2.5–4.7 V, and the current densities were from 80 to 480 mA/g.

The galvanostatic intermittent titration (GITT) was performed during discharge process to estimate the resistance values. At first, the formation cycles were also carried out; and then the samples were cycled in the required voltage range. The cell was discharged at a constant current during 30 min succeeded by the current interruption (relaxation time) for 60 min, which is sufficient to achieve an equilibrium voltage value. The steps were repeated four times within a discharge, i.e., full discharge time was 120 min. From the GITT results, we estimated the resistance values by the following procedure. At first, the voltage was measured at the end of each discharge step ($U1$). Then, the cut-off voltage was measured immediately after the current interruption ($U2$) and at the end of relaxation time ($U3$). The cell resistances were calculated from the voltage differences between $U2$ and $U1$ ($R_{ohm.}$), and $U3$ and $U2$ ($R_{pol.}$) according to Equations (1) and (2), respectively:

$$R_{ohm.} = \frac{U2 - U1}{I} \tag{1}$$

$$R_{pol.} = \frac{U3 - U2}{I} \tag{2}$$

where I-discharge current (A).

3. Results and Discussion

3.1. Chemical and Structural Analyses

The SEM images of the carbonate precursors for the Li-rich materials of different compositions are presented in Figure S1. All precursors were spheric-like agglomerates. The agglomerates became larger in the samples with a larger manganese content. The average agglomerate size of the LMR20 carbonate precursor was 8 μm, but there were also smaller agglomerates (size range of 0.5–1 μm). The average agglomerate size for the LMR35 precursor was about 8 μm, and there were also small agglomerates 1–2 μm in size. The same average value for the LMR50 precursor was 5–12 μm, whereas it was 15–16 μm for LMR65. The narrowest size distribution was observed for the LMR65 carbonate precursor. The average primary particle sizes for the LMR65 and LMR50 precursors were 400–500 and 40–50 nm, respectively. The same values for LMR35 and LMR20 were less than 50 nm.

The cathode materials, obtained by a solid-state reaction with lithium hydroxide monohydrate and following annealing, maintained the shape and size of the carbonate agglomerates (Figure 1). The primary particle sizes varied in the ranges of 250 nm–2 μm, 300–800 nm, 400 nm–1 μm, and 400 nm–2.5 μm for LMR20, LMR35, LMR50, and LMR65, respectively.

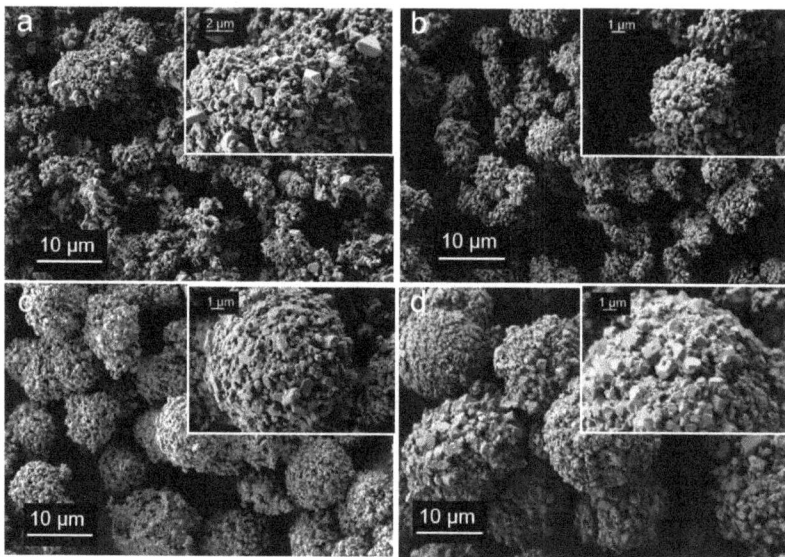

Figure 1. SEM micrographs of the cathode materials at different magnifications: (**a**) LMR20, (**b**) LMR35, (**c**) LMR50, and (**d**) LMR65.

The differential size distributions for the cathode materials are shown in Figure S2. The numeric values d10, d50, d90 are listed in Table 2. As is observed, the agglomerates increase with an increase in the manganese content in the material composition. At the same time, the size distributions become narrower. Width of agglomerate size distribution characterized by (d90-d10)/d50 value is very close to LMR50 and LMR65.

Table 2. The agglomerate size distribution for the Li-rich samples.

	LMR20	LMR35	LMR50	LMR65
d10, μm	3.50	4.56	6.23	8.75
d50, μm	8.11	8.54	10.88	16.97
d90, μm	15.97	15.26	17.64	27.15
(d90-d10)/d50	1.54	1.25	1.05	1.08

The cathode material compositions determined by ICP-OES were close to the targeted compositions (Table S1).

Most of the peaks in the XRD patterns can be described by both the trigonal structure with $R\bar{3}m$ space group and the monoclinic structure with $C2/m$ space group. In the range of a 20–30° 2θ in the LR35, LR50, LR65 diffractograms, the broadened, low-intensity peaks were observed, characteristic of a superlattice monoclinic phase with ordering of some lithium ions in the transition metal layers (Figure 2). The intensities of these peaks increased in the order of LMR35, LMR50, LMR65. No superstructural peaks were observed in the X-ray pattern of LMR20, which may indicate a small number of the layers with ordered Li ions. We used the model of the solid solution based on the monoclinic phase for the Rietveld refinement (Table 3).

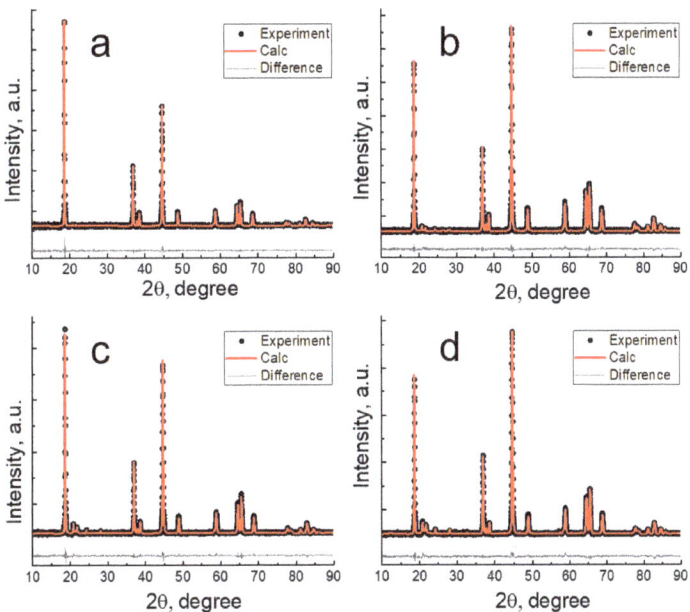

Figure 2. XRD patterns for the cathode materials with different compositions: (**a**) LMR20, (**b**) LMR35, (**c**) LMR50, (**d**) LMR65.

Table 3. Crystallographic data and the unit cell parameters from the Rietveld refinement ($C2/m$ space group) for the cathode materials.

	LMR20	LMR35	LMR50	LMR65
Rwp, %	1.76	2.84	2.10	2.40
Rp, %	1.28	2.22	1.60	1.81
GOOF, %	3.19	5.05	4.58	5.59
a, Å	4.9441(4)	4.9363(4)	4.9317(4)	4.9305(3)
b, Å	8.5637(8)	8.5502(6)	8.5422(8)	8.5401(6)
c, Å	5.0245(14)	5.0218(4)	5.0216(4)	5.0233(4)
β, Å	109.226(3)	109.264(3)	109.283(2)	109.303(2)
V, Å^3	200.87(7)	200.08(3)	199.68(4)	199.63(3)

The dependence of the peak broadening on the diffraction angle was described by the Williamson–Hall approach. The larger was the manganese content and the smaller were the nickel and cobalt concentrations in the composition of the cathode materials, the smaller were the unit cell parameters.

3.2. Electrochemical Characterization

Three different protocols of the electrochemical tests were used to study the samples. **Protocol 1** included the formation cycles consisting of the successive cycles by two samples in the ranges of 2.5–4.3, 2.5–4.5, 2.5–4.6, and 2.5–4.7 V at the current density of 20 mA/g; **protocol 2** contained the same successive formation cycles at the current density of 80 mA/g; and **protocol 3** constituted the cycling without formation.

A comparison of the formation cycles (LMR50 sample) at different current densities (**protocols 1 and 2**) is presented in Figure 3.

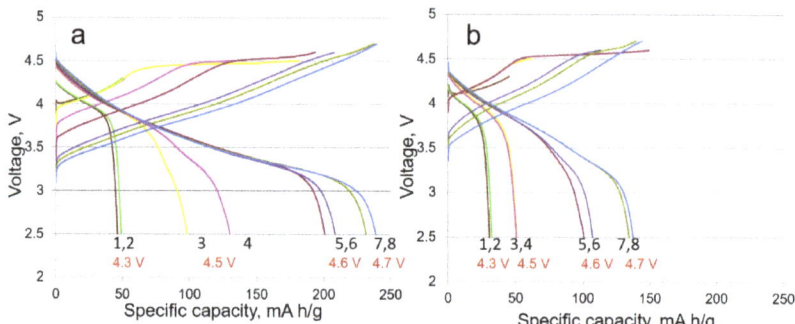

Figure 3. Charge–discharge curves of the formation cycles for LMR50 sample at the current densities of (**a**) 20 mA/g (**protocol 1**) and (**b**) 80 mA/g (**protocol 2**).

As is seen, the formation cycles at the current density of 80 mA/g have a negative effect on the discharge capacity value compared with the formation cycles at the current density of 20 mA/g. The discharge capacity was about 240 mAh/g at the second formation cycle to 4.7 V (the eighth cycle in total, Figure 3) at the current density of 20 mA/g, whereas the discharge capacity was only 140 mAh/g at 80 mA/g in the same cycle. Notice also that the highest irreversible capacity at 20 mA/g was observed at the first formation cycle at 2.5–4.5 V (the third cycle in total), and the highest irreversible capacity at 80 mA/g (**protocol 2**) was observed in the later cycles (first cycle at 2.5–4.6 V, i.e., the fifth cycle in total). This fact can indicate that the activation of the materials is kinetically hindered.

The differential capacity curves (dQ/dV) for the third, fifth, seventh, and eighth cycles performed according to **protocol 1** and **protocol 2** for the sample LMR50 are shown in Figure 4.

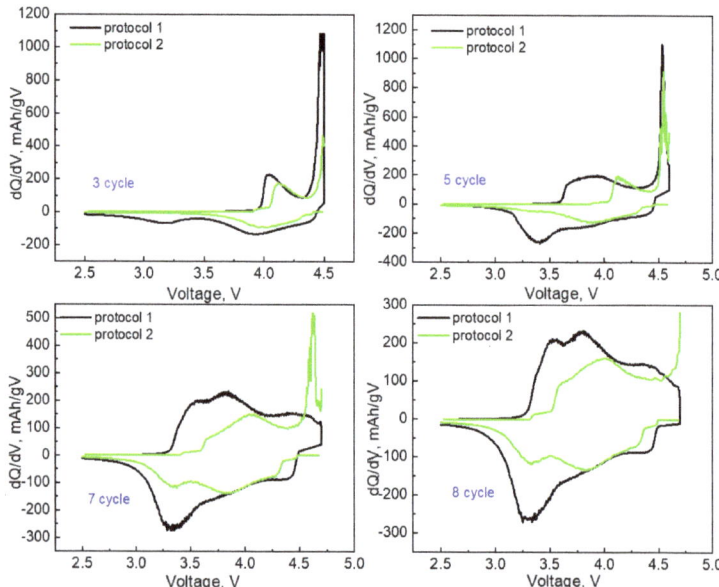

Figure 4. dQ/dV curves of LMR50 sample with different formation protocols.

The peak in the anode curve in the region of 4.5–4.6 V is responsible for the oxidation of O^{2-} [34]. This peak appeared later for the samples prepared by **protocol 2**, which correlated with the irreversible capacities due to a partial oxygen loss. The activation also led to the formation of an electrochemically active manganese-containing phase, which manifested itself by additional cathodic and anodic peaks in the range of 3.2–3.4 V [13,35]. These peaks appeared later in case of **protocol 2**. The samples prepared according to this protocol demonstrated worse performance in the course of further cycling.

All the cathode materials were cycled at the current density of 80 mA/g according to the three different protocols described above. The materials formed by **protocol 2** worked no more than 20 cycles, and their discharge capacities at this current density were about 20–30 mAh/g (the data are not represented graphically).

The cycling profiles at 80 mA/g for the oxides tested by **protocol 1** and **protocol 3** are compared in Figure 5. LMR65 had very poor electrochemical performance, so its cycling results are not shown in all the graphs. The specific capacity and energy of LMR50 and LMR35 gradually increased in the first 10–20 cycles (**protocol 3**), whereas for LMR20, these values gradually decreased from the first cycle (Figure 5c,d). The samples after the formation cycles according to **protocol 1** (Figure 5a,b) demonstrated considerably higher specific capacities and energies than the samples without this preliminary procedure. Notice also that LMR35 showed better cyclability compared with the other two samples. In our opinion, the formation cycles with successive voltage increase give a possibility to smooth the inherent transformation of the initial structure in the course of activation. The confirmation of this assumption is the difference in the behavior of LMR50 and LMR35 samples. The structure transformation in LMR35 occurred more gradually (Figure 5c,d), which led to a better cyclability of this material.

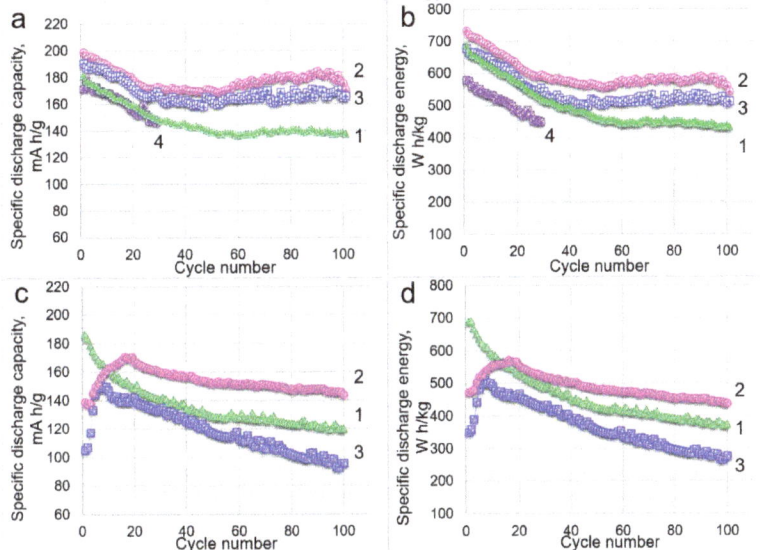

Figure 5. Cycling behavior of the cathode materials with different compositions cycled according to (**a,b**) **protocol 1** and (**c,d**) **protocol 3** at the current density of 80 mA/g. Designations: 1—LMR20, 2—LMR35, 3—LMR50, and 4—LMR65.

The dQ/dV curves for the first, second, and 100th cycles for the materials without formation and the 100th cycle for the samples cycled by **protocol 1** are shown in Figure 6.

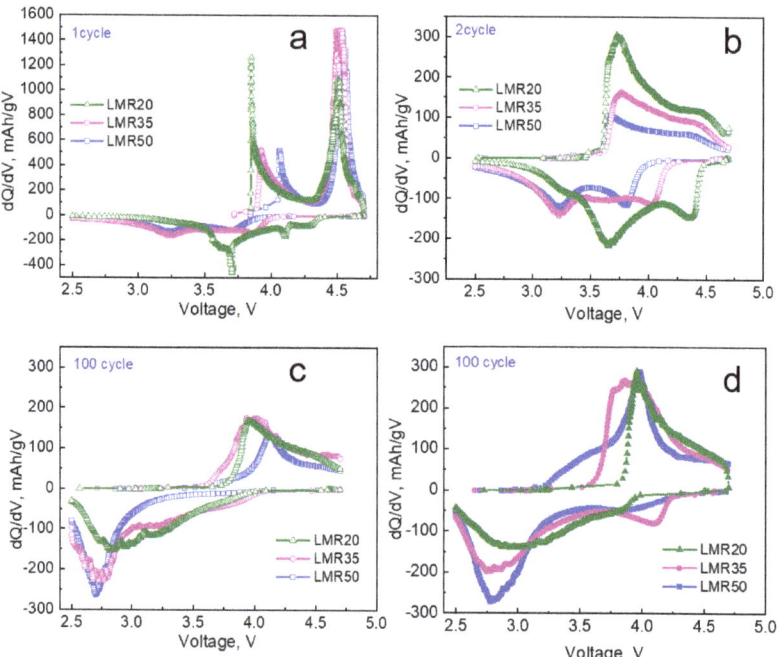

Figure 6. dQ/dV curves for the samples cycled (**a–c**) without formation and (**d**) by **protocol 1** at the current density of 80 mA/g.

Apparently, the oxygen oxidation in the samples without formation cycles occurred only in the first cycle (Figure 6a), the anodic peak in the range of 4.5–4.6 V was not observed in the second cycle (Figure 6b). At the same time, the structure transformation continued in the further cycles, as evidenced by increasing the discharge capacity and appearance of the new cathodic peak at 3.2–3.4 V that was described above. This can indicate that an increase in the capacity in the further 10–20 cycles is provided by the manganese-containing species. We did not observe the similar behavior for the sample LMR20 with the lowest manganese concentration from the samples studied. This sample showed only a small shoulder in the area of 3.2–3.4 V to the second cycle and a very broad low-intensity cathodic peak in the area of 2.8–3.2 V to the 100th cycle.

The dQ/dV curves for the 100th cycle for the same materials without the formation cycles (**protocol 3**) and formed by **protocol 1** are shown in Figures 6c and 6d, respectively. The intensities of the peaks were higher for the samples after preliminary slow formation cycles at 20 mA/g (Figure 6d), which correlated with the higher capacities of these samples. The additional peak in the range of 4.0–4.1 V was observed in the cathodic curve of LMR50 and especially in that of LMR35, where its value was considerably larger (Figure 6d). This peak was attributed in the literature to reversible oxygen activity [36]. Therefore, the LMR35 oxide most likely showed better cyclability and larger values of the discharge capacity (energy) due to better reversibility of the oxygen redox process.

The materials tested by **protocol 1** were also cycled at 20 mA/g (Figure 7).

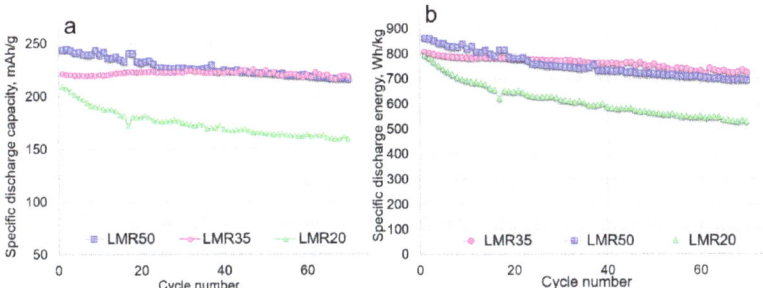

Figure 7. Cycling behavior of cathode materials at 20 mA/g: (**a**) discharge capacity vs. cycle number, (**b**) discharge energy vs. cycle number.

As was discussed above, LMR65 had very poor electrochemical performance and its data were not shown in the Figure 7. Only 8–10 cycles could be obtained for this material at 20 mA/g. LMR50 showed the highest initial capacity and energy values from the samples under study, but the capacities of LMR35 and LMR50 became comparable to the 20th cycle (Figure 7a,b). At the same time, the energy values for LMR35 material became higher after 20 cycles due to a lower voltage decay. It is significant that the capacity fade for LMR35 material to the 70th cycle was only 1%, which is comparable or better than in the literature data for the Li-rich cathode materials [37–39]. The energy fade for LMR35 was 10% to the same cycle because of the voltage decay. The energy fades for LMR50 and LMR20 were 20 and 34%, respectively.

The preliminary formation cycles according to **protocol 1** also positively affected the further cycling at a low rate of 20 mA/g, as was observed for the cycling at the current density of 80 mA/g. However, the formation cycles performed at the high current density (**protocol 2**) worsened the electrochemical performance of the materials compared with the samples cycled both according to **protocol 1** and **protocol 3**. Apparently, the formation cycles at the low current density led to the structure capable to reversibly oxidize/reduce oxygen.

To study the oxidation/reduction processes taking place during charge/discharge, three compositions (LMR20, LMR35, LMR50) after slow formation cycles (**protocol 1**) were cycled in the three different voltage ranges of 2.5–4.3, 3.0–4.8 (Figure 8), and 2.5–4.7 V (Figure 5a,b). Note that a comparison of the cycling behavior in the voltage ranges of 2.5–4.7 and 2.5–4.8 V showed the minimal difference in the capacity values and capacity retentions for all the samples. The range of 2.5–4.3 V reflects the effect of an increase in the resistance due to a deep discharge. The range of 3.0–4.8 V shows the effect of deep charges, which may contribute to the structural transformations. During cycling in the wide voltage range of 2.5–4.7 V both factors may contribute to the capacity fade.

The comparison of the cycling profiles in the different voltage ranges showed that the capacity fade was not maximal in the widest voltage range of 2.5–4.7 V, as might be expected. However, we observed the maximum of the capacity fade in the 3.0–4.8 V voltage range. The capacity retention for all the materials to the 110th cycle was only about 50% for the 3.0–4.8 V voltage range (Figure 8c,d). The capacity retentions to the same cycle in the ranges of 2.5–4.7 and 2.5–4.3 V were, respectively, 76, 90, and 87% and 92, 89, and 85% for LMR20, LMR35, and LMR50. It should be noted, that the capacity retentions for LMR35 and LMR50 were somewhat lower in the range of 2.5–4.3 V than those in the voltage range of 2.5–4.7 V. The sample LMR20, on the contrary, showed the better cyclability in the range of 2.5–4.3 V. This correlates with the lowest manganese content in this sample, which formally corresponds to the lowest content of the monoclinic phase (or the ordered layers containing transition metals and Li ions). The voltage decay is maximal for all the cathode materials in the 2.5–4.7 V voltage range due to formation of the electrochemically active phase with a lower redox potential and an increase in the polarization resistance in the course of cycling.

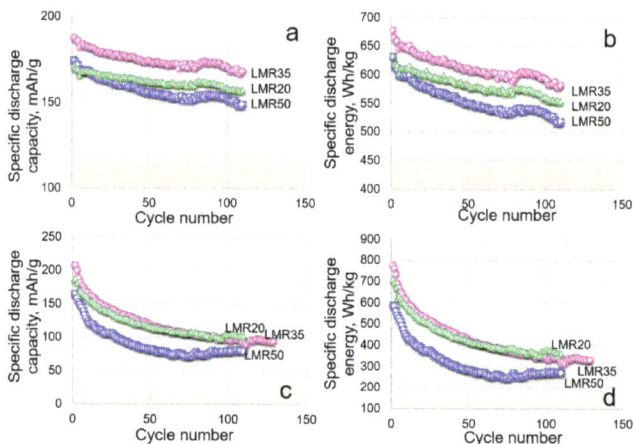

Figure 8. Cycling behavior of the materials in the range of (**a,b**) 2.5–4.3 V and (**c,d**) 3.0–4.8 V at 80 mA/g.

The charge–discharge profiles and the dQ/dV curves in the 100th cycle for the three different voltage ranges are shown in Figure 9.

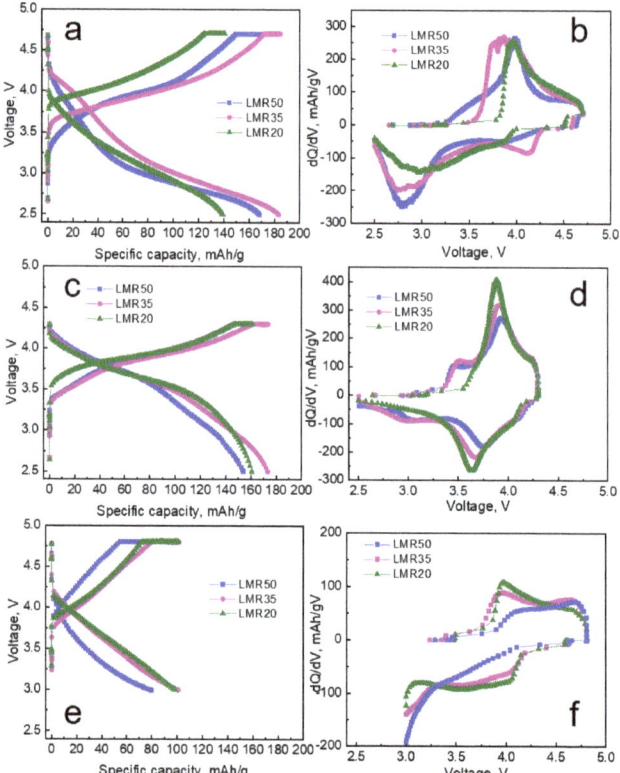

Figure 9. (**a,c,e**) Charge—discharge profiles and (**b,d,e**) dQ/dV curves in the 100th cycle for the different compositions in the voltage ranges of 2.5–4.7, 2.5–4.3, and 3.0–4.8 V, respectively; the current density is equal to 80 mA/g.

The redox peaks in the dQ/dV curves (Figure 9b,d) for all the samples cycled in the ranges of 2.5–4.3 and 2.5–4.7 V were observed at different potentials. The cathodic peak for the samples cycled in the range of 2.5–4.7 V was broadened and shifted to lower potentials, whereas that for the 2.5–4.3 V range slightly changed with cycling. The LMR35 sample had the additional cathodic peak at 4.1–4.2 V in the range of 2.5–4.7 V. The discharge capacities obtained in the range of 3.0–4.8 V in the 100th cycle for all the samples were considerably lower (Figure 9e) than those in the voltage ranges of 2.5–4.3 and 2.5–4.7 V. The redox process corresponding to the low-voltage species could not occur in this range, and the related cathodic peaks were not displayed in the dQ/dV plots (Figure 9f). Therefore, to carry out the cycling tests at the lower potential limit higher than 2.5 V is impractical.

The polarization resistances were calculated for all the cathode materials in the different voltage ranges from the GITT data. As a rule, the polarization resistance characterizes the processes associated with the surface concentration changes due to the hindered diffusion of lithium ions from the particle surface to the bulk [40].

The polarization resistances calculated for the cell discharge vs. the voltage values are shown in Figure 10. In the voltage range of 2.5–4.3 V, the resistance increased in the course of discharge in each cycle for all the samples. In the range of 2.5–4.7 V, the character of the curves was more complex and varied depending on the cycle number for all the materials. In this range, we also observed an increase in the resistance during the first cycle. However, from the 45th cycle, the character of the resistance dependence changed within a cycle showing an increase at both the beginning and the end of the cycle. At the same time, the resistance values were lower in the wide voltage range than those in the range of 2.5–4.3 V. The least changes within a cycle were observed for LMR35, which demonstrated the better cyclability.

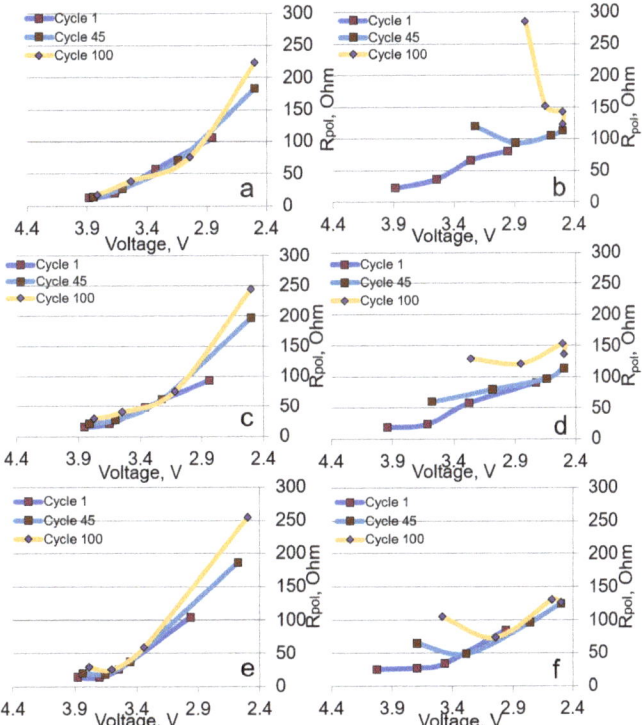

Figure 10. Polarization resistance values on discharge during cycling in the ranges of 2.5–4.3 V ((**a**) LMR50, (**c**) LMR35, (**e**) LMR20) and 2.5–4.7 V ((**b**) LMR50, (**d**) LMR35, (**f**) LMR20).

The rate capabilities in the range of 2.5–4.7 V were studied for all the materials with the different compositions (Figure S3). The LMR35 sample showed a significantly lower discharge capacity fade at the high current densities compared to the other samples.

4. Conclusions

The effect of the different Li-rich compositions on the properties and electrochemical performance of the materials was investigated; the Li/(Mn + Ni + Co) ratios were varied from 1.2 to 1.65. The oxides were synthesized by coprecipitation of TM carbonates followed by a solid-state reaction with a lithium source. Different schemes of the electrochemical tests were used to study these materials. Some of the tests included so-called formation cycles— eight cycles with gradual increasing of the upper voltage limit at the different current densities. In addition, we studied also the electrochemical behavior of the materials in the three voltage ranges, namely, 2.5–4.3, 2.5–4.7, and 3.0–4.8 V. It was found that successive formation cycles at the low current density (20 mA/g, 0.1C) are necessary to obtain the high discharge capacities and stable cycling of the materials under study. The electrochemical tests in the three different voltage ranges were performed after preliminary formation cycles described above. The results of this study showed that the largest capacity fade and the lowest capacities during cycling are observed in the range of 3.0–4.8 V, whereas the cycling in the widest voltage range of 2.5–4.7 V demonstrates very good cyclability with the high discharge capacities. This behavior of the materials under study may be explained by the formation of the electrochemically active phase with the redox processes at a low potential as a result of the structural transformations during the first cycles. These structural changes, although leading to a voltage decay, provide an additional discharge capacity. In addition, the structures formed by the slow preliminary formation revealed a greater reversibility of the anionic redox process, which also might contribute to the discharge capacity. Comparing the 2.5–4.3 V and 2.5–4.7 V ranges, it may be noted that the polarization resistance in the 2.5–4.3 V range increased significantly within one discharge cycle, as opposite to that when the samples cycled in the 2.5–4.7 V range. In this voltage range, the resistance values were lower (except LMR50) than those the 2.5–4.3 V range. The LMR35 sample (Li/(Mn + Ni + Co) = 1.35) showed the best electrochemical properties among all the other samples, likely due to a lower polarization resistance, a more gradual structure transformation, and a greater reversibility of the oxygen redox process.

Supplementary Materials: The following supporting information can be downloaded at: https://www.mdpi.com/article/10.3390/nano12224054/s1, Figure S1: Micrographs of carbonate precursors for cathode materials with different magnifications: (a,b) LMR20, (c,d) LMR35, (e,f) LMR50, (g,h) LMR65; Figure S2: Differential agglomerate distribution of cathode materials with different compositions: (a) LMR20, (b) LMR35, (c) LMR50, (d) LMR65; Figure S3: Rate capabilities of the cathode materials; Table S1: Determined by ICP-OES compositions of cathode materials.

Author Contributions: Conceptualization, L.P. and E.M.; methodology, L.P. and E.M.; synthesis, A.M. and Y.P.; SEM studies, A.M.; XRD analysis, A.G. and A.M.; electrochemical measurements, A.R., Y.K., and Y.P.; data analysis, E.M., L.P., and A.M.; writing—original draft preparation, L.P. and E.M.; writing—review and editing, L.P. and E.M.; visualization, L.P. and E.M.; supervision, E.M. and I.E.; project administration, E.M.; funding acquisition, E.M. All authors have read and agreed to the published version of the manuscript.

Funding: This research was funded by the Russian Science Foundation, grant number 20-13-00423.

Acknowledgments: The compound's characterization was performed with the financial support from Ministry of Science and Higher Education of the Russian Federation using the equipment of the Joint Research Center of IGIC RAS (SEM, ICP-OES) and the Center for molecular composition studies of INEOS RAS (Powder XRD).

Conflicts of Interest: The authors declare no conflict of interest.

References

1. Poudel, M.B.; Kim, A.R.; Ramakrishan, S.; Logeshwaran, N.; Ramasamy, S.K.; Kim, H.J.; Yoo, D.J. Integrating the Essence of Metal Organic Framework-Derived ZnCoTe–N–C/MoS2 Cathode and ZnCo-NPS-N-CNT as Anode for High-Energy Density Hybrid Supercapacitors. *Compos. B Eng.* **2022**, *247*, 110339. [CrossRef]
2. Zhang, M.; Kitchaev, D.A.; Lebens-Higgins, Z.; Vinckeviciute, J.; Zuba, M.; Reeves, P.J.; Grey, C.P.; Whittingham, M.S.; Piper, L.F.J.; van der Ven, A.; et al. Pushing the Limit of 3d Transition Metal-Based Layered Oxides That Use Both Cation and Anion Redox for Energy Storage. *Nat. Rev. Mater.* **2022**, *7*, 522–540. [CrossRef]
3. House, R.A.; Marie, J.-J.; Pérez-Osorio, M.A.; Rees, G.J.; Boivin, E.; Bruce, P.G. The Role of O2 in O-Redox Cathodes for Li-Ion Batteries. *Nat. Energy* **2021**, *6*, 781–789. [CrossRef]
4. Assat, G.; Iadecola, A.; Foix, D.; Dedryvère, R.; Tarascon, J.-M. Direct Quantification of Anionic Redox over Long Cycling of Li-Rich NMC via Hard X-Ray Photoemission Spectroscopy. *ACS Energy Lett.* **2018**, *3*, 2721–2728. [CrossRef]
5. Li, Q.; Ning, D.; Wong, D.; An, K.; Tang, Y.; Zhou, D.; Schuck, G.; Chen, Z.; Zhang, N.; Liu, X. Improving the Oxygen Redox Reversibility of Li-Rich Battery Cathode Materials via Coulombic Repulsive Interactions Strategy. *Nat. Commun.* **2022**, *13*, 1123. [CrossRef]
6. Zhang, J.; Cheng, F.; Chou, S.; Wang, J.; Gu, L.; Wang, H.; Yoshikawa, H.; Lu, Y.; Chen, J. Tuning Oxygen Redox Chemistry in Li-Rich Mn-Based Layered Oxide Cathodes by Modulating Cation Arrangement. *Adv. Mater.* **2019**, *31*, 1901808. [CrossRef]
7. Seo, D.-H.; Lee, J.; Urban, A.; Malik, R.; Kang, S.; Ceder, G. The Structural and Chemical Origin of the Oxygen Redox Activity in Layered and Cation-Disordered Li-Excess Cathode Materials. *Nat. Chem.* **2016**, *8*, 692–697. [CrossRef]
8. Li, X.; Qiao, Y.; Guo, S.; Xu, Z.; Zhu, H.; Zhang, X.; Yuan, Y.; He, P.; Ishida, M.; Zhou, H. Direct Visualization of the Reversible O^{2-}/O$^-$ Redox Process in Li-Rich Cathode Materials. *Adv. Mater.* **2018**, *30*, 1705197. [CrossRef]
9. Assat, G.; Tarascon, J.-M. Fundamental Understanding and Practical Challenges of Anionic Redox Activity in Li-Ion Batteries. *Nat. Energy* **2018**, *3*, 373–386. [CrossRef]
10. Merz, M.; Ying, B.; Nagel, P.; Schuppler, S.; Kleiner, K. Reversible and Irreversible Redox Processes in Li-Rich Layered Oxides. *Chem. Mater.* **2021**, *33*, 9534–9545. [CrossRef]
11. Zhang, H.; Liu, H.; Piper, L.F.J.; Whittingham, M.S.; Zhou, G. Oxygen Loss in Layered Oxide Cathodes for Li-Ion Batteries: Mechanisms, Effects, and Mitigation. *Chem. Rev.* **2022**, *122*, 5641–5681. [CrossRef] [PubMed]
12. Teufl, T.; Strehle, B.; Müller, P.; Gasteiger, H.A.; Mendez, M.A. Oxygen Release and Surface Degradation of Li- and Mn-Rich Layered Oxides in Variation of the Li$_2$MnO$_3$ Content. *J. Electrochem. Soc.* **2018**, *165*, A2718–A2731. [CrossRef]
13. Zheng, J.; Shi, W.; Gu, M.; Xiao, J.; Zuo, P.; Wang, C.; Zhang, J.-G. Electrochemical Kinetics and Performance of Layered Composite Cathode Material Li[Li$_{0.2}$Ni$_{0.2}$Mn$_{0.6}$]O$_2$. *J. Electrochem. Soc.* **2013**, *160*, A2212–A2219. [CrossRef]
14. Thackeray, M.M. Manganese Oxides for Lithium Batteries. *Prog. Solid State Chem.* **1997**, *25*, 1–71. [CrossRef]
15. Chai, K.; Zhang, J.; Li, Q.; Wong, D.; Zheng, L.; Schulz, C.; Bartkowiak, M.; Smirnov, D.; Liu, X. Facilitating Reversible Cation Migration and Suppressing O$_2$ Escape for High Performance Li-Rich Oxide Cathodes. *Small* **2022**, *18*, 2201014. [CrossRef]
16. Wang, T.; Zhang, C.; Li, S.; Shen, X.; Zhou, L.; Huang, Q.; Liang, C.; Wang, Z.; Wang, X.; Wei, W. Regulating Anion Redox and Cation Migration to Enhance the Structural Stability of Li-Rich Layered Oxides. *ACS Appl. Mater Interfaces* **2021**, *13*, 12159–12168. [CrossRef]
17. Viji, M.; Budumuru, A.K.; Hebbar, V.; Gautam, S.; Chae, K.H.; Sudakar, C. Influence of Morphology and Compositional Mixing on the Electrochemical Performance of Li-Rich Layered Oxides Derived from Nanoplatelet-Shaped Transition Metal Oxide–Hydroxide Precursors. *Energy Fuels* **2021**, *35*, 4533–4549. [CrossRef]
18. Guo, L.; Tan, X.; Mao, D.; Zhao, T.; Song, L.; Liu, Y.; Kang, X.; Wang, H.; Sun, L.; Chu, W. Improved Electrochemical Activity of the Li2MnO3-like Superstructure in High-Nickel Li-Rich Layered Oxide Li1.2Ni0.4Mn0.4O2 and Its Enhanced Performances via Tungsten Doping. *Electrochim. Acta* **2021**, *370*, 137808. [CrossRef]
19. Bian, X.; Zhang, R.; Yang, X. Effects of Structure and Magnetism on the Electrochemistry of the Layered Li1+x(Ni0.5Mn0.5)1−XO2 Cathode Material. *Inorg. Chem.* **2020**, *59*, 17535–17543. [CrossRef]
20. Wei, Z.; Shi, Z.; Wen, X.; Li, X.; Qiu, B.; Gu, Q.; Sun, J.; Han, Y.; Luo, H.; Guo, H.; et al. Eliminating Oxygen Releasing of Li-Rich Layered Cathodes by Tuning the Distribution of Superlattice Domain. *Mater. Today Energy* **2022**, *27*, 101039. [CrossRef]
21. Shukla, A.K.; Ramasse, Q.M.; Ophus, C.; Duncan, H.; Hage, F.; Chen, G. Unravelling Structural Ambiguities in Lithium- and Manganese-Rich Transition Metal Oxides. *Nat. Commun.* **2015**, *6*, 8711. [CrossRef] [PubMed]
22. Genevois, C.; Koga, H.; Croguennec, L.; Ménétrier, M.; Delmas, C.; Weill, F. Insight into the Atomic Structure of Cycled Lithium-Rich Layered Oxide Li1.20Mn0.54Co0.13Ni0.13O2 Using HAADF STEM and Electron Nanodiffraction. *J. Phys. Chem. C* **2015**, *119*, 75–83. [CrossRef]
23. Yin, W.; Grimaud, A.; Rousse, G.; Abakumov, A.M.; Senyshyn, A.; Zhang, L.; Trabesinger, S.; Iadecola, A.; Foix, D.; Giaume, D.; et al. Structural Evolution at the Oxidative and Reductive Limits in the First Electrochemical Cycle of Li1.2Ni0.13Mn0.54Co0.13O2. *Nat. Commun.* **2020**, *11*, 1252. [CrossRef]
24. Chen, C.-J.; Pang, W.K.; Mori, T.; Peterson, V.K.; Sharma, N.; Lee, P.-H.; Wu, S.; Wang, C.-C.; Song, Y.-F.; Liu, R.-S. The Origin of Capacity Fade in the Li$_2$MnO$_3$·LiMO_2 (*M* = Li, Ni, Co, Mn) Microsphere Positive Electrode: An *Operando* Neutron Diffraction and Transmission X-Ray Microscopy Study. *J. Am. Chem. Soc.* **2016**, *138*, 8824–8833. [CrossRef]
25. Xie, H.; Cui, J.; Yao, Z.; Ding, X.; Zhang, Z.; Luo, D.; Lin, Z. Revealing the Role of Spinel Phase on Li-Rich Layered Oxides: A Review. *Chem. Eng. J.* **2022**, *427*, 131978. [CrossRef]

26. Bettge, M.; Li, Y.; Gallagher, K.; Zhu, Y.; Wu, Q.; Lu, W.; Bloom, I.; Abraham, D.P. Voltage Fade of Layered Oxides: Its Measurement and Impact on Energy Density. *J. Electrochem. Soc.* **2013**, *160*, A2046–A2055. [CrossRef]

27. Croy, J.R.; Gallagher, K.G.; Balasubramanian, M.; Long, B.R.; Thackeray, M.M. Quantifying Hysteresis and Voltage Fade in XLi2MnO3•(1-x)LiMn0.5Ni0.5O 2 Electrodes as a Function of Li2MnO3 Content. *J. Electrochem. Soc.* **2014**, *161*, A318–A325. [CrossRef]

28. Pechen, L.; Makhonina, E.; Volkov, V.; Rumyantsev, A.; Koshtyal, Y.; Politov, Y.; Pervov, V.; Eremenko, I. Investigation of Capacity Fade and Voltage Decay in Li-Rich Cathode Materials with Different Phase Composition. In Proceedings of the 11th International Conference on Nanomaterials - Research & Application, Brno, Czech Republic, 16–18 October 2019; pp. 144–149.

29. Makhonina, E.V.; Pechen, L.S.; Volkov, V.V.; Rumyantsev, A.M.; Koshtyal, Y.M.; Dmitrienko, A.O.; Politov, Y.A.; Pervov, V.S.; Eremenko, I.L. Synthesis, Microstructure, and Electrochemical Performance of Li-Rich Layered Oxide Cathode Materials for Li-Ion Batteries. *Russ. Chem. Bull.* **2019**, *68*, 301–312. [CrossRef]

30. Fu, F.; Yao, Y.; Wang, H.; Xu, G.-L.; Amine, K.; Sun, S.-G.; Shao, M. Structure Dependent Electrochemical Performance of Li-Rich Layered Oxides in Lithium-Ion Batteries. *Nano Energy* **2017**, *35*, 370–378. [CrossRef]

31. Riekehr, L.; Liu, J.; Schwarz, B.; Sigel, F.; Kerkamm, I.; Xia, Y.; Ehrenberg, H. Effect of Pristine Nanostructure on First Cycle Electrochemical Characteristics of Lithium–Nickel–Cobalt–Manganese-Oxide Cathode Ceramics for Lithium Ion Batteries. *J. Power Sources* **2016**, *306*, 135–147. [CrossRef]

32. Makhonina, E.; Pechen, L.; Medvedeva, A.; Politov, Y.; Rumyantsev, A.; Koshtyal, Y.; Volkov, V.; Goloveshkin, A.; Eremenko, I. Effects of Mg Doping at Different Positions in Li-Rich Mn-Based Cathode Material on Electrochemical Performance. *Nanomaterials* **2022**, *12*, 156. [CrossRef] [PubMed]

33. Pechen, L.S.; Makhonina, E.V.; Medvedeva, A.E.; Rumyantsev, A.M.; Koshtyal, Y.M.; Politov, Y.A.; Goloveshkin, A.S.; Eremenko, I.L. Effect of Dopants on the Functional Properties of Lithium-Rich Cathode Materials for Lithium-Ion Batteries. *Russ. J. Inorg. Chem.* **2021**, *66*, 682–694. [CrossRef]

34. Farahmandjou, M.; Zhao, S.; Lai, W.-H.; Sun, B.; Notten Peter, H.L.; Wang, G. Oxygen Redox Chemistry in Lithium-Rich Cathode Materials for Li-Ion Batteries: Understanding from Atomic Structure to Nano-Engineering. *Nano Mater. Sci.* **2022**; *in press*. [CrossRef]

35. Yu, X.; Lyu, Y.; Gu, L.; Wu, H.; Bak, S.-M.; Zhou, Y.; Amine, K.; Ehrlich, S.N.; Li, H.; Nam, K.-W.; et al. Understanding the Rate Capability of High-Energy-Density Li-Rich Layered Li 1.2 Ni 0.15 Co 0.1 Mn 0.55 O 2 Cathode Materials. *Adv. Energy Mater.* **2014**, *4*, 1300950. [CrossRef]

36. Shanmugam, V.; Natarajan, S.; Lobo, L.S.; Mathur, A.; Sahana, M.B.; Sundararajan, G.; Gopalan, R. Surface Oxygen Vacancy Engineering and Physical Protection by In-Situ Carbon Coating Process of Lithium Rich Layered Oxide. *J. Power Sources* **2021**, *515*, 230623. [CrossRef]

37. Jo, Y.N.; Prasanna, K.; Park, S.J.; Lee, C.W. Characterization of Li-Rich XLi2MnO3·(1−x)Li[MnyNizCo1−y−z]O2 as Cathode Active Materials for Li-Ion Batteries. *Electrochim. Acta* **2013**, *108*, 32–38. [CrossRef]

38. Zhang, S.; Gu, H.; Tang, T.; Du, W.; Gao, M.; Liu, Y.; Jian, D.; Pan, H. Insight into the Synergistic Effect Mechanism between the Li2MO3 Phase and the LiMO2 Phase (M = Ni, Co, and Mn) in Li- and Mn-Rich Layered Oxide Cathode Materials. *Electrochim. Acta* **2018**, *266*, 66–77. [CrossRef]

39. Amalraj, F.; Kovacheva, D.; Talianker, M.; Zeiri, L.; Grinblat, J.; Leifer, N.; Goobes, G.; Markovsky, B.; Aurbach, D. Integrated Materials XLi[Sub 2]MnO[Sub 3]·(1−x)LiMn[Sub 1/3]Ni[Sub 1/3]Co[Sub 1/3]O[Sub 2] (X=0.3, 0.5, 0.7) Synthesized. *J. Electrochem. Soc.* **2010**, *157*, A1121. [CrossRef]

40. Galkin, V.V.; Lanina, E.V.; Shel'deshov, N.V. Dependence of the Electrochemical Characteristics of Lithium-Ion Battery in the Initial State and after Degradation of the Structural Parameters of the Positive Electrode. *Electrochem. Energetics* **2013**, *13*, 103–112. [CrossRef]

nanomaterials

MDPI

Review

Recent Advances in g-C₃N₄ for the Application of Perovskite Solar Cells

Jian Yang [1,2], Yuhui Ma [2], Jianping Yang [1], Wei Liu [1,*] and Xing'ao Li [1,*]

[1] New Energy Technology Engineering Laboratory of Jiangsu Province, Institute of Advanced Materials, School of Science, Nanjing University of Posts and Telecommunications (NJUPT), Nanjing 210023, China
[2] Department of Mathematics and Physics, Nanjing Institute of Technology, Nanjing 211167, China
* Correspondence: liuwei@njupt.edu.cn (W.L.); lixa@njupt.edu.cn (X.L.)

Abstract: In this study, graphitic carbon nitride (g-C₃N₄) was extensively utilized as an electron transport layer or interfacial buffer layer for simultaneously realizing photoelectric performance and stability improvement of perovskite solar cells (PSCs). This review covers the different g-C₃N₄ nanostructures used as additive and surface modifier layers applied to PSCs. In addition, the mechanism of reducing the defect state in PSCs, including improving the crystalline quality of perovskite, passivating the grain boundaries, and tuning the energy level alignment, were also highlighted in this review. Currently, the power conversion efficiency of PSCs based on modified g-C₃N₄ has been increased up to 22.13%, and its unique two-dimensional (2D) package structure has enhanced the stability of PSCs, which can remain stable in the dark for over 1500 h. Finally, the potential challenges and perspectives of g-C₃N₄ incorporated into perovskite-based optoelectronic devices are also included in this review.

Keywords: g-C₃N₄; perovskite solar cells; additive; surface modifier layer

Citation: Yang, J.; Ma, Y.; Yang, J.; Liu, W.; Li, X. Recent Advances in g-C₃N₄ for the Application of Perovskite Solar Cells. *Nanomaterials* **2022**, *12*, 3625. https://doi.org/10.3390/nano12203625

Academic Editor: Iván Mora-Seró

Received: 24 September 2022
Accepted: 13 October 2022
Published: 16 October 2022

Publisher's Note: MDPI stays neutral with regard to jurisdictional claims in published maps and institutional affiliations.

1. Introduction

Fossil fuels created a huge amount of pollution in the environment. Solar energy, as one of the main sources of clean and renewable energy, can solve both environmental pollution and energy demand. Perovskite solar cells (PSCs) are a new type of photovoltaic device that can directly convert solar energy into electrical energy [1,2]. Compared with silicon-based solar cells, the manufacturing cost of PSCs is lower, and their photoelectric characteristics are more prominent [3–5]. Moreover, the PSCs can be attached on the flexible substrate, significantly expanding its application scenarios [6,7]. Currently, the lab-scale power conversion efficiency (PCE) has been certified to boost efficiency up to 25.5% [8]. For realizing the commercialization of PSCs as soon as possible [9,10], several urgent breakthroughs are required, such as obtaining a higher PCE, longer-term stability, and eco-friendliness [11–17].

Generally, the structure of PSCs are composed of a hole transport layer (HTL), a perovskite layer, and an electron transport layer (ETL), along with counter electrodes [18,19]. The defect-induced recombination of photo-generated carriers significantly impacts the extraction and transportation of charges in PSCs [20–23], which severely diminishes the performance of PSCs, in including short-circuit current density (Jsc), PCE, current hysteresis and stability, etc. [24,25]. For PSCs, defects are mainly located at the interfaces of ETL/perovskite/HTL, as well as at the grain boundaries (GBs) of perovskite films [26–28]. Many routes have been explored to reduce defects in PSCs, including additive and interface engineering, etc. [29–40]. Meanwhile, it is worth noting that the stability and eco-friendliness of the materials used in additive or interface engineering should also be taken into account.

Due to the polymeric feature, the surface of g-C₃N₄ abounds with N, -NH₂, -NH-, and other groups, which are facile functionalized by surface modification [41–43]. Thus, g-C₃N₄ has been utilized in PSCs as an additive or interface engineering material, as depicted in Figure 1a [44,45]. Additionally, different nanostructures (bulk, nanosheets, nanoparticle and quantum dots, etc.) of g-C$_3$N$_4$ reveal diversified photoelectric characteristics [46,47]. In this work, we summarized the role of g-C$_3$N$_4$ in PSCs. The main role are as follows: first, facilitating electron transport or perovskite growth via adjusting the energy level or roughness of ETL, corresponding. Second, decreasing the deep electron defect state of PSCs via improving the crystalline quality of perovskite. Moreover, we looked into the development trend of applying g-C$_3$N$_4$ in perovskite-based optoelectronic devices.

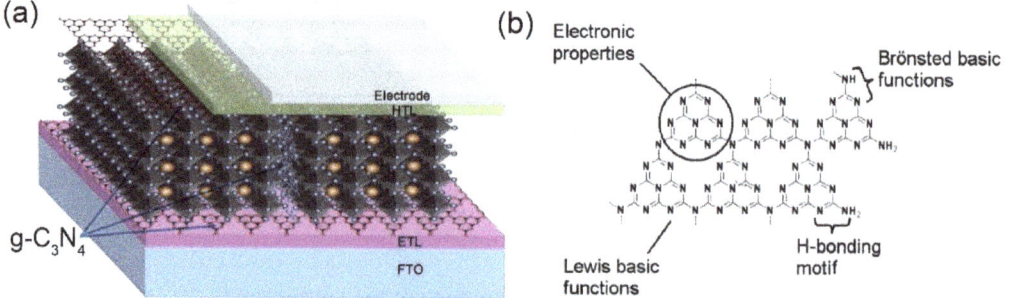

Figure 1. (a) g-C$_3$N$_4$ as an additive or interface engineering material in PSCs, (b) surface groups and properties of tri-s-triazine-based g-C$_3$N$_4$ (reproduced with permission [42]; copyright 2020, American Chemical Society).

2. g-C$_3$N$_4$ as an Additive in Perovskite Films

g-C$_3$N$_4$ is a metal–free, non–toxic, and high–yield polymeric semiconductor with an around 2.7 eV bandgap. More importantly, it has excellent thermal and chemical stability [46,48–50]. Consequently, g–C$_3$N$_4$ has been broadly used in pollutant degradation, sensing, optoelectronic devices, and other fields [51–60]. Figure 1b plotted the molecule structure diagrams of g–C$_3$N$_4$, based on tri–s–triazine connection patterns. The hexagon triazine ring is comprised of sp^2 hybridized N and C atoms, with hydrogen bonds between the –NH$_2$ groups and the N edge atoms, and they are linked at the end with a C–N bond, creating an extended network–like planar structure.

2.1. Pure g–C$_3$N$_4$ Nanosheets as an Additive

g–C$_3$N$_4$ nanosheets are π–conjugated nanomaterials with two–dimensional structure, and a larger specific surface area, thus conducting the separation of photo–generated charges [61,62]. In 2018, Jiang et al. reported pure g–C$_3$N$_4$ nanosheets mixed into a perovskite precursor solution as additives [63], suppressing nucleation and slowing down the growth of perovskite during the crystallization process; this results in the g–C$_3$N$_4$:CH$_3$NH$_3$PbI$_3$ films have larger grain sizes and a lower defect density (n$_t$). In 2019, Liao et al. used a method similar to that of Jiang et al. to add g–C$_3$N$_4$ nanosheets into perovskite films [64]. Importantly, they revealed the location of g–C$_3$N$_4$ in perovskite. Meanwhile, the mechanisms of defect passivation and charge extraction by g–C$_3$N$_4$ were clarified. From Figure 2a, it can be seen that g-C$_3$N$_4$ was uniformly anchored at the surface of the GBs, and the dangling Pb^{2+} can coordinate with the N atom in g-C$_3$N$_4$, retarding the crystallization of perovskite [65,66]. Moreover, the conductive g-C$_3$N$_4$ network condensed at the GBs can act as an efficient carrier shuttle, facilitating the electron transport. From the TRPL spectra, as presented in Figure 2b, for g-C$_3$N$_4$ additive CH$_3$NH$_3$PbI$_3$ films, the PL lifetime reduce to 17 ns, which is less than that of the CH$_3$NH$_3$PbI$_3$ films, indicating ultrafast photo–excited carrier transport due to the addition of g–C$_3$N$_4$ [67].

Yang et al. added g–C$_3$N$_4$ into carbon-based PSCs [68]. Additionally, the insulating layer was prepared on the surface of the ETL by spin-coating Al$_2$O$_3$ [69–71]. Figure 2c presents the fabrication procedure of the device. From the J-V curve, as shown in Figure 2d, it can be noted that the Jsc has barely changed after incorporating the Al_2O_3 layer, which suggests that the conductive g-C_3N_4 network at the GBs can provide electrons via a carrier shuttle. This can be proved from the TRPL spectra, as shown in Figure 2e. In the field of wearable electronics, g-C_3N_4 nanosheets were applied into flexible tin-based PSCs [72]. Figure 2f plots the device configuration: PDMS/hc-PEDOT:PSS/PEDOT:PSS/FASnI$_3$/C$_{60}$/BCP/Ag. The network structure of g-C_3N_4 indicates a better lattice match with formamidine cation, which is more conducive to the crystallization of FASnI$_3$ films. The illustration of bonding and passivation between FASnI$_3$ and g-C_3N_4 is shown in Figure 2g. Finally, the stabilized PCE of 8.56% was obtained.

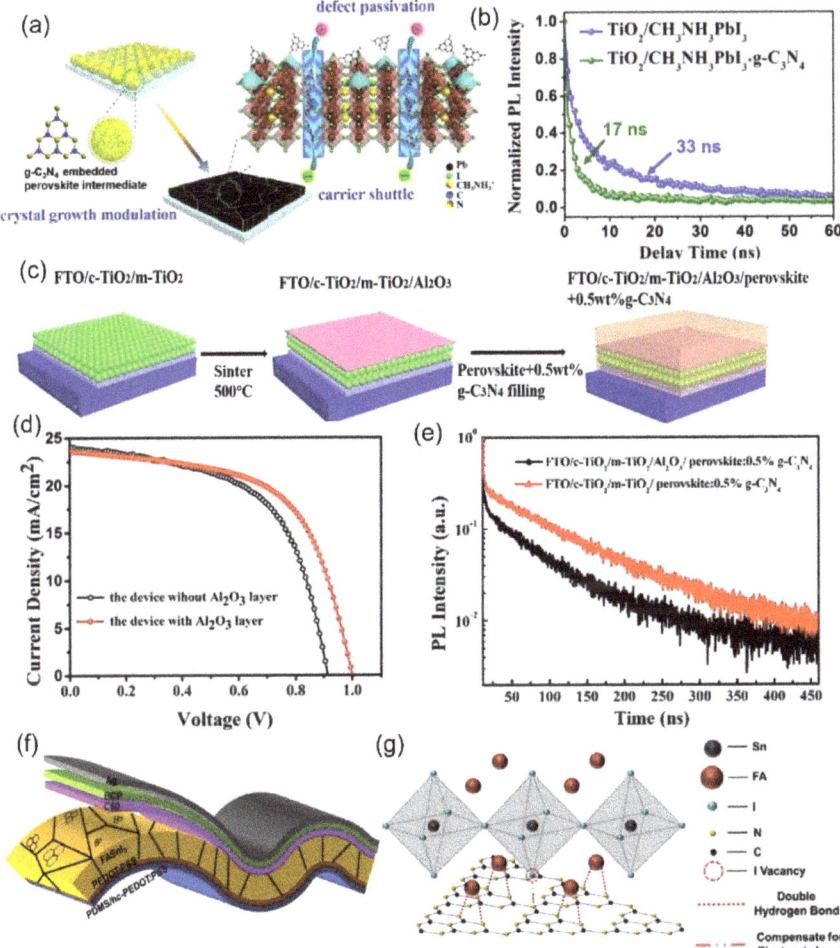

Figure 2. (**a**) Schematic illustration of the functions of g–C_3N_4 as an additive in perovskite films. (**b**) TRPL lifetime of g–C_3N_4:$CH_3NH_3PbI_3$ and $CH_3NH_3PbI_3$ films (reproduced with permission [64]; copyright 2019, Royal Society of Chemistry). (**c**) Fabrication procedure of a device with FTO/c–TiO_2/m–TiO_2/Al_2O_3/g–C_3N_4:$CH_3NH_3PbI_3$/carbon. (**d**) J–V curves of the devices and (**e**) TRPL spectra of perovskite films (reproduced with permission [68]; copyright 2019, Elsevier). (**f**) Device configuration: PDMS/hc–PEDOT:PSS/PEDOT:PSS/FASnI$_3$/C$_{60}$/BCP/Ag. (**g**) Illustration of bonding and passivation between FASnI$_3$ and g–C_3N_4 (reproduced with permission [72]; copyright 2021, John Wiley and Sons).

2.2. Functionalized g–C₃N₄ Nanosheets as an Additive

The pure g–C_3N_4 nanosheets tend to agglomerate in the organic or aqueous environment due to the robust van der Waals interactions, resulting in lower dispersibility [49,73]. However, the functionalized g–C_3N_4 nanosheets show good dispersity in liquid, mainly caused by the electrostatic repulsion of the charged groups [42]. Here, there are two main methods to achieve functionalized g–C_3N_4 nanosheets—doped, and surface modified [74,75].

In 2019, Cao et al. reported that iodine–doped g–C_3N_4 (g–CNI) was added into triple cation perovskite films as an additive [76]. Due to the fact that doped iodine can coordinate with dangling Pb^{2+} at GBs, the trap states in PSCs were effectively passivated [77,78]. Therefore, they achieved high–quality perovskite films with fewer trap states [65,79,80]. Figure 3a displays the mechanism of g–CNI modified PSCs. From the XPS spectra, as shown in Figure 3b, for perovskite with g–CNI, the I 3d signal is higher than that of the ref., suggesting the incorporation of g–CNI into the perovskite [81]. After adding g–CNI, the n_t reduced to 1.07×10^{16} cm^3 from 1.43×10^{16} cm^3, with the maximum PCE of up to 18.28%.

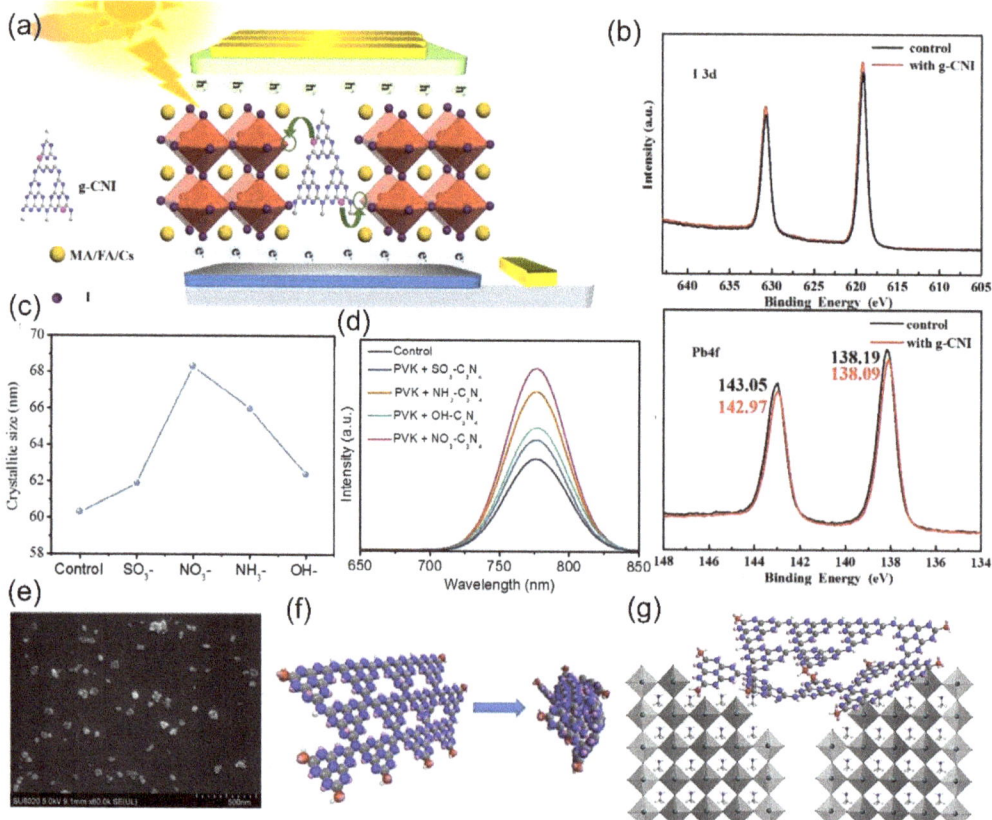

Figure 3. (**a**) Mechanism illustration of g–CNI modified PSCs. (**b**) XPS spectra of Pb 4f and I 3d in perovskite (reproduced with permission [76]; copyright 2019, Royal Society of Chemistry). (**c**) Average crystallite size. (**d**) PL spectra of perovskite films with different functionalized g–C_3N_4 (reproduced with permission [41]; copyright 2019, John Wiley and Sons). (**e**) FESEM images of U–g–C_3N_4 nanoparticles. (**f**) Scheme of g–C_3N_4 fragments coils into U–g–C_3N_4 nanoparticles. (**g**) Scheme of the U–g–C_3N_4 self–recognizing grain boundaries of $CH_3NH_3PbI_3$ films (reproduced with permission [82]; copyright 2019, Elsevier).

Nanomaterials **2022**, *12*, 3625

Li et al. reported that surface–modified g–C$_3$N$_4$ with various organic groups (–NO$_3$, –NH$_3$, –SO$_3$ and –OH) was mixed into perovskite films [41], which can improve crystalline quality and passivate defects state at the GBs. Specifically, for NO$_3$–C$_3$N$_4$–based perovskite, the average crystallite size of up to 68 nm was achieved, as exhibited in Figure 3c, leading to a decent photovoltaic performance, which can be proved by the PL spectra, as shown in Figure 3d. It is worth noting that the PL intensity of NO$_3$-C$_3$N$_4$–based perovskite film is the strongest compared to the other perovskite films. This can be ascribed to the better crystallinity of perovskite with the NO$_3$–C$_3$N$_4$ addition, as well as a reduction in the trap states density. As a result, for p–i–n PSCs based on NO$_3$–C$_3$N$_4$, the best PCE obtained was up to 20.08%.

2.3. Ultrafine g–C$_3$N$_4$ Nanoparticles as an Additive

In 2019, Liu et al. reported that g–C$_3$N$_4$ nanoparticles were introduced into perovskite films as an additive [82]. Here, the ultrafine size of g–C$_3$N$_4$ nanoparticles is about 20–50 nm, which were successfully synthesized with exfoliated g–C$_3$N$_4$ nanosheets. The surface of ultrafine g–C$_3$N$_4$ nanoparticles (U–g–C$_3$N$_4$) is rich in O–H or N–H groups, which can easily bond with N–H bonds on CH$_3$NH$_3$PbI$_3$ GBs [83]. Thus, it can self–recognize GBs and adhere to them, decreasing the deep electron trap state. The FESEM images of U–g–C$_3$N$_4$ nanoparticles are plotted in Figure 3e. Figure 3f presents the scheme of U–g–C$_3$N$_4$ nanoparticles from g–C$_3$N$_4$ fragments. Figure 3g describes the scheme of the U–g–C$_3$N$_4$ self–recognizing CH$_3$NH$_3$PbI$_3$ GBs. Finally, the champion PCE of U–g–C$_3$N$_4$ based planar PSCs is up to 15.8%.

3. g–C$_3$N$_4$ as a Surface Modifier Layer

3.1. g–C$_3$N$_4$ Quantum Dots (g–CNQD) as Modifier Layer

g–C$_3$N$_4$ quantum dots are a type of zero–dimensional nanomaterial in which electrons and holes cannot move freely [84]. The tiny particle size creates its unique size effect, macroscopic quantum tunnel effect, edge effect [85], etc. In 2020, Chen et al. prepared g–CNQD via acid etching and hydrothermal cure [86]; the diameter of g–CNQD were about 5–10 nm, and they were added into an SnO$_2$ colloid precursor, forming nanocomposite ETL (G–SnO$_2$). Figure 4a exhibits the position of g–CNQD in SnO$_2$. Tiny and conductive g–CNQD could reorganize the electronic density distribution of SnO$_2$. The charge density difference between G–SnO$_2$ and SnO$_2$ is displayed in Figure 4b, which was obtained using the density functional theory. It can be seen that the vacancies surrounding the three Sn atoms interact with g–C$_3$N$_4$, and an obvious charge redistribution occurs around the oxygen vacancy, resulting in the elimination of the trap state defects [87]. Importantly, g–CNQD can effectively adjust the Fermi level of ETL, promoting electron transport [88,89]. The energy–level diagrams are shown in Figure 4c. For G–SnO$_2$ based planar PSCs, the PCE is up to 22.13% with a Voc of 1.176 V. They also exhibit excellent long–term stability under ambient conditions, e.g., about 60% humidity [90].

Liu et al. published a similar report on g–CNQD. The g–CNQD was synthesized with urea and sodium citrate [91], the diameter was about 10–30 nm, and it was well monodispersed. In this work, g–CNQD were intercalated into the ETL/perovskite layer, which facilitates the formation of high–quality perovskite films due to the smoother surface of ETL [92,93]. Figure 4d presents the schematic illustration of the g–CNQD–based device. The roughness of the ETL was evaluated by AFM, as shown in Figure 4e. After intercalating g–CNQD, the root mean square roughness reduced from 17.5 nm to 12.8 nm, suggesting that a smoother ETL was obtained, which is conducive to perovskite growth [94]. From the SEM images and XRD spectra of the perovskite films as shown in Figure 4f,g, it can be seen that the g–CNQD based perovskite films have purer phases, fewer GBs, and lower trap states. Finally, the maximum PCE is noted, up to 21.23% under full air–processing, and without apparent current hysteresis.

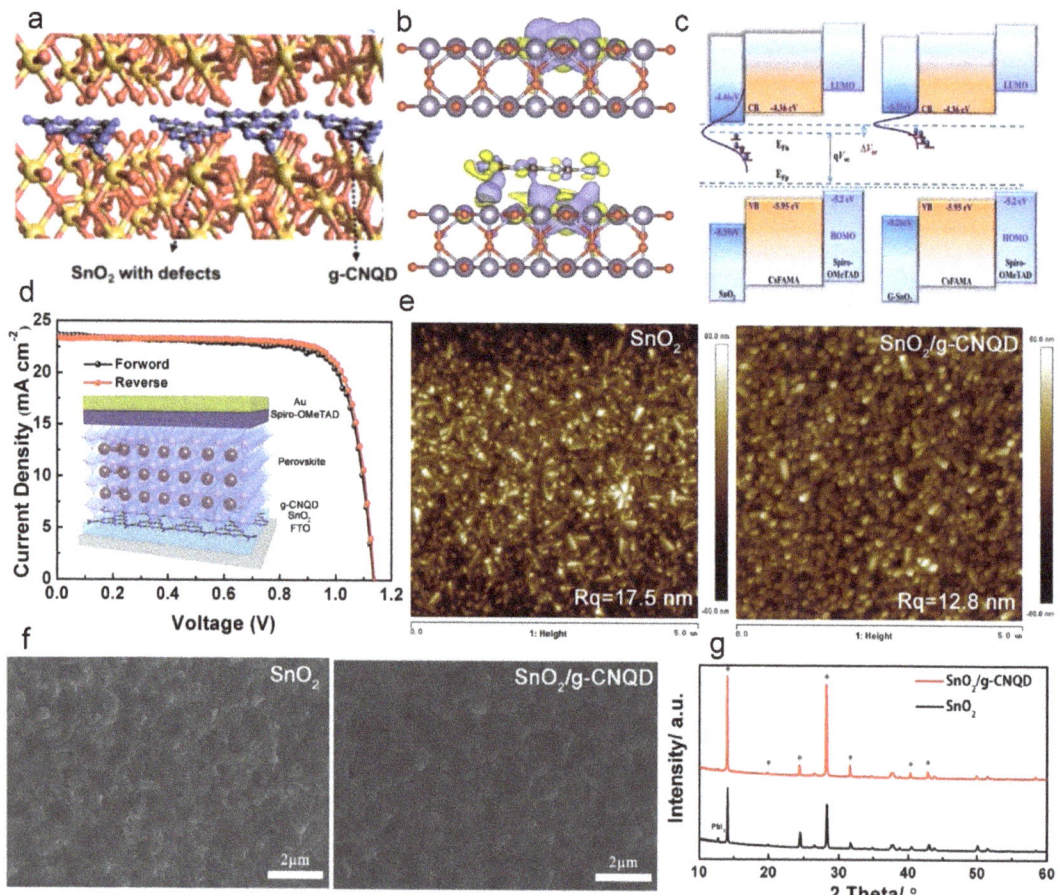

Figure 4. (**a**) The position of G–CNQD in SnO$_2$. (**b**) The side view for the charge density difference of SnO$_2$ (above) and G–SnO$_2$ (below) with oxygen vacancy; the yellow and cyan areas indicate electron depletion and accumulation, respectively. (**c**) Energy band alignment of the devices (reproduced with permission [86]; copyright 2020, Royal Society of Chemistry). (**d**) Schematic illustration of the devices with a g–CNQD layer and a J–V curve. (**e**) AFM images of pristine SnO$_2$ film and SnO$_2$/g–CNQD films. (**f**) SEM images and (**g**) XRD spectra of perovskite films based on different ETL. The asterisks indicate the main peaks of perovskites structure (reproduced with permission [91]; copyright 2020, Elsevier).

3.2. g–C$_3$N$_4$ Nanosheets as a Modified Layer

In 2020, Liu et al. used multilayer g–C$_3$N$_4$ to simultaneously modify the upper and lower interfaces of perovskite [95]. For the perovskite/HTL interface, the dangling Pb^{2+} can coordinate with lone–pair electrons on g–C$_3$N$_4$, reducing the defects state at the perovskite film surface. For the ETL/perovskite interface, the Gibbs free energy of SnO$_2$ surface was decreased, which facilitates the preparation of flat and non–pinhole perovskite films [96,97]. Figure 5a displays the schematic diagram of dual–modified PSCs with g–C$_3$N$_4$. The improvement of perovskite films can be seen from SEM, as shown in Figure 5b. A maximum PCE of 19.67% was obtained for planar PSCs with longer–term stability. The main reason for this was that g–C$_3$N$_4$ could reduce the surface defects of perovskite films, thus decreasing the migration of iodide and ion mobilization within the perovskite lattices.

In 2021, Yang et al. adopted g–C$_3$N$_4$ nanosheets as a modified layer [98], and the work function of ETL was finely tuned, as shown in Figure 5c, resulting in the enhancement of Voc from 1.01 to 1.11 V, and diminishing the current hysteresis of PSCs. Therefore, the maximum PCE was boosted to 19.55% from its initial 15.81%. Yang et al. reported new buried layers for efficient perovskite [99], which are composed of a mixture of g–C$_3$N$_4$ and SnO$_2$. Due to the fact that amine–rich g–C$_3$N$_4$ can promote the prenucleation of the Pb–related intermediates, the vertical crystallization of perovskite films were obviously optimized, exhibiting superior carrier transmission characteristics. Figure 5d presents the schematic illustration of vertical carrier transportation via buried manipulation.

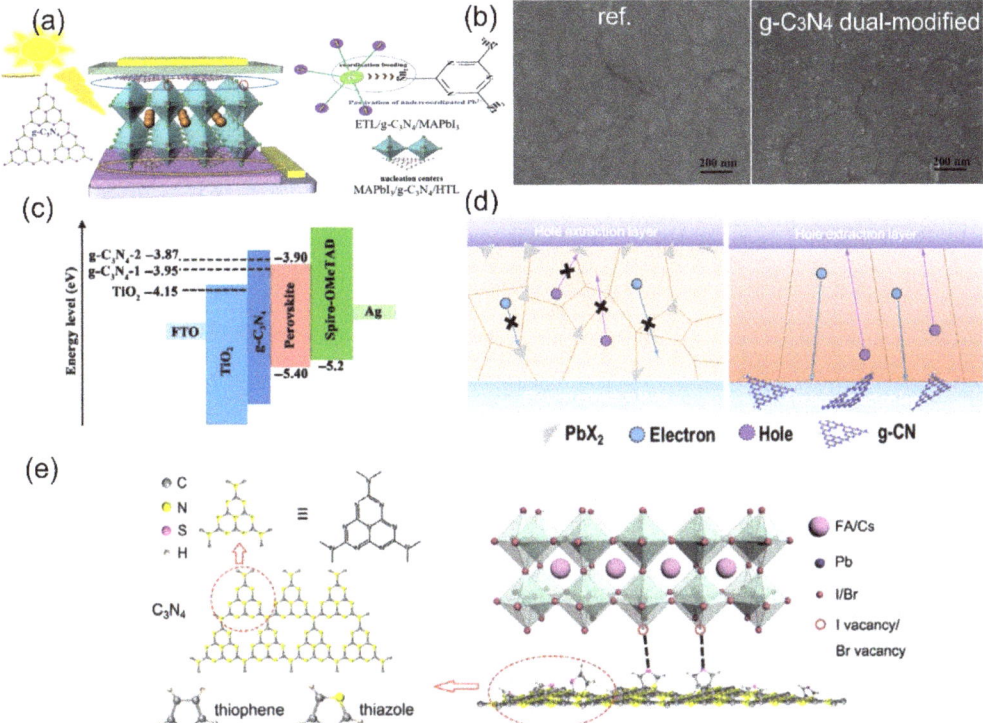

Figure 5. (**a**) Schematic diagram of g–C$_3$N$_4$ dual–modified PSCs. (**b**) SEM of perovskite films (reproduced with permission [95]; copyright 2018, Royal Society of Chemistry). (**c**) Energy band alignment of the device. The E$_F$ level of the ETL, without and with g–C$_3$N$_4$, are represented with a dotted line (reproduced with permission [98]; copyright 2021, Springer Nature). (**d**) Schematic of the vertical carrier transportation of the perovskite films via buried manipulation (reproduced with permission [99]; copyright 2021, John Wiley and Sons). (**e**) Schematic illustration of the possible interface defect sites (reproduced with permission [100]; copyright 2021, John Wiley and Sons).

3.3. Functionalized g–C$_3$N$_4$ as a Modified Layer

In 2019, Cruz et al. prepared thiazole–modified g–C$_3$N$_4$ via exfoliation treatment [101], then intercalated it in p–i–n PSCs as ETL. The charge recombination in the interface was suppressed due to the enhanced energy level of electronic interface [88]. Finally, the Voc of 1.09 V and the Jsc of 20.17 mA/cm^2 were achieved. In 2021, Wang et al. prepared the functionalized g–C$_3$N$_4$ with thiophene or thiazole by thermal treatment [100]. Then, the functionalized g–C$_3$N$_4$ were intercalated between ETL and the perovskite layer. Notably, there was a well–matched energy level in the device. Due to the strong chemical affinity between Pb^{2+} and N or S atoms [67], the defect state in the device was efficiently passivated. Figure 5e shows the possible interface defect sites in perovskite, as well as the passivation of thiophene or thiazole. For the thiazole $g–C_3N_4$based device, the maximum PCE increased to 19.23% from pristine 13.42%, with the Voc increasing from 1.02 V. to 1.11 V.

4. Conclusions and Future Perspectives

In this paper, we summarize the recent progress of $g–C_3N_4$ application in PSCs. $g–C_3N_4$ is an eco–friendly polymeric semiconductor with a suitable bandgap. Recently, it has been widely applied in PSCs, which can reduce defect states in PSCs, and simultaneously enhance the PCE and long–term stability of PSCs. Specifically, different nanostructures of $g–C_3N_4$ (i.e., nanosheets, nanoparticles, and QDs) used as additive and surface modifier layers have been discussed in detail. The performance of the devices is listed in Table 1.

Pure $g–C_3N_4$ nanosheets are a kind of two–dimensional nanomaterial with large N, $–NH_2$, $–NH$, and other groups, and the N atom in $g–C_3N_4$ can coordinate a bond with dangling Pb^{2+}, increasing the perovskite grain size. Moreover, when it functions as an additive, the conductive $g–C_3N_4$ network acts as an efficient carrier shuttle, facilitating electron transport. $g–C_3N_4$ nanoparticles are formed by the self–coiling of the exfoliated $g–C_3N_4$ nanosheets, with an ultra–fine size (20–50 nm), and the surface of the $g–C_3N_4$ nanoparticles contains abundant O–H or N–H groups. Therefore, $g–C_3N_4$ nanoparticles can magically self–recognize GBs in perovskite films, decreasing the deep electron trap state of PSCs. Functionalized $g–C_3N_4$ nanosheets can also be achieved from nanosheets, but the former shows better dispersity in organic solvent. They exhibit a similar effect as additives. Specifically, when used as a modifier layer, they can adjust the energy level alignment of PSCs, making the electron transport more efficient. g–CNQD has a tiny particle size, which can adjust the energy level or roughness of the ETL, facilitating electron transport or perovskite growth, corresponding. These features give the PSCs potential for consideration as next–generation photovoltaic devices, as the most promising means of simultaneously solving environmental pollution and energy demand. The passivation of the perovskite layer is mainly used to improve the PCE and the long–term stability of PSCs. $g–C_3N_4$ can be added in different solutions to prepare efficient PSCs with long–term stability, which have the potential to compete with other conventional silicon cells in the future. Meanwhile, $g–C_3N_4$ shows great promise in solving external and internal concerns, including packaging, additional technology, and reducing charge recombination. Thus, the $g–C_3N_4$ can increase the grain size of perovskite and passivate the interface defect, making it more conducive to charge extraction. However, the development of the $g–C_3N_4$ nanostructure is still in its early stages. Additional methods for improving the efficiency and stability of PSCs require further exploration. We expect to develop new additives exhibiting eco–friendliness, long–term stability, and compatibility with flexible substrates, as well as other new strategies to improve the performance of perovskite. This effort should be closely connected to application of dedicated defect passivation strategies to produce high–performance and enduring stable PSCs.

Table 1. Recent development of g-C$_3$N$_4$ based PSCs photovoltaic performance.

Structure	PCE (%)	Voc (V)	Jsc (mA·cm^{-2})	FF (%)	Ref.	Year
ITO/PTAA/NO$_3$-g-C$_3$N$_4$:CsFAMAPbI$_{3-x}$Br$_x$/PCBM/BCP/Ag	20.08	1.11	22.84	79.20	[41]	2019
FTO/c-TiO$_2$/g-C$_3$N$_4$:MAPbI$_3$/spiro-OMeTAD/MoO$_3$/Ag	19.49	1.07	24.31	74.0	[63]	2018
FTO/c-TiO$_2$/g-C$_3$N$_4$:MAPbI$_3$/spiro-OMeTAD/Au	21.10	1.16	23.00	79.0	[64]	2019
FTO/c-TiO$_2$/m-TiO$_2$/g-C$_3$N$_4$:CsPbBr$_3$/carbon	8.00	1.277	7.80	80.32	[66]	2021
FTO/c-TiO$_2$/m-TiO$_2$/Al$_2$O$_3$/g-C$_3$N$_4$:MAPbI$_3$/carbon	14.34	1.00	23.80	60.1	[68]	2019
PDMS/hc-PEDOT:PSS/PEDOT:PSS/g-C$_3$N$_4$:FASnI$_3$/C$_{60}$/BCP/Ag	8.56	0.621	20.68	66.68	[72]	2021
FTO/TiO$_2$/G-CNI:CsFAMAPbI$_{3-x}$Br$_x$/spiro-OMeTAD/Au	18.28	1.07	22.97	74.0	[76]	2019
FTO/c-TiO$_2$/U-g-C$_3$N$_4$:MAPbI$_3$/spiro-OMeTAD/Au	15.80	1.10	23.20	62.0	[82]	2019
ITO/CNQDs:SnO$_2$/CsFAMAPbI$_{3-x}$Br$_x$/Spiro-MeOTAD/Au	22.13	1.18	24.03	78.3	[86]	2020
FTO/SnO$_2$/CNQDs/(FA/MA/Cs)PbI$_{3-(x+y)}$Br$_x$Cl$_y$/spiro-OMeTAD/Au	21.23	1.14	23.39	79.6	[91]	2020
FTO/SnO$_2$/g-C$_3$N$_4$/MAPbI$_3$/g-C$_3$N$_4$/Sspiro-OMeTAD/Au	19.67	1.14	21.45	80.7	[95]	2020
FTO/c-TiO$_2$/m-TiO$_2$/g-C$_3$N$_4$ nanosheets/MAPbI$_3$/Carbon	11.37	1.02	16.91	66	[97]	2021
FTO/c-TiO$_2$/g-C$_3$N$_4$/MAPbI$_3$/spiro-OMeTAD/Ag	19.55	1.11	23.69	74.0	[98]	2021
ITO/g-C$_3$N$_4$:SnO$_2$/FA$_{0.85}$MA$_{0.11}$Cs$_{0.04}$PbI$_{2.67}$Br$_{0.33}$:xPbI$_2$/spiro-MeOTAD/Au	21.54	1.19	23.21	78	[99]	2021
ITO/PTAA/MAPbI$_3$/PC$_{60}$BM/CMB-vTA/AZO/Ag	17.15	1.09	20.17	78.03	[100]	2019
FTO/TiO$_2$/thiazole-C$_3$N$_4$/(FAPbI$_3$)$_{0.875}$(CsPbBr$_3$)$_{0.125}$/spiro-OMeTAD/Ag	19.23	1.11	22.50	77	[101]	2021

Author Contributions: Conceptualization, J.Y. (Jian Yang); resources, W.L. and X.L.; writing—original draft preparation, J.Y. (Jianping Yang); writing—review and editing, W.L.; visualization, Y.M.; supervision, X.L. All authors have read and agreed to the published version of the manuscript.

Funding: This work was supported by the Ministry of Education of China (IRT1148), the National Natural Science Foundation of China (52172205, 51872145, U1732126, 11804166, 51602161, 51372119), the China Postdoctoral Science Foundation (2022M711685), the Priority Academic Program Development of Jiangsu Higher Education Institutions (YX03001), the Natural Science Research Project of Jiangsu Universities (22KJB510035), and the Start-up Fund of Nanjing University of Posts and Telecommunications (NUPTSF, NY221029).

Institutional Review Board Statement: Not applicable.

Informed Consent Statement: Not applicable.

Data Availability Statement: No new data were created or analyzed in this study. Data sharing is not applicable to this article.

Conflicts of Interest: The authors declare no conflict of interest.

References

1. Zhang, H.; Lu, Y.; Han, W.; Zhu, J.; Zhang, Y.; Huang, W. Solar energy conversion and utilization: Towards the emerging photo-electrochemical devices based on perovskite photovoltaics. *Chem. Eng. J.* **2020**, *393*, 124766. [CrossRef]
2. Kovalenko, M.V.; Protesescu, L.; Bodnarchuk, M.I. Properties and potential optoelectronic applications of lead halide perovskite nanocrystals. *Science* **2017**, *358*, 745–750. [CrossRef] [PubMed]
3. Wang, F.; Yang, M.; Zhang, Y.; Du, J.; Han, D.; Yang, L.; Fan, L.; Sui, Y.; Sun, Y.; Meng, X.; et al. Constructing m-TiO$_2$/a-WO$_x$ hybrid electron transport layer to boost interfacial charge transfer for efficient perovskite solar cells. *Chem. Eng. J.* **2020**, *402*, 126303. [CrossRef]
4. Chen, H.; Ye, F.; Tang, W.; He, J.; Yin, M.; Wang, Y.; Xie, F.; Bi, E.; Yang, X.; Grätzel, M.; et al. A solvent- and vacuum-free route to large-area perovskite films for efficient solar modules. *Nature* **2017**, *550*, 92–95. [CrossRef]
5. Li, H.; Chen, C.; Jin, J.; Bi, W.; Zhang, B.; Chen, X.; Xu, L.; Liu, D.; Dai, Q.; Song, H. Near-infrared and ultraviolet to visible photon conversion for full spectrum response perovskite solar cells. *Nano Energy* **2018**, *50*, 699–709. [CrossRef]
6. Han, G.S.; Jung, H.S.; Park, N.-G. Recent cutting-edge strategies for flexible perovskite solar cells toward commercialization. *Chem. Commun.* **2021**, *57*, 11604–11612. [CrossRef]
7. Tang, G.; Yan, F. Recent progress of flexible perovskite solar cells. *Nano Today* **2021**, *39*, 101155. [CrossRef]
8. Kim, M.; Jeong, J.; Lu, H.; Lee, T.K.; Eickemeyer, F.T.; Liu, Y.; Choi, I.W.; Choi, S.J.; Jo, Y.; Kim, H.-B.; et al. Conformal quantum dot–SnO$_2$ layers as electron transporters for efficient perovskite solar cells. *Science* **2022**, *375*, 302–306. [CrossRef]
9. Cheng, Y.; Ding, L. Pushing commercialization of perovskite solar cells by improving their intrinsic stability. *Energy Environ. Sci.* **2021**, *14*, 3233–3255. [CrossRef]
10. Rong, Y.G.; Hu, Y.; Mei, A.Y.; Tan, H.R.; Saidaminov, M.I.; Seok, S.I.; McGehee, M.D.; Sargent, E.H.; Han, H.W. Challenges for commercializing perovskite solar cells. *Science* **2018**, *361*, eaat8235. [CrossRef]
11. Wu, T.; Qin, Z.; Wang, Y.; Wu, Y.; Chen, W.; Zhang, S.; Cai, M.; Dai, S.; Zhang, J.; Liu, J.; et al. The Main Progress of Perovskite Solar Cells in 2020–2021. *Nano-Micro Lett.* **2021**, *13*, 152. [CrossRef]
12. Wang, D.; Wright, M.; Elumalai, N.K.; Uddin, A. Stability of perovskite solar cells. *Sol. Energy Mater. Sol. Cells* **2016**, *147*, 255–275. [CrossRef]
13. Babayigit, A.; Ethirajan, A.; Muller, M.; Conings, B. Toxicity of organometal halide perovskite solar cells. *Nat. Mater.* **2016**, *15*, 247–251. [CrossRef]
14. Kim, M.-c.; Ham, S.-Y.; Cheng, D.; Wynn, T.A.; Jung, H.S.; Meng, Y.S. Advanced Characterization Techniques for Overcoming Challenges of Perovskite Solar Cell Materials. *Adv. Energy Mater.* **2021**, *11*, 2001753. [CrossRef]
15. Fan, R.; Huang, Y.; Wang, L.; Li, L.; Zheng, G.; Zhou, H. The Progress of Interface Design in Perovskite-Based Solar Cells. *Adv. Energy Mater.* **2016**, *6*, 1600460. [CrossRef]
16. Arjun, V.; Muthukumaran, K.P.; Ramachandran, K.; Nithya, A.; Karuppuchamy, S. Fabrication of efficient and stable planar perovskite solar cell using copper oxide as hole transport material. *J. Alloys Compd.* **2022**, *923*, 166285. [CrossRef]
17. Li, S.; Zhang, X.; Xue, X.; Wu, Y.; Hao, Y.; Zhang, C.; Liu, Y.; Dai, Z.; Sun, Q.; Hao, Y. Importance of tin (II) acetate additives in sequential deposited fabrication of Sn-Pb-based perovskite solar cells. *J. Alloys Compd.* **2022**, *904*, 164050. [CrossRef]
18. Zhao, Z.; Sun, W.; Li, Y.; Ye, S.; Rao, H.; Gu, F.; Liu, Z.; Bian, Z.; Huang, C. Simplification of device structures for low-cost, high-efficiency perovskite solar cells. *J. Mater. Chem. A* **2017**, *5*, 4756–4773. [CrossRef]
19. Correa-Baena, J.-P.; Abate, A.; Saliba, M.; Tress, W.; Jesper Jacobsson, T.; Grätzel, M.; Hagfeldt, A. The rapid evolution of highly efficient perovskite solar cells. *Energy Environ. Sci.* **2017**, *10*, 710–727. [CrossRef]

20. Ma, Y.; Deng, K.; Gu, B.; Cao, F.; Lu, H.; Zhang, Y.; Li, L. Boosting Efficiency and Stability of Perovskite Solar Cells with CdS Inserted at TiO_2/Perovskite Interface. *Adv. Mater. Interfaces* **2016**, *3*, 1600729. [CrossRef]
21. Taheri, S.; Ahmadkhan kordbacheh, A.; Minbashi, M.; Hajjiah, A. Effect of defects on high efficient perovskite solar cells. *Opt. Mater.* **2021**, *111*, 110601. [CrossRef]
22. Ke, W.; Fang, G.; Wan, J.; Tao, H.; Liu, Q.; Xiong, L.; Qin, P.; Wang, J.; Lei, H.; Yang, G.; et al. Efficient hole-blocking layer-free planar halide perovskite thin-film solar cells. *Nat. Commun.* **2015**, *6*, 6700. [CrossRef]
23. Li, B.; Ferguson, V.; Silva, S.R.P.; Zhang, W. Defect Engineering toward Highly Efficient and Stable Perovskite Solar Cells. *Adv. Mater. Interfaces* **2018**, *5*, 1800326. [CrossRef]
24. Tavakoli, M.M.; Yadav, P.; Tavakoli, R.; Kong, J. Surface Engineering of TiO_2 ETL for Highly Efficient and Hysteresis-Less Planar Perovskite Solar Cell (21.4%) with Enhanced Open-Circuit Voltage and Stability. *Adv. Energy Mater.* **2018**, *8*, 1800794. [CrossRef]
25. Montoya, D.M.; Pérez-Gutiérrez, E.; Barbosa-Garcia, O.; Bernal, W.; Maldonado, J.-L.; Percino, M.J.; Meneses, M.-A.; Cerón, M. Defects at the interface electron transport layer and alternative counter electrode, their impact on perovskite solar cells performance. *Sol. Energy* **2020**, *195*, 610–617. [CrossRef]
26. Dong, Y.; Li, W.; Zhang, X.; Xu, Q.; Liu, Q.; Li, C.; Bo, Z. Highly Efficient Planar Perovskite Solar Cells Via Interfacial Modification with Fullerene Derivatives. *Small* **2016**, *12*, 1098–1104. [CrossRef]
27. Jain, S.M.; Phuyal, D.; Davies, M.L.; Li, M.; Philippe, B.; De Castro, C.; Qiu, Z.; Kim, J.; Watson, T.; Tsoi, W.C.; et al. An effective approach of vapour assisted morphological tailoring for reducing metal defect sites in lead-free, $(CH_3NH_3)_3Bi_2I_9$ bismuth-based perovskite solar cells for improved performance and long-term stability. *Nano Energy* **2018**, *49*, 614–624. [CrossRef]
28. Ali, J.; Li, Y.; Gao, P.; Hao, T.; Song, J.; Zhang, Q.; Zhu, L.; Wang, J.; Feng, W.; Hu, H.; et al. Interfacial and structural modifications in perovskite solar cells. *Nanoscale* **2020**, *12*, 5719–5745. [CrossRef]
29. Han, T.H.; Tan, S.; Xue, J.; Meng, L.; Lee, J.W.; Yang, Y. Interface and Defect Engineering for Metal Halide Perovskite Optoelectronic Devices. *Adv. Mater.* **2019**, *31*, 1803515. [CrossRef]
30. Qiu, L.; Ono, L.K.; Jiang, Y.; Leyden, M.R.; Raga, S.R.; Wang, S.; Qi, Y. Engineering Interface Structure to Improve Efficiency and Stability of Organometal Halide Perovskite Solar Cells. *J. Phys. Chem. B* **2018**, *122*, 511–520. [CrossRef] [PubMed]
31. Bai, Y.; Meng, X.; Yang, S. Interface Engineering for Highly Efficient and Stable Planar p-i-n Perovskite Solar Cells. *Adv. Energy Mater.* **2018**, *8*, 1701883. [CrossRef]
32. Vasilopoulou, M.; Fakharuddin, A.; Coutsolelos, A.G.; Falaras, P.; Argitis, P.; Yusoff, A.; Nazeeruddin, M.K. Molecular materials as interfacial layers and additives in perovskite solar cells. *Chem. Soc. Rev.* **2020**, *49*, 4496–4526. [CrossRef]
33. Boopathi, K.M.; Mohan, R.; Huang, T.-Y.; Budiawan, W.; Lin, M.-Y.; Lee, C.-H.; Ho, K.-C.; Chu, C.-W. Synergistic improvements in stability and performance of lead iodide perovskite solar cells incorporating salt additives. *J. Mater. Chem. A* **2016**, *4*, 1591–1597. [CrossRef]
34. Gong, X.; Li, M.; Shi, X.-B.; Ma, H.; Wang, Z.-K.; Liao, L.-S. Controllable Perovskite Crystallization by Water Additive for High-Performance Solar Cells. *Adv. Funct. Mater.* **2015**, *25*, 6671–6678. [CrossRef]
35. Kang, Y.F.; Wang, A.R.; Li, R.; Song, Y.L.; Dong, Q.F. A Review: Flexible Perovskite Solar Cells towards High Mechanical Stability. *Acta Polym. Sin.* **2021**, *52*, 920–937. [CrossRef]
36. Jiang, Q.; Zhao, Y.; Zhang, X.; Yang, X.; Chen, Y.; Chu, Z.; Ye, Q.; Li, X.; Yin, Z.; You, J. Surface passivation of perovskite film for efficient solar cells. *Nat. Photonics* **2019**, *13*, 460–466. [CrossRef]
37. Zhao, X.; Tao, L.; Li, H.; Huang, W.; Sun, P.; Liu, J.; Liu, S.; Sun, Q.; Cui, Z.; Sun, L.; et al. Efficient Planar Perovskite Solar Cells with Improved Fill Factor via Interface Engineering with Graphene. *Nano Lett.* **2018**, *18*, 2442–2449. [CrossRef]
38. Ioakeimidis, A.; Choulis, S.A. Nitrobenzene as Additive to Improve Reproducibility and Degradation Resistance of Highly Efficient Methylammonium-Free Inverted Perovskite Solar Cells. *Materials* **2020**, *13*, 3289. [CrossRef]
39. Chu, L.; Ahmad, W.; Liu, W.; Yang, J.; Zhang, R.; Sun, Y.; Yang, J.; Li, X.A. Lead-Free Halide Double Perovskite Materials: A New Superstar Toward Green and Stable Optoelectronic Applications. *Nano-Micro Lett.* **2019**, *11*, 16. [CrossRef]
40. Li, Y.; Wang, D.; Yang, L.; Yin, S. Preparation and performance of perovskite solar cells with two dimensional MXene as active layer additive. *J. Alloys Compd.* **2022**, *904*, 163742. [CrossRef]
41. Li, Z.; Wu, S.; Zhang, J.; Yuan, Y.; Wang, Z.; Zhu, Z. Improving Photovoltaic Performance Using Perovskite/Surface-Modified Graphitic Carbon Nitride Heterojunction. *Sol. RRL* **2019**, *4*, 1900413. [CrossRef]
42. Majdoub, M.; Anfar, Z.; Amedlous, A. Emerging Chemical Functionalization of g-C_3N_4: Covalent/Noncovalent Modifications and Applications. *ACS Nano* **2020**, *14*, 12390–12469. [CrossRef] [PubMed]
43. Pu, Y.-C.; Fan, H.-C.; Liu, T.-W.; Chen, J.-W. Methylamine lead bromide perovskite/protonated graphitic carbon nitride nanocomposites: Interfacial charge carrier dynamics and photocatalysis. *J. Mater. Chem. A* **2017**, *5*, 25438–25449. [CrossRef]
44. Jia, C.; Yang, L.; Zhang, Y.; Zhang, X.; Xiao, K.; Xu, J.; Liu, J. Graphitic Carbon Nitride Films: Emerging Paradigm for Versatile Applications. *ACS Appl. Mater. Interfaces* **2020**, *12*, 53571–53591. [CrossRef]
45. Javad, S.; Guoxiu, W. Progress and prospects of two-dimensional materials for membrane-based osmotic power generation. *Nano Res. Energy* **2022**, *1*, e9120008. [CrossRef]
46. Thomas, A.; Fischer, A.; Goettmann, F.; Antonietti, M.; Müller, J.-O.; Schlögl, R.; Carlsson, J.M. Graphitic carbon nitride materials: Variation of structure and morphology and their use as metal-free catalysts. *J. Mater. Chem.* **2008**, *18*, 4893–4908. [CrossRef]
47. Ismael, M. A review on graphitic carbon nitride (g-C_3N_4) based nanocomposites: Synthesis, categories, and their application in photocatalysis. *J. Alloys Compd.* **2020**, *846*, 156446. [CrossRef]

48. Cao, Q.; Kumru, B.; Antonietti, M.; Schmidt, B.V.K.J. Graphitic carbon nitride and polymers: A mutual combination for advanced properties. *Mater. Horiz.* **2020**, *7*, 762–786. [CrossRef]
49. Niu, X.; Yi, Y.; Bai, X.; Zhang, J.; Zhou, Z.; Chu, L.; Yang, J.; Li, X. Photocatalytic performance of few-layer graphitic g-C$_3$N$_4$: Enhanced by interlayer coupling. *Nanoscale* **2019**, *11*, 4101–4107. [CrossRef]
50. Lu, D.; Fang, P.; Wu, W.; Ding, J.; Jiang, L.; Zhao, X.; Li, C.; Yang, M.; Li, Y.; Wang, D. Solvothermal-assisted synthesis of self-assembling TiO$_2$ nanorods on large graphitic carbon nitride sheets with their anti-recombination in the photocatalytic removal of Cr(vi) and rhodamine B under visible light irradiation. *Nanoscale* **2017**, *9*, 3231–3245. [CrossRef] [PubMed]
51. Wang, A.; Wang, C.; Fu, L.; Wong-Ng, W.; Lan, Y. Recent Advances of Graphitic Carbon Nitride-Based Structures and Applications in Catalyst, Sensing, Imaging, and LEDs. *Nano-Micro Lett.* **2017**, *9*, 47. [CrossRef] [PubMed]
52. Niu, P.; Zhang, L.; Liu, G.; Cheng, H.-M. Graphene-Like Carbon Nitride Nanosheets for Improved Photocatalytic Activities. *Adv. Funct. Mater.* **2012**, *22*, 4763–4770. [CrossRef]
53. Lu, D.; Wang, H.; Zhao, X.; Kondamareddy, K.K.; Ding, J.; Li, C.; Fang, P. Highly Efficient Visible-Light-Induced Photoactivity of Z-Scheme g-C$_3$N$_4$/Ag/MoS$_2$ Ternary Photocatalysts for Organic Pollutant Degradation and Production of Hydrogen. *ACS Sustain. Chem. Eng.* **2017**, *5*, 1436–1445. [CrossRef]
54. Ou, M.; Tu, W.; Yin, S.; Xing, W.; Wu, S.; Wang, H.; Wan, S.; Zhong, Q.; Xu, R. Amino-Assisted Anchoring of CsPbBr$_3$ Perovskite Quantum Dots on Porous g-C$_3$N$_4$ for Enhanced Photocatalytic CO$_2$ Reduction. *Angew. Chem. Int. Ed.* **2018**, *57*, 13570–13574. [CrossRef]
55. Mamba, G.; Mishra, A.K. Graphitic carbon nitride (g-C$_3$N$_4$) nanocomposites: A new and exciting generation of visible light driven photocatalysts for environmental pollution remediation. *Appl. Catal. B-environ.* **2016**, *198*, 347–377. [CrossRef]
56. Ansari, M.S.; Banik, A.; Qureshi, M. Morphological tuning of photo-booster g-C$_3$N$_4$ with higher surface area and better charge transfers for enhanced power conversion efficiency of quantum dot sensitized solar cells. *Carbon* **2017**, *121*, 90–105. [CrossRef]
57. Yuan, Z.; Tang, R.; Zhang, Y.; Yin, L. Enhanced photovoltaic performance of dye-sensitized solar cells based on Co$_9$S$_8$ nanotube array counter electrode and TiO$_2$/g-C$_3$N$_4$ heterostructure nanosheet photoanode. *J. Alloys Compd.* **2017**, *691*, 983–991. [CrossRef]
58. Xie, F.; Dong, G.; Wu, K.; Li, Y.; Wei, M.; Du, S. In situ synthesis of g-C$_3$N$_4$ by glass-assisted annealing route to boost the efficiency of perovskite solar cells. *J. Colloid Interface Sci.* **2021**, *591*, 326–333. [CrossRef]
59. Gao, Q.; Sun, S.; Li, X.; Zhang, X.; Duan, L.; Lu, W. Enhancing Performance of CdS Quantum Dot-Sensitized Solar Cells by Two-Dimensional g-C$_3$N$_4$ Modified TiO$_2$ Nanorods. *Nanoscale Res. Lett.* **2016**, *11*, 463. [CrossRef]
60. Zou, J.; Liao, G.; Wang, H.; Ding, Y.; Wu, P.; Hsu, J.-P.; Jiang, J. Controllable interface engineering of g-C$_3$N$_4$/CuS nanocomposite photocatalysts. *J. Alloys Compd.* **2022**, *911*, 165020. [CrossRef]
61. Sheng, Y.; Zhao, A.; Yu, L.; Yuan, S.; Di, Y.; Liu, C.; Dong, L.; Gan, Z. Highly Efficient Charge Transfer between Perovskite Nanocrystals and g-C$_3$N$_4$ Nanosheets. *Phys. Status Solidi (B)* **2020**, *257*, 2000198. [CrossRef]
62. Wang, K.; Liu, J.; Yin, J.; Aydin, E.; Harrison, G.T.; Liu, W.; Chen, S.; Mohammed, O.F.; De Wolf, S. Defect Passivation in Perovskite Solar Cells by Cyano-Based π-Conjugated Molecules for Improved Performance and Stability. *Adv. Funct. Mater.* **2020**, *30*, 2002861. [CrossRef]
63. Jiang, L.-L.; Wang, Z.-K.; Li, M.; Zhang, C.-C.; Ye, Q.-Q.; Hu, K.-H.; Lu, D.-Z.; Fang, P.-F.; Liao, L.-S. Passivated Perovskite Crystallization via g-C$_3$N$_4$ for High-Performance Solar Cells. *Adv. Funct. Mater.* **2018**, *28*, 1705875. [CrossRef]
64. Liao, J.-F.; Wu, W.-Q.; Zhong, J.-X.; Jiang, Y.; Wang, L.; Kuang, D.-B. Enhanced efficacy of defect passivation and charge extraction for efficient perovskite photovoltaics with a small open circuit voltage loss. *J. Mater. Chem. A* **2019**, *7*, 9025–9033. [CrossRef]
65. Li, X.; Bi, D.; Yi, C.; Décoppet, J.-D.; Luo, J.; Zakeeruddin, S.M.; Hagfeldt, A.; Grätzel, M. A vacuum flash-assisted solution process for high-efficiency large-area perovskite solar cells. *Science* **2016**, *353*, 58–62. [CrossRef]
66. Liu, W.W.; Liu, Y.C.; Cui, C.Y.; Niu, S.T.; Niu, W.J.; Liu, M.C.; Liu, M.J.; Gu, B.; Zhang, L.Y.; Zhao, K.; et al. All-inorganic CsPbBr$_3$ perovskite solar cells with enhanced efficiency by exploiting lone pair electrons via passivation of crystal boundary using carbon nitride (g-C$_3$N$_4$) nanosheets. *Mater. Today Energy* **2021**, *21*, 100782. [CrossRef]
67. Zhu, P.; Gu, S.; Luo, X.; Gao, Y.; Li, S.; Zhu, J.; Tan, H. Simultaneous Contact and Grain-Boundary Passivation in Planar Perovskite Solar Cells Using SnO$_2$-KCl Composite Electron Transport Layer. *Adv. Energy Mater.* **2019**, *10*, 1903083. [CrossRef]
68. Yang, Z.-L.; Zhang, Z.-Y.; Fan, W.-L.; Hu, C.-s.; Zhang, L.; Qi, J.-J. High-performance g-C$_3$N$_4$ added carbon-based perovskite solar cells insulated by Al$_2$O$_3$ layer. *Sol. Energy* **2019**, *193*, 859–865. [CrossRef]
69. Xiong, Y.; Zhu, X.; Mei, A.; Qin, F.; Liu, S.; Zhang, S.; Jiang, Y.; Zhou, Y.; Han, H. Bifunctional Al$_2$O$_3$ Interlayer Leads to Enhanced Open-Circuit Voltage for Hole-Conductor-Free Carbon-Based Perovskite Solar Cells. *Sol. RRL* **2018**, *2*, 1800002. [CrossRef]
70. Han, G.S.; Chung, H.S.; Kim, B.J.; Kim, D.H.; Lee, J.W.; Swain, B.S.; Mahmood, K.; Yoo, J.S.; Park, N.-G.; Lee, J.H.; et al. Retarding charge recombination in perovskite solar cells using ultrathin MgO-coated TiO$_2$ nanoparticulate films. *J. Mater. Chem. A* **2015**, *3*, 9160–9164. [CrossRef]
71. Xia, Z.; Zhang, C.; Feng, Z.; Wu, Z.; Wang, Z.; Chen, X.; Huang, S. Synergetic Effect of Plasmonic Gold Nanorods and MgO for Perovskite Solar Cells. *Nanomaterials* **2020**, *10*, 1830. [CrossRef] [PubMed]
72. Rao, L.; Meng, X.; Xiao, S.; Xing, Z.; Fu, Q.; Wang, H.; Gong, C.; Hu, T.; Hu, X.; Guo, R.; et al. Wearable Tin-Based Perovskite Solar Cells Achieved by a Crystallographic Size Effect. *Angew. Chem. Int. Ed. Engl.* **2021**, *60*, 14693–14700. [CrossRef] [PubMed]
73. Gillan, E.G. Synthesis of Nitrogen-Rich Carbon Nitride Networks from an Energetic Molecular Azide Precursor. *Chem. Mater.* **2000**, *12*, 3906–3912. [CrossRef]

74. Sriram, B.; Baby, J.N.; Hsu, Y.F.; Wang, S.F.; George, M.; Veerakumar, P.; Lin, K.C. Electrochemical sensor-based barium zirconate on sulphur-doped graphitic carbon nitride for the simultaneous determination of nitrofurantoin (antibacterial agent) and nilutamide (anticancer drug). *J. Electroanal. Chem.* **2021**, *901*, 115782. [CrossRef]

75. Rakibuddin, M.; Kim, H.; Khan, M.E. Graphite-like carbon nitride (C_3N_4) modified N-doped $LaTiO_3$ nanocomposite for higher visible light photocatalytic and photo-electrochemical performance. *Appl. Surf. Sci.* **2018**, *452*, 400–412. [CrossRef]

76. Cao, W.; Lin, K.; Li, J.; Qiu, L.; Dong, Y.; Wang, J.; Xia, D.; Fan, R.; Yang, Y. Iodine-doped graphite carbon nitride for enhancing photovoltaic device performance via passivation trap states of triple cation perovskite films. *J. Mater. Chem. C* **2019**, *7*, 12717–12724. [CrossRef]

77. Niu, T.; Lu, J.; Munir, R.; Li, J.; Barrit, D.; Zhang, X.; Hu, H.; Yang, Z.; Amassian, A.; Zhao, K.; et al. Stable High-Performance Perovskite Solar Cells via Grain Boundary Passivation. *Adv. Mater.* **2018**, *30*, e1706576. [CrossRef]

78. Chen, S.; Pan, Q.; Li, J.; Zhao, C.; Guo, X.; Zhao, Y.; Jiu, T. Grain boundary passivation with triazine-graphdiyne to improve perovskite solar cell performance. *Sci. China Mater.* **2020**, *63*, 2465–2476. [CrossRef]

79. Wei, J.; Wang, X.; Sun, X.; Yang, Z.; Moreels, I.; Xu, K.; Li, H. Polymer assisted deposition of high-quality $CsPbI_2Br$ film with enhanced film thickness and stability. *Nano Res.* **2020**, *13*, 684–690. [CrossRef]

80. Zeng, J.; Bi, L.; Cheng, Y.; Xu, B.; Jen, A.K.Y. Self-assembled monolayer enabling improved buried interfaces in blade-coated perovskite solar cells for high efficiency and stability. *Nano Res. Energy* **2022**, *1*, e9120004. [CrossRef]

81. Ye, S.; Rao, H.; Zhao, Z.; Zhang, L.; Bao, H.; Sun, W.; Li, Y.; Gu, F.; Wang, J.; Liu, Z.; et al. A Breakthrough Efficiency of 19.9% Obtained in Inverted Perovskite Solar Cells by Using an Efficient Trap State Passivator Cu(thiourea)I. *J. Am. Chem. Soc.* **2017**, *139*, 7504–7512. [CrossRef]

82. Wei, X.; Liu, X.; Liu, H.; Yang, S.; Zeng, H.; Meng, F.; Lei, X.; Liu, J. Exfoliated graphitic carbon nitride self-recognizing $CH_3NH_3PbI_3$ grain boundaries by hydrogen bonding interaction for improved perovskite solar cells. *Sol. Energy* **2019**, *181*, 161–168. [CrossRef]

83. Kang, B.; Biswas, K. Preferential $CH_3NH_3^+$ Alignment and Octahedral Tilting Affect Charge Localization in Cubic Phase $CH_3NH_3PbI_3$. *J. Phys. Chem. C* **2017**, *121*, 8319–8326. [CrossRef]

84. Hao, Q.; Jia, G.; Wei, W.; Vinu, A.; Wang, Y.; Arandiyan, H.; Ni, B.-J. Graphitic carbon nitride with different dimensionalities for energy and environmental applications. *Nano Res.* **2019**, *13*, 18–37. [CrossRef]

85. Chen, X.; Liu, Q.; Wu, Q.; Du, P.; Zhu, J.; Dai, S.; Yang, S. Incorporating Graphitic Carbon Nitride (g-C_3N_4) Quantum Dots into Bulk-Heterojunction Polymer Solar Cells Leads to Efficiency Enhancement. *Adv. Funct. Mater.* **2016**, *26*, 1719–1728. [CrossRef]

86. Chen, J.; Dong, H.; Zhang, L.; Li, J.; Jia, F.; Jiao, B.; Xu, J.; Hou, X.; Liu, J.; Wu, Z. Graphitic carbon nitride doped SnO_2 enabling efficient perovskite solar cells with PCEs exceeding 22%. *J. Mater. Chem. A* **2020**, *8*, 2644–2653. [CrossRef]

87. Grimme, S.; Antony, J.; Ehrlich, S.; Krieg, H. A consistent and accurate ab initio parametrization of density functional dispersion correction (DFT-D) for the 94 elements H-Pu. *J. Chem. Phys.* **2010**, *132*, 154104. [CrossRef]

88. Wang, S.; Sakurai, T.; Wen, W.; Qi, Y. Energy Level Alignment at Interfaces in Metal Halide Perovskite Solar Cells. *Adv. Mater. Interfaces* **2018**, *5*, 1800260. [CrossRef]

89. Jena, A.K.; Ishii, A.; Guo, Z.; Kamarudin, M.A.; Hayase, S.; Miyasaka, T. Cesium Acetate-Induced Interfacial Compositional Change and Graded Band Level in $MAPbI_3$ Perovskite Solar Cells. *ACS Appl. Mater. Interfaces* **2020**, *12*, 33631–33637. [CrossRef]

90. Singh, A.N.; Kajal, S.; Kim, J.; Jana, A.; Kim, J.Y.; Kim, K.S. Interface Engineering Driven Stabilization of Halide Perovskites against Moisture, Heat, and Light for Optoelectronic Applications. *Adv. Energy Mater.* **2020**, *10*, 2000768. [CrossRef]

91. Liu, P.; Sun, Y.; Wang, S.; Zhang, H.; Gong, Y.; Li, F.; Shi, Y.; Du, Y.; Li, X.; Guo, S.-s.; et al. Two dimensional graphitic carbon nitride quantum dots modified perovskite solar cells and photodetectors with high performances. *J. Power Sources* **2020**, *451*, 227825. [CrossRef]

92. Ameri, M.; Ghaffarkani, M.; Ghahrizjani, R.T.; Safari, N.; Mohajerani, E. Phenomenological morphology design of hybrid organic-inorganic perovskite solar cell for high efficiency and less hysteresis. *Sol. Energy Mater. Sol. Cells* **2020**, *205*, 110251. [CrossRef]

93. Zeng, W.; Liu, X.; Guo, X.; Niu, Q.; Yi, J.; Xia, R.; Min, Y. Morphology Analysis and Optimization: Crucial Factor Determining the Performance of Perovskite Solar Cells. *Molecules* **2017**, *22*, 520. [CrossRef]

94. Zheng, L.; Zhang, D.; Ma, Y.; Lu, Z.; Chen, Z.; Wang, S.; Xiao, L.; Gong, Q. Morphology control of the perovskite films for efficient solar cells. *Dalton Trans.* **2015**, *44*, 10582–10593. [CrossRef]

95. Liu, Z.; Wu, S.; Yang, X.; Zhou, Y.; Jin, J.; Sun, J.; Zhao, L.; Wang, S. The dual interfacial modification of 2D g-C_3N_4 for high-efficiency and stable planar perovskite solar cells. *Nanoscale Adv.* **2020**, *2*, 5396–5402. [CrossRef]

96. Cao, J.; Tang, G.; You, P.; Wang, T.; Zheng, F.; Zhao, J.; Yan, F. Enhanced Performance of Planar Perovskite Solar Cells Induced by Van Der Waals Epitaxial Growth of Mixed Perovskite Films on WS_2 Flakes. *Adv. Funct. Mater.* **2020**, *30*, 2002358. [CrossRef]

97. Jin, J.; Wu, S.; Yang, X.; Zhou, Y.; Li, Z.; Cao, Q.; Chi, B.; Li, J.; Zhao, L.; Wang, S. Improve the efficiency of perovskite solar cells through the interface modification of g-C_3N_4 nanosheets. *Mater. Lett.* **2021**, *304*, 130685. [CrossRef]

98. Yang, J.; Chu, L.; Hu, R.; Liu, W.; Liu, N.; Ma, Y.; Ahmad, W.; Li, X.a. Work function engineering to enhance open-circuit voltage in planar perovskite solar cells by g-C_3N_4 nanosheets. *Nano Res.* **2021**, *14*, 2139–2144. [CrossRef]

99. Yang, X.; Li, L.; Wu, J.; Hu, Q.; Wang, Y.; Russell, T.P.; Tu, Y.; Zhu, R. Optimizing Vertical Crystallization for Efficient Perovskite Solar Cells by Buried Composite Layers. *Sol. RRL* **2021**, *5*, 2100457. [CrossRef]

100. Wang, L.; Fu, L.; Li, B.; Li, H.; Pan, L.; Chang, B.; Yin, L. Thiazole-Modified C_3N_4 Interfacial Layer for Defect Passivation and Charge Transport Promotion in Perovskite Solar Cells. *Sol. RRL* **2021**, *5*, 2000720. [CrossRef]
101. Cruz, D.; Garcia Cerrillo, J.; Kumru, B.; Li, N.; Dario Perea, J.; Schmidt, B.; Lauermann, I.; Brabec, C.J.; Antonietti, M. Influence of Thiazole-Modified Carbon Nitride Nanosheets with Feasible Electronic Properties on Inverted Perovskite Solar Cells. *J. Am. Chem. Soc.* **2019**, *141*, 12322–12328. [CrossRef]

 *nanomaterials*

Review

Recent Advances of Transition Metal Chalcogenides as Cathode Materials for Aqueous Zinc-Ion Batteries

Ying Liu and Xiang Wu *

School of Materials Science and Engineering, Shenyang University of Technology, Shenyang 110870, China
* Correspondence: wuxiang05@sut.edu.cn

Abstract: In recent years, advances in lithium-ion batteries (LIBs) have pushed the research of other metal-ion batteries to the forefront. Aqueous zinc ion batteries (AZIBs) have attracted much attention owing to their low cost, high capacity and non-toxic characteristics. Among various cathodes, transition metal chalcogenides (TMCs) with a layered structure are considered as suitable electrode materials. The large layer spacing facilitates the intercalation/de-intercalation of Zn^{2+} between the layers. In this mini-review, we summarize a variety of design strategies for the modification of TMCs. Then, we specifically emphasize the zinc storage capacity of the optimized electrodes. Finally, we propose the challenges and future prospects of cathode materials for high-energy AZIBs.

Keywords: aqueous zinc ion battery; transition metal chalcogenides; layered structure; cathode; energy storage mechanism

Citation: Liu, Y.; Wu, X. Recent Advances of Transition Metal Chalcogenides as Cathode Materials for Aqueous Zinc-Ion Batteries. *Nanomaterials* **2022**, *12*, 3298. https://doi.org/10.3390/nano12193298

Academic Editor: Henrich Frielinghaus

Received: 4 September 2022
Accepted: 20 September 2022
Published: 22 September 2022

Publisher's Note: MDPI stays neutral with regard to jurisdictional claims in published maps and institutional affiliations.

1. Introduction

The growing energy crisis has driven the unprecedented development of renewable clean energy [1–3]. To date, lithium-ion batteries (LIBs) are the most widely used energy storage devices. However, the scarcity of lithium resources, the inflammability of electrolytes, and high operating environment requirements limit their growth [4–7]. Rechargeable aqueous zinc ion batteries (AZIBs) are a new generation of safety batteries. They possess certain advantages in terms of abundant zinc reserves, low anode potential (−0.763 V vs. SHE) and high theoretical capacity (820 mAh g^{-1}) [8–11]. Therefore, AZIBs have become one of the candidates to replace LIBs. However, zinc anodes undergo dissolution–precipitation reactions with several adverse reactions, such as dendrite growth, corrosion, and by-product formation. They are inevitable during repeated plating and stripping and seriously damage the cycle life of the cells [12]. Also, divalent zinc ions possess stronger electrostatic interactions than monovalent lithium ions [13,14]. Therefore, the choice of a suitable intercalation material is the crucial to break through this challenge.

In recent years, many efforts are devoted to the exploration of cathode materials, including Prussian blue analogs, vanadium-based and manganese-based compounds, and transition metal chalcogenides (TMCs) [15–19]. Among them, Prussian blue analogs are featured by a high voltage window, but their crystal structure is unstable and prone to phase transformation [20]. The inherent low electrical conductivity and poor structural stability of V- and Mn-based materials lead to their slow electrochemical kinetics [21–23]. The electrical conductivity of TMCs is superior to that of oxides. Additionally, TMCs are characterized by a unique layer structure with large layer spacing. Their high specific surface area can provide many active sites and reduce ion transfer paths [24]. In previous reports, Naveed et al. designed the VS_2 nanosheet materials as cathodes, which maintain a capacity of 138.3 mAh g^{-1} at 0.1 A g^{-1} after 500 cycles with a retention of 94.38% [25]. Kang's group summarized the Zn storage ability of various oxides, sulfides and borides [26]. The results show that the activated MnS electrode is a potential cathode material with both high capacity and stable cycling performance. However, bulk MoS_2 and WS_2 materials are virtually incapable of storing zinc ions. Therefore, it is essential to improve the electrochemical performance of TMCs by effective tuning strategies.

In recent years, there have been numerous reports on TMCs-based cathode materials. Herein, we first summarize several feasible strategies for optimizing the electrode structure. Then, we discuss the electrochemical performance of TMCs cathodes for AZIBs. Lastly, we overview the advances in cathode materials and present the current challenges and prospects for constructing advanced cathodes of AZIBs.

2. The Electrochemical Performance of TMCs Cathodes

Two-dimensional (2D) TMCs are composed of transition metals (M = V, Ni, Mo, W and Mn, etc.) and chalcogen elements (X = S, Se, Te) with tunable electrical properties from semiconductors to metals. They is widely studied in the field of energy storage and conversion field [24,27,28]. In addition, graphene-like 2D layered TMCs are highly advantageous for battery applications because of their large specific surface area, which can significantly increase the contact area between the active material and the electrolyte [29]. Figure 1 illustrates the crystal structures of various TMCs materials. Their non-bonding properties allow the insertion of atoms, ions and molecules. The design strategies of TMCs materials mainly include defect engineering, hybridization, phase modulation, and in situ electrochemical oxidation. The main focus is to widen the interlayer space, improve the electrode conductivity, and accelerate the electrochemical kinetic process. We will categorize the strategies of structural design and zinc storage capabilities of various cathode materials in the following sections.

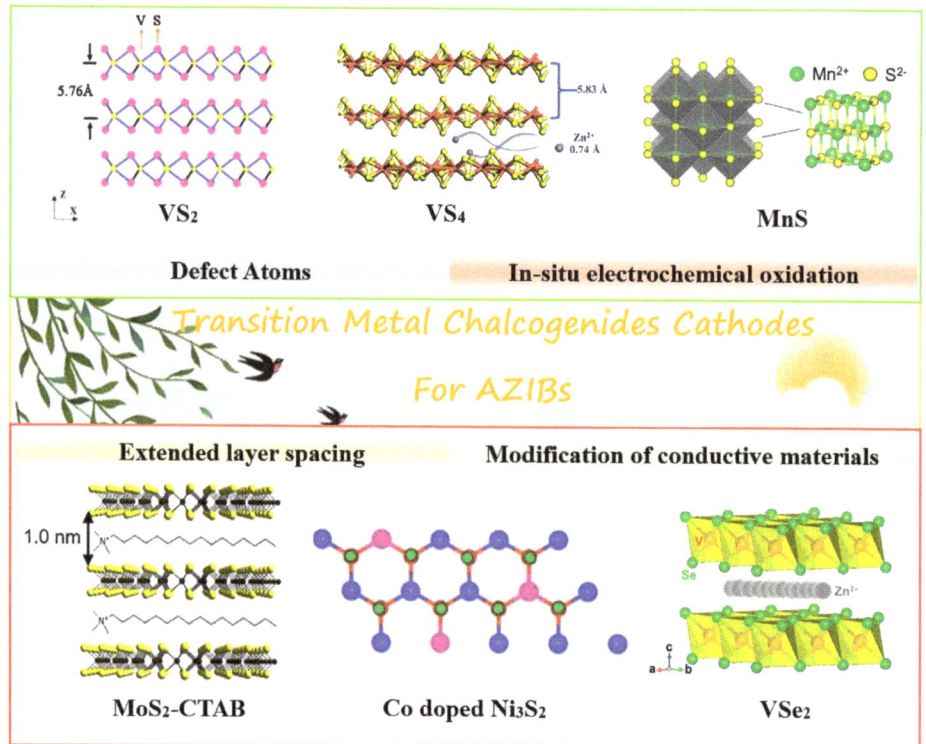

Figure 1. The classification of typical TMCs cathodes and the strategies of modification.

2.1. VS$_2$ and VS$_4$

The hexagonal-structured vanadium disulfide (VS$_2$) is a member of the TMC family. It owns a layered structure with a layer spacing of 5.76 Å [30]. The S-V-S layers rely on weak van der Waals interactions (Figure 2a). Hence, VS$_2$ materials have become an attractive host for Zn ion insertion/extraction. Nevertheless, VS$_2$ is unstable in aqueous solutions, leading to severe capacity decay during cycling. From the aspect of modulating the material structure, coating is considered to be effective strategy. Pu et al. prepared rose-shaped VS$_2$ encapsulated with a hydrophilic VOOH coating using a one-pot hydrothermal route [31]. The assembled cells maintained 82% capacity after 400 cycles. The rate capability and long cycle life of the optimized sample is significantly improved compared to the VS$_2$ sample, which is attributed to the O–H in VOOH. From Figure 2b, its presence not only enhances the wetting of the electrode and electrolyte, but also prevents the dissolution of the main materials. Fan's group prepared ultrathin VS$_2$ nanosheets grown on graphene sheets (rGO–VS$_2$) by a solvothermal strategy [32]. The rGO offers a large specific surface area and excellent electrical conductivity. In addition, the close contact of VS$_2$ nanosheets with rGO can effectively prevent the dissolution and corrosion of the host materials. It ensures the high stability of the electrode in long-term cycling. Thus, Zn/rGO–VS$_2$ cells can deliver a large specific capacity (238 mAh g^{-1} at 0.1 A g^{-1}) and excellent rate performance (190 mAh g^{-1} at 5 A g^{-1}, Figure 2c). After 1000 cycles of charging and discharging, it can still maintain 93.3% of the initial capacity, as shown in Figure 2d.

The N dopant provides high affinity for the transition metals, so it is possible to form a strong coupling between the host material and the N-doped carbon. This contributes to accelerate the interfacial electron transfer and reduces cycling-induced stress and volume changes. Liu and co-workers prepared spun VS$_2$ materials on a N-doped carbon layer (VS$_2$@N-C) by an in situ hybridization strategy [33]. This strategy ensures a strong interfacial interaction between the active material and the N-doped carbon. It promotes the enhancement of electrochemical kinetics and cycling stability of the cathode. The optimal electrode possesses a capacity of 203 mAh g^{-1} at 0.05 A g^{-1}. Based on the Zn ion insertion and extraction mechanism, the cells can obtain a capacity of 144 mAh g^{-1} after 600 cycles. Liu's group synthesized 1T-VS$_2$ colloidal nanospheres assembled from nanoflakes [34]. By controlling the charge cutoff voltage, a number of the Zn ions were trapped in the interlayer of the structure after the initial charge/discharge cycle. These "dead Zn" act as "pillars" to ensure the stability of the layered structure of VS$_2$. After 2000 cycles, the cells maintained the capacity retention of 86.7%. This unique layered structure increases the conductivity due to the presence of carbon and oxygen groups on the surface, facilitating the penetration of the aqueous electrolyte and providing more active sites.

In order to further improve the stability of VS$_2$ electrodes in an aqueous electrolyte. Yang et al. employed an in situ electrochemical oxidation approach to enhance the interlayer space of vanadium disulfide (VS$_2$NH$_3$) hollow spheres [35]. This large layer spacing (1.21 nm) is favorable to the improvement of zinc ion storage capacity. The VS$_2$NH$_3$ samples transform into a porous structured V$_2$O$_5$·nH$_2$O phase during the first charging cycle (Figure 2e). It enhances the active sites and is conducive to a rapid electrochemical kinetic process. This derived electrode maintains a high capacity at 3 A g^{-1} even after 2000 cycles. Similarly, Du and co-workers proposed the formation of VS$_2$/VO$_x$ heterostructures by in situ electrochemical induction [36]. When the sample is charged to 1.8 V, the morphology of the composite changes from rose-like shape to sheet-like one. This structure can withstand the volume expansion caused by repeated cycles. Compared to the pure Zn–VS$_2$ cell, the Zn–VS$_2$/VO$_x$ one demonstrates a cycling stability of 3000 cycles at 1 A g^{-1} with an improved working potential of 0.25 V (Figure 3a). This strategy of combining highly conductive sulfides and excellent chemically stable oxides leads to an enhancement of the Zn^{2+} storage capacity of the VS$_2$/VO$_x$ cathode.

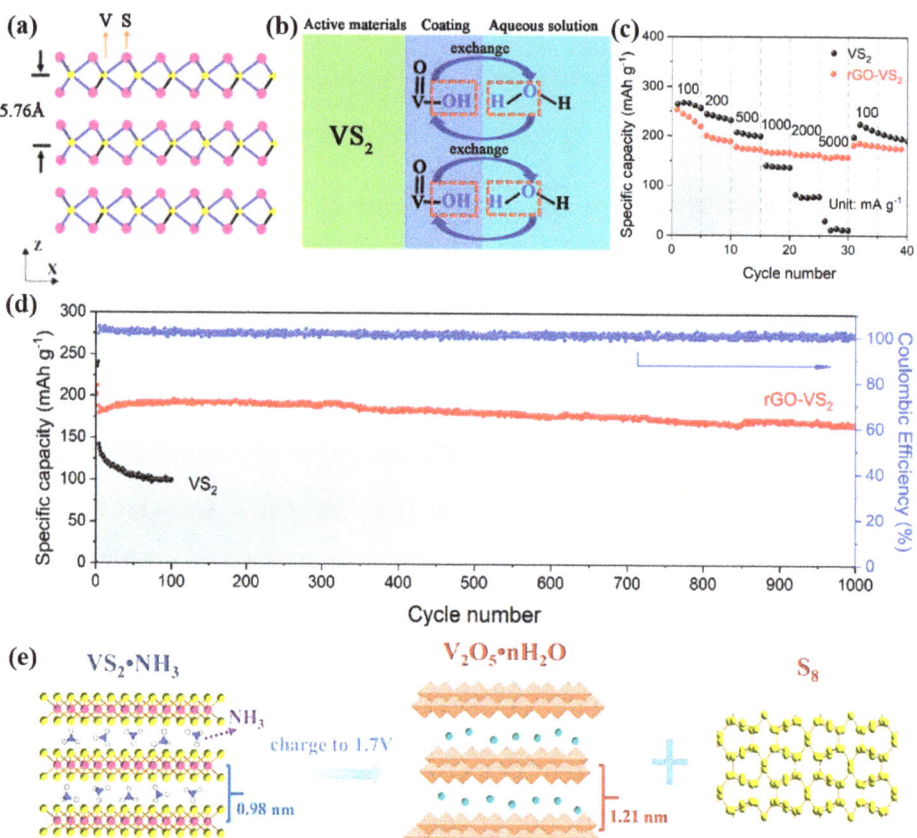

Figure 2. The electrochemical performance of VS$_2$ electrodes. (**a**) Atomic structure of VS$_2$ materials. (**b**) The working mechanism of cathode. Reproduced with permission [31]. Copyright 2019, Elsevier B.V. (**c**) Rate capability of rGO–VS$_2$ and VS$_2$. (**d**) Cycling performance of these two electrodes. Reproduced with permission [32]. Copyright 2020, Elsevier B.V. (**e**) Zn^{2+} storage mechanism. Reproduced with permission [35]. Copyright 2021, Elsevier Inc.

VS$_4$ is a material with a one-dimensional (1D) atomic chain structure. When compared with the VS$_2$ material, it possesses many S atoms with layer spacing up to 5.83 Å, as shown in Figure 3b. It indicates that VS$_4$ may show excellent zinc storage capacity. Zhu et al. prepared VS$_4$ materials as cathodes via a hydrothermal route [37]. Density functional theory (DFT) calculations demonstrate that the electrode is capable of storing zinc ions up to a maximum specific capacity of 262 mAh g^{-1}. Then, the cell can deliver a specific capacity of 310 mAh g^{-1} at 0.1 A g^{-1}. It is higher than the theoretical capacity, which could be due to the additional absorption capacity. Furthermore, the absence of additional by-product generation suggests that the energy storage of material follows a zinc-ion embedding/de-embedding mechanism. The construction of heterostructures is also an attractive strategy. In theory, the intrinsic zinc storage capacity of the cathode can be effectively enhanced by building a heterostructure with sufficient interfaces and grain boundaries. Fang's group designed VS$_4$/V$_2$O$_3$ heterostructures with a high specific surface area [38]. The assembled battery shows a capacity of 163 mAh g^{-1} at 0.1 A g^{-1}. As a contrast, the single electrode presents inferior electrochemical performance. This demonstrates that the optimized heterogeneous material can boost the energy storage capacity.

It is also an effective strategy for improving electrochemical kinetics by compositing with highly conductive materials. Qin and co-workers synthesized VS_4 composite material immobilized on reduced graphene oxide (VS_4@rGO) as a cathode [39]. This synergistic effect enables the VS_4@rGO electrode to reach a capacity of 180 mAh g^{-1} (1 A g^{-1}) after 165 cycles with a capacity retention of 93.3%. Nevertheless, the above-mentioned electrochemical performance is still unsatisfactory. Chen and co-workers optimized the morphology of the VS_4@rGO composites to achieve a specific capacity of 450 mAh g^{-1} at a current density of 0.5 A g^{-1} when used as cathodes [40]. Moreover, the capacity of 313.8 mAh g^{-1} was maintained at high current densities (10 A g^{-1}). It indicates that the batteries possess an excellent rate capability. It is noteworthy that a new phase $Zn_3(OH)_2V_2O_7 \cdot 2H_2O$ (ZVO) appears during charging. The following reactions may occur in the electrode material:

$$VS_4 + xZn^{2+} + 2xe^- \leftrightarrow Zn_xVS_4 \tag{1}$$

$$VS_4 + 11H_2O + 3Zn^{2+} \rightarrow Zn_3(OH)_2V_2O_7 \cdot 2H_2O + 8S + 16H^+ + 10e^- \tag{2}$$

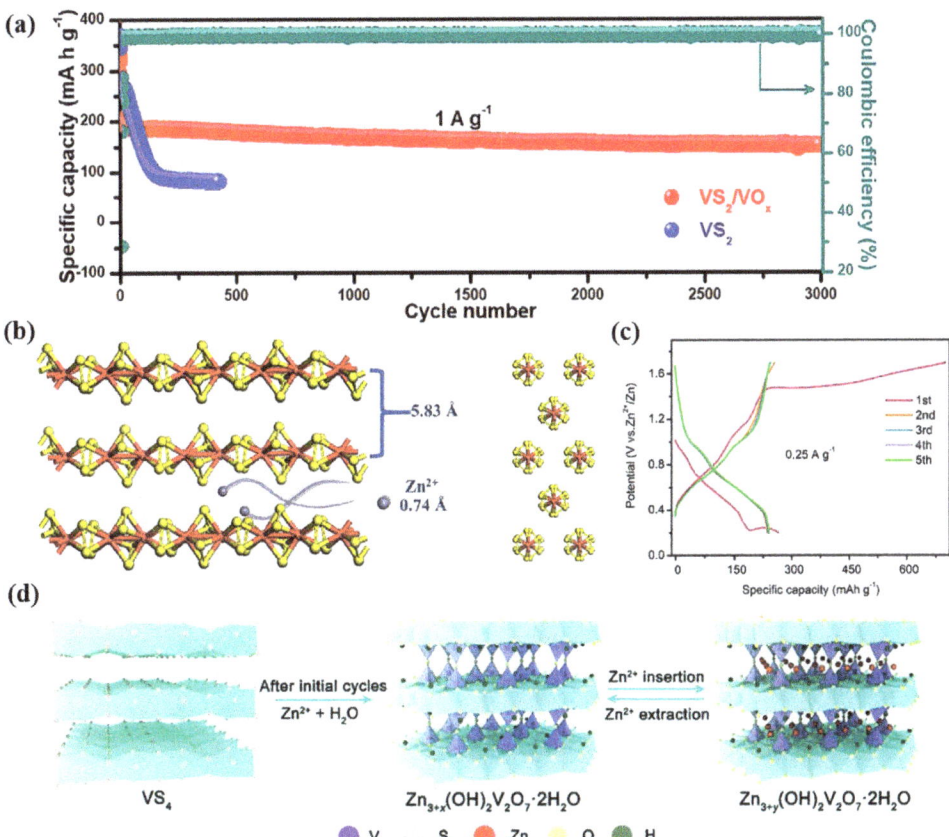

Figure 3. The energy storage mechanism and cycling performance. (**a**) The comparison of cycling performance at 1 A g^{-1}. Reproduced with permission [36]. Copyright 2020 Wiley-VCH GmbH. (**b**) The crystal structure of VS_4 materials. Reproduced with permission [37]. Copyright 2020, The Royal Society of Chemistry. (**c**) The charge–discharge curves at 0.25 A g^{-1}. (**d**) Schematic of the energy storage mechanism of Zn–VS_4/CNTs batteries. Reproduced with permission [41]. Copyright 2021, Elsevier B.V. and Science Press.

After that, it transforms into the ZnV_3O_8 phase during the long cycles, which is associated with the subsequent capacity decay. Gao et al. reported a flower-like VS_4/CNTs cathode with an abundant mesoporous structure, which effectively shortens the diffusion path of zinc ions [41]. In Figure 3c, when the first cycle is charged to 1.7 V, the charging curve undergoes a slow upward trend, which implies a phase transition process. Figure 3d further confirms that the mechanism of the phase change reaction of VS_4 with zinc pyrovandate ($Zn_{3+x}(OH)_2V_2O_7 \cdot 2H_2O$). The results show that the Zn–VS_4/CNTs batteries possess a reversible capacity of 265 mAh g^{-1} (0.25 A g^{-1}) and a good rate performance in the potential range from 0.2 to 1.7 V. Although the energy storage capacity has been significantly improved by modification of the electrode material, the inevitable phase change during the reaction process still hinders the cycle life. This may be related to the high charging voltage.

2.2. MoS$_2$

MoS_2 is a typical 2D-layered structure bound by weak van der Waals forces [42–44]. However, the ionic radius of hydrated Zn^{2+} is 0.43 nm, which places high demands on the interlayer space of the host materials. To enhance the reaction kinetics of Zn ion insertion and extraction, Li et al. extended the interlayer spacing of the (002) plane of MoS_2 nanosheets from 0.62 nm to 0.70 nm [45]. Due to the addition of glucose, an amorphous carbon layer is wrapped on the surface of MoS_2. This facilitates the alleviation of volume expansion and promotes charge transfer. From Figure 4a, the specific capacity of the batteries can be maintained at 164.5 mAh g^{-1} after 600 cycles. In Figure 4b, the charge storage mechanism can be described as follow:

$$\text{Cathode: } xZn^{2+} + x2e^- + MoS_2 \leftrightarrow Zn_xMoS_2 \tag{3}$$

$$\text{Anode: } Zn^{2+} + 2e^- \leftrightarrow Zn \tag{4}$$

In addition, a flexible solid-state Zn/E-MoS_2 cell was further assembled using the starch/polyacrylamide (PAM) polymer electrolyte. Under different mechanical strengths, the cell still can maintain a stable charge/discharge process.

Due to the diversity of coordination of Mo and S atoms, MoS_2 can show a semiconductor phase with a triangular prismatic structure (2H phase) and a metallic phase with an octahedral structure (1T phase). 1T-phase MoS_2 possesses higher electrical conductivity and better hydrophilicity than 2H-phase ones [46,47]. Therefore, the material is also a promising electrode for zinc storage. Huang et al. synthesized a 1T-phase MoS_2 nanosheet grown directly on reduced graphene oxide (rGO) scaffolds [48]. The addition of the rGO scaffold can serve to stabilize the 1T phase and reduce the possibility of phase transition during zinc ions insertion/extraction. In addition, it can improve the electrical conductivity, thus shortening the diffusion path of zinc ions. The initial discharge capacity of the 1T-MoS_2/rGO heterogeneous electrode is 108.3 mAh g^{-1}, and the cell maintains a capacity retention of 88% after repeated charge/discharges of 1000 times.

Tang's group synthesized N-doped 1T MoS_2 nanoflowers assembled from ultrathin nanosheets by a one-step hydrothermal sulfidation of Mo-based organic framework (MOF) precursors [49]. The introduction of defects effectively widens the interlayer spacing and increases the number of sulfur vacancies as well as the hydrophilicity of the sample. Zn/N-doped 1T MoS_2 batteries deliver a capacity of 149.6 mAh g^{-1} (0.1 A g^{-1}). The capacity retention is up to 89.1% after 1000 cycles at 3 A g^{-1}. Additionally, the electrochemical performance was studied for the difference in area mass loading of the electrodes. The area capacity shows an outstanding performance when the area loading reaches 1.701 mg cm^{-2}. This implies that the electrode presents excellent rate capacity even at high loadings. Liu and co-workers reported MoS_2 nanosheets with different phase contents as cathode materials [50]. Among them, the MoS_2 nanosheet electrode with 1T phase of content around 70% presents favorable long-term cycling stability. This indicates that the presence of metallic 1T phase favors the ion and charge transfer.

Nanomaterials **2022**, *12*, 3298

Apart from composite with conductive materials, combination with organic molecules also promotes the increase of zinc ion storage capacity. Yao et al. designed a 2D $MoS_2/C_{19}H_{42}N^+$ (CTAB) organic–inorganic superlattice structure (MoS_2–CTAB) as a cathode [51]. This unique structure can significantly enlarge the interlayer spacing (1.0 nm) of the host materials (Figure 4c). In addition, the stable electrode structure can accommodate the expansion and contraction of Zn^{2+} within the host structure. The loading mass of the active material is a very important parameter for the evaluation of the specific capacity and energy density of the cell. The areal capacity of the battery increases with the area mass loading in a potential window of 0.2–1.3 V. Figure 4d demonstrates the optimal adsorption position of Zn ions at the pure MoS_2 and modified MoS_2 electrodes by DFT calculations. In the former structure, the Zn ion prefers to adsorb at the top site of the Mo atom with a corresponding energy of -0.31 eV, but the charge accumulation of adsorbed Zn with adjacent S atoms suggests a strong electrostatic interaction between Zn and the host material. In the latter one, the adsorption energy of Zn ion at the same site is -0.25 eV, and the distance between Zn and its three neighboring S atoms remains almost unchanged relative to the original electrode.

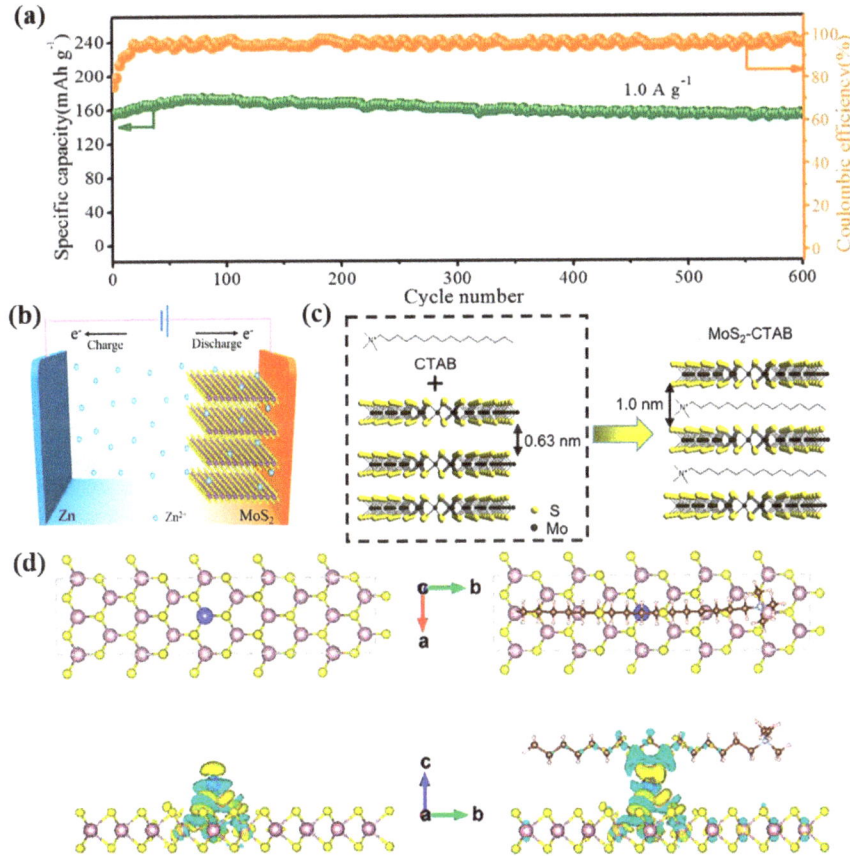

Figure 4. The modification of MoS_2 materials and DFT calculation. (**a**) Cycling stability of Zn/E–MoS_2 batteries at 1.0 A g^{-1}. (**b**) Schematics of the mechanism of batteries. Reproduced with permission [45]. Copyright 2018, Elsevier B.V. (**c**) Schematic of the MoS_2–CTAB superlattice nanosheets and the d-spacing of the samples. (**d**) The charge density of pristine MoS_2 and MoS_2–CTAB nanosheets during the insertion of Zn^{2+}. Reproduced with permission [51]. Copyright 2022, American Chemical Society.

2.3. MnS

In recent years, many efforts have been made in Mn-based oxide cathodes [52]. For instance, Minakshi et al. compared the cathodic behavior of electrolytic manganese dioxide (EMD) and chemically prepared battery-grade manganese dioxide (BGM) in a lithium hydroxide (LiOH) electrolyte [53]. The EMD cell demonstrated stable discharge/charge cycles compared to the BGM. Wang's group designed nanocrystal line structures of MnO_2 materials with particle sizes typically less than 10 nm [54]. This structure confers some electrode/electrolyte contact interfaces. Therefore, the Zn/MnO_2 cell delivers a capacity of 260 mAh g^{-1} at 1.3 C.

However, their rate capability and cycle stability cannot meet the current high-capacity energy storage requirements. In addition, Zn ions show strong electrostatic interactions with the Mn oxide lattice, leading to large energy barriers for Zn^{2+} migration [55]. Chen et al. reported the transformation of α-MnS materials into high-performance manganese oxide cathodes (MnS–EDO) by in situ electrochemical oxidation [56]. Compared to α-MnO_2, this electrode generated more defects and vacancies after structural reconfiguration. This indicates a rapid electrochemical kinetic process. The Zn/MnS–EDO cell shows a high specific capacity of 335.7 mAh g^{-1} with capacity retention close to 100% and reversible rate performance (Figure 5a). In addition, it can undergo a repeated charge/discharge process of 4000 cycles, as shown in Figure 5b.

To further enhance the electrical conductivity of MnS, Ma and co-workers designed a MnS and rGO composite material. This synergistic effect effectively improves the Zn storage capacity of the electrode material [57]. MnS possesses various phase types: α-MnS, β-MnS and γ-MnS. Both β- and γ-phases are sub-stable and readily transform to the stable rock salt structure α-MnS (Figure 5c). Jiang et al. fabricated flexible zinc ion microcells with MnS as cathodes and guar gels as the quasi-solid electrolyte by etching soft templates on various substrates [58]. The cells prepared on PET substrates deliver an area-specific capacity of 178 μAh cm^{-2}. After 1000 cycles, they can maintain a capacity of 150 mAh g^{-1} at 1 A g^{-1} in Figure 5d. Moreover, the area energy density can reach 322 μWh cm^{-2} at a power density of 120 μW cm^{-2}. In addition, this quasi-solid-state cell shows excellent flexibility with almost no significant capacity degradation when the device is bent at multiple angles.

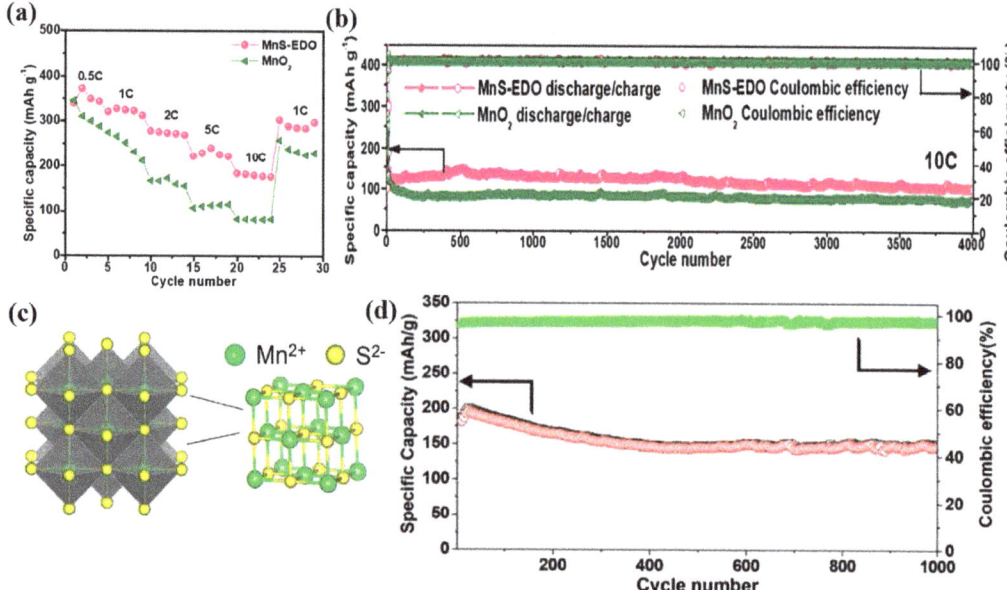

Figure 5. The long-term cycle performance of MnS electrodes. (**a**) Rate capability of the two electrode.

(**b**) The long-term cycling at 10 C. Reproduced with permission [56]. Copyright 2020, Elsevier Ltd. (**c**) Crystal structure model of α-MnS; (**d**) Cyclic stability in 2 M ZnSO$_4$ and 0.1 M MnSO$_4$ aqueous electrolyte. Reproduced with permission [58]. Copyright 2021, Wiley-VCH GmbH.

2.4. VSe$_2$

Among numerous TMCs, vanadium diselenide (VSe$_2$) presents a typical layered structure with a sandwich-like Se–V–Se connected by van der Waals interactions [59]. The materials possess a layer spacing of 6.11 Å, which can provide sufficient transport channels and active sites for the intercalated ions [60]. Moreover, the strong electronic coupling force between adjacent V^{4+}–V^{4+} endows it with metallic properties. Thus, it shows great potential in terms of energy storage [61,62]. For instance, Alshareef's group explored the energy storage capacity of VSe$_2$ materials in different alkali metal batteries [63]. The morphology of VSe$_2$ nanosheets can be modulated using *N*-methylpyrrolidone (NMP) solvent, and the electrochemical performance of the samples is further improved by in situ carbon coating. The electrodes can provide a specific capacity of 768 mAh g^{-1} (lithium storage) and 571 mAh g^{-1} (sodium storage), respectively.

Recently, VSe$_2$ materials also show attractive performance in zinc storage. Wu et al. designed an ultrathin VSe$_2$ nanosheet as a cathode [64]. The assembled cells maintain an initial capacity of 80.8% after 500 cycles. This durable cycling stability is attributed to their fast Zn^{2+} diffusion kinetic process and durable cycling stability. The local charge density map in Figure 6a shows a decrease in charge around the Zn ion and an increase in charge around the Se and some V sites. This demonstrates that the intercalated Zn is bonded to the Se ligand. From Figure 6b, the migration barrier for the optimal diffusion pathway of Zn ions is 0.91 eV, which corresponds to a fast ion migration rate. Based on the Zn ion insertion and extraction mechanism (Figure 6c), the cell possesses an energy density of 107.3 Wh kg^{-1} at a power density of 81.2 W kg^{-1}. Cai and co-workers synthesized homogeneous flower-shaped VSe$_2$ spheres using MXene as a support [65]. The specific capacity of Zn–VSe$_2$/MXene cells was higher than the initial capacity after 2000 cycles at 5 A g^{-1} in the voltage window of 0.2–1.6 V (Figure 6d). This may be attributed to the generation of a Zn$_{0.25}$V$_2$O$_5$ (ZVO) phase. With repeated discharge/charging, the generation and accumulation of ZVO phase provides a continuous capacity contribution to the battery.

2.5. Ni$_3$S$_2$

Nickel sulfide (Ni$_3$S$_2$) is a promising electrode material with the advantages of high activity and theoretical capacity, excellent stability, and low cost [66,67]. However, Ni$_3$S$_2$ materials perform inferiorly in cycling stability and specific capacity when they are used as a cathode for AZIBs. Structural defects are an approach to modulate the crystal structure of the materials [68]. In addition, the introduction of defects can increase the electrical conductivity and carrier density, effectively improving the electrochemical activity of the electrode materials [69].

For instance, Tong et al. doped highly reactive Co ions in Ni$_3$S$_2$ nanocones by atomic layer deposition (ALD) and the hydrothermal method [70]. A Co12-Ni$_3$S$_2$/NF (C12NS) sample was used as the cathode; the hydrogel electrolyte is immersed in 5 M KOH/0.1 M Zn(AC)$_2$ solution to assemble a quasi-solid-state flexible cell. The reactivity of the electrodes is increased due to the Co doping and sulfation. The specific capacity of the activated electrode is about four times higher than that before activation. This is mainly due to the adsorption of OH radicals from the alkaline electrolyte on the sulfur sites of the Co-doped Ni$_3$S$_2$. The reaction mechanism can be described as:

$$\text{Cathode: Co-Ni}_3\text{S}_2 + 2\text{OH}^- \leftrightarrow \text{Co-Ni}_3\text{S}_2(\text{OH})_2 + 2\text{e}^- \tag{5}$$

$$\text{Co-Ni}_3\text{S}_2(\text{OH})_2 + 3\text{OH}^- \leftrightarrow \text{Co-Ni}_3\text{S}_2\text{O}_2\text{OH} + 2\text{H}_2\text{O} + 3\text{e}^- \tag{6}$$

$$\text{Anode: Zn[(OH)}_4]^{2-} + 2\text{e}^- \leftrightarrow \text{Zn} + 4\text{OH}^- \tag{7}$$

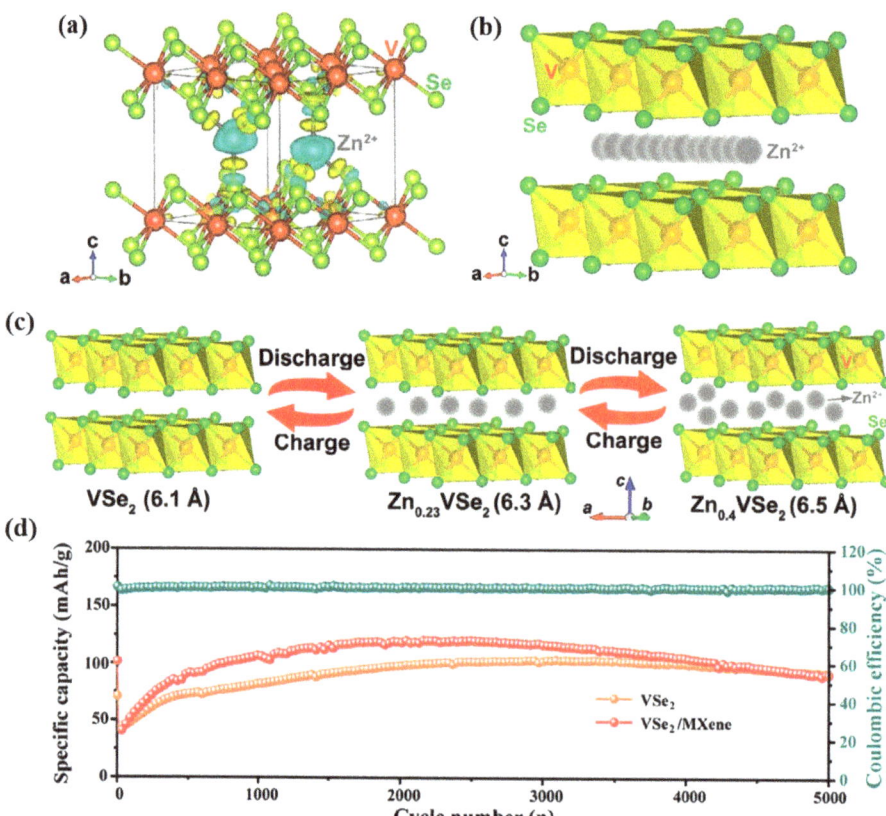

Figure 6. The kinetic process and cycle capability of a VSe$_2$-based Zn battery. (**a**) The schematic of charge density map after zinc-ion intercalation. (**b**) The optimal diffusion pathway of zinc ions. (**c**) Schematic of the two-step Zn^{2+} intercalation/de-intercalation process. Reproduced with permission [64]. Copyright 2020, Wiley-VCH GmbH. (**d**) Long-cycles performance at 5 A g^{-1}. Reproduced with permission [65]. Copyright 2022, Elsevier Inc.

During the continuous adsorption of hydroxide radicals, sulfur and oxygen coexist. The formed oxygen–sulfur bonds enhance the electrochemical performance of the electrode material, as shown in Figure 7a. The Co-doped electrodes present a low energy barrier. It indicates that they can achieve a fast Zn ion migration (Figure 7b). In Figure 7c, the capacity retention of Co12-Ni$_3$S$_2$/NF electrode after 5000 cycles is up to 90%.

In addition to the defect strategy, Wang's group prepared Ni/Ni$_3$S$_2$ nanocomposites with large specific surface areas using Ni–ZIF MOFs as precursors through a simple medium temperature solid–gas phase reaction [71]. The assembled batteries with 6 M KOH-0.6 M ZnO electrolyte possess an excellent rate capability (Figure 7d). They can maintain a capacity retention of 83% after 1000 cycles at 20 A g^{-1} (Figure 7e). Additionally, an energy density of 379 Wh kg^{-1} can be obtained at 340 W kg^{-1} at a high output voltage (1.7 V). Table 1 summarizes the electrochemical performance of TMC cathodes. It can be observed that these materials present some disadvantages, mainly in terms of limited voltage window, low specific capacity and inferior cycle life. In addition, the modification tactics lead to a large difference in the electrochemical performance of the electrodes. This demonstrates that an appropriate strategy can enhance the structural stability as well as the cycle life of the cells. The VS$_4$ composite shows significant advantages with a cycle life of 3500 cycles with a stable charge/discharge process. It can be noted that the electrolytes used in these batteries are varied. The high concentration of electrolyte contributes to the increase in capacity. Therefore, the choice of electrolyte is also crucial in future work to affect the electrochemical performance of the Zn/TMCs batteries.

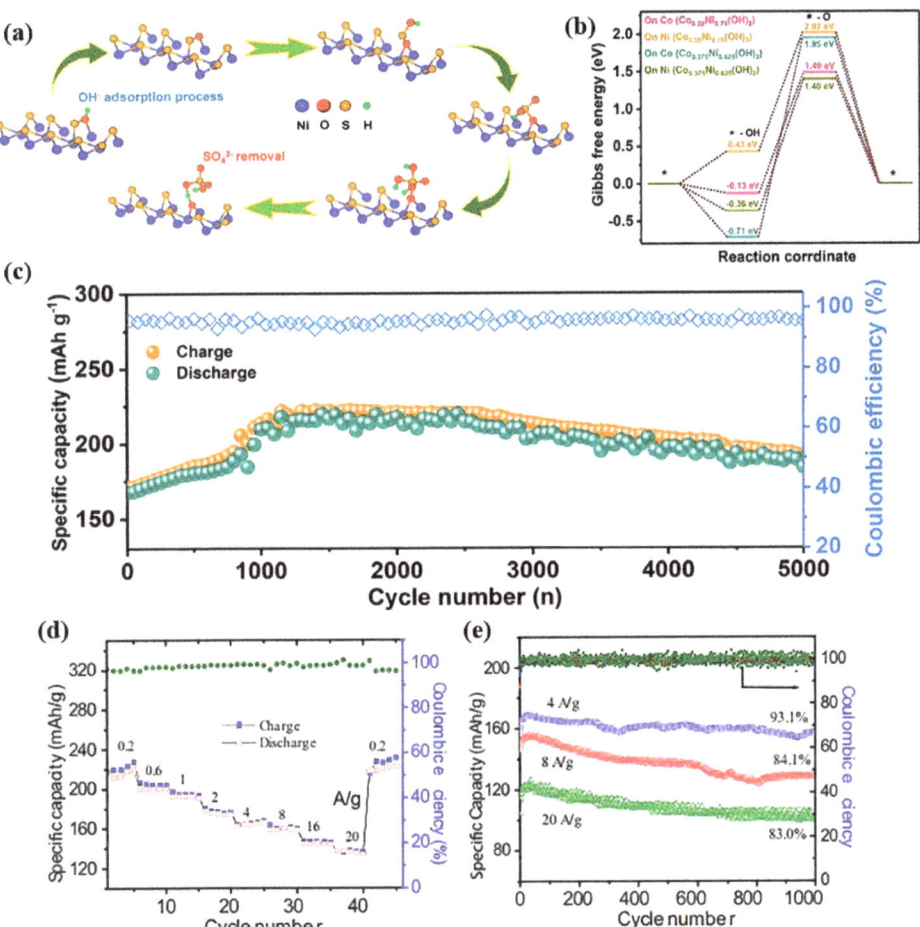

Figure 7. The electrochemical performance of Zn/Ni_3S_2 batteries. (**a**) Schematic of nickel sulfide-adsorbing hydroxide radicals and being oxidized to form nickel hydroxide and sulfate radicals. (**b**) Gibbs free energies of Zn/Ni_3S_2 batteries. (**c**) The cycling stability of the flexible ZC12NS battery. Reproduced with permission [70]. Copyright 2021, Elsevier B.V. and Science Press. (**d**) Rate capability. (**e**) The long-term performance. Reproduced with permission [71]. Copyright 2021, Elsevier B.V.

Table 1. The electrochemical performance of TMC cathodes.

Cathodes	Anodes	Voltage (V)	Electrolyte	Capacity (mAh g^{-1})	Cycle Stability (mAh g^{-1}, Cycles, A g^{-1})	Ref.
VS$_2$@VOOH	Zinc foil	0.4–1.0	3 M ZnSO$_4$	165 (0.1 A g^{-1})	91.4, 400, 2.5	[31]
VS$_2$ and N-doped carbon	Zinc foil	0.2–1.8	3 M Zn(CF$_3$SO$_3$)$_2$	203 (0.05 A g^{-1})	144, 600, 1	[33]
VS$_2$/VO$_x$	Zinc foil	0.1–1.8	25 M ZnCl$_2$	260 (0.1 A g^{-1})	75%, 3000, 1	[36]

Table 1. *Cont.*

Cathodes	Anodes	Voltage (V)	Electrolyte	Capacity (mAh g^{-1})	Cycle Stability (mAh g^{-1}, Cycles, A g^{-1})	Ref.
VS$_4$	Zinc foil	0.2–1.6	1 M ZnSO$_4$	310 (0.1 A g^{-1})	110, 500, 2.5	[37]
VS$_4$/V$_2$O$_3$	Zinc foil	0.3–1.2	3 M Zn(CF$_3$SO$_3$)$_2$	163 (0.1 A g^{-1})	-	[38]
VS$_4$@rGO	Zinc foil	0–1.8	1 M Zn(CF$_3$SO$_3$)$_2$	450 (0.5 A g^{-1})	82%, 3500, 10	[40]
MoS$_2$	Deposited zinc on carbon cloth	0.5–1.5	2 M ZnSO$_4$	202.6 (0.1 A g^{-1})	164.5, 600, 1	[45]
MoS$_2$-CTAB	Zn anode plated on carbon paper	0.2–1.3	3 M ZnSO$_4$	181.8 (0.1 A g^{-1})	92.8%, 2100, 10	[51]
MnS	zinc powder	0.9–1.95	2 M ZnSO$_4$ and 0.1 M MnSO$_4$	297 (0.1 A g^{-1})	150, 1000, 1	[58]
VSe$_2$	Zinc foil	0.2–1.6	2 M ZnSO$_4$	131.8 (0.1 A g^{-1})	80.8%, 500, 0.1	[64]
Ni/Ni$_3$S$_2$	Zinc foil	1.3–1.9	6 M KOH and 0.6 M ZnO	220 (0.2 A g^{-1})	93.1%, 1000, 4	[71]

3. Summary and Outlook

AZIBs have attracted a considerable attention as an alternative to LIBs with the features of being inexpensive, environmentally friendly, resource-rich, and having high theoretical capacity. However, current cathode materials are still hampered by their inferior conductivity, slow kinetics, structural instability, and dissolution of active substances. Among them, TMCs possess a unique layered structure and a large layer spacing and high conductivity, which facilitate the transfer of ion carriers. First, the high specific surface area provides many electrochemically active sites and short ion transfer paths. Secondly, its excellent electrical conductivity ensures fast electron transfer. Finally, the open-layer structure favors the embedding of electrolyte ions and reduces the volume change. We summarized the research advances of TMCs as electrode materials in recent years. The main focus was on their modification strategies and the improvement of zinc storage capacity. Reasonable modification strategies are beneficial for the improvement of energy storage capability. Nevertheless, the goal of commercialization is still not reached. More efforts may be required to try to shorten this gap.

The electrochemical performance of TMCs materials, such as specific capacity and long cycle performance, still needs to be improved. There are several strategies for considering directions for this: (1) introducing defects (vacancies or doping) in electrode materials. It can add many active sites and conductivity to effectively increase the capacity of the battery. (2) Combining high conductivity and specific surface area materials such as MOFs and rGO. MOFs possess rich pore structures. The calcination of MOFs precursors is an effective method to prepare metal compound/carbon composites. (3) Expanded layer spacing. The increase of layer spacing is beneficial to the rapid shuttling of zinc ions, which can effectively improve the electrochemical kinetics.

Although some TMCs have been employed for energy storage, there are few materials with desired zinc storage capability. Therefore, it is still necessary to explore the unexploited cathode materials (Figure 8). We should place our eyes on the study of materials such as WS$_2$, MoSe$_2$, etc., and focus on the design of effective modification strategies. Additionally, the mechanism of electrochemical energy storage is still unclear. Advanced in situ characterization techniques are beneficial to explore the evolution of the structure of electrode materials during charging and discharging as well as elemental valence changes.

In addition, the structural stability of electrode materials is closely associated with their electrochemical performance. The cycling process may generate irreversible phase changes and side reactions, which lead to structural collapse or capacity loss. The modulation of heterogeneous interfaces may suppress the occurrence of side reactions. Finally, flexible electronic devices have attracted much attention because of their high safety, portability, and wearable features. There are few reports related to flexible devices for AZIBs. They should focus on flexible substrates and cathodes with certain mechanical flexibility, and stable solid-state electrolytes with high ionic conductivity.

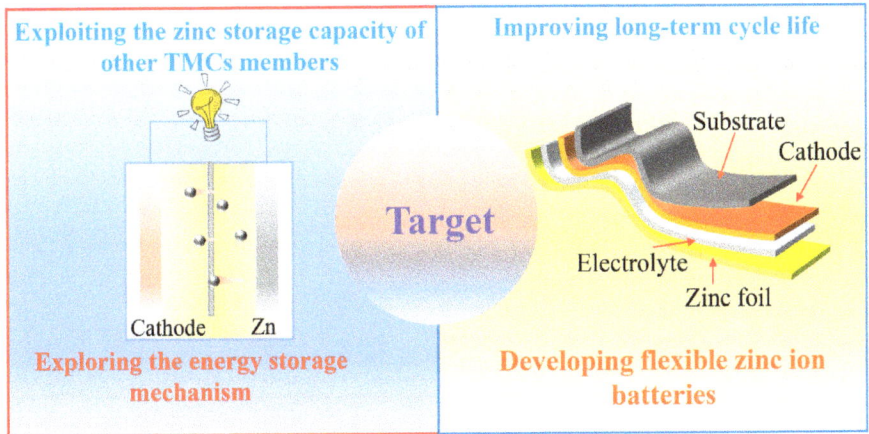

Figure 8. The future target of TMCs cathodes for AZIBs.

Author Contributions: Conceptualization, Y.L.; methodology, Y.L.; software, Y.L.; validation, Y.L.; formal analysis, Y.L.; investigation, Y.L.; resources, X.W.; data curation, Y.L. and X.W.; writing—original draft preparation, Y.L.; writing—review and editing, X.W.; visualization, Y.L.; supervision, X.W.; project administration, X.W.; funding acquisition, X.W. All authors have read and agreed to the published version of the manuscript.

Funding: This research was funded by National Natural Science Foundation of China (No. 52172218).

Conflicts of Interest: The authors declare no conflict of interest.

References

1. Goodenough, J.B. Energy storage materials: A perspective. *Energy Storage Mater.* **2015**, *1*, 158–161. [CrossRef]
2. Zhao, D.P.; Dai, M.Z.; Zhao, Y.; Liu, H.Q.; Liu, Y.; Wu, X. Improving electrocatalytic activities of $FeCo_2O_4@FeCo_2S_4@PPy$ electrodes by surface/interface regulation. *Nano Energy* **2020**, *72*, 104715. [CrossRef]
3. Ruan, P.; Liang, S.; Lu, B.; Fan, H.J.; Zhou, J. Design strategies for high-energy-density aqueous zinc batteries. *Angew. Chem. Int. Ed.* **2022**, *61*, e202200598. [CrossRef]
4. Liu, C.; Wu, X.; Wang, B. Performance modulation of energy storage devices: A case of Ni-Co-S electrode materials. *Chem. Eng. J.* **2020**, *392*, 123651. [CrossRef]
5. Tarascon, J.M.; Armand, M. Issues and challenges facing rechargeable lithium batteries. *Nature* **2001**, *414*, 359–367. [CrossRef]
6. Yang, D.; Zhou, Y.; Geng, H.; Liu, C.; Lu, B.; Rui, X.; Yan, Q. Pathways towards high energy aqueous rechargeable batteries. *Coord. Chem. Rev.* **2020**, *424*, 213521. [CrossRef]
7. Ju, Z.; Zhao, Q.; Chao, D.; Hou, Y.; Pan, H.; Sun, W.; Yuan, Z.; Li, H.; Ma, T.; Su, D.; et al. Energetic Aqueous Batteries. *Adv. Energy Mater.* **2022**, *12*, 2201074. [CrossRef]
8. Liu, Y.; Liu, Y.; Wu, X. Toward long-life aqueous zinc ion batteries by constructing stable zinc anodes. *Chem. Rec.* **2022**, e202200088. [CrossRef]
9. Li, C.; Xie, X.; Liang, S.; Zhou, J. Issues and future perspective on zinc metal anode for rechargeable aqueous zinc-ion batteries. *Energy Environ. Mater.* **2020**, *3*, 146–159. [CrossRef]
10. Liu, Y.; Wu, X. Hydrogen and sodium ions co-intercalated vanadium dioxide electrode materials with enhanced zinc ion storage capacity. *Nano Energy* **2021**, *86*, 106124. [CrossRef]
11. Song, M.; Tan, H.; Chao, D.; Fan, H.J. Recent advances in Zn-ion batteries. *Adv. Funct. Mater.* **2018**, *28*, 1802564. [CrossRef]
12. Yang, J.; Yin, B.; Sun, Y.; Pan, H.; Sun, W.; Jia, B.; Zhang, S.; Ma, T. Zinc Anode for Mild Aqueous Zinc-Ion Batteries: Challenges, Strategies, and Perspectives. *Nano-Micro Lett.* **2022**, *14*, 42. [CrossRef]
13. Liu, Y.; Liu, Y.; Wu, X.; Cho, Y.R. Enhanced electrochemical performance of Zn/VO_x batteries by a carbon-encapsulation strategy. *ACS Appl. Mater. Interfaces* **2022**, *14*, 11654–11662. [CrossRef]

14. Li, H.F.; Ma, L.T.; Han, C.P.; Wang, Z.F.; Liu, Z.X.; Tang, Z.J.; Zhi, C.Y. Advanced rechargeable zinc-based batteries: Recent progress and future perspectives. *Nano Energy* **2019**, *62*, 550–587. [CrossRef]
15. Liu, Y.; Wu, X. Review of vanadium-based electrode materials for rechargeable aqueous zinc ion batteries. *J. Energy Chem.* **2021**, *56*, 223–237. [CrossRef]
16. Guo, X.; Li, J.; Jin, X.; Han, Y.; Lin, Y.; Lei, Z.; Wang, S.; Qin, L.; Jiao, S.; Cao, R. A hollow-structured manganese oxide cathode for stable Zn-MnO$_2$ batteries. *Nanomaterials* **2018**, *8*, 301. [CrossRef]
17. Liu, B. Transition metal dichalcogenides for high-performance aqueous zinc ion batteries. *Batteries* **2022**, *8*, 62. [CrossRef]
18. Trócoli, R.; La Mantia, F. An aqueous zinc-ion battery based on copper hexacyanoferrate. *ChemSusChem* **2015**, *8*, 481–485. [CrossRef]
19. Liu, Y.; Hu, P.F.; Liu, H.Q.; Wu, X.; Zhi, C.Y. Tetragonal VO$_2$ hollow nanospheres as robust cathode material for aqueous zinc ion batteries. *Mater. Today Energy* **2020**, *17*, 100431. [CrossRef]
20. Yang, Q.; Mo, F.; Liu, Z.; Ma, L.; Li, X.; Fang, D.; Chen, S.; Zhang, S.; Zhi, C. Activating C-coordinated iron of iron hexacyanoferrate for Zn hybrid-ion batteries with 10000-cycle lifespan and superior rate capability. *Adv. Mater.* **2019**, *31*, 1901521. [CrossRef]
21. Liu, Y.; Liu, Y.; Yamauchi, Y.; Alothman, Z.A.; Kaneti, Y.V.; Wu, X. Enhanced zinc ion storage capability of V$_2$O$_5$ electrode materials with hollow interior cavities. *Batt. Supercap.* **2021**, *4*, 1867–1873. [CrossRef]
22. Zhong, Y.; Xu, X.; Veder, J.P.; Shao, Z. Self-recovery chemistry and cobalt-catalyzed electrochemical deposition of cathode for boosting performance of aqueous zinc-ion batteries. *iScience* **2020**, *23*, 100943. [CrossRef] [PubMed]
23. Liu, Y.; Liu, Y.; Wu, X.; Cho, Y.R. High performance aqueous zinc battery enabled by potassium ion stabilization. *J. Colloid Interface Sci.* **2022**, *628*, 33–40. [CrossRef] [PubMed]
24. Tan, C.; Cao, X.; Wu, X.J.; He, Q.; Yang, J.; Zhang, X.; Chen, J.; Zhao, W.; Han, S.; Nam, G.H.; et al. Recent advances in ultrathin two-dimensional nanomaterials. *Chem. Rev.* **2017**, *117*, 6225–6331. [CrossRef]
25. Naveed, A.; Yang, H.; Shao, Y.; Yang, J.; Yanna, N.; Liu, J.; Shi, S.; Zhang, L.; Ye, A.; He, B.; et al. A Highly reversible Zn anode with intrinsically safe organic electrolyte for long-cycle-life batteries. *Adv. Mater.* **2019**, *31*, 1900668. [CrossRef]
26. Liu, W.; Hao, J.; Xu, C.; Mou, J.; Dong, L.; Jiang, F.; Kang, Z.; Wu, J.; Jiang, B.; Kang, F. Investigation of zinc ion storage of transition metal oxides, sulfides, and borides in zinc ion battery systems. *Chem. Commun.* **2017**, *53*, 6872–6874. [CrossRef]
27. Liu, J.; Zhou, W.; Zhao, R.; Yang, Z.; Li, W.; Chao, D.; Qiao, S.Z.; Zhao, D. Sulfur-based aqueous batteries: Electrochemistry and strategies. *J. Am. Chem. Soc.* **2021**, *143*, 15475–15489. [CrossRef]
28. Gao, M.R.; Xu, Y.F.; Jiang, J.; Yu, S.H. Nanostructured metal chalcogenides: Synthesis, modification, and applications in energy conversion and storage devices. *Chem. Soc. Rev.* **2013**, *42*, 2986–3017. [CrossRef]
29. Gong, J.; Li, H.; Zhang, K.; Zhang, Z.; Cao, J.; Shao, Z.; Tang, C.; Fu, S.; Wang, Q.; Wu, X. Zinc-ion storage mechanism of polyaniline for rechargeable aqueous zinc-ion batteries. *Nanomaterials* **2022**, *12*, 1438. [CrossRef]
30. He, P.; Yan, M.; Zhang, G.; Sun, R.; Chen, L.; An, Q.; Mai, L. Layered VS$_2$ nanosheet-based aqueous Zn ion battery cathode. *Adv. Energy Mater.* **2017**, *7*, 1601920. [CrossRef]
31. Pu, X.; Song, T.; Tang, L.; Tao, Y.; Cao, T.; Xu, Q.; Liu, H.; Wang, Y.; Xia, Y. Rose-like vanadium disulfide coated by hydrophilic hydroxyvanadium oxide with improved electrochemical performance as cathode material for aqueous zinc-ion batteries. *J. Power Source* **2019**, *437*, 226917. [CrossRef]
32. Chen, T.; Zhu, X.; Chen, X.; Zhang, Q.; Li, Y.; Peng, W.; Zhang, F.; Fan, X. VS$_2$ nanosheets vertically grown on graphene as high-performance cathodes for aqueous zinc-ion batteries. *J. Power Source* **2020**, *477*, 228652. [CrossRef]
33. Liu, J.; Peng, W.; Li, Y.; Zhang, F.; Fan, X. A VS$_2$@N-doped carbon hybrid with strong interfacial interaction for high-performance rechargeable aqueous Zn-ion batteries. *J. Mater. Chem. C* **2021**, *9*, 6308–6315. [CrossRef]
34. Tan, Y.; Li, S.; Zhao, X.; Wang, Y.; Shen, Q.; Qu, X.; Liu, Y.; Jiao, L. Unexpected role of the interlayer "dead Zn^{2+}" in strengthening the nanostructures of VS$_2$ cathodes for high-performance aqueous Zn-ion storage. *Adv. Energy Mater.* **2022**, *12*, 2104001. [CrossRef]
35. Yang, M.; Wang, Z.; Ben, H.; Zhao, M.; Luo, J.; Chen, D.; Lu, Z.; Wang, L.; Liu, C. Boosting the zinc ion storage capacity and cycling stability of interlayer-expanded vanadium disulfide through in-situ electrochemical oxidation strategy. *J. Colloid Interface Sci.* **2022**, *607*, 68–75. [CrossRef]
36. Yu, D.; Wei, Z.; Zhang, X.; Zeng, Y.; Wang, C.; Chen, G.; Shen, Z.X.; Du, F. Boosting Zn^{2+} and NH$_4^{4+}$ storage in aqueous media via in-situ electrochemical induced VS$_2$/VO$_x$ heterostructures. *Adv. Funct. Mater.* **2020**, *31*, 2008743. [CrossRef]
37. Zhu, Q.; Xiao, Q.; Zhang, B.; Yan, Z.; Liu, X.; Chen, S.; Ren, Z.; Yu, Y. VS$_4$ with a chain crystal structure used as an intercalation cathode for aqueous Zn-ion batteries. *J. Mater. Chem. A* **2020**, *8*, 10761–10766. [CrossRef]
38. Ding, J.; Gao, H.; Liu, W.; Wang, S.; Wu, S.; Fang, S.; Cheng, F. Operando constructing vanadium tetrasulfide-based heterostructures enabled by extrinsic adsorbed oxygen for enhanced zinc ion storage. *J. Mater. Chem. A* **2021**, *9*, 11433–11441. [CrossRef]
39. Qin, H.; Yang, Z.; Chen, L.; Chen, X.; Wang, L. A high-rate aqueous rechargeable zinc ion battery based on the VS$_4$@rGO nanocomposite. *J. Mater. Chem. A* **2018**, *6*, 23757–23765. [CrossRef]
40. Chen, K.; Li, X.; Zang, J.; Zhang, Z.; Wang, Y.; Lou, Q.; Bai, Y.; Fu, J.; Zhuang, C.; Zhang, Y.; et al. Robust VS$_4$@rGO nanocomposite as a high-capacity and long-life cathode material for aqueous zinc-ion batteries. *Nanoscale* **2021**, *13*, 12370–12378. [CrossRef]
41. Gao, S.; Ju, P.; Liu, Z.; Zhai, L.; Liu, W.; Zhang, X.; Zhou, Y.; Dong, C.; Jiang, F.; Sun, J. Electrochemically induced phase transition in a nanoflower vanadium tetrasulfide cathode for high-performance zinc-ion batteries. *J. Energy Chem.* **2022**, *69*, 356–362. [CrossRef]

42. Zhao, C.; Yu, C.; Zhang, M.; Sun, Q.; Li, S.; Banis, M.N.; Han, X.; Dong, Q.; Yang, J.; Wang, G.; et al. Enhanced sodium storage capability enabled by super wide-interlayer-spacing MoS$_2$ integrated on carbon fibers. *Nano Energy* **2017**, *41*, 66–74. [CrossRef]

43. Liu, Y.; Zhao, D.; Liu, H.; Umar, A.; Wu, X. High performance hybrid supercapacitor based on hierarchical MoS$_2$/Ni$_3$S$_2$ metal chalcogenide. *Chin. Chem. Lett.* **2019**, *30*, 1105–1110. [CrossRef]

44. Liang, Y.; Feng, R.; Yang, S.; Ma, H.; Liang, J.; Chen, J. Rechargeable Mg batteries with graphene-like MoS$_2$ cathode and ultrasmall Mg nanoparticle anode. *Adv. Mater.* **2011**, *23*, 640–643. [CrossRef]

45. Li, H.; Yang, Q.; Mo, F.; Liang, G.; Liu, Z.; Tang, Z.; Ma, L.; Liu, J.; Shi, Z.; Zhi, C. MoS$_2$ nanosheets with expanded interlayer spacing for rechargeable aqueous Zn-ion batteries. *Energy Storage Mater.* **2019**, *19*, 94–101. [CrossRef]

46. Jiao, Y.; Mukhopadhyay, A.; Ma, Y.; Yang, L.; Hafez, A.M.; Zhu, H. Ion transport nanotube assembled with vertically aligned metallic MoS$_2$ for high rate lithium-ion batteries. *Adv. Energy Mater.* **2018**, *8*, 1702779. [CrossRef]

47. Lei, Z.; Zhan, J.; Tang, L.; Zhang, Y.; Wang, Y. Recent development of metallic (1T) phase of molybdenum disulfide for energy conversion and storage. *Adv. Energy Mater.* **2018**, *8*, 1703482. [CrossRef]

48. Huang, M.; Mai, Y.; Fan, G.; Liang, X.; Fang, Z.; Jie, X. Toward fast zinc-ion storage of MoS$_2$ by tunable pseudocapacitance. *J. Alloys Compd.* **2021**, *871*, 159541. [CrossRef]

49. Sheng, Z.; Qi, P.; Lu, Y.; Liu, G.; Chen, M.; Gan, X.; Qin, Y.; Hao, K.; Tang, Y. Nitrogen-doped metallic MoS$_2$ derived from a metal-organic framework for aqueous rechargeable zinc-ion batteries. *ACS Appl. Mater. Interfaces* **2021**, *13*, 34495–34506. [CrossRef]

50. Liu, J.; Xu, P.; Liang, J.; Liu, H.; Peng, W.; Li, Y.; Zhang, F.; Fan, X. Boosting aqueous zinc-ion storage in MoS$_2$ via controllable phase. *Chem. Eng. J.* **2020**, *389*, 124405. [CrossRef]

51. Yao, Z.; Zhang, W.; Ren, X.; Yin, Y.; Zhao, Y.; Ren, Z.; Sun, Y.; Lei, Q.; Wang, J.; Wang, L.; et al. A volume self-regulation MoS$_2$ superstructure cathode for stable and high mass-loaded Zn-ion storage. *ACS Nano* **2022**, *16*, 12095–12106. [CrossRef] [PubMed]

52. Liu, Y.; Wu, X. Strategies for constructing manganese-based oxide electrode materials for aqueous rechargeable zinc-ion batteries. *Chin. Chem. Lett.* **2022**, *33*, 1236–1244. [CrossRef]

53. Minakshi, M.; Singh, P.; Issa, T.B.; Thurgate, S.; Marco, R.D. Lithium insertion into manganese dioxide electrode in MnO$_2$/Zn aqueous battery. *J. Power Source* **2004**, *138*, 319–322. [CrossRef]

54. Sun, W.; Wang, F.; Hou, S.; Yang, C.; Fan, X.; Ma, Z.; Gao, T.; Han, F.; Hu, R.; Zhu, M.; et al. Zn/MnO$_2$ Battery Chemistry With H$^+$ and Zn^{2+} Coinsertion. *J. Am. Chem. Soc.* **2017**, *139*, 9775–9778. [CrossRef] [PubMed]

55. Chao, D.; Zhou, W.; Ye, C.; Zhang, Q.; Chen, Y.; Gu, L.; Davey, K.; Qiao, S.Z. An electrolytic Zn-MnO$_2$ battery for high-voltage and scalable energy storage. *Angew. Chem. Int. Ed.* **2019**, *58*, 7823–7828. [CrossRef] [PubMed]

56. Chen, X.; Li, W.; Xu, Y.; Zeng, Z.; Tian, H.; Velayutham, M.; Shi, W.; Li, W.; Wang, C.; Reed, D.; et al. Charging activation and desulfurization of MnS unlock the active sites and electrochemical reactivity for Zn-ion batteries. *Nano Energy* **2020**, *75*, 104869. [CrossRef]

57. Ma, S.C.; Sun, M.; Sun, B.Y.; Li, D.; Liu, W.L.; Ren, M.M.; Kong, F.G.; Wang, S.J.; Guo, Z.X. In situ preparation of manganese sulfide on reduced graphene oxide sheets as cathode for rechargeable aqueous zinc-ion battery. *J. Solid State Chem.* **2021**, *299*, 122166. [CrossRef]

58. Jiang, K.; Zhou, Z.; Wen, X.; Weng, Q. Fabrications of high-performance planar zinc-ion microbatteries by engraved soft templates. *Small* **2021**, *17*, 2007389. [CrossRef]

59. Lv, R.; Robinson, J.A.; Schaak, R.E.; Sun, D.; Sun, Y.; Mallouk, T.E.; Terrones, M. Transition metal dichalcogenides and beyond: Synthesis, properties, and applications of single- and few-layer nanosheets. *Acc. Chem. Res.* **2015**, *48*, 56–64. [CrossRef]

60. Xu, K.; Chen, P.; Li, X.; Wu, C.; Guo, Y.; Zhao, J.; Wu, X.; Xie, Y. Ultrathin nanosheets of vanadium diselenide: A metallic two-dimensional material with ferromagnetic charge-density-wave behavior. *Angew. Chem. Int. Ed.* **2013**, *52*, 10477. [CrossRef]

61. Yang, C.; Feng, J.; Lv, F.; Zhou, J.; Lin, C.; Wang, K.; Zhang, Y.; Yang, Y.; Wang, W.; Li, J.; et al. Metallic graphene-like VSe$_2$ ultrathin nanosheets: Superior potassium-ion storage and their working mechanism. *Adv. Mater.* **2018**, *30*, 1800036. [CrossRef]

62. Wu, Y.; Chen, H.; Zhang, L.; Li, Q.; Xu, M.; Bao, S.J. A rough endoplasmic reticulum-like VSe$_2$/rGO anode for superior sodium-ion capacitors. *Inorg. Chem. Front.* **2019**, *6*, 2935–2943. [CrossRef]

63. Ming, F.; Liang, H.; Lei, Y.; Zhang, W.; Alshareef, H.N. Solution synthesis of VSe$_2$ nanosheets and their alkali metal ion storage performance. *Nano Energy* **2018**, *53*, 11–16. [CrossRef]

64. Wu, Z.; Lu, C.; Wang, Y.; Zhang, L.; Jiang, L.; Tian, W.; Cai, C.; Gu, Q.; Sun, Z.; Hu, L. Ultrathin VSe$_2$ nanosheets with fast ion diffusion and robust structural stability for rechargeable zinc-ion battery cathode. *Small* **2020**, *16*, 2000698. [CrossRef]

65. Cai, S.; Wu, Y.; Chen, H.; Ma, Y.; Fan, T.; Xu, M.; Bao, S.J. Why does the capacity of vanadium selenide based aqueous zinc ion batteries continue to increase during long cycles? *J. Colloid Interface Sci.* **2022**, *615*, 30–37. [CrossRef]

66. Lin, Y.; Chen, G.; Wan, H.; Chen, F.; Liu, X.; Ma, R. 2D free-standing nitrogen-doped Ni-Ni$_3$S$_2$ @carbon nanoplates derived from metal-organic frameworks for enhanced oxygen evolution reaction. *Small* **2019**, *15*, 1900348. [CrossRef]

67. Liu, Y.; Hu, P.; Liu, H.; Song, J.; Umar, A.; Wu, X. Toward a high performance asymmetric hybrid capacitor by electrode optimization. *Inorg. Chem. Front.* **2019**, *6*, 2824–2831. [CrossRef]

68. Zhang, Y.; Tao, L.; Xie, C.; Wang, D.; Zou, Y.; Chen, R.; Wang, Y.; Jia, C.; Wang, S. Defect engineering on electrode materials for rechargeable batteries. *Adv. Mater.* **2020**, *32*, 1905923. [CrossRef]

69. Liu, Z.; Sun, H.; Qin, L.; Cao, X.; Zhou, J.; Pan, A.; Fang, G.; Liang, S. Interlayer doping in layered vanadium oxides for low-cost energy storage: Sodium-ion batteries and aqueous zinc-ion batteries. *ChemNanoMat* **2020**, *6*, 1553–1566. [CrossRef]

70. Tong, X.; Li, Y.; Pang, N.; Zhou, Y.; Wu, D.; Xiong, D.; Xu, S.; Wang, L.; Chu, P.K. Highly active cobalt-doped nickel sulfide porous nanocones for high-performance quasi-solid-state zinc-ion batteries. *J. Energy Chem.* **2022**, *66*, 237–249. [CrossRef]

71. Wang, Q.; Liu, T.; Chen, Y.; Wang, Q. The Ni/Ni$_3$S$_2$ nanocomposite derived from Ni-ZIF with superior energy storage performance as cathodes for asymmetric supercapacitor and rechargeable aqueous zinc ion battery. *J. Alloys Compd.* **2022**, *891*, 161935. [CrossRef]

Article

Graphene-Modified Co-B-P Catalysts for Hydrogen Generation from Sodium Borohydride Hydrolysis

Xinlei Jia [1,†], Zhen Sang [1,†], Lixian Sun [1,2,*], Fen Xu [1,*], Hongge Pan [3,*], Chenchen Zhang [1], Riguang Cheng [1], Yuqian Yu [1], Haopan Hu [1], Li Kang [1] and Yiting Bu [1]

1   School of Material Science & Engineering, Guangxi Key Laboratory of Information Materials, Guangxi Collaborative Innovation Center of Structure and Property for New Energy and Materials, Guilin University of Electronic Technology, Guilin 541004, China
2   School of Mechanical & Electrical Engineering, Guilin University of Electronic Technology, Guilin 541004, China
3   School of New Energy Science and Technology, Xi'an Technological University, Xi'an 710021, China
*   Correspondence: sunlx@guet.edu.cn (L.S.); xufen@guet.edu.cn (F.X.); hgpan@zju.edu.cn (H.P.)
†   These authors contributed equally to this work.

Abstract: Sodium borohydride ($NaBH_4$) is considered a good candidate for hydrogen generation from hydrolysis because of its high hydrogen storage capacity (10.8 wt%) and environmentally friendly hydrolysis products. However, due to its sluggish hydrogen generation (HG) rate in the water, it usually needs an efficient catalyst to enhance the HG rate. In this work, graphene oxide (GO)-modified Co-B-P catalysts were obtained using a chemical in situ reduction method. The structure and composition of the as-prepared catalysts were characterized, and the catalytic performance for $NaBH_4$ hydrolysis was measured as well. The results show that the as-prepared catalyst with a GO content of 75 mg (Co-B-P/75rGO) exhibited an optimal catalytic efficiency with an HG rate of 12087.8 mL min^{-1} g^{-1} at 25 °C, far better than majority of the findings that have been reported. The catalyst had a good stability with 88.9% of the initial catalytic efficiency following 10 cycles. In addition, Co-, B-, and P-modified graphene showed a synergistic effect improving the kinetics and thermodynamics of $NaBH_4$ hydrolysis with a lower activation energy of 28.64 kJ mol^{-1}. These results reveal that the GO-modified Co-B-P catalyst has good potential for borohydride hydrolysis applications.

Keywords: $NaBH_4$; graphene oxide; catalytic activity; hydrolysis

Citation: Jia, X.; Sang, Z.; Sun, L.; Xu, F.; Pan, H.; Zhang, C.; Cheng, R.; Yu, Y.; Hu, H.; Kang, L.; et al. Graphene-Modified Co-B-P Catalysts for Hydrogen Generation from Sodium Borohydride Hydrolysis. *Nanomaterials* 2022, 12, 2732. https://doi.org/10.3390/nano12162732

Academic Editor: Ioannis V. Yentekakis

Received: 4 July 2022
Accepted: 6 August 2022
Published: 9 August 2022

Publisher's Note: MDPI stays neutral with regard to jurisdictional claims in published maps and institutional affiliations.

1. Introduction

Since the first industrial revolution, the overconsumption of fossil energy has created issues of air pollution and energy storage [1]. Therefore, the development of new renewable green and efficient energy has become an urgent matter for the future development of society and the economy. Hydrogen is expected to be a fossil energy alternative, which relies on its outstanding features of the nonemission of pollutants and high efficiency [2]. In general, there are several ways, such as photocatalysis, biomass decomposition, chemical hydrides hydrolysis, to produce hydrogen [3,4]. In the method described above, hydrogen-rich compound hydrolysis, such as $NaBH_4$ [5] and ammonia borane (NH_3BH_3) [6], has been considered as a convenient, economical, and efficient way to produce hydrogen.

$NaBH_4$ is rich in hydrogen (10.8 wt%), environmentally friendly, safe, and non-flammable, which can be employed for producing hydrogen through hydrolysis reactions [7]. The hydrolysis reaction occurs through the following reactions:

$$NaBH_4 + 2H_2O \rightarrow NaBO_2 + 4H_2 + heat \ (217 \ kJ \ mol^{-1}) \tag{1}$$

During this process, four moles of H_2 can be produced by one mole of $NaBH_4$. In particular, $NaBH_4$ and water each provide 50% of the hydrogen. [8]. In addition, the byproduct $NaBO_2$ can be collected to reproduce $NaBH_4$, which shows a sustainable development value. However, the $NaBH_4$ hydrolysis reaction exhibits sluggish kinetics in the solutions. Previous reports have proved that selecting an appropriate catalyst can significantly improve the HG rate. Noble metal-based catalysts (Pt [9,10], Ru [11,12], and Pd [13,14]) have shown positive catalytic performance. However, their scarce storage and high price limit the related practical applications. Transition metal catalysts (Co [15,16], Ni [17], and Co-Ni [18]) with inferior cost and relatively good catalytic activity have been broadly investigated for hydrogen production from $NaBH_4$. In addition, transition metals combined with heteroatoms, such as boron (B) and phosphorus (P), could further enhance the catalytic activity [19,20]. For example, Patel et al. reported that the transition-metal borides (e.g., CoB, NiB) exhibited superior catalytic activities due to the mutual electronic interaction between boron and transition metals (Co or Ni), thus preventing them from oxidation and protecting the active metal center. [21]. Chen et al. prepared cobalt–phosphorus (Co-P) catalysts and investigated their catalytic efficiency in alkaline sodium borohydride solutions. The Co-P catalyst showed favorable hydrolysis performance with a low activation energy, which was attributed to the improvement of the catalytic performance by the appropriate amount of P doping [22]. So far, catalysts including Co-P [22], Ni-B [23], Co-W-B [24], Co-Ni-B [25], Co-B-P [26], etc., have been extensively researched and have shown good catalytic performance. Although these catalysts possess preferable catalytic activity, they usually show a low cycle stability. To address the above issue, selecting a suitable matrix, such as MOFs [27], porous carbon [28], MWCNTs [29], SiO_2 [30], and γ-Al_2O_3 [31], which possesses a high specific surface area to support the active metals, can effectively improve the catalytic performance. Recently, graphene with excellent physical and chemical characteristics has been researched, making it an ideal carrier material to support metal clusters [32,33]. The large specific surface area can not only improve the distribution of metal clusters, thereby exposing more catalytically active sites for catalysis reaction, but it can also suppress the aggregation issue during the catalytic process, thus presenting a superior catalytic performance.

In this study, we successfully prepared the graphene modified Co-B-P catalysts through chemical in situ reduction. The structural characteristics and catalytic efficiency of the Co-B-P/xrGO (x = 25, 50, 75, 100) catalysts were studied. The Co-B-P/75rGO catalyst exhibited an optimal catalytic performance with an average HG rate of 12,087.8 mL min^{-1} g^{-1}. In addition, the effects of the GO content and heteroatom types on the catalytic activity of $NaBH_4$ hydrolysis were also studied. The excellent hydrogen generation performance is attributed to the fact that the large specific surface area of graphene oxide can better disperse Co-B-P clusters and thus expose more active sites. Meanwhile the elemental B and P doping exhibits a synergistic catalytic effect. This is because the presence of GO increases the specific surface area for uniform dispersion of Co-B-P clusters on the surface of the GO, thus exposing more catalytically active sites for the hydrolysis reaction.

2. Materials and Methods

2.1. Materials

High purity flake graphite (300 mesh), sulfuric acid, hydrochloric acid, hydrogen peroxide, sodium nitrate, potassium permanganate, sodium borohydride, sodium hydroxide, cobalt chloride hexahydrate, and sodium hypophosphite monohydrate were obtained from Alfa Aesar Co., Ltd. (Tianjin, China). The chemicals used were analytical reagent. All experiments used ultrapure water.

2.2. Synthesis of GO

We prepared the GO materials using a modification of the Hummers method [34]. First, concentrated H_2SO_4 (60 mL), $NaNO_3$ (2 g), and flake graphite (2 g) were mixed at 5 °C to obtain a solution. Subsequently, 12 g of $KMnO_4$ was slowly added to the above solution, and the solution was heated to 35 °C for 7 h with magnetic stirring. Then, 200 mL ice water and 15 mL H_2O_2 were added in turn to the mixed solution, until the mixed solution changed from brown to bright yellow. Next, the mixed solution was repeatedly washed with hydrochloric acid and deionized water until the pH was 7 to obtain the GO solution. Finally, the GO was obtained after freeze-drying for 72 h.

2.3. Catalyst Preparation

The Co-B-P/75rGO was obtained through chemical in situ reduction synthesis. First, GO (75 mg), $CoCl_2 \cdot 6H_2O$ (5 mmol), and $NaH_2PO_2 \cdot H_2O$ (30 mmol) were dispersed into 20 mL ultrapure water with sonication for 30 min. Next, 20 mL solution containing an appropriate amount of $NaBH_4$ (30 mmol) was slowly dropped into the reaction solution with intense agitation. After being aged in an ice water bath for 10 h, the Co-B-P/75rGO catalysts were obtained after washing with water, washing with ethanol, and drying. For comparison, a Co-B-P cluster without GO was prepared under the same conditions. In addition, we controlled the addition of GO to be 25, 50, and 100 mg, and the obtained corresponding composites were labeled as Co-B-P/xrGO (x = 25, 50, and 100), respectively. The comparison samples of CoB, CoP, and CoBP without GO were prepared under the same conditions.

2.4. Catalyst Characterization

The Co-B-P/xrGO catalyst structures were analyzed by X-ray diffraction (XRD). The chemical structures of the catalyst were characterized by Fourier transform infrared (FTIR) spectroscopy. The elemental valence states of the Co-B-P/xrGO catalysts were determined by X-ray photoelectron spectroscopy (XPS). The morphologies of the Co-B-P/xrGO catalysts were determined by scanning electron microscopy (SEM). The degree of graphitization of the Co-B-P/xrGO catalysts was analyzed by Raman microscope (Raman spectra). The specific surface areas of the Co-B-P/xrGO catalysts were calculated by the Brunauer–Emmett–Teller (BET) method. The bulk elemental composition of Co, B, and P in the as-prepared Co-B-P/xrGO catalysts was measured via inductive coupled plasma–optical emission spectroscopy (ICP-OES).

2.5. Hydrogen Generation Measurement

The catalytic efficiency of Co-B-P/xrGO in alkaline $NaBH_4$ solution was evaluated by a laboratory fabricated self-assembled drainage device [35]. The amount of hydrogen produced was determined by the volume of water drained, and the hydrogen generation rate was calculated through tracking the volume of water expelled at regular periods. Firstly, 0.1 g of Co-B-P/xrGO was added to a dry 125 mL wide-mouth flask. Then, 10 mL of a solution (1.5 wt% $NaBH_4$ and 5 wt% NaOH) was placed into a wide-mouth flask through a 10 mL capacity syringe, and an appropriate amount of sodium hydroxide solution inhibited the $NaBH_4$ self-hydrolysis reaction. The hydrogen generation efficiency of the catalyst hydrolysis was tested at different temperatures, and the reaction activation energy (Ea) was evaluated by the exponential law of reaction rate. After the hydrolysis test, the catalyst was washed with water and vacuum dried for 10 h. Then, the catalyst was tested for durability by adding 10 mL of fresh $NaBH_4$ solution as described above.

3. Results and Discussion

3.1. Catalyst Characterization

The Co-B-P/xrGO was prepared through the chemical in situ reduction method (Figure 1) [36]. In a typical procedure, GO material was prepared by a modified Hummers method and distributed in ultrapure water under ultrasonic conditions. Subsequently, Co^{2+} was anchored on the GO surface by the electrostatic adsorption. After adding the $NaH_2PO_2 \cdot H_2O$ and $NaBH_4$ solution, GO was reduced to rGO, and Co-B-P clusters formed on the rGO surface [37].

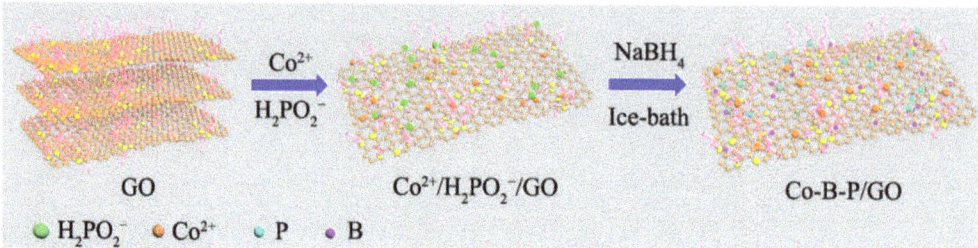

Figure 1. The illustration of the synthetic route of Co-B-P/xrGO.

The microscopic morphology and nanostructure of the catalysts were characterized through SEM. The prepared pure Co-B-P alloy catalyst was agglomerated in a granular morphology (Figure 2a), which was ascribed to the exothermic nature of the catalyst during the preparation process. Figure 2b shows that the GO was successfully synthesized by the modified Hummers method with a typical pleated-sheet morphology. To overcome the aggregation issue, GO with a typical pleated structure can act as a matrix material to disperse the Co-B-P clusters [38]. For exploring the effect of the GO addition on the catalytic performance, catalysts with different contents of GO were prepared. Figure 2c–f show the morphologies of Co-B-P/xrGO (x = 25, 50, 75, and 100), respectively. All the SEM images showed that the metal clusters were tightly anchored to the surface of the reduced graphene. Increasing the content of GO means the larger specific surface area can be used to provide a larger space for the dispersion of Co-B-P clusters. The Co-B-P clusters tended to grow uniformly on the surface of the reduced graphene. However, when the addition content was 100 mg, the redundant reduced graphene wrapped around leading to the aggregation issue of Co-B-P clusters. Among them, Co-B-P/75rGO exhibited an optimal morphology with Co-B-P clusters tightly and uniformly anchored on the surface of the reduced graphene. This structure can expose more active sites for the catalytic reaction, which was verified in subsequent hydrolysis catalysis measurements [29]. In addition, the EDX spectra (Figure 2g–l) showed that the Co, B, P, O, and C elements were uniformly dispersed in the Co-B-P/75rGO catalyst.

The XRD patterns and Raman spectra of Co-B-P/xrGO were measured as shown in Figure 3a. A broad diffraction peak near $2\theta = 45°$ corresponded to the Co-B and Co-P phases, indicating that the as-prepared catalysts were a typical amorphous structure [39,40], and the addition of GO would not affect the amorphous structure of the catalyst. The peaks around 26.0° belonged to the (002) plane of reduced graphene, indicating that the GO was reduced. The short-range ordered and long-range disordered amorphous structures are generally considered to have an unsaturated surface coordination, which has been proved to be beneficial for catalytic hydrolysis [21]. The characteristic peaks of the D-band and G-band were observed near 1350 and 1580 cm^{-1}, as shown in Figure 3b. The ratio of the strength of the D band to the G band represents the disorder of the carbon-based hybrid material [41]. Experimental results showed that with the addition of GO, the I_D/I_G values of all catalysts were greater than 1.00; in particular, Co-B-P/75rGO (I_D/I_G) reached 1.28. The I_D/I_G value indicated that Co-B-P/75rGO had more defects, which can anchor more metal and metal-like clusters to improve the catalytic performance. The following performance test experiments also confirmed this conclusion.

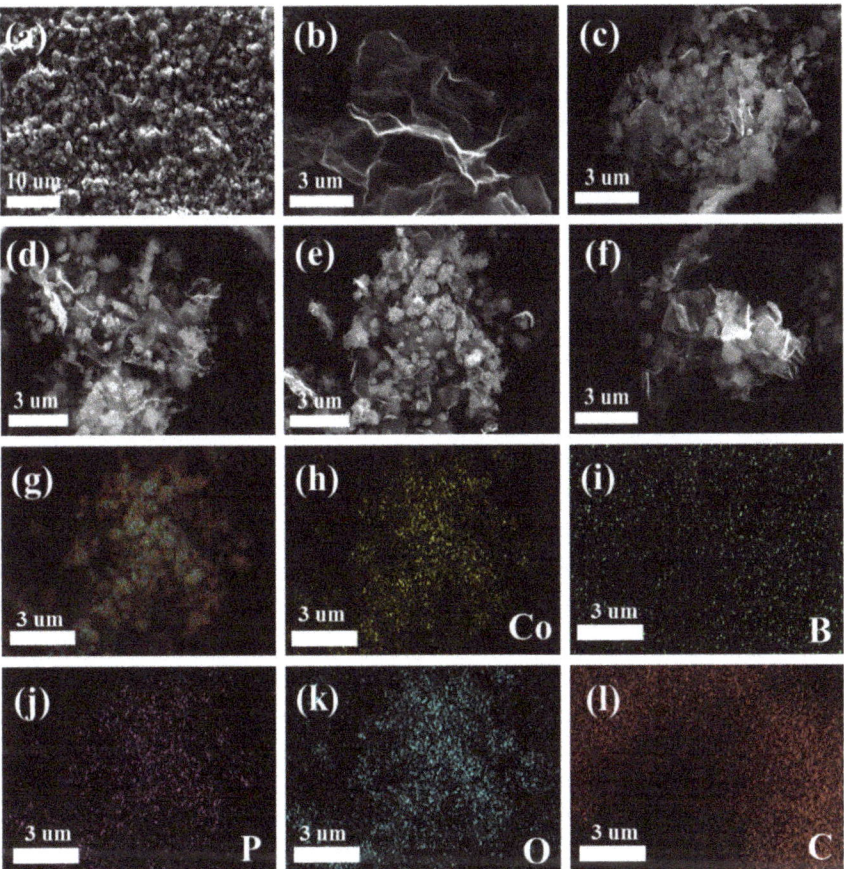

Figure 2. SEM images of (**a**) Co-B-P, (**b**) GO, (**c–f**) Co-B-P/xrGO (x = 25, 50, 75, and 100), and EDX mapping images of (**g**) Co-B-P/75rGO, (**h**) Co, (**i**) B, (**j**) P, (**k**) O, and (**l**) C in Co-B-P/xrGO.

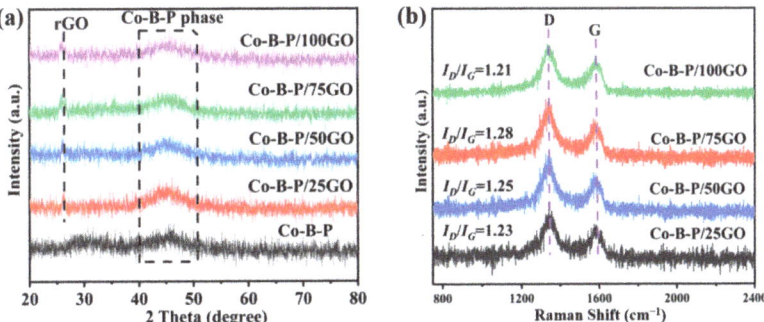

Figure 3. (**a**) XRD patterns of Co-B-P and Co-B-P/xrGO catalysts; (**b**) Raman spectra of Co-B-P/xrGO catalysts (x = 25, 50, 75, and 100).

The chemical structures of GO and the Co-B-P/75rGO were characterized by FTIR (Figure 4). For the spectra of GO, the peak of the -OH stretching vibration of water molecules appeared at 3431 cm^{-1} [42]. The characteristic peaks at 1736, 1630, and 1089 cm^{-1} were observed for the -COOH stretching vibration, C=C bond skeleton vibration, and C-O-C vibration of GO, respectively [42]. The considerable numbers of oxygen-containing groups contained in the GO were produced during the oxidation of the graphite with a strong oxidizer, which can easily absorb metal ions. The FTIR spectrum of Co-B-P/75rGO was similar to the GO; yet, the peak near 1736 cm^{-1} disappeared. We ascribed this to the addition of $H_2PO_2^-$ and BH_4^-, which acted as reducing agents to reduce the GO to reduced graphene (rGO) [39]. These experimental results show that the Co-B-P/75rGO catalyst was successfully synthesized.

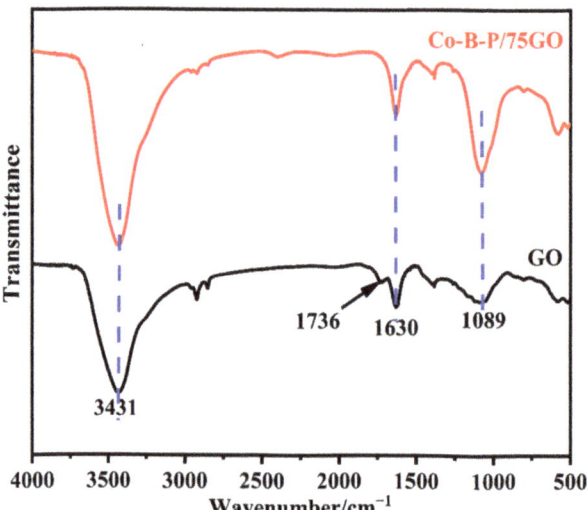

Figure 4. FTIR spectra of GO and CO-B-P/75rGO.

The surface interactions and electronic states of the Co-B-P/75rGO catalyst were investigated by XPS. In the XPS spectrum of Co 2p (Figure 5b), the two major peaks at 781.4 and 797.2 eV were Co 2p 3/2 and Co 2p 1/2, respectively [43], while two satellite peaks were observed at 786.5 and 803.1 eV, indicating the presence of elemental Co and the oxidized state of Co in the catalyst [44]. The C 1s spectrum (Figure 5c) showed three peaks located at 284.8, 285.9, and 288.7 eV, which belonged to the C-C/C=C, C-O, and O=C-O groups of rGO, respectively [45]. The peaks of B 1s at 187.7 and 191.2 eV were attributed to boron in the elemental and oxidized states, respectively. The elemental boron was positively shifted by 1.2 eV compared to the pure boron (186.5 eV) binding energy [46]. This was due to the transfer of electrons from boron to cobalt, filling the empty d-orbitals of cobalt (Figure 5d). In the O 1s XPS spectrum, two peaks located at 531.6 and 533.0 eV were ascribed to -C=O and -C-O, respectively (Figure 5e). In addition, two distinctive characteristic peaks near 129.5 and 132.9 eV in the full spectrum of element P (Figure 5f) were attributed to the presence of P^0 and P-O, respectively [29]. Due to the high electronegativity of P, the binding energy of P^0 was negatively shifted by 0.7 eV compared to pure P (130.2 eV) [47]. Apparently, as shown in Figure 2, the binding energy of cobalt in Co-B-P/75rGO was positively shifted by 0.3 eV compared to that in Co-B/75rGO.These experimental results suggest that there was an interaction between Co, B, and P, which is more favorable for catalysis.

Nanomaterials **2022**, *12*, 2732

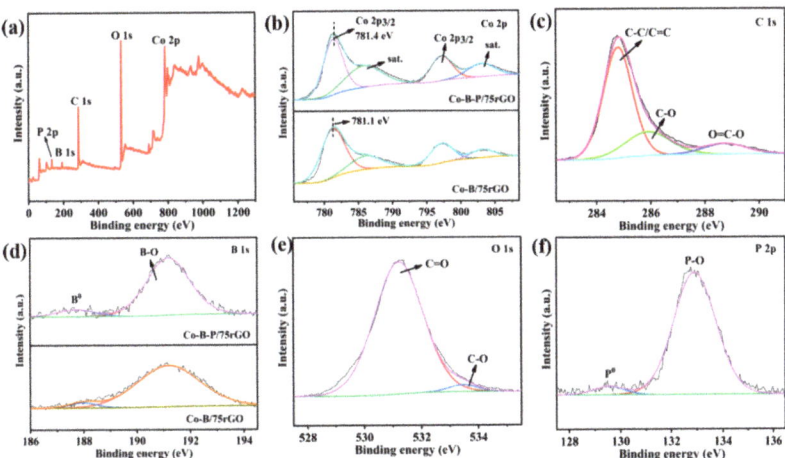

Figure 5. XPS analysis of Co-B-P/75rGO: (**a**) full spectrum, (**b**) Co 2p, (**c**) C 1s, (**d**) B 1s, (**e**) O 1s, and (**f**) P 2p.

The specific surface area and surface pore characteristics of the catalysts were tested by an Autosorb-iQ analyzer. According to the IUPAC classification, both curves in Figure 6a show hysteresis back loops, which were apparently type IV isotherms, indicating that both catalysts had a mesoporous characteristic [48]. The mesoporous channels are beneficial to the diffusion and contact between catalyst and reactant [49]. In addition, according to Table 1, the specific surface area of the Co-B-P/75rGO catalyst increased from 3 m^2/g to 89 m^2/g as the GO was added. Compared with pure Co-B-P, the total pore volume of Co-B-P/75rGO was increased, and the average pore diameter of 12.0 nm decreased to 9.0 nm. The addition of GO significantly increased the specific surface area for uniform distribution of Co-B-P clusters; thus, the composite catalyst offered more active sites for catalyzing the hydrolysis.

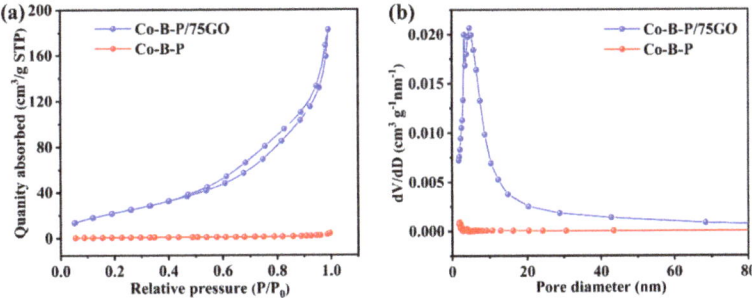

Figure 6. Nitrogen sorption isotherms (**a**) and pore-size distributions (**b**) for the Co-B-P and Co-B-P/75rGO catalysts.

Table 1. Textural parameters of the Co-B-P and Co-B-P/75rGO catalysts.

Catalyst	Specific Surface Area (m^2 g^{-1})	Pore Volume (cm^3 g^{-1})	Average Pore Diameter (nm)
Co-B-P	3	0.01	12.0
Co-B-P/75rGO	89	0.28	9.0

3.2. Effect of Different Types of Catalysts

In order to evaluate the properties of the catalyst, performance tests with different comparison samples were carried out. Figure 7 shows the hydrogen production per unit time of sodium borohydride hydrolysis catalyzed by the GO, Co-P, Co-B, Co-B-P, and Co-B-P/75rGO catalysts, and the magnitude of the slope represents the different superior and inferior performances. The experimental results showed that pure GO had less catalytic performance when used for $NaBH_4$ hydrolysis. Moreover, the combination of Co elements with heteroatoms (e.g., B and P) presented better catalytic performance than pure Co-based catalysts, which is due to the addition of heteroatoms forming electronic interactions with Co, thereby enhancing the catalytic behavior [21]. Based on the above conclusion, the Co-B-P catalyst with two heteroatoms exhibited a better performance than Co-B and Co-P catalysts because of the synergistic effect between Co, B, and P. Moreover, after combining Co-B-P with GO, the Co-B-P/rGO catalyst presented the optimal catalytic activity and has a higher competitive advantage over previously reported catalysts (Table 2). This is because the presence of GO increased the specific surface area for the uniform dispersion of the Co-B-P clusters on the surface of GO, thus exposing more catalytically active sites for the hydrolysis reaction [38]. Therefore, our further research was based on the Co-B-P/rGO catalyst.

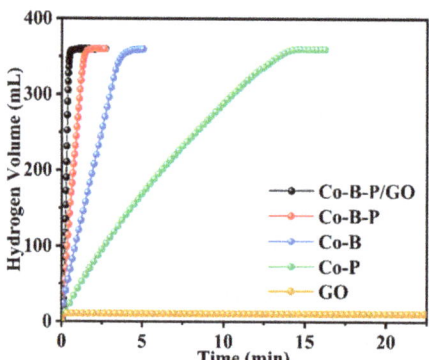

Figure 7. Hydrogen volume versus time for GO, Co-P, Co-B, Co-B-P, and Co-B-P/rGO (batch system, 25 °C, 1.5 wt% $NaBH_4$ + 5 wt% NaOH, 0.1 g catalyst).

Table 2. The Co-B-P/75rGO catalyst was compared with those previously reported in the literature.

Sample	Maximum HG Rate (mL min^{-1} g^{-1})	E_a (kJ mol^{-1})	Number of Cycles	Cyclic Stability	References
Co@3DGO	4394	37.42	5	54.0%	[50]
Co@GO	5955	64.87	5	73.0%	[51]
Co-P	1647.9	47.0	5	31.0%	[39]
CoO-Co$_2$P	3940	27.4	4	60.0%	[52]
Co@N MGC-500	3575	35.2	20	82.5%	[53]
Cu-Co-P/γ-Al$_2$O$_3$	1115	47.8	6	66.0%	[54]
Co-P/CNTs-Ni foam	2430	49.94	8	74.0%	[55]
Co-B-10CNTs	12,000	23.5	5	64.0%	[29]
Co-O-P	4850	63	5	78.0%	[56]
Co-B-50GO	14,430	26.2	5	81.5%	[40]
Co-B-P/75rGO	12,087.8	28.64	10	88.9%	This work

3.3. Effect of GO Amount

The appropriate amount of carrier plays a crucial role in the synthesis of catalysts. The effects on Co-B-P/rGO catalysts with different amounts of GO (25, 50, 75, and 100 mg) for the catalytic activity of $NaBH_4$ were also investigated. The hydrolysis of $NaBH_4$

experiments showed that, with an increase in the amount of GO, the HG rate first increased and then decreased. The Co-B-P/75rGO sample with 75 mg GO presented an optimal catalytic performance with the HG rate of 12,087.8 mL min^{-1} g^{-1} (Figure 8a,b). Previous studies have proved that metal clusters play a major role in the hydrolysis $NaBH_4$ reaction. The chemical composition of the prepared catalysts with different amounts of GO were determined by ICP–OES (Table 3). The results showed that the Co-B-P/75rGO catalyst had the highest Co content (61.79%), which also corresponded to the results of the hydrolysis experiment. Combined with the textural and surface morphology analysis, there were two factors for the superior performance of the Co-B-P/75rGO. First, the optimal content of GO supplied sufficient specific surface area for uniform distribution of Co-B-P/75rGO and provided more active sites for catalysis reaction [57]. Meanwhile, B and P heteroatoms doping led to a higher electron density in the active site of the catalyst, thus exhibiting a better catalytic performance.

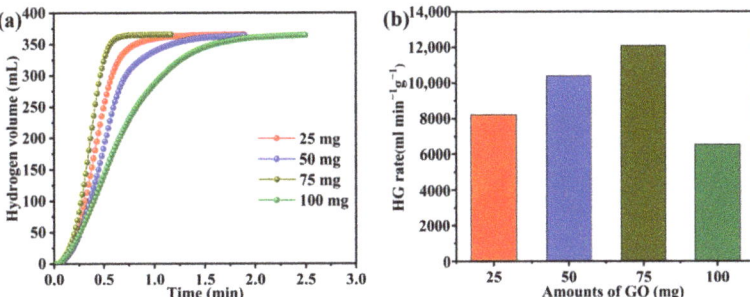

Figure 8. Hydrogen volume versus reaction time for the as-prepared catalysts (**a**); the histogram of the H_2 generation rate versus the additive amount of GO (**b**) (batch system, 25 °C, 1.5 wt% $NaBH_4$ + 5 wt% NaOH, 0.1 g catalyst).

Table 3. The chemical composition of the prepared catalysts with different amounts of GO were determined by ICP–OES.

Catalyst	Amount of Co (wt%)	Amount of B (wt%)	Amount of P (wt%)
Co-B-P/25GO	35.80	0.04	16.48
Co-B-P/50GO	40.2	1.02	12.62
Co-B-P/75GO	61.79	2.51	5.50
Co-B-P/100GO	34.65	0.72	14.34

3.4. Effect of Catalyst Amount

In order to investigate the relationship between the catalyst amount and catalytic performance, four groups of different masses of Co-B-P/75rGO (25, 50, 75, and 100 mg) were tested for hydrolysis performance (Figure 9a). Each test reached the theoretical capacity of hydrogen volume, and the HG rate became increasingly faster with the increase in catalyst dosage. A linear relationship between the two can be seen in Figure 9b. This indicates that the Co-B-P/75rGO catalyst's catalyzing hydrogen production from $NaBH_4$ was characterized by first-order reaction kinetics.

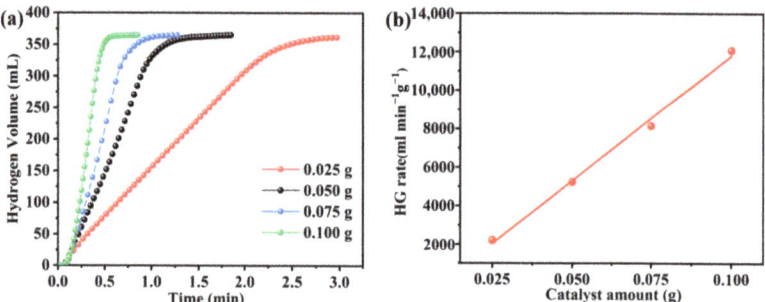

Figure 9. (**a**) Effect of catalyst loadings on the HG rate (batch system, 25 °C, 1.5 wt% NaBH$_4$ + 5 wt% NaOH); (**b**) HG rate versus catalyst dosage.

3.5. Effect of NaBH$_4$ Concentration

The effect of NaBH$_4$ concentration on hydrogen generation was studied under the condition of 0.1 g Co-B-P/75rGO catalyst and 25 °C (Figure 10a). The generated hydrogen volume was gradually increased to the theoretical capacity after increasing the NaBH$_4$ concentration from 0.5 wt% to 2.0 wt%. In addition, Figure 10b shows that the HG rate remained nearly identical as the NaBH$_4$ concentration increased. The insignificant change in HG rate indicated that the concentration of NaBH$_4$ did not affect the HG reaction, showing zero-order reaction kinetics [58].

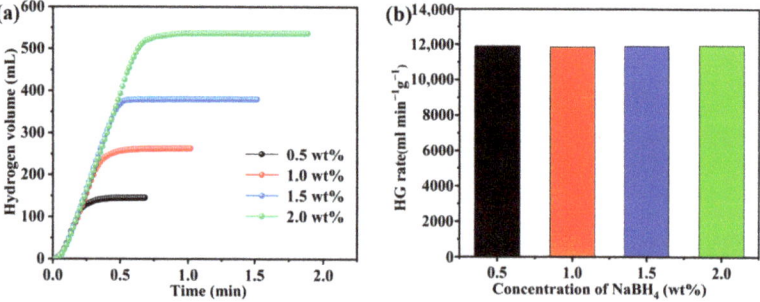

Figure 10. (**a**) Effect of NaBH$_4$ concentrations on the HG rate (batch system, 25 °C, 5 wt% NaOH, 0.1 g of catalyst); (**b**) HG rate versus NaBH$_4$ concentration.

3.6. Kinetic Studies at Different Temperatures

The HG rate of the Co-B-P/75rGO catalyst was measured under standard conditions. The temperature was controlled from 15 °C to 55 °C with 10 °C as a gradient. Figure 11a shows that high temperature had a significant promotion effect on the rate of hydrogen production. The total HG volume reached theoretical capacity at different temperatures. The formula is shown below:

$$k = k_0 \cdot \exp\left(\frac{E_a}{RT}\right) \tag{2}$$

where k_0 is the rate constant (mL min^{-1} g^{-1}), E_a is the activation energy (kJ mol^{-1}), T is the reaction temperature (K), and R is the gas constant (8.314 kJ mol^{-1} K^{-1}). Figure 11b shows the Arrhenius plot of ln k and the reciprocal of the absolute temperature (1/T). According to the slope of the fitting line, the Ea of the hydrolysis reaction in this study was calculated to be 28.64 kJ mol^{-1}, which is lower than most previous reports in the literature (Table 2). The favorable catalytic activity was ascribed to the presence of GO, which promoted the uniform dispersion of Co-B-P clusters and exposed more catalytically active sites for the hydrolysis. Meanwhile the synergistic effect of the GO and Co-B-P clusters was also conducive to the hydrolysis activity of $NaBH_4$.

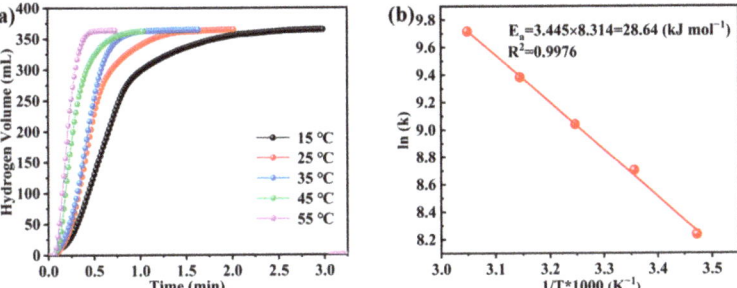

Figure 11. (**a**) Hydrogen generation kinetics curves and (**b**) Arrhenius plot obtained using 1.5 wt% $NaBH_4$ and 1.0 wt% NaOH solution and employing Co-B-P/75rGO as a catalyst at different solution temperatures.

3.7. Reusability Performance

The cycle stability of catalysts is critical in practical applications. Therefore, $NaBH_4$ was hydrolyzed 10 times with Co-B-P/75rGO catalyst in the same conditions. Figure 12 shows the variation in the catalytic hydrogen production efficiency of the Co-B-P/75rGO catalyst with the number of cycles. It can be observed that the HG rate decreased slightly as the cycle time increased. The HG rate still maintained 88.9% of the initial rate after 10 cycles, which shows better stability compared to other previously reported cobalt-based catalysts (Table 2). The decline in the catalytic activity may be due to the active clusters being reunited during each cycle. In addition, the produced boride byproducts (such as $B_\alpha O_\beta (OH)_\gamma$ and $B_x O_y \cdot nH_2O$) were adsorbed on the catalyst surface during the catalysis process, thereby decreasing the HG rate [59].

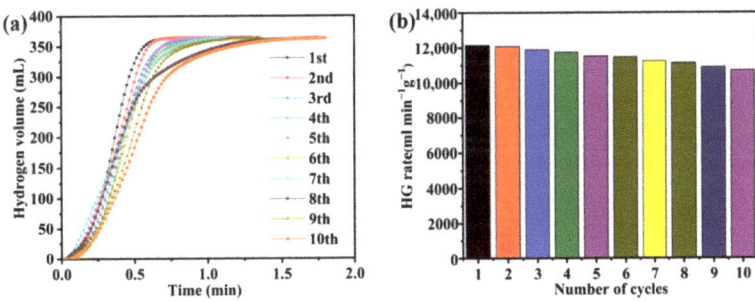

Figure 12. (**a**) Reusability of Co-B-P/75rGO with 0.1 g catalyst and 1.5 wt% $NaBH_4$ + 5 wt% NaOH solution at 25 °C; (**b**) HG rate bar chart of catalyst used 10 times.

4. Conclusions

In summary, a series of Co-B-P/xrGO catalysts were achieved using a chemical in situ reduction method and were employed for $NaBH_4$ hydrolysis. The experimental results showed that Co-B-P/xrGO had a strong effect on the catalytic behaviors, in which the Co-B-P/75rGO presented an optimal HG rate (12,087.8 mL min^{-1} g^{-1}) and lower activation energy (28.64 kJ mol^{-1}). The satisfied catalytic performances were due to the uniform dispersion of clusters and the synergistic catalytic effect between Co, B, and P. In addition, the repeatability test results showed that 88.9% of the initial catalytic efficiency could be maintained following 10 cycles, indicating that the catalyst had a good cycle stability. The above findings suggest that the Co-B-P/75rGO catalyst has great promise for producing hydrogen via chemical hydrate hydrolysis.

Author Contributions: Conceptualization, X.J. and Z.S.; methodology, X.J. and Z.S.; software, X.J., Z.S., H.H., Y.Y. and C.Z.; validation, X.J. and Z.S.; formal analysis, X.J., L.S. and F.X.; investigation, X.J., L.S. and F.X.; resources, L.S. and F.X.; data curation, X.J., Z.S., L.K., R.C. and Y.B.; writing—original draft preparation, X.J. and Z.S.; writing—review and editing, L.S., F.X. and H.P.; visualization, X.J. and Z.S.; supervision, L.S. and F.X.; project administration, L.S. and F.X.; funding acquisition, L.S. and F.X. All authors have read and agreed to the published version of the manuscript.

Funding: This work was supported by the National Key Research and Development Program of China (2018YFB1502103, 2018YFB1502105), the National Natural Science Foundation of China (51971068, U20A20237 and 51871065), the Scientific Research and Technology Development Program of Guangxi (AA19182014, AD17195073, AA17202030-1, 2021AB17045), Guangxi Bagui Scholar Foundation, Guangxi Collaborative Innovation Centre of Structure and Property for New Energy and Materials, Guangxi Advanced Functional Materials Foundation and Application Talents Small Highlands, Chinesisch-Deutsche Kooperationsgruppe (GZ1528), and the Science Research and Technology Development Project of Guilin (20210102-4, 20210216-1).

Data Availability Statement: Not applicable.

Conflicts of Interest: The authors declare no conflict of interest.

References

1. Liang, Z.Q.; Yao, Z.D.; Li, R.H.; Xiao, X.Z.; Ye, Z.C.; Wang, X.C.; Qi, J.C.; Bi, J.P.; Fan, X.L.; Kou, H.Q.; et al. Regulating local chemistry in ZrCo-based orthorhombic hydrides via increasing atomic interference for ultra-stable hydrogen isotopes storage. *J. Energy Chem.* **2022**, *69*, 397–405. [CrossRef]
2. Cai, W.T.; Yang, Y.Z.; Tao, P.J.; Ouyang, L.Z.; Wang, H. Correlation between structural stability of LiBH$_4$ and cation electronegativity in metal borides: An experimental insight for catalyst design. *Dalton Trans.* **2018**, *47*, 4987–4993. [CrossRef] [PubMed]
3. Schapbach, L.; Züttel, A. Hydrogen-storage materials for mobile applications. *Nature* **2001**, *414*, 265–270.
4. Zhu, Y.Y.; Ouyang, L.Z.; Zhong, H.; Liu, J.W.; Wang, H.; Shao, H.Y.; Huang, Z.G.; Zhu, M. Closing the loop for hydrogen storage: Facile regeneration of NaBH$_4$ from its hydrolytic product. *Angew. Chem. Int. Ed.* **2020**, *132*, 8701–8707. [CrossRef]
5. Hashimi, A.S.; Nohan, M.A.N.M.; Chin, S.X.; Zakaria, S.; Chia, C.H. Copper Nanowires as Highly Efficient and Recyclable Catalyst for Rapid Hydrogen Generation from Hydrolysis of Sodium Borohydride. *Nanomaterials* **2020**, *9*, 1153. [CrossRef]
6. Wang, Y.; Zou, K.L.; Wang, D.; Meng, W.; Qi, N.; Cao, Z.Q.; Zhang, K.; Chen, H.H.; Li, G.D. Highly efficient hydrogen evolution from the hydrolysis of ammonia borane solution with the Co-Mo-B/NF nanocatalyst. *Renew. Energy* **2020**, *154*, 453–460. [CrossRef]
7. Zhong, H.; Ouyang, L.Z.; Ye, J.S.; Liu, J.W.; Wang, H.; Yao, X.D.; Zhu, M. An one-step approach towards hydrogen production and storage through regeneration of NaBH$_4$. *Energy Storage Mater.* **2017**, *7*, 222–228. [CrossRef]
8. Lee, J.; Kong, K.Y.; Jung, C.R.; Cho, E.; Yoon, S.P.; Han, J.; Lee, T.G.; Nam, S.W. A structured Co–B catalyst for hydrogen extraction from NaBH$_4$ solution. *Catal. Today* **2007**, *120*, 305–310. [CrossRef]
9. Huang, Y.K.; An, C.H.; Zhang, Q.Y.; Zang, L.; Shao, H.X.; Liu, Y.F.; Zhang, Y.; Yuan, H.T.; Wang, C.Y.; Wang, Y.J. Cost-effective mechanochemical synthesis of highly dispersed supported transition metal catalysts for hydrogen storage. *Nano Energy* **2021**, *80*, 105535. [CrossRef]
10. Zhang, H.M.; Zhang, L.; Rodriguez-Perez, I.A.; Miao, W.K.; Chen, K.L.; Wang, W.F.; Li, Y.; Han, S.M. Carbon nanospheres supported bimetallic Pt-Co as an efficient catalyst for NaBH$_4$ hydrolysis. *Appl. Surf. Sci.* **2021**, *540*, 148296. [CrossRef]
11. Liu, H.Y.; Ning, H.L.; Peng, S.G.; Yu, Y.H.; Ran, C.; Chen, Y.M.; Ma, J.Y.; Xie, J.P. Surface tailored Ru catalyst on magadiite for efficient hydrogen generation. *Colloids Surf. A* **2021**, *631*, 127627. [CrossRef]
12. Zou, Y.C.; Nie, M.; Huang, Y.M.; Wang, J.Q.; Liu, H.L. Kinetics of NaBH$_4$ hydrolysis on carbon-supported ruthenium catalysts. *Int. J. Hydrogen Energy* **2011**, *36*, 12343–12351. [CrossRef]
13. Zhao, Y.C.; Ning, Z.; Tian, J.N.; Wang, H.W.; Liang, X.Y.; Nie, S.L.; Yu, Y.; Li, X.X. Hydrogen generation by hydrolysis of alkaline NaBH$_4$ solution on Co–Mo–Pd–B amorphous catalyst with efficient catalytic properties. *J. Power Sources* **2012**, *207*, 120–126. [CrossRef]
14. Zhang, B.P.; Xia, G.L.; Sun, D.L.; Fang, F.; Yu, X.B. Magnesium hydride nanoparticles self-assembled on graphene as anode material for high-performance lithium-ion batteries. *ACS Nano* **2018**, *12*, 3816–3824. [CrossRef] [PubMed]
15. Ma, H.; Ji, W.Q.; Zhao, J.Z.; Liang, J.; Chen, J. Preparation, characterization and catalytic NaBH$_4$ hydrolysis of Co-B hollow spheres. *J. Alloys Compd.* **2009**, *474*, 584–589. [CrossRef]
16. Zhang, X.W.; Zhang, Q.; Xu, B.; Liu, X.Q.; Zhang, K.M.; Fan, A.Y.; Jiang, W.D. Efficient Hydrogen Generation from the NaBH$_4$ Hydrolysis by Cobalt-Based Catalysts: Positive Roles of Sulfur-Containing Salts. *ACS Appl. Mater. Inter.* **2020**, *12*, 9376–9386. [CrossRef]

17. Lin, H.J.; Xu, C.; Gao, M.; Ma, Z.L.; Meng, Y.Y.; Li, L.Q.; Hu, X.H.; Zhu, Y.F.; Pan, S.P.; Li, W. Hydrogenation properties of five-component $Mg_{60}Ce_{10}Ni_{20}Cu_{5 \times 5}$ (X=Co, Zn) metallic glasses. *Intermetallics* **2019**, *108*, 94–99. [CrossRef]
18. Didehban, A.; Zabihi, M.; Shahrouzi, J.R. Experimental studies on the catalytic behavior of alloy and core-shell supported Co-Ni bimetallic nano-catalysts for hydrogen generation by hydrolysis of sodium borohydride. *Int. J. Hydrogen Energy* **2018**, *43*, 20645–20660. [CrossRef]
19. Eom, K.S.; Cho, K.W.; Kwon, H.S. Effects of electroless deposition conditions on microstructures of cobalt-phosphorous catalysts and their hydrogen generation properties in alkaline sodium borohydride solution. *J. Power Sources* **2008**, *180*, 484–490. [CrossRef]
20. Li, H.; Yang, P.F.; Chu, D.S.; Li, H.X. Selective maltose hydrogenation to maltitol on a ternary Co-P-B amorphous catalyst and the synergistic effects of alloying B and P. *Appl. Catal. A-Gen.* **2007**, *325*, 34–40. [CrossRef]
21. Patel, N.; Fernandes, R.; Miotello, A. Hydrogen generation by hydrolysis of $NaBH_4$ with efficient Co-P-B catalyst: A kinetic study. *J. Power Sources* **2009**, *188*, 411–420. [CrossRef]
22. Zhang, X.W.; Zhao, J.Z.; Cheng, F.Y.; Liang, J.; Tao, Z.L.; Chen, J. Electroless-deposited Co–P catalysts for hydrogen generation from alkaline $NaBH_4$ solution. *Int. J. Hydrogen Energy* **2010**, *35*, 8363–8369. [CrossRef]
23. Ekinci, A.; Cengiz, E.; Kuncan, M.; Sahin, Ö. Hydrolysis of sodium borohydride solutions both in the presence of Ni-B catalyst and in the case of microwave application. *Int. J. Hydrogen Energy* **2020**, *45*, 34749–34760.
24. Sun, H.M.; Meng, J.; Jiao, L.F.; Cheng, F.Y.; Chen, J. A review of transition-metal boride/phosphide-based materials for catalytic hydrogen generation from hydrolysis of boron-hydrides. *Inorg. Chem. Front.* **2018**, *5*, 760–772. [CrossRef]
25. Li, L.; Wang, Y.J.; Wang, Y.P.; Ren, Q.L.; Jiao, L.F.; Yuan, H.T. Effect of Ni Content in $Co_{1-x}Ni_xB$ Catalysts on Hydrogen Generation during Hydrolysis. *Acta Phys.-Chim. Sin.* **2010**, *26*, 1575–1578. [CrossRef]
26. Şahin, Ö.; Karakaş, D.E.; Kaya, M.; Saka, C. The effects of plasma treatment on electrochemical activity of Co-B-P catalyst for hydrogen production by hydrolysis of $NaBH_4$. *J. Energy Inst.* **2017**, *90*, 466–475. [CrossRef]
27. Zhang, X.Y.; Sun, X.W.; Xu, D.Y.; Tao, X.M.; Dai, P.; Guo, Q.J.; Liu, X. Synthesis of MOF-derived Co@C composites and application for efficient hydrolysis of sodium borohydride. *Appl. Surf. Sci.* **2019**, *469*, 764–769. [CrossRef]
28. Zhang, P.F.; Chen, N.Q.; Chen, D.; Yang, S.Z.; Liu, X.F.; Wang, L.; Wu, P.W.; Phillip, N.; Yang, J.; Dai, S. Ultra-Stable and High-Cobalt-Loaded Cobalt@Ordered Mesoporous Carbon Catalysts: All-in-One Deoxygenation of Ketone into Alkylbenzene. *ChemCatChem* **2018**, *10*, 3299–3304. [CrossRef]
29. Shi, L.M.; Chen, Z.; Jian, Z.Y.; Guo, F.H.; Gao, C.L. Carbon nanotubes-promoted Co-B catalysts for rapid hydrogen generation via $NaBH_4$ hydrolysis. *Int. J. Hydrogen Energy* **2019**, *44*, 19868–19877. [CrossRef]
30. Yang, C.C.; Chen, M.S.; Chen, Y.W. Hydrogen generation by hydrolysis of sodium borohydride on CoB/SiO_2 catalyst. *Int. J. Hydrogen Energy* **2011**, *36*, 1418–1423. [CrossRef]
31. Ye, W.; Zhang, H.; Xu, D.; Ma, L.; Yi, B.L. Hydrogen generation utilizing alkaline sodium borohydride solution and supported cobalt catalyst. *J. Power Sources* **2007**, *164*, 544–548. [CrossRef]
32. Zhang, R.Z.; Zheng, J.L.; Chen, T.W.; Ma, G.S.; Zhou, W. RGO-wrapped Ni-P hollow octahedrons as noble-metal-free catalysts to boost the hydrolysis of ammonia borane toward hydrogen generation. *J. Alloys Compd.* **2018**, *763*, 538–545. [CrossRef]
33. Yao, Q.; Lu, Z.H.; Huang, W.; Chen, X.S.; Zhu, J. Highly Pt-like activity of Ni-Mo/graphene catalyst for hydrogen evolution from hydrolysis of ammonia borane. *J. Mater. Chem. A* **2016**, *4*, 8579–8583. [CrossRef]
34. Hummers, W.S.; Offeman, R.E. Preparation of Graphitic Oxide. *J. Am. Chem. Soc.* **1958**, *208*, 1334–1339. [CrossRef]
35. Fernandes, R.; Patel, N.; Miotello, A. Efficient catalytic properties of Co-Ni-P-B catalyst powders for hydrogen generation by hydrolysis of alkaline solution of $NaBH_4$. *Int. J. Hydrogen Energy* **2009**, *34*, 2893–2900. [CrossRef]
36. Zhao, X.; Xu, D.Y.; Liu, K.; Dai, P.; Gao, J. Remarkable enhancement of PdAg/rGO catalyst activity for formic acid dehydrogenation by facile boron-doping through $NaBH_4$ reduction. *Appl. Surf. Sci.* **2020**, *512*, 145746. [CrossRef]
37. Wang, W.Y.; Liu, P.L.; Wu, K.; Tan, S.; Li, W.S.; Yang, Y.Q. Preparation of hydrophobic reduced graphene oxide supported Ni-B-P-O and Co-B-P-O catalysts and their high hydrodeoxygenation activities. *Green Chem.* **2016**, *18*, 984–988. [CrossRef]
38. Shi, L.M.; Xie, W.; Jian, Z.Y.; Liao, X.M.; Wang, Y.J. Graphene modified Co-B catalysts for rapid hydrogen production from $NaBH_4$ hydrolysis. *Int. J. Hydrogen Energy* **2019**, *44*, 17954–17962. [CrossRef]
39. Wang, Y.; Qi, K.Z.; Wu, S.W.; Cao, Z.Q.; Zhang, K.; Lu, Y.S.; Liu, H.X. Preparation, characterization and catalytic sodium borohydride hydrolysis of nanostructured cobalt-phosphorous catalysts. *J. Power Sources* **2015**, *284*, 130–137. [CrossRef]
40. Dai, H.B.; Liang, Y.; Wang, P. Effect of trapped hydrogen on the induction period of cobalt-tungsten-boron/nickel foam catalyst in catalytic hydrolysis reaction of sodium borohydride. *Catal. Today* **2011**, *170*, 27–32. [CrossRef]
41. Bharath, G.; Anwer, S.; Mangalaraja, R.V.; Alhseinat, E.; Banat, F.; Ponpandian, N. Sunlight-Induced photochemical synthesis of Au nanodots on α-Fe_2O_3@Reduced graphene oxide nanocomposite and their enhanced heterogeneous catalytic properties. *Sci. Rep.* **2018**, *8*, 5718. [CrossRef] [PubMed]
42. Riahi, K.Z.; Sdiri, N.; Ennigrou, H.; Horchani-Naifer, K. Investigations on electrical conductivity and dielectric properties of graphene oxide nanosheets synthetized from modified Hummer's method. *J. Mol. Struct.* **2020**, *1216*, 128304. [CrossRef]
43. Bai, Y.J.; Zhang, H.J.; Liu, L.; Xu, H.T.; Wang, Y. Tunable and Specific Formation of C@NiCoP Peapods with Enhanced HER Activity and Lithium Storage Performance. *Chem. Eur. J.* **2016**, *22*, 1021–1029. [CrossRef] [PubMed]
44. Zou, Y.J.; Yin, Y.; Gao, Y.B.; Xiang, C.L.; Chu, H.L.; Qiu, S.J.; Yan, E.H.; Xu, F.; Sun, L.X. Chitosan-mediated Co-Ce-B nanoparticles for catalyzing the hydrolysis of sodium borohydride. *Int. J. Hydrogen Energy* **2018**, *43*, 4912–4921. [CrossRef]

45. Fan, G.; Huang, W.; Wang, C. In situ synthesis of Ru/RGO nanocomposites as a highly efficient catalyst for selective hydrogenation of halonitroaromatics. *Nanoscale* **2013**, *5*, 6819–6825. [CrossRef]

46. Moddeman, W.E.; Burke, A.R.; Bowling, W.C.; Foose, D.S. Surface oxides of boron and $B_{12}O_2$ as determined by XPS. *Surf. Interface Anal.* **1989**, *14*, 224–232. [CrossRef]

47. Guang, H.L.; Zhu, S.L.; Liang, Y.Q.; Wu, S.L.; Li, Z.Y.; Luo, S.Y.; Cui, Z.D.; Inoue, A. Highly efficient nanoporous CoBP electrocatalyst for hydrogen evolution reaction. *Rare Metals* **2021**, *40*, 1031–1039. [CrossRef]

48. Shi, L.M.; Zhang, G. Improved Low-Temperature Activity of CuO-CeO_2-ZrO_2 Catalysts for Preferential Oxidation of CO in H_2-Rich Streams. *Catal. Lett.* **2016**, *146*, 1449–1456. [CrossRef]

49. Tong, D.G.; Han, X.; Chu, W.; Chen, H.; Ji, X.Y. Preparation of mesoporous Co-B catalyst via self-assembled triblock copolymer templates. *Mater. Lett.* **2007**, *61*, 4679–4682. [CrossRef]

50. Wang, J.; Ke, D.D.; Li, Y.; Zhang, H.M.; Wang, C.X.; Zhao, X.; Yuan, Y.J.; Han, S.M. Efficient hydrolysis of alkaline sodium borohydride catalyzed by cobalt nanoparticles supported on three-dimensional graphene oxide. *Mater. Res. Bull.* **2017**, *95*, 204–210. [CrossRef]

51. Zhang, H.M.; Feng, X.L.; Cheng, L.; Hou, X.W.; Li, Y.; Han, S.M. Non-noble Co anchored on nanoporous graphene oxide, as an efficient and long-life catalyst for hydrogen generation from sodium borohydride. *Colloid Surf. A* **2019**, *563*, 112–119. [CrossRef]

52. Liu, H.Y.; Shi, Q.Y.; Yang, Y.M.; Yu, Y.N.; Zhang, Y.; Zhang, M.S.; Wei, L.; Lu, Y.H. CoO-Co_2P composite nanosheets as highly active catalysts for sodium borohydride hydrolysis to generate hydrogen. *Funct. Mater. Lett.* **2020**, *13*, 2051025. [CrossRef]

53. Li, J.H.; Hong, X.Y.; Wang, Y.L.; Luo, Y.M.; Huang, P.R.; Li, B.; Zhang, K.X.; Zou, Y.J.; Sun, L.X.; Xu, F.; et al. Encapsulated cobalt nanoparticles as a recoverable catalyst for the hydrolysis of sodium borohydride. *Energy Storage Mater.* **2020**, *27*, 187–197. [CrossRef]

54. Li, Z.; Wang, L.N.; Zhang, Y.; Xie, G.W. Properties of Cu-Co-P/γ-Al_2O_3 catalysts for efficient hydrogen generation by hydrolysis of alkaline $NaBH_4$ solution. *Int. J. Hydrogen Energy* **2017**, *42*, 5749–5757. [CrossRef]

55. Wang, F.H.; Zhang, Y.J.; Wang, Y.N.; Luo, Y.M.; Chen, Y.N.; Zhu, H. Co-P nanoparticles supported on dandelion-like CNTs-Ni foam composite carrier as a novel catalyst for hydrogen generation from $NaBH_4$ methanolysis. *Int. J. Hydrogen Energy* **2018**, *43*, 8805–8814. [CrossRef]

56. Wei, L.; Dong, X.L.; Yang, Y.M.; Shi, Q.Y.; Lu, Y.H.; Liu, H.Y.; Yu, Y.N.; Zhang, M.H.; Qi, M.; Wang, Q. Co-O-P composite nanocatalysts for hydrogen generation from the hydrolysis of alkaline sodium borohydride solution. *Int. J. Hydrogen Energy* **2020**, *45*, 10745–10753. [CrossRef]

57. Cui, Z.; Guo, Y.; Ma, J. In situ synthesis of graphene supported Co-Sn-B alloy as an efficient catalyst for hydrogen generation from sodium borohydride hydrolysis. *Int. J. Hydrogen Energy* **2016**, *41*, 1592–1599. [CrossRef]

58. Wang, L.N.; Li, Z.; Zhang, P.P.; Wang, G.X.; Xie, G.W. Hydrogen generation from alkaline NaBH4 solution using Co-Ni-Mo-P/γ-Al_2O_3 catalysts. *Int. J. Hydrogen Energy* **2016**, *41*, 1468–1476. [CrossRef]

59. Li, Z.; Li, H.L.; Wang, L.N.; Liu, T.Y.; Zhang, T.; Wang, G.X.; Xie, G.W. Hydrogen generation from catalytic hydrolysis of sodium borohydride solution using supported amorphous alloy catalysts (Ni-Co-P/γ-Al_2O_3). *Int. J. Hydrogen Energy* **2014**, *39*, 14935–14941. [CrossRef]

nanomaterials

MDPI

Article

Bimetallic Pt-Ni Nanoparticles Confined in Porous Titanium Oxide Cage for Hydrogen Generation from NaBH$_4$ Hydrolysis

Yuqian Yu [1,†], Li Kang [1,†], Lixian Sun [1,2,*], Fen Xu [1,*], Hongge Pan [3,*], Zhen Sang [1], Chenchen Zhang [1], Xinlei Jia [1], Qingli Sui [1], Yiting Bu [1], Dan Cai [1], Yongpeng Xia [1], Kexiang Zhang [1] and Bin Li [1]

1   Guangxi Key Laboratory of Information Materials, Guangxi Collaborative Innovation Center for Structure and Properties for New Energy and Materials, School of Material Science and Engineering, Guilin University of Electronic Technology, Guilin 541004, China; yuyuqian997@163.com (Y.Y.); kangli000hello@163.com (L.K.); 15755028821@163.com (Z.S.); Zhang_linba_3760@163.com (C.Z.); jia15292079581@163.com (X.J.); sqlguet123@163.com (Q.S.); ytb1172701255@163.com (Y.B.); dancai1985@guet.edu.cn (D.C.); ypxia@guet.edu.cn (Y.X.); kxzhang@guet.edu.cn (K.Z.); li_bin@guet.edu.cn (B.L.)
2   School of Mechanical & Electrical Engineering, Guilin University of Electronic Technology, Guilin 541004, China
3   School of New Energy Science and Technology, Xi'an Technological University, Xi'an 710021, China
*   Correspondence: sunlx@guet.edu.cn (L.S.); xufen@guet.edu.cn (F.X.); hgpan@zju.edu.cn (H.P.)
†   These authors contributed equally to this work.

Abstract: Sodium borohydride (NaBH$_4$), with a high theoretical hydrogen content (10.8 wt%) and safe characteristics, has been widely employed to produce hydrogen based on hydrolysis reactions. In this work, a porous titanium oxide cage (PTOC) has been synthesized by a one-step hydrothermal method using NH$_2$-MIL-125 as the template and L-alanine as the coordination agent. Due to the evenly distributed PtNi alloy particles with more catalytically active sites, and the synergistic effect between the PTOC and PtNi alloy particles, the PtNi/PTOC catalyst presents a high hydrogen generation rate (10,164.3 mL·min^{-1}·g^{-1}) and low activation energy (28.7 kJ·mol^{-1}). Furthermore, the robust porous structure of PTOC effectively suppresses the agglomeration issue; thus, the PtNi/PTOC catalyst retains 87.8% of the initial catalytic activity after eight cycles. These results indicate that the PtNi/PTOC catalyst has broad applications for the hydrolysis of borohydride.

Keywords: hydrogen generation; porous titanium oxide cage; PtNi nanoparticles; sodium borohydride hydrolysis

check for updates

Citation: Yu, Y.; Kang, L.; Sun, L.; Xu, F.; Pan, H.; Sang, Z.; Zhang, C.; Jia, X.; Sui, Q.; Bu, Y.; et al. Bimetallic Pt-Ni Nanoparticles Confined in Porous Titanium Oxide Cage for Hydrogen Generation from NaBH$_4$ Hydrolysis. *Nanomaterials* **2022**, *12*, 2550. https://doi.org/10.3390/nano12152550

Academic Editor: Yuichi Negishi

Received: 29 June 2022
Accepted: 21 July 2022
Published: 25 July 2022

Publisher's Note: MDPI stays neutral with regard to jurisdictional claims in published maps and institutional affiliations.

1. Introduction

The overconsumption of traditional fossil fuels has brought in severe energy shortages and environmental pollution issues, such as the greenhouse effect [1]. To solve the above issues, hydrogen energy, as an efficient and sustainable energy, is considered to be a promising alternative to fossil energy [2,3]. The extensive development and use of hydrogen energy is conducive to the pursuit of carbon neutrality and emission peak. Hydrogen produced by the hydrolysis of sodium borohydride (NaBH$_4$) has been regarded as one of the most promising hydrogen production methods due to the advantages of high theoretical hydrogen production density (10.8 wt%), low hydrogen release temperature, controllable reaction process, high hydrogen purity, and environmental friendliness [4]. However, the slow hydrolysis reaction limits the wide use of hydrogen.

In order to accelerate the reaction kinetics of hydrolysis NaBH$_4$, a variety of catalysts such as Co [5–8], Ni [9–11], Rh [12], Pd [13], Ru [14,15], and Pt [16,17] have been comprehensively studied. Although Pt-based catalysts are one of the most active catalysts, the scarce storage and expensive price are the main obstacles to their large-scale application. Therefore, increasing the utilization efficiency of Pt remains the focus of the search. Previous studies have demonstrated that combining Pt with non-noble metals (such as Co [18,19],

Ni [20–22], and Fe [23,24]) could significantly improve the utilization efficiency of the catalysts. For example, Shumin Han et al. synthesized a carbon nanosphere (CNS)-supported ultrafine bimetallic Pt-Co nanoparticle (CNSs@$Pt_{0.1}Co_{0.9}$) catalyst for $NaBH_4$ catalysis. The as-prepared CNSs@$Pt_{0.1}Co_{0.9}$ catalyst exhibited excellent performance in kinetic and thermodynamic tests [25]. Younghun Kim et al. designed a magnetic core and multi-shelled silica/titania-supported bimetallic (Pt/Ni NPs Fe_3O_4@SiO_2@TiO_2) catalyst for catalyzing the hydrolysis of $NaBH_4$ [20]. Jong-Sung Yu et al. uniformly deposited PtFe hydroxide by in situ hydrolysis of urea, followed by the preparation of a carbon-supported PtFe catalyst in ethylene glycol, and the catalyst exhibited excellent electrocatalytic performance [24].

Recently, Ni combined with precious metals, such as Pt-Ni and Ru-Ni, have been confirmed to be effective catalysts for hydrogen production from hydrolysis $NaBH_4$ [26–28]. However, these catalysts exhibit poor catalytic activity due to the accumulation of metal nanoparticles (NPs) during the reaction. Strategies including structural and morphology control, as well as the addition of suitable carries can effectively inhibit the agglomeration problem [29–31]. In addition, the introduction of support material not only is conducive to the distribution of metal NPs but also improves the metal properties through geometric and electronic effects. Anelia Kakanakova-Georgieva et al., employing theoretical calculations, demonstrated that the porous structural material and the synergistic effect between metal NPs with support materials played an important role in the activity of the catalyst [32,33]. Porous hollow structures assembled from nanosheets with large surface areas could provide a unique microenvironment both on the inside and outside through species channels for guest shuttling [34]. Among numerous porous materials, metal-organic frameworks (MOFs) with tunable metal ions and organic ligands are extensively searched in the fields of energy storage and catalysis [35]. Furthermore, they also act as a self-sacrificing template in preparing the porous hollow materials. For example, pioneering studies used the Ti-MOFs as carriers to improve the catalytic performance of metal catalysts for hydrogen production [36,37]. Therefore, reasonably designed porous structural carries for the dispersion of metal NPs enables the achievement of a satisfactory catalytic performance.

Herein, PtNi NPs were confined in a porous titanium oxide cage (PTOC) derived from NH_2-MIL-125 (Ti) by a facile hydrothermal method and used for the hydrogen production of hydrolysis $NaBH_4$. The synthesized catalysts exhibit good catalytic activity with a high hydrogen generation rate (10,164.3 mL·min^{-1}·g^{-1}) and low activation energy (28.7 kJ·mol^{-1}). In addition, the robust porous structure of PTOC benefits from the distribution of PtNi alloy particles and suppresses the agglomeration issue; thus, the PtNi/PTOC nanocomposite catalyst retains 87.8% of the initial catalytic activity after eight cycles.

2. Materials and Methods

2.1. Materials

All chemicals were of analytical grade and used without further purification. 2-aminoterephthalic acid, sodium borohydride ($NaBH_4$), and nickel nitrate hexahydrate ($Ni(NO_3)_2$·$6H_2O$) with a purity of 99% were purchased from Alfa Aesar Co., Ltd. (Tianjin, China). Chloroplatinic acid hexahydrate (H_2PtCl_6·$6H_2O$), titanium (IV) isopropoxide, L-alanine, and dimethylformamide (DMF)were purchased from Aladdin Reagent (Shanghai, China). All experiments were performed using DMF and anhydrous CH_3OH as solvents.

2.2. Synthesis of NH_2-MIL-125

The preparation of NH_2-MIL-125 nanocrystals followed a previously reported process [38]. Using DMF and ethanol as organic reaction solvents, 2-aminoterephthalic acid (500 mg, 2.76 mmol) was dissolved in a mixture solvent (10 mL) of 1 mL of CH_3OH (1 mL) and DMF (9 mL). Subsequently, 0.76 mmol of titanium isopropoxide was slowly added to the mixture under ultrasound. The solution was then placed in a 25 mL Teflon-lined reactor and heated at 150 °C for 72 h. After, the mixture was cooled to room temperature and the yellow powder was recovered by centrifugation. To remove impurities, the collected powder was washed sequentially with DMF, ethanol, and deionized water and dried at 80 °C for 12 h.

2.3. Synthesis of PTOC

NH$_2$-MIL-125 (10 mg) was sonicated and dispersed in 5 mL of anhydrous ethanol. Next, 47.5 mg of L-alanine was added to the mixture and stirred for 6 h. The solution was then placed in a reaction vessel containing 25 mL of Teflon liner and heated at 176 °C for 36 h. The white precipitate was recovered by centrifugation, washed with ethanol, and dried under vacuum at 80 °C for 12 h.

2.4. Preparation of PtNi/PTOC

Amino acid molecules (generally a class of mild reducing agents) are used to prepare metal NPs. Herein, NH$_2$-MIL-125 (10 mg) was sonicated and dispersed in 5 mL of anhydrous ethanol. Then, L-alanine (47.5 mg) was added into the mixture and stirred for 6 h. H$_2$PtCl$_6$·6H$_2$O (1.0 mg, 2.3 μmol) and Ni(NO$_3$)$_2$·6H$_2$O (3 mg, 10.3 μmol) were added sequentially and stirred for 1 h. Then, the mixture was placed in a 25 mL Teflon pan and heated at 176 °C for 36 h. After centrifugation, the mixture was dried with ethanol at 80 °C for 12 h. For the comparison, Ni(NO$_3$)$_2$·6H$_2$O was not added in the preparation process of Pt/PTOC, and H$_2$PtCl$_6$·6H$_2$O was not added to Ni/PTOC; the other steps were consistent with the preparation process of PtNi/PTOC.

2.5. Characterization

The morphology of the PtNi/PTOC catalyst was analyzed by scanning electron microscopy (SEM, Quanta 200, FEI, Hillsboro, OR, USA) under a vacuum environment and 30 kV AC voltage. The test sample was dispersed on conductive material and stuck on a small sample holder. Excess powder was blown off with gas to avoid contaminating the cavity. The morphology and elemental composition of the catalyst was analyzed using a transmission electron microscope (TEM, JEOL 2010, JEOL, Tokyo, Japan) and dispersive X-ray detector (EDX) with an informal resolution of 0.12 nm and a point resolution of 0.25 nm. The powder was put into an anhydrous ethanol solution, shaken well with ultrasonic waves, and dropped onto the microgrid support film to obtain the sample to be tested. The chemical structure of the catalyst was characterized by Fourier-transform infrared (FT-IR) spectroscopy (Nicolet 6700, Waltham, MA, USA) in the wavenumber range of 400–4000 cm^{-1}. The fine powder of the sample was uniformly dispersed in potassium bromide in the ratio of 1:100 (m$_{catalyst}$:m$_{KBr}$) and the transparent flakes were obtained by the tablet method at the pressure of 5 MPa for 30 s. The crystal structure was analyzed by X-ray diffraction (XRD, 1820, Philips, Amsterdam, The Netherlands), with a scan angle from 5° to 90°, a step size set to 0.02, a working voltage of 40 kV, and a working current of 40 mA. The sample preparation was carried out as follows: the powder sample was evenly distributed in the sample holder and compacted with the glass plate. The sample surface was required to be smooth and flush with the glass surface. The nitrogen-desorption isotherms of the PtNi/PTOC catalysts were investigated using a QuantachromeAutosorb-iQ2 adsorber. The specific surface area of PtNi/PTOC catalyst was determined using a fully automated ratio meter and porosity analyzer. The samples were degassed in a glass tube at 150 °C for 10 h and then analyzed in liquid nitrogen. The pore size of a pore of the PtNi/PTOC catalyst was determined by the BJH method. X-ray photoelectron spectroscopy (XPS; Thermo Electron ESCALAB 250, Waltham, MA, USA), was mainly used qualitatively and semi-quantitatively through the analysis of catalysts, the valence state, species class, and surface content. The sample was pressed on aluminum foil, the excitation light source was Al Kα (hv= 1486.6 eV), and the final XPS was calibrated by C 1s (284.8 eV).

2.6. Hydrogen Production Testing

The catalytic hydrogen generation experiments were measured on a self-built hydrogen generation device [39]. The volume of hydrogen produced was determined by the equivalent displacement of water. First, 0.1 g of catalyst was added in a 125 mL conical flask. Next, 10 mL of the solution containing 1.5 wt% $NaBH_4$ and 5 wt% NaOH was injected into the conical flask. The produced gas was collected in a container filled with water after flowing through a condenser and dryer to remove water vapor. The volume of produced H_2 was measured by the water displacement method. The water was displaced into a 1 L flask through a tube connected with a gas-gathering container and weighted by an electronic balance (UX2200H, Shimadzu Corporation, Kyoto, Japan). A computer connected to the electronic balance was used to record water quality automatically. The hydrogen released per gram of catalyst per unit time (HGR) was calculated through the display on the computer. After one hydrolysis test was completed, the catalyst was immediately washed and dried for 12 h. Subsequently, a fresh 10 mL of the 1.5 wt% $NaBH_4$ and 5 wt% NaOH solution was added to repeat the above measurements.

The hydrogen generation rate (HGR) was calculated according to the following equation:

$$HGR = \frac{V_{H_20}(\text{mL})}{t(\text{min}) \times m(\text{g})}$$

where V_{H_20} is the volume of drained water, m is the total mass of the catalyst, and t is the total reaction time in minutes [40].

3. Results and Discussion

3.1. Catalyst Characterization

In this paper, PtNi/PTOC was synthesized by a simple hydrothermal method and wet-reduction method. Figure 1 shows a schematic diagram of the preparation of PtNi/PTOC (PTOC). First, a round cake of NH_2-MIL-125 (Ti) was obtained using 2-aminoterephthalic acid as organic ligands and titanium (IV) isopropoxide as a metal precursor. Then, PTOC with a porous hollow structure was formed into an alcoholic thermal process at 176 °C under auxiliary amino acid molecules L-alanine. Lastly, PtNi precursors were reduced to Pt_3Ni NPs by the L-alanine. The formation of PTOC involved the Kirkendall effect of Ti ion dissolution and recrystallization. The Ti(iv) ions were firstly dissolved from NH_2-MIL-125(Ti) nanocrystals by coordination of Ti(iv) with amino acids (l-alanine), leading to the formation of sheet-like titanium oxide NPs on the NH_2-MIL-125(Ti) nanocrystals. With this continuous transformation, successive shells of titanium oxides were generated and transferred into the completed porous cages.

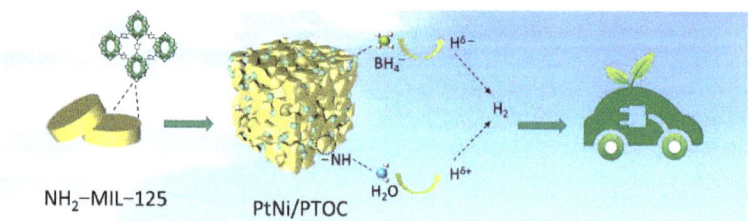

Figure 1. The illustration of the synthetic route of PtNi/PTOC.

The morphology of NH_2-MIL-125 and PTOC were characterized by SEM; as can be seen in Figure 2a,b, the NH_2-MIL-125 exhibits a round cake with a smooth surface, and the size is around 300–500 nm. After the auxiliary of the amino acid molecule L-alanine under hydrothermal circumstances, the cage structure of PTOC remains, with multi-channel interlacing on the surface (Figure 2c,d). Due to the alcoholization of NH_2-MIL-125, the nanosheets were assembled into a cage structure. L-alanine is commonly used as a mild reducing agent for the preparation of metal NPs [41,42]. After the hydrothermal reaction, PtNi precursors were reduced to Pt_3Ni NPs and confined in PTOC. As shown in Figure 2e,f, compared with the PTOC sample, Pt-Ni/PTOC still retained the unique "nanocage" structure.

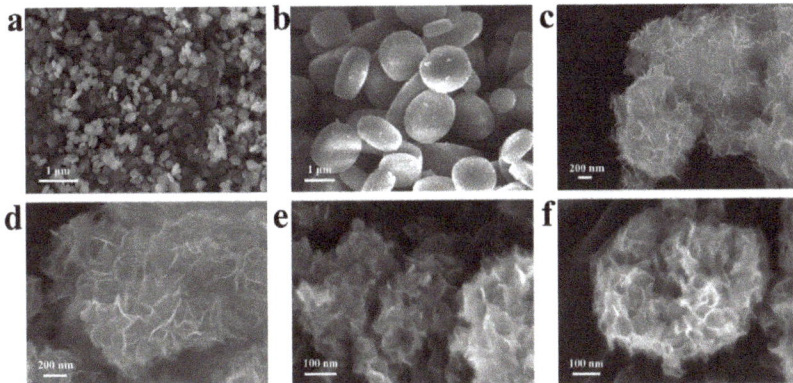

Figure 2. SEM images of (**a,b**) NH$_2$-MIL-125; (**c,d**) PTOC and (**e,f**) PtNi/PTOC catalyst.

As shown in Figure 3, HRTEM analysis showed that PtNi NPs were uniformly distributed in the PTOC nanocages with average particle sizes of 1.68 nm (Figure 3b). Further, high-resolution HRTEM analysis revealed that the d-spacing of 0.223 nm is between the (111) crystal faces of Pt (0.227 nm) and Ni (0.204 nm), indicating the formation of PtNi alloy NPs (Figure 3c) [43]. The HAADF-STEM image also confirmed the formation of uniformly distributed PtNi NPs (Figure 3d). EDX analysis showed that the as-prepared nanocomposites consisted of Ti, N, Pt, and Ni (Figure 3e–h). These results indicate that the 3D structure of PtNi NPs encapsulated by PTOC nanocages have been successfully prepared.

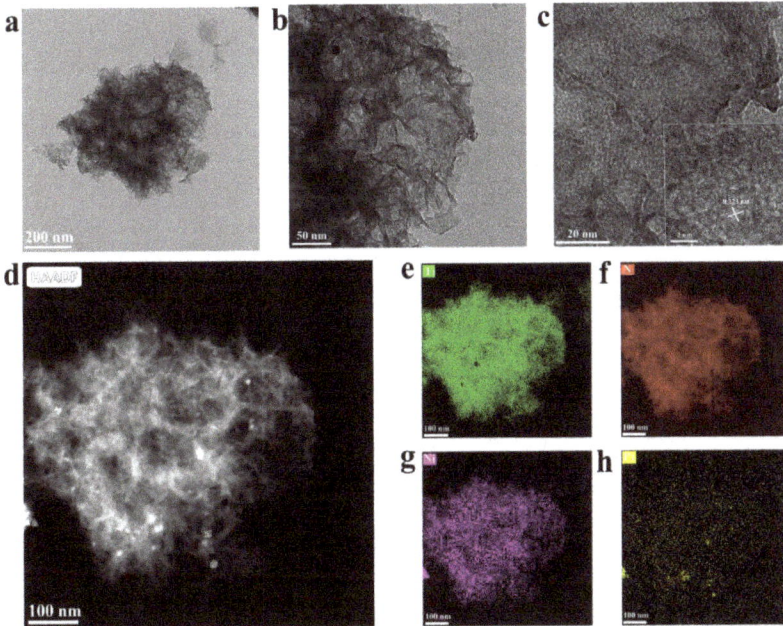

Figure 3. TEM images of (**a,b**) PtNi/PTOC catalyst; (**c**) HRTEM; (**d**) HADDF-STEM (**e–h**); EDX images of PtNi/PTOC.

FT-IR measurements (Figure 4a) were carried out to detect the functional groups of the as-prepared catalysts. The experimental results showed that all the prepared catalysts contained benzene rings and amino groups (the characteristic peaks at 3428 cm^{-1} and 1630 cm^{-1}). The existence of amino groups stably binds the metal NPs due to the strong chelation/complexation effect between the metal and amine groups [44]. Therefore, our results indicate that PTOC precursors are beneficial for the distribution of metal NPs. The XRD spectrum (Figure 4b) had four peaks at 2θ = 25°, 48°, 55°, and 62°, which corresponded to the (101), (200), (211), and (213) crystal planes of anatase TiO$_2$ (PDF, No. 21-1272). The diffraction peaks of layered titanate H$_2$Ti$_8$O$_{17}$ also appeared (PDF No. 36-0656), indicating that the PTOC had a two-component titanium oxide porous cage. In addition, the reflected signals of the Pt/PTOC and Ni/PTOC samples matched well with metallic Pt (PDF No. 87-0647) and Ni (PDF No. 65-0380), respectively. The diffraction peaks of PtNi alloy laid between the corresponding characteristic peaks of Pt and Ni, which further reflected the well-alloyed PtNi nanoparticles [45].

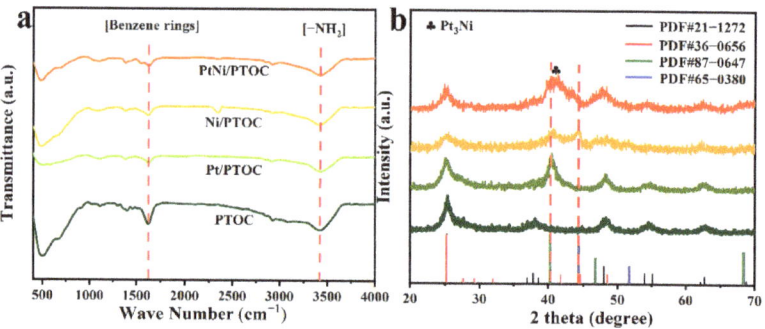

Figure 4. (**a**) FTIR spectra of catalysts. (**b**) XRD patterns of catalysts.

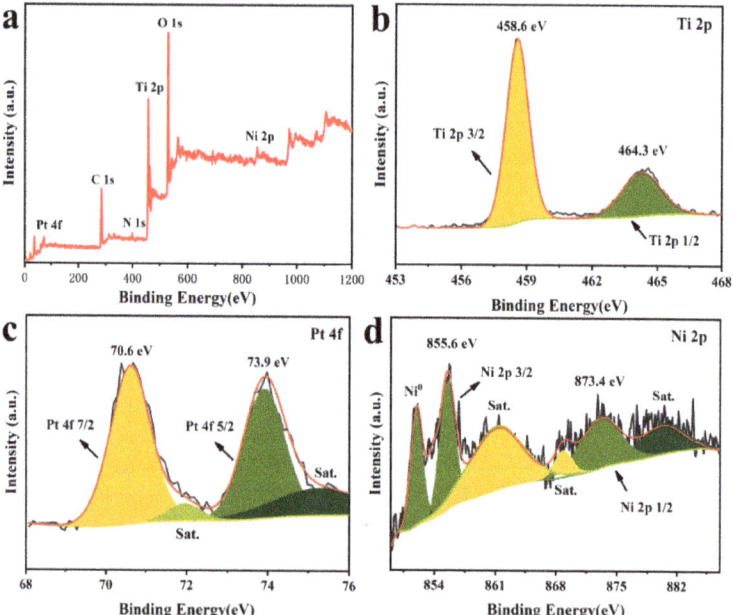

Figure 5. X-ray photoelectron spectra of Pt-Ni/PTOC catalyst (**a**) full spectrum; (**b**) Ti 2p; (**c**) Pt 4f; (**d**) Ni 2p.

The surface interactions and electronic state of PtNi/PTOC were investigated using XPS. Figure 5a shows the whole XPS pattern of PtNi/PTOC. The signals generated by the PTOC corresponded to C 1s, N 1s, O 1s, Ti 2p, Pt 4f, and Ni 2p. The narrow range spectra of Ti 2p is depicted in Figure 5b, which also proves the presence of PTOC. In Figure 5c, it can be seen that the Pt 4f region of core level binding energies is deconvoluted into two sets of spin-orbit doublet peaks. The Pt 4f spectrum exhibited two peaks at 71.4 and 74.7 eV and were assigned to Pt 4f7/2 and Pt 4f5/2, respectively, suggesting the presence of Pt^0. Two peaks at 71.9 and 75.3 eV corresponded to the satellite peaks of Pt. The binding energy located at 855.6 and 873.4 eV belonged to the Ni 2p3/2, and Ni 2p1/2, respectively. The binding energies at 861.38, 868.92, and 873.4 eV corresponded to the satellite peaks. In addition, the binding energy around 852.05 eV is attributed to the Ni^0 peak, which confirmed the existence of metallic Ni in PtNi/PTOC (Figure 5d). The strong interaction between Pt and Ni within the catalyst may lead to an increased oxidation resistance, which is beneficial to the catalysis activity and durability [46].

As shown in Figure 6a, PtNi/PTOC exhibited a typical IV-type isotherm with obvious hysteresis loops with a high specific surface area of approximately 206.2 $m^2 \cdot g^{-1}$. In addition, PTOC showed that similar isotherms with the specific surface area decreased from 206.2 $m^2 \cdot g^{-1}$ to 145.8 $m^2 \cdot g^{-1}$, which is due to the addition of PtNi NPs. From the IV-type isotherms with obvious hysteresis loops, the main pore size distribution of two materials is mesopores. Pore-size distribution curves showed that the size of the pore in PtNi/PTOC ranged from 3.5 nm to 10.0 nm (Figure 6b). The rich mesopores are conducive to the penetration of the electrolyte and the transport of electrons, thereby enhancing the catalytic activity of the material.

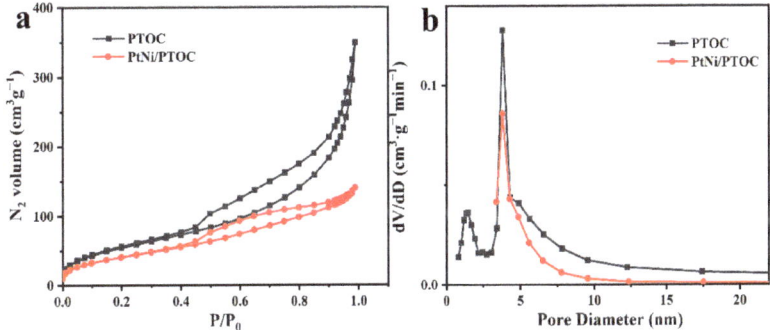

Figure 6. N_2 desorption/adsorption isotherm (**a**) and pore-size distributions (**b**) for PTOC and PtNi/PTOC.

3.2. Effect of Different Types of Catalysts

The effect of different catalysts on the hydrolysis of $NaBH_4$ under alkaline conditions was investigated. As shown in Figure 7, PtNi/PTOC exhibited optimal performance with hydrogen release rate (HGR) of 10,164.3 $mL \cdot min^{-1}$ at 25 °C, which is higher than Pt/PTOC and Ni/PTOC. Compared to most of the previously reported results, PtNi/PTOC also exhibited a good catalytic activity (Table 1) [25,47–52]. According to Figure 7b, the magnitude of the catalytic performance was PtNi/PTOC > Pt/PTOC > Ni/PTOC, while PTOC and NH_2-MIL-125 had no catalytic activity. The experimental results show that the synergistic effect between Pt and Ni enhanced the catalytic activity more than the single Pt or Ni-based catalyst, thereby promoting the rapid release of hydrogen from $NaBH_4$. Furthermore, the evenly distributed PtNi alloy particles with more catalytically active sites simultaneously enhanced the hydrolysis activity.

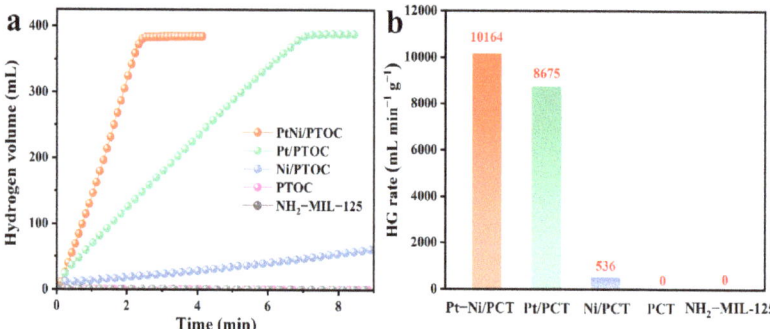

Figure 7. Hydrogen volume versus time (**a**) and HG rate bar chart of NH$_2$-MIL-125, PTOC, Pt/PTOC, Ni/PTOC, and (**b**) PtNi/PTOC (reaction conditions: batch system, 25 °C, 1.5 wt% NaBH$_4$ + 5 wt% NaOH, 0.1 g catalyst).

Table 1. Comparison of catalyst systems, reaction temperatures, HGR, *Ea* values, and number of cycles for NaBH$_4$ hydrolysis catalyzed by various catalysts.

Sample	Tempera-ture (°C)	HG Rate (mL·min^{-1}·g$_M$$^{-1}$)	E_a (kJ·mol^{-1})	Number of Cycles	Cyclic Stability	Ref.
CNSs@Pt$_{0.1}$Co$_{0.9}$	30	8943.0	38.0	5	85.2%	[25]
Pt/MWCNTs	30	16.9	46.2	5	80.0%	[47]
Pt/CeO$_2$-Co$_7$Ni$_2$O$_x$	25	7834.8	47.4	5	85.0%	[48]
PtPd/GO	25	3940.0	29.4	4	60.0%	[49]
Pt/Si$_3$N$_4$	80	13,000.0	35.2	5	82.5%	[50]
NiCoP NA/Ti	30	3016.8	52.7	8	70.0%	[51]
RuNi/Ti$_3$C$_{2×2}$	30	1649.0	34.7	4	50%	[52]
PtNi/PTOC	29	10,164.3	28.7	8	87.8%	This work

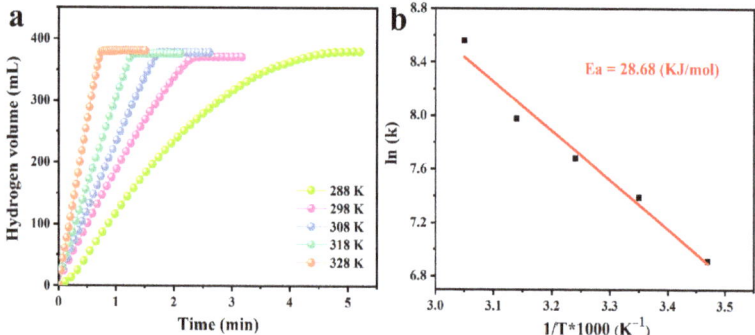

Figure 8. (**a**) Hydrogen generation kinetics curves and (**b**) Arrhenius plot obtained using 1.5 wt% NaBH$_4$ and 1.0 wt% NaOH solution and employing PtNi/PTOC as a catalyst at different solution temperatures.

In order to measure the activation energy (E_a) of the hydrolysis reaction, hydrolysis tests were carried out using different temperatures with the other parameters unchanged, controlling the reaction temperature from 15 to 55 °C (Figure 8a) with a gradient of 10 °C. As expected, all of the tests reached the theoretical hydrogen quantity, and the hydrogen release rate increased with the increase in reaction temperature, which belongs to the first-order reaction [53–55]. According to the Arrhenius slope calculation, the activation energy of Pt-Ni/PTOC is 28.7 kJ·mol^{-1} (Figure 8b), which was lower than most of the catalysts that have been reported (Table 1). The synergistic effect between PtNi NPs and PTOC may be the main factor for the decrease in the E_a value. The small particle size of the PtNi NPs is well supported on the pores of PTOC, thus avoiding excessive losses and agglomeration during hydrolysis. Moreover, the porous hollow structure promotes the interaction mass transfer between the catalyst and $NaBH_4$ in the pores. Therefore, these results show that PtNi/PTOC has good kinetic properties for catalyzing $NaBH_4$ hydrogen release.

3.3. Stability of PtNi/PTOC

The stability of the catalyst is a key index to the actual application of hydrogen generation from $NaBH_4$ hydrolysis. Figure 9 shows that the PtNi/PTOC catalyst was tested eight times under conventional conditions (25 °C). The catalytic activity of the hydrogen evolution of $NaBH_4$ decreased slightly and maintained the initial catalytic activity of 87.8% after eight cycles. The excellent cycling performance may be related to the hollow porous structure of PTOC, not only providing a large surface area for the distribution of PtNi alloy particles but also suppressing the agglomeration issues.

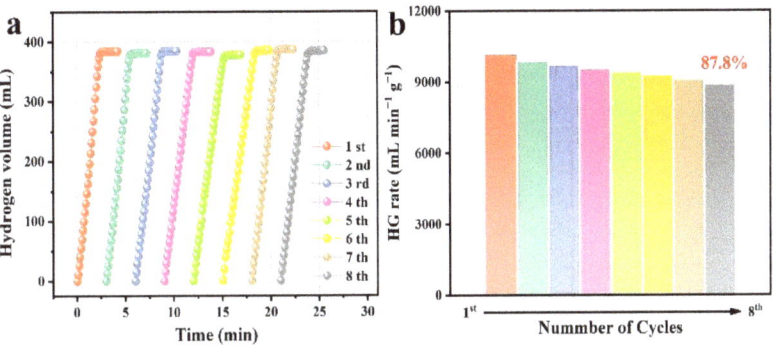

Figure 9. (**a**) Reusability of PtNi/PTOC with 0.1 g catalyst and 1.5 wt% $NaBH_4$ + 5 wt% NaOH solution at 25 °C; (**b**) HG rate bar chart of catalyst used 8 times.

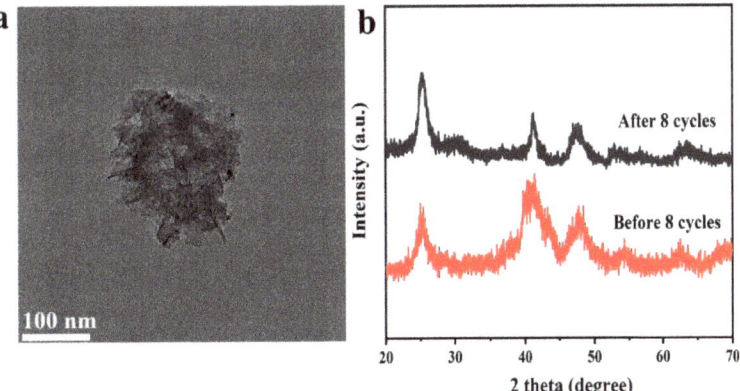

Figure 10. (**a**) TEM images of the PtNi/PTOC catalyst after 8 cycles; (**b**) XRD patterns of the PtNi/PTOC catalyst before cycling and after 8 cycles.

To verify the structural stability of the PtNi/PTOC catalyst, TEM (Figure 10a) and XRD characterizations were carried out after the cyclability test (Figure 10b). The TEM images of PtNi/PTOC after cycle tests show that the material maintained the nanocage structure with numerous sheets, indicating the stable structural integrity of the catalyst. In addition, compared to the original PtNi/PTOC sample, there was no significant agglomeration, which is favorable for the catalytic reaction. The XRD spectra of the obtained products showed that two XRD spectra were well-matched, and the peak of the catalyst became sharp after cycle tests, indicating the increased crystallinity of the catalyst. According to the Scherrer formula, ($D = K\lambda/\beta Cos\theta$), the size of PtNi/PTOC catalyst increased from 1.68 to 2.32 nm after eight cycle tests, which was one of the reasons for the decay of catalytic activity. The stable structure and high catalytic activity of metal NPs are promising for the hydrolysis of borohydride.

4. Conclusions

In this work, ultra-small PtNi NPs were confined in a porous titanium oxide cage (PTOC) derived from NH_2-MIL-125 (Ti) by a facile hydrothermal method and used for the hydrogen production of hydrolysis $NaBH_4$. At a room temperature of 25 °C, the hydrogen production rate of PtNi/PTOC reached 10,164.3 mL·min^{-1}·g_M^{-1}, and the activation energy was 28.7 kJ·mol^{-1}. After eight cycles of testing, 87.8% of the initial test performance was maintained. Such excellent performance can be attributed to the following: (i) The porous and hollow structure of PTOC creates a unique microenvironment between its interior and exterior, which provides more reaction channels. (ii) PTOC with a high surface area enables the even distribution of PtNi alloy particles, thus exhibiting a large number of active sites. (iii) The synergistic effect between PTOC and PtNi alloy particles can improve the reactivity. (iv) The robust porous structure maintains the integrity of the catalyst and suppresses the aggregation of nanoparticles. The catalyst has the advantages of a simple operation and economic efficiency and shows promise for producing hydrogen for fuel-cell vehicles.

Author Contributions: Conceptualization, Y.Y. and L.K.; methodology, Y.Y. and L.K.; software, Y.Y., L.K., Z.S. and C.Z.; validation, Y.Y. and L.K.; formal analysis, C.Z., L.S. and F.X.; investigation, Y.Y., L.K., L.S. and F.X.; resources, L.S. and F.X.; data curation, Y.Y., L.K., Z.S. X.J., Q.S., Y.B., D.C., Y.X., K.Z. and B.L.; writing—original draft preparation, Y.Y., L.K. and C.Z; writing—review and editing, L.S., F.X. and H.P.; visualization, Y.Y. and L.K.; supervision, L.S., F.X. and H.P; project administration, L.S. and F.X.; funding acquisition, L.S. and F.X. All authors have read and agreed to the published version of the manuscript.

Funding: This work was supported by the National Key Research and Development Program of China (2018YFB1502103 and 2018YFB1502105), the National Natural Science Foundation of China (51971068, U20A20237 and 51871065), the Scientific Research and Technology Development Program of Guangxi (AA19182014, AD17195073 and AA17202030-1), Guangxi Bagui Scholar Foundation, Guangxi Collaborative Innovation Centre of Structure and Property for New Energy and Materials, Guangxi Advanced Functional Materials Foundation and Application Talents Small Highlands, Chinesisch-Deutsche Kooperationsgruppe (GZ1528), Science research and Technology Development project of Guilin (20210216-1) and Science Research and Technology Development Project of Guilin (20210102-4).

Institutional Review Board Statement: Not applicable.

Informed Consent Statement: Not applicable.

Data Availability Statement: Not applicable.

Conflicts of Interest: The authors declare no conflict of interest.

References

1. Chen, M.; Xiao, X.; Wang, X.; Lu, Y.; Zhang, M.; Zheng, J.; Chen, L. Self-templated carbon enhancing catalytic effect of ZrO_2 nanoparticles on the excellent dehydrogenation kinetics of MgH_2. *Carbon* **2020**, *166*, 46–55. [CrossRef]
2. Zhang, M.; Xiao, X.; Luo, B.; Liu, M.; Chen, M.; Chen, L. Superior de/hydrogenation performances of MgH_2 catalyzed by 3D flower-like TiO_2@C nanostructures. *J. Energy Chem.* **2020**, *46*, 191–198. [CrossRef]
3. Chen, W.; Xiao, X.; He, J.; Dong, Z.; Wang, X.; Chen, M.; Chen, L. A dandelion-like amorphous composite catalyst with outstanding performance for sodium borohydride hydrogen generation. *Int. J. Hydrogen Energy* **2021**, *46*, 10809–10818. [CrossRef]
4. Zhu, Y.; Ouyang, L.; Zhong, H.; Liu, J.; Wang, H.; Shao, H.; Huang, Z.; Zhu, M. Closing the Loop for Hydrogen Storage: Facile Regeneration of $NaBH_4$ from its Hydrolytic Product. *Angew. Chem. Int. Ed.* **2020**, *59*, 8623–8629. [CrossRef]

5. Huang, Y.; An, C.; Zhang, Q.; Zang, L.; Shao, H.; Liu, Y.; Yuan, H.; Wang, C.; Wang, Y. Cos-effective mechanochemical synthesis of highly dispersed supported transition metal catalysts for hydrogen storage. *Nano Energy* **2021**, *80*, 105535. [CrossRef]
6. Min, J.; Jeffery, A.; Kim, Y.; Jung, N. Electrochemical Analysis for Demonstrating CO Tolerance of Catalysts in Polymer Electrolyte Membrane Fuel Cells. *Nanomaterials* **2019**, *9*, 1425. [CrossRef] [PubMed]
7. Bu, Y.; Liu, J.; Chu, H.; Wei, S.; Yin, Q.; Kang, L.; Luo, X.; Sun, L.; Xu, F.; Huang, P.; et al. Catalytic Hydrogen Evolution of $NaBH_4$ Hydrolysis by Cobalt Nanoparticles Supported on Bagasse-Derived Porous Carbon. *Nanomaterials* **2021**, *11*, 3259. [CrossRef]
8. Ren, Y.; Wang, J.; Hu, W.; Wen, H.; Qiu, Y.; Tang, P.; Chen, M.; Wang, P. Hierarchical Nanostructured Co-Mo-B/$CoMoO_{4-x}$ Amorphous Composite for the Alkaline Hydrogen Evolution Reaction. *ACS Appl. Mater. Interfaces* **2021**, *13*, 42605–42612. [CrossRef]
9. Shao, H.; Huang, G.; Liu, Y.; Guo, Y.; Wang, Y.N. Thermally stable Ni MOF catalyzed MgH_2 for hydrogen storage. *Int. J. Hydrogen Energy* **2021**, *46*, 37977–37985. [CrossRef]
10. Gao, H.; Shao, Y.; Shi, R.; Liu, Y.; Zhu, J.; Liu, J.; Zhu, Y.; Zhang, J.; Li, L.; Hu, X. Effect of Few-Layer $Ti_3C_2T_x$ Supported Nano-Ni via Self-Assembly Reduction on Hydrogen Storage Performance of MgH_2. *ACS Appl. Mater. Interfaces* **2020**, *12*, 47684–47694. [CrossRef]
11. Liu, W.; Zhi, H.; Yu, X. Recent progress in phosphorus based anode materials for lithium/sodium ion batteries. *Energy Storage Mater.* **2019**, *16*, 290–322. [CrossRef]
12. Larichev, Y.V.; Netskina, O.V.; Komova, O.V.; Simagina, V.I. Comparative XPS study of Rh/Al_2O_3 and Rh/TiO_2 as catalysts for $NaBH_4$ hydrolysis. *Int. J. Hydrogen Energy* **2010**, *35*, 6501–6507. [CrossRef]
13. Liu, S.; Chen, X.; Wu, Z.J.; Zheng, X.C.; Peng, Z.K.; Liu, P. Chitosan-reduced graphene oxide hybrids encapsulated Pd (0) nanocatalysts for H_2 generation from ammonia borane. *Int. J. Hydrogen Energy* **2019**, *44*, 23610–23619. [CrossRef]
14. Konuş, N.; Karataş, Y.; Gulcan, M. In situ formed ruthenium (0) nanoparticles supported on TiO_2 catalyzed hydrogen generation from aqueous ammonia-borane solution at room temperature under air. *Synth. React. Inorg. Met. Org. Chem.* **2016**, *46*, 534–542. [CrossRef]
15. Zhang, J.; Lin, F.; Yang, L.; He, Z.; Huang, X.; Zhang, D.; Dong, H. Ultrasmall Ru nanoparticles supported on chitin nanofibers for hydrogen production from $NaBH_4$ hydrolysis. *Chin. Chem. Lett.* **2020**, *31*, 2019–2022. [CrossRef]
16. Dai, P.; Zhao, X.; Xu, D.; Wang, C.; Tao, X.; Liu, X.; Gao, J. Preparation, characterization, and properties of Pt/Al_2O_3/cordierite monolith catalyst for hydrogen generation from hydrolysis of sodium borohydride in a flow reactor. *Int. J. Hydrogen Energy* **2019**, *44*, 28463–28470. [CrossRef]
17. Ro, G.; Kim, Y. H_2 generation using Pt nanoparticles encapsulated in Fe_3O_4@SiO_2@TiO_2 multishell particles. *Colloids Surf. A* **2019**, *577*, 48–52. [CrossRef]
18. Chen, J.; Gu, Y.; Kong, D.; Xiang, S.; Wang, P.; Zhang, H.; Zhang, S. Preparation of Colloidal Pt/Co Bimetallic Nanoparticle Catalysts and Their Catalytic Activity for Hydrogen Generation from Hydrolysis Reaction of $NaBH_4$. *Rare Met. Mater. Eng.* **2014**, *43*, 2209–2214.
19. Kotkondawar, A.V.; Rayalu, S. Enhanced H_2 production from dehydrogenation of sodium borohydride over the ternary $Co_{0.97}Pt_{0.03}$/CeO_x nanocomposite grown on CGO catalytic support. *RSC Adv.* **2020**, *10*, 38184–38195. [CrossRef]
20. Ro, G.; Hwang, D.K.; Kim, Y. Hydrogen generation using Pt/Ni bimetallic nanoparticles supported on Fe_3O_4@SiO_2@TiO_2 multi-shell microspheres. *J. Ind. Eng. Chem.* **2019**, *79*, 364–369. [CrossRef]
21. Guo, H.; Chen, C.; Chen, K.; Cai, H.; Chang, X.; Liu, S.; Wang, C. High performance carbon-coated hollow $Ni_{12}P_5$ nanocrystals decorated on GNS as advanced anodes for lithium and sodium storage. *J. Mater. Chem. A* **2017**, *5*, 22316–22324. [CrossRef]
22. Aranishi, K.; Singh, A.K.; Xu, Q. Dendrimer-Encapsulated Bimetallic Pt-Ni Nanoparticles as Highly Efficient Catalysts for Hydrogen Generation from Chemical Hydrogen Storage Materials. *Chem. Cat. Chem.* **2013**, *5*, 2248–2252. [CrossRef]
23. Li, W.; Zhao, W.; Zhou, W.; Zhou, X.; Sun, L.; Zhang, H.; Lu, L.; Zhang, S. Fabrication and catalytic activity of Pt/Ni/Fe trimetallic nanoparticles for hydrogen generation from $NaBH_4$. *Chem. J. Chin. Univ. Chin.* **2014**, *35*, 2164–2169.
24. Yang, D.S.; Kim, M.S.; Song, M.Y.; Yu, J.S. Highly efficient supported PtFe cathode electrocatalysts prepared by homogeneous deposition for proton exchange membrane fuel cell. *Int. J. Hydrogen Energy* **2012**, *37*, 13681–13688. [CrossRef]
25. Zhang, H.; Zhang, L.; Rodríguez-Pérez, I.A.; Miao, W.; Chen, K.; Wang, W.; Han, S. Carbon nanospheres supported bimetallic Pt-Co as an efficient catalyst for $NaBH_4$ hydrolysis. *Appl. Surf. Sci.* **2021**, *540*, 14829. [CrossRef]
26. Qi, X.; Li, X.; Chen, B.; Lu, H.; Wang, L.; He, G. Highly active nanoreactors: Patchlike or thick Ni coating on Pt nanoparticles based on confined catalysis. *ACS Appl. Mat. Interfaces* **2016**, *8*, 1922–1928. [CrossRef] [PubMed]
27. Ghosh, S.K.; Mandal, M.; Kundu, S.; Nath, S.; Pal, T. Bimetallic Pt-Ni nanoparticles can catalyze reduction of aromatic nitro compounds by sodium borohydride in aqueous solution. *Appl. Catal. A Gen.* **2004**, *268*, 61–66. [CrossRef]
28. Hannauer, J.; Demirci, U.B.; Geantet, C.; Herrmann, J.M.; Miele, P. Transition metal-catalyzed dehydrogenation of hydrazine borane $N_2H_4BH_3$ via the hydrolysis of BH_3 and the decomposition of N_2H_4. *Int. J. Hydrogen Energy* **2012**, *37*, 10758–10767. [CrossRef]
29. Zhang, H.; Feng, X.; Cheng, L.; Hou, X.; Li, Y.; Han, S. Non-noble Co anchored on nanoporous graphene oxide, as an efficient and long-life catalyst for hydrogen generation from sodium borohydride. *Colloids Surf. A* **2019**, *563*, 112–119. [CrossRef]
30. Xia, G.; Zhang, L.; Fang, F.; Sun, D.; Guo, Z.; Liu, H.; Yu, X. General synthesis of transition metal oxide ultrafine nanoparticles embedded in hierarchically porous carbon nanofibers as advanced electrodes for lithium storage. *Adv. Funct. Mater.* **2016**, *26*, 6188–6196. [CrossRef]

31. Song, Q.; Wang, W.D.; Hu, X.; Dong, Z. Ru nanoclusters confined in porous organic cages for catalytic hydrolysis of ammonia borane and tandem hydrogenation reaction. *Nanoscale* **2019**, *11*, 21513–21521. [CrossRef] [PubMed]
32. Kakanakova-Georgieva, A.; Gueorguiev, G.; Sangiovanni, D.; Suwannaharn, N.; Ivanov, I.; Cora, I.; Pécz, B.; Nicotra, G.; Giannazzo, F. Nanoscale phenomena ruling deposition and intercalation of AlN at the graphene/SiC interface. *Nanoscale* **2020**, *12*, 19470–19476. [CrossRef] [PubMed]
33. Freitas, R.; Brito Mota, F.; Castilho, C.; Kakanakova-Georgieva, A.; Gueorguiev, G. Spin-orbit-induced gap modification in buckled honeycomb XBi and XBi$_3$ (X = B, Al, Ga, and In) sheets. *J. Phys.-Condens. Matter* **2015**, *27*, 485306. [CrossRef]
34. Chen, Y.; Shi, J. Chemistry of mesoporous organosilica in nanotechnology: Molecularly organic-inorganic hybridization into frameworks. *Adv. Mater.* **2016**, *28*, 3235–3272. [CrossRef] [PubMed]
35. Li, Y.; Xie, L.; Li, Y.; Zheng, J.; Li, X. Metal-Organic-Framework-Based Catalyst for Highly Efficient H$_2$ Generation from Aqueous NH$_3$BH$_3$ Solution. *Chem. Eur. J.* **2009**, *15*, 8951–8954. [CrossRef]
36. Yan, Y.; Li, C.; Wu, Y.; Gao, J.; Zhang, Q. From isolated Ti-oxo clusters to infinite Ti-oxo chains and sheets: Recent advances in photoactive Ti-based MOFs. *J. Mater. Chem. A* **2020**, *8*, 15245–15270. [CrossRef]
37. Karthik, P.; Shaheer, A.M.; Vinu, A.; Neppolian, B. Amine Functionalized Metal-Organic Framework Coordinated with Transition Metal Ions: D-d Transition Enhanced Optical Absorption and Role of Transition Metal Sites on Solar Light Driven H$_2$ Production. *Small* **2020**, *16*, 1902990. [CrossRef] [PubMed]
38. Gu, Z.; Chen, L.; Wang, H.; Guo, Y.; Xu, M.; Zhang, Y.; Duan, C. Light control of charge transfer in metal/semiconductor heterostructures for efficient hydrogen evolution: Optical transition versus SPR. *Int. J. Hydrogen Energy* **2017**, *42*, 26713–26722. [CrossRef]
39. Xu, F.; Zhang, X.; Sun, L.; Yu, F.; Li, P.; Chen, J.; Wu, Y.; Cao, L.; Xu, C.; Yang, X.; et al. Hydrogen generation of a novel Al-NaMgH$_3$ composite reaction with water. *Int. J. Hydrogen Energy* **2017**, *42*, 30535–30542. [CrossRef]
40. Yang, F.; Zou, Y.; Xiang, C.; Xu, F.; Sun, L. Synthesis of "needle-cluster" NiCo$_2$O$_4$ carbon nanofibers and loading of Co-B nanoparticles for hydrogen production through the hydrolysis of NaBH$_4$. *J. Alloys Compd.* **2022**, *911*, 165069. [CrossRef]
41. Zhou, H.; Kang, M.; Xie, B.; Wen, P.; Zhao, N. L-Alanine mediated controllable synthesis: Ultrathin Co$_3$O$_4$ nanosheets@ Ni foam for high performance supercapacitors. *J. Alloys Compd.* **2021**, *874*, 160030. [CrossRef]
42. Shen, Y.; Zhu, C.; Chen, B.; Chen, J.; Fang, Q.; Wang, J. Novel photocatalytic performance of nanocage-like MIL-125-NH$_2$ induced by adsorption of phenolic pollutants. *Environ. Sci. Nano* **2020**, *7*, 1525–1538. [CrossRef]
43. Wang, W.; Hong, X.; Yao, Q.; Lu, Z.H. Bimetallic Ni-Pt nanoparticles immobilized on mesoporous N-doped carbon as a highly efficient catalyst for complete hydrogen evolution from hydrazine borane. *J. Mater. Chem. A* **2020**, *8*, 13694–13701. [CrossRef]
44. Georgieva, J.; Valova, E.; Mintsouli, I.; Sotiropoulos, S.; Tatchev, D.; Armyanov, S.; Malet, L. Pt (Ni) electrocatalysts for methanol oxidation prepared by galvanic replacement on TiO$_2$ and TiO$_2$-C powder supports. *J. Electroanal. Chem.* **2015**, *754*, 65–74. [CrossRef]
45. Liu, Y.; Chen, H.; Tian, C.; Geng, D.; Wang, D.; Bai, S. One-Pot Synthesis of Highly Efficient Carbon-Supported Polyhedral Pt$_3$Ni Alloy Nanoparticles for Oxygen Reduction Reaction. *Electrocatalysis* **2019**, *10*, 613–620. [CrossRef]
46. Li, M.; Zhu, Y.; Song, N.; Wang, C.; Lu, X. Fabrication of Pt nanoparticles on nitrogen-doped carbon/Ni nanofibers for improved hydrogen evolution activity. *J. Colloid Interface Sci.* **2018**, *514*, 199–207. [CrossRef]
47. Huff, C.; Quach, Q.; Long, J.M.; Abdel-Fattah, T.M. Nanocomposite catalyst derived from ultrafine platinum nanoparticles and carbon nanotubes for hydrogen generation. *ECS. J. Solid State Technol.* **2020**, *9*, 101008. [CrossRef]
48. Wu, C.; Zhang, J.; Guo, J.; Sun, L.; Ming, J.; Dong, H.; Yang, X. Ceria-induced strategy to tailor Pt atomic clusters on cobalt–nickel oxide and the synergetic effect for superior hydrogen generation. *ACS. Sustain. Chem. Eng.* **2018**, *6*, 7451–7457. [CrossRef]
49. Lu, L.; Shu, H.; Ruan, Z.; Ni, J.; Zhang, H. Preparation of Graphene-supported Pt-Pd Catalyst and Its Catalytic Activity and Mechanism for Hydrogen Generation Reaction. *Chem. J. Chin. Chi.* **2018**, *39*, 949–955. [CrossRef]
50. Lale, A.; Wasan, A.; Kumar, R.; Miele, P.; Demirci, U.B.; Bernard, S. Organosilicon polymer-derived mesoporous 3D silicon carbide, carbonitride and nitride structures as platinum supports for hydrogen generation by hydrolysis of sodium borohydride. *Int. J. Hydrogen Energy* **2016**, *41*, 15477–15488. [CrossRef]
51. Li, K.; Ma, M.; Xie, L.; Yao, Y.; Kong, R.; Du, G.; Sun, X. Monolithically integrated NiCoP nanosheet array on Ti mesh: An efficient and reusable catalyst in NaBH$_4$ alkaline media toward on-demand hydrogen generation. *Int. J. Hydrogen Energy* **2017**, *42*, 19028–19034. [CrossRef]
52. Li, X.; Zeng, C.; Fan, G. Ultrafast hydrogen generation from the hydrolysis of ammonia borane catalyzed by highly efficient bimetallic RuNi nanoparticles stabilized on Ti$_3$C$_2$X$_2$ (X = OH and/or F). *Int. J. Hydrogen Energy* **2015**, *40*, 3883–3891. [CrossRef]
53. Pornea, A.M.; Abebe, M.W.; Kim, H. Ternary NiCoP urchin like 3D nanostructure supported on nickel foam as a catalyst for hydrogen generation of alkaline NaBH$_4$. *Chem. Phys.* **2019**, *516*, 152–159. [CrossRef]
54. Wang, Y.; Li, G.; Wu, S.; Wei, Y.; Meng, W.; Xie, Y.; Zhang, X. Hydrogen generation from alkaline NaBH$_4$ solution using nanostructured Co-Ni-P catalysts. *Int. J. Hydrogen Energy* **2017**, *42*, 16529–16537. [CrossRef]
55. Zhou, Y.; Fang, C.; Fang, Y.; Zhu, F.; Liu, H.; Ge, H. Hydrogen generation mechanism of BH$_4^-$ spontaneous hydrolysis: A sight from ab initio calculation. *Int. J. Hydrogen Energy* **2016**, *41*, 22668–22676. [CrossRef]

Article

Micro/Nano Energy Storage Devices Based on Composite Electrode Materials

Yanqi Niu [1,2,3,*], Deyong Shang [1,2,3] and Zhanping Li [1,2,3]

1   School of Mechanical, Electronic & Information Engineering, China University of Mining and Technology (Beijing), Beijing 100083, China; sdycumtb@163.com (D.S.); ky202110@cumtb.edu.cn (Z.L.)
2   Institute of Intelligent Mining & Robotics, China University of Mining and Technology (Beijing), Beijing 100083, China
3   Key Laboratory of Intelligent Mining and Robotics, Ministry of Emergency Management, Beijing 100083, China
*   Correspondence: niuyanqi@cumtb.edu.cn

Abstract: It is vital to improve the electrochemical performance of negative materials for energy storage devices. The synergistic effect between the composites can improve the total performance. In this work, we prepare $\alpha\text{-}Fe_2O_3@MnO_2$ on carbon cloth through hydrothermal strategies and subsequent electrochemical deposition. The $\alpha\text{-}Fe_2O_3@MnO_2$ hybrid structure benefits electron transfer efficiency and avoids the rapid decay of capacitance caused by volume expansion. The specific capacitance of the as-obtained product is 615 mF cm^{-2} at 2 mA cm^{-2}. Moreover, a flexible supercapacitor presents an energy density of 0.102 mWh cm^{-3} at 4.2 W cm^{-3}. Bending tests of the device at different angles show excellent mechanical flexibility.

Keywords: $\alpha\text{-}Fe_2O_3@MnO_2$; electrode materials; electrochemical performance; flexibility

Citation: Niu, Y.; Shang, D.; Li, Z. Micro/Nano Energy Storage Devices Based on Composite Electrode Materials. *Nanomaterials* **2022**, *12*, 2202. https://doi.org/10.3390/nano12132202

Academic Editor: Diego Cazorla-Amorós

Received: 1 June 2022
Accepted: 24 June 2022
Published: 27 June 2022

Publisher's Note: MDPI stays neutral with regard to jurisdictional claims in published maps and institutional affiliations.

1. Introduction

Supercapacitors (SCs) have attracted much attention from researchers as an innovative type of energy storage device [1–4]. Compared with traditional capacitors, SCs shows the advantages of superior cycle stability, outstanding power density and fast charging/discharging [5–7]. Recently, electronic devices have progressively high requirements for long-term endurance. However, SCs is severely limited with low energy density [8–10]. According to the present research results, one of the most valid ways to settle this issue is to increase the specific capacity of electrode [11]. Therefore, designing electrodes with high specific capacitance is the primary task to broaden the application range of SCs.

Currently, the research on positive and negative materials is unevenly developed and research on negative electrodes is relatively little, which makes it difficult to increase the energy density of SCs. Commonly used negative materials are carbon (AC, CNTs and rGO), transition metal oxides (such as Fe_3O_4, $\alpha\text{-}Fe_2O_3$, MoO_3 and Mn_3O_4) and a small amount of metal nitride [12–17]. Among them, $\alpha\text{-}Fe_2O_3$ is considered to have the highest potential and is the most widely used anode material, because of its high redox activity, large theoretical specific capacitance and environmental protection [18]. Nonetheless, the weak conductivity of $\alpha\text{-}Fe_2O_3$ electrodes leads low practical specific capacitance and poor electrochemical stability [19,20]. Manganese dioxide (MnO_2) has gained extensive attention in the construction of supercapacitors due to its high oxidation activity [21]. At present, preparing nanocomposite materials utilizing the synergistic effect of two materials not only promotes redox reactions, but also enhance device energy density [22]. $Co_3O_4@MnO_2$, $SnO_2@MnO_2$, $ZnO@MnO_2$, $CuO@MnO_2$ and $\alpha\text{-}Fe_2O_3@MnO_2$ nanostructures were compounded to achieve both excellent cyclic stability and high capacitance [23–26].

Seol et al. prepared two types of SCs (EDLC and PC) using activated carbon and graphene/Mn_3O_4 nanocomposite. The performance degradation of EDLC was negligible after 100,000 cycles, while PC was less than 10% after 25,000 cycles [27]. Both devices demonstrate excellent cyclic stability and durability. Sarkar et al. fabricated α-Fe_2O_3/MnO_2 nano-heterostructure with a specific capacitance of 750 mFcm^{-2} at 2 mV s^{-1} [28]. However, in practice, these composites, because of loose contact, might impact their electrochemical performance. Thus, it is necessary to construction α-Fe_2O_3-based materials with unique nanostructures and excellent electrochemical performance. By combining two materials with high oxidative activity, the synthesis of ordered nanostructures will help to construct electrode materials with excellent specific capacitance. The main objective of our research is that by compounding nanomaterials, the advantages of both can be fully exploited and the electrochemical performance can be effectively enhanced.

Herein, we synthesized α-Fe_2O_3 nanorods structures through a hydrothermal route. Then, a MnO_2 film is coated on α-Fe_2O_3 surface by subsequent electrochemical deposition. When utilized as negative material for SCs, α-Fe_2O_3@MnO_2 electrode shows a specific capacitance of 615 mF cm^{-2} at 2 mA cm^{-2}. After 10,000 cycles, it maintains 92.3% of the initial capacitance. Finally, a flexible supercapacitor possesses the maximum energy density is 0.102 mWh cm^{-3} at 4.2 W cm^{-3}. The results under different angles bending tests demonstrated that the device possesses excellent mechanical flexibility.

2. Experimental Section

Material Preparation

The α-Fe_2O_3 sample was synthesized via a hydrothermal method. In total, 0.808 g $Fe(NO_3)_3 \cdot 9H_2O$, 0.2841 g Na_2SO_4 and 0.5 g PVP were dissolved into 45 mL deionized water. Then, a clean carbon cloth (2.5 × 2.5 cm^2) and the above mixed solution was transferred into an 80 mL autoclave and kept 110 °C for 9 h. Finally, the as-synthesized samples were annealed at 350 °C for 2 h (2 °C min^{-1}). An α-Fe_2O_3@MnO_2 sample was prepared by subsequent electrochemical deposition. In total, 2.4509 g $C_4H_6MnO_4 \cdot 4H_2O$ and 1.4204 g Na_2SO_4 was used as electrolyte. The α-Fe_2O_3 product was used as the working electrode, Ag/AgCl as the reference electrode and Pt foil as the counter one, with deposition at 1 V constant potential for 30 s. The $NiCo_2S_4$ sample was prepared from a homogeneous solution of 0.4 g $Ni(NO_3)_2 \cdot 6H_2O$, 1 g $Co(NO_3)_2 \cdot 6H_2O$, 0.5 g urea, 0.1 g NH_4F and 60 mL deionized water, heated with nickel foam at 140 °C for 12 h. It was then combined with 0.5 g $Na_2S \cdot 9H_2O$ and 60 mL deionized water at 140 °C for 6 h. α-Fe_2O_3, α-Fe_2O_3@MnO_2 and $NiCo_2S_4$ mass loading is 2, 2.3 and 1.2 mg cm^{-2}, respectively.

A supercapacitor was assembled with PVA-KOH gel as the electrolyte, $NiCo_2S_4$ as the positive electrode and α-Fe_2O_3@MnO_2 as the negative electrode. The preparation process of PVA-KOH gel electrolyte is as follows: stir 2 g KOH with 2 mL distilled water, mix well and set aside for later use. In a 20 mL beaker, add 2 g polyvinyl alcohol (PVA) and 20 mL deionized water, and stir at 80 °C until transparent. Finally, drop the KOH solution into the PVA solution at a constant speed, and stir at a constant temperature until it becomes a clear and transparent gel.

The crystal structure and the elemental compositions of the products were investigated by an X-ray diffractometer (XRD, Shimadzu-7000, Kyoto, Japan, CuKα, 40 kV) and X-ray photoelectron spectrometer (XPS, Amsterdam, Holland,). The morphology and microstructure of the sample is characterized by scanning electron microscope (SEM, Gemini 300-71-31, Berlin, Germany).

In a three-electrode system, the as-prepared electrode was measured through an electrochemical workstation (Shanghai Chenhua). Electrochemical performance methods include cyclic voltammetry (CV), galvanostatic charge-discharge (GCD) and electrochemical impedance spectroscopy (EIS). The as-synthesized materials were used as the working electrode, Pt foil as the counter electrode and Ag/AgCl as the reference electrode.

3. Results and Discussion

Figure 1 presents the growth process of α-Fe_2O_3@MnO_2 products on carbon cloth. Firstly, α-Fe_2O_3 nanorods are obtained via a facile hydrothermal approach. Afterwards, a

Nanomaterials **2022**, *12*, 2202

layer of MnO_2 film is deposited by subsequent electrochemical deposition on the nanorod-shaped α-Fe_2O_3 surface.

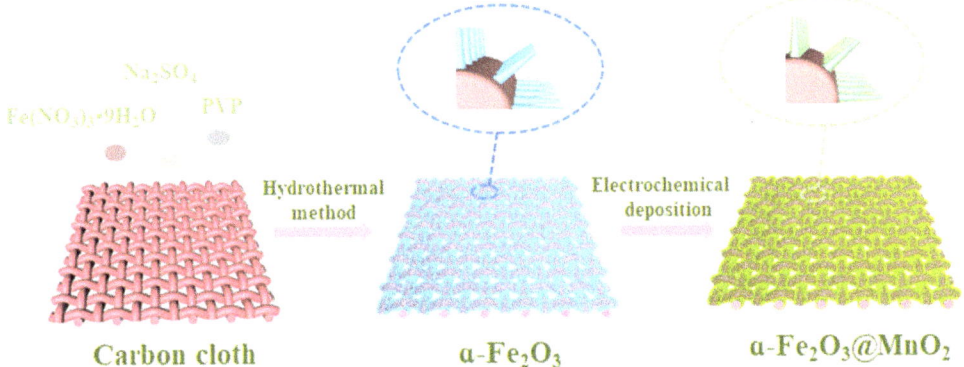

Figure 1. Synthesis schematic of the products.

First, the crystal structure of the obtained product is studied by XRD. Figure 2a shows the XRD patterns of α-Fe_2O_3 and α-Fe_2O_3@MnO_2 composites. A typical peak of the carbon cloth can be clearly observed. The peaks at 2θ values of $33.4°$, $35.8°$, $49.7°$, $54.4°$, $64.3°$ and $72.4°$ can be indexed to (104), (110), (024), (116), (300) and (1010) planes of α-Fe_2O_3 phases, respectively (PDF No. 84-0308). Those at $28.7°$, $37.6°$, $41.1°$, $47.2°$ and $72.6°$ match well with (310), (121), (420), (510) and (631) planes of MnO_2 (PDF No. 72-1982). The shape and sharpness of the diffraction peaks in figure reveal that the products possess high crystallinity.

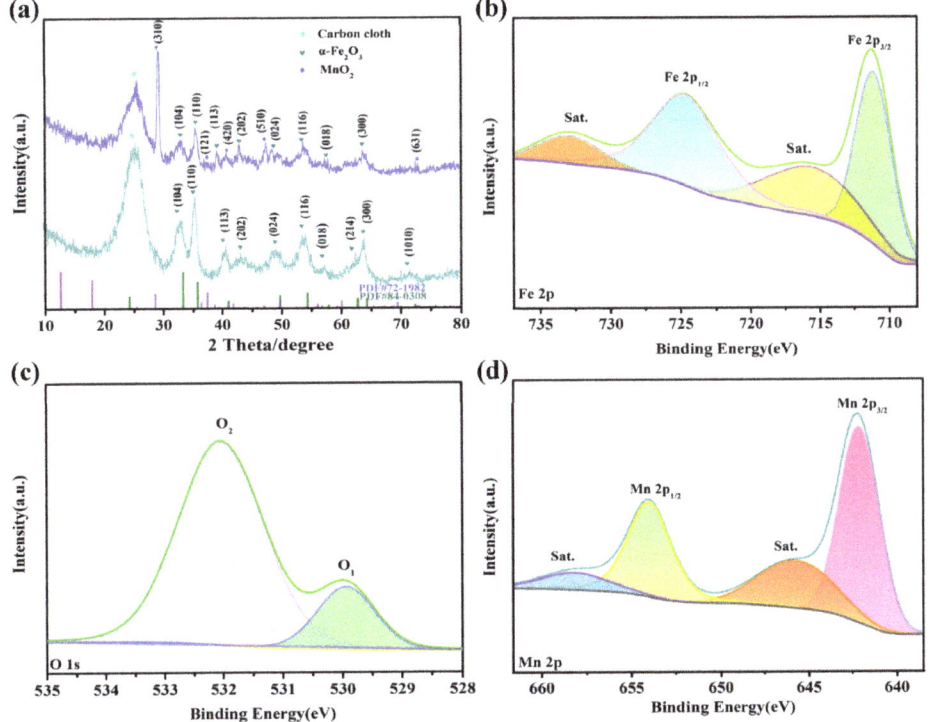

Figure 2. Structural characterization using (**a**) XRD patterns and (**b**–**d**) XPS spectra.

Then, XPS is used to investigate the α-Fe_2O_3@MnO_2 materials surface element composition. In Fe 2p spectra, the characteristic peaks of Fe $2p_{3/2}$ and Fe $2p_{1/2}$ at 711.2 eV and 724.8 eV, respectively (Figure 2b). Additionally, two shake-up satellite peaks (Sat.) at 716 eV and 732.9 eV are determined. This indicates that Fe^{3+} exists in composite product [29]. Figure 2c depicts the two main peaks of O 1s spectra located at 529.9 eV and 532 eV [30]. Binding energies at 529.9 eV, labeled as O_1, denote metal oxygen [31]. Another O_2 peak located at 532 eV is due to some degree of hydrolysis on the product surface [32]. For Mn 2p spectra (Figure 2d), four peaks at 642.2 eV, 645.8 eV, 653.9 eV and 658.1 eV are from Mn $2p_{3/2}$, Sat., Mn $2p_{1/2}$ and Sat., respectively [33].

Figure 3a indicates that α-Fe_2O_3 shows a short rod-like structure. In addition, it can be found that many nanorods homogeneously grown on carbon cloth with uniform size and shape, and the cross-section of nanorods is rough. The high magnification image (Figure 3b) shows the as-synthesized products average length is 100 nm. Figure 3c presents a thin MnO_2 film covers α-Fe_2O_3, and still maintains the shape of nanorods. From Figure 3d, the cross-section of α-Fe_2O_3@MnO_2 nanorods becomes smooth.

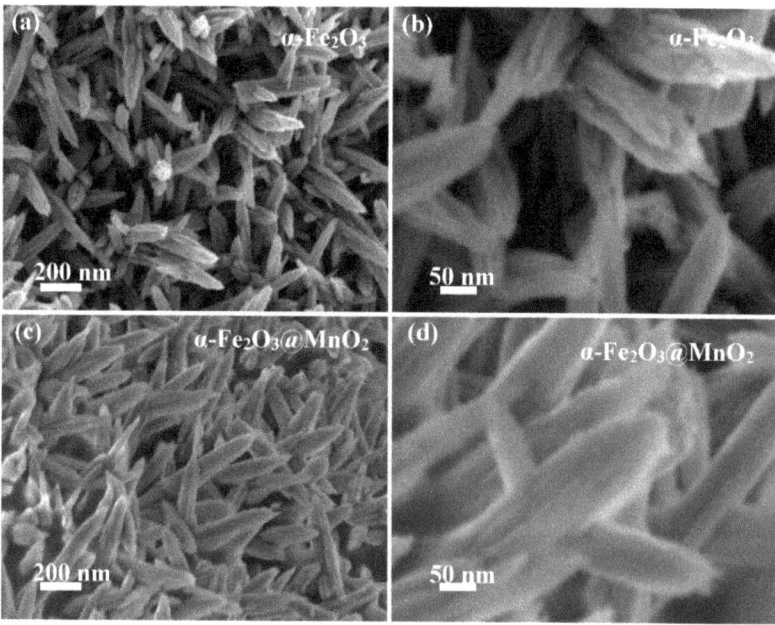

Figure 3. SEM images of the samples. (**a,c**) single materials (**b,d**) conposite materials.

Next, we analyzed several as-obtained electrode electrochemical performances by CV, GCD and EIS. Figure 4a shows CV curves of α-Fe_2O_3, MnO_2 and α-Fe_2O_3@MnO_2 materials. Evidently, α-Fe_2O_3@MnO_2 delivers a large CV area in -1–0 V, reflecting its good energy storage effect in this range. At 8 mA cm^{-2} (Figure 4b), the GCD curves obvious that α-Fe_2O_3@MnO_2 product with long discharge times, which can be correlative to the synergistic effect between α-Fe_2O_3 and MnO_2 materials. Figure 4c presents CV curves of α-Fe_2O_3@MnO_2 from 5 to 40 mV s^{-1}. The shape of CV curves almost the same as the scan rate increased, indicating excellent reversibility of electrode. In Figure 4d, the GCD curves of α-Fe_2O_3@MnO_2 materials are measured from 2 to 10 mA cm^{-2}. Areal capacitance (C_a) is obtained by GCD, and the equation is shown below:

$$Ca = I \int Vdt/V \qquad (1)$$

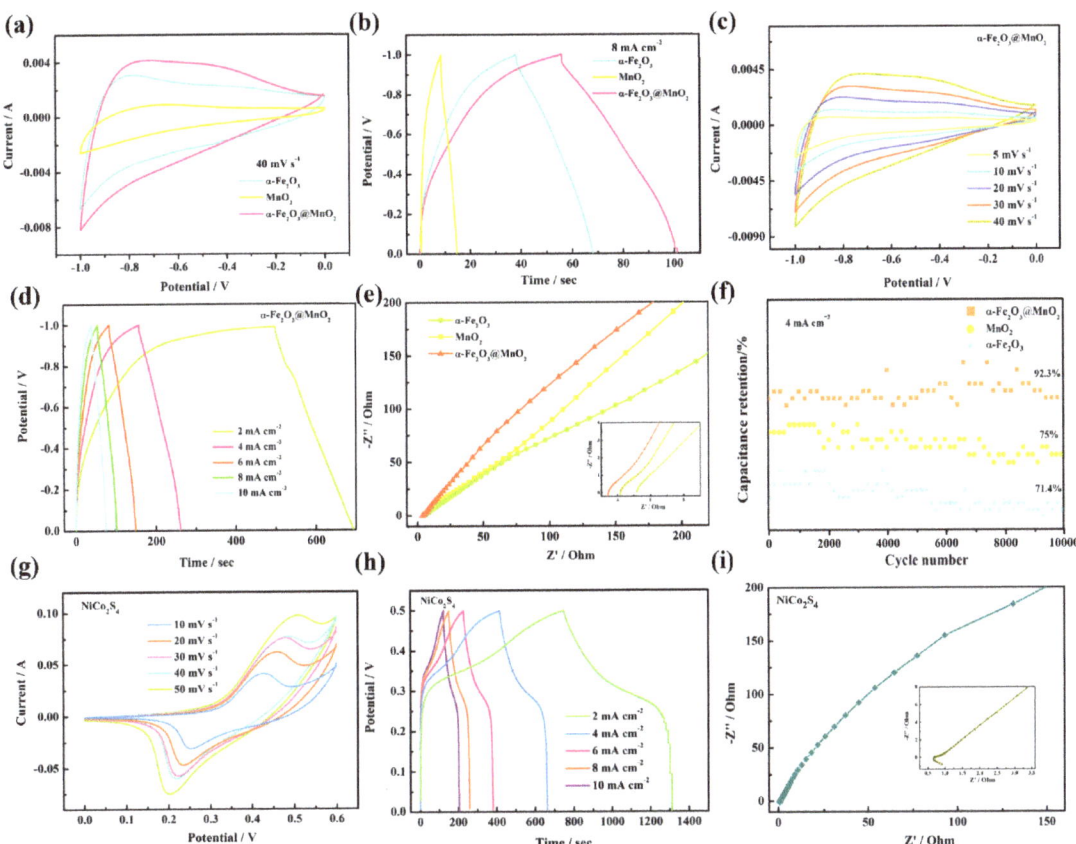

Figure 4. Electrochemical performance. (**a**) CV curves. (**b**) GCD curves. (**c**) CV curves of α-Fe$_2$O$_3$@MnO$_2$. (**d**) GCD curves of α-Fe$_2$O$_3$@MnO$_2$. (**e**) Nyquist plots. (**f**) Cycling performance at 4 mA cm^{-2}. (**g**) NiCo$_2$S$_4$ CV curves. (**h**) NiCo$_2$S$_4$ GCD curves. (**i**) NiCo$_2$S$_4$ Nyquist plots.

In Equation (1), I is current density, $\int$ Vdt stands for the integral area of discharge curve and V is the constant discharge voltage range (V). The α-Fe$_2$O$_3$@MnO$_2$ electrode delivers 615 mF cm^{-2} specific capacitance at 2 mA cm^{-2}

EIS is a significant factor in assessing the electrochemical kinetics of products. The sample is tested over a frequency range of 0.01 Hz to 100 kHz (Figure 4e). In the low frequency region, the slope of the straight line shows the ion diffusion resistance. Among the three samples, α-Fe$_2$O$_3$@MnO$_2$ sample presents the largest slope, which expresses fast diffusion of ions in electrolyte [34]. The intersection with the real axis represents the equivalent resistance (Rs) [35]. α-Fe$_2$O$_3$, MnO$_2$ and α-Fe$_2$O$_3$@MnO$_2$ electrodes Rs value is 5.1 Ω, 4.1 Ω and 3.3 Ω, respectively. According to above analysis, α-Fe$_2$O$_3$@MnO$_2$ shows the largest slope and smallest Rs, so the conductivity of composite material is better than α-Fe$_2$O$_3$ and MnO$_2$.

At the end, the cyclic stability is investigated at 4 mA cm^{-2}. Figure 4f indicates that the capacitance of α-Fe$_2$O$_3$@MnO$_2$ is only reduced by 7.7% after 10,000 cycles, while α-Fe$_2$O$_3$ and MnO$_2$ products present only 71.4% and 75% of the initial capacitance. This phenomenon is due to the MnO$_2$ film covering the α-Fe$_2$O$_3$ nanorods, which can help alleviate the volume expansion during long cycle measurements. Similarly, the positive NiCo$_2$S$_4$ is also studied by the same methods. Figure 4g presents the CV curves of NiCo$_2$S$_4$

sample. Redox peaks and shapes, confirming its pseudocapacitive material. Five symmetrical GCD curves shows an obvious platform (Figure 4h), which indicates their Faradaic redox behavior [36]. At 2 mA cm^{-2}, the specific capacitance is 720.8 mF cm^{-2}. Nyquist plots of NiCo$_2$S$_4$ products are shown in Figure 4i; the value of Rs is 0.9 Ω.

To further explore the α-Fe$_2$O$_3$@MnO$_2$ electrodes for practical applications, a flexible supercapacitor is assembled. From Figure 5a, the voltage windows of α-Fe$_2$O$_3$@MnO$_2$ and NiCo$_2$S$_4$ are -1–0 V and 0–0.6 V, respectively. Figure 5b shows CV curves from 1.1 V to 1.5 V with a sweep rate of 100 mV s^{-1}, demonstrating the device can maintain operate stably within 1.5 V. It can be seen that with the decrease of voltage, the area becomes small. Figure 5c depicts all CV curves at different scan rates keep similar shapes, revealing outstanding rate performance of device. GCD curves from 1 to 8 mA cm^{-2} possess the same charging and discharging time (Figure 5d). The specific capacitance of the device at 1 mA cm^{-2} is 37.8 mF cm^{-2} and it still delivers 15.6 mF cm^{-2} at 8 mA cm^{-2}. The equivalent resistance value of the device is 1.9 Ω, as shown in Figure 5e.

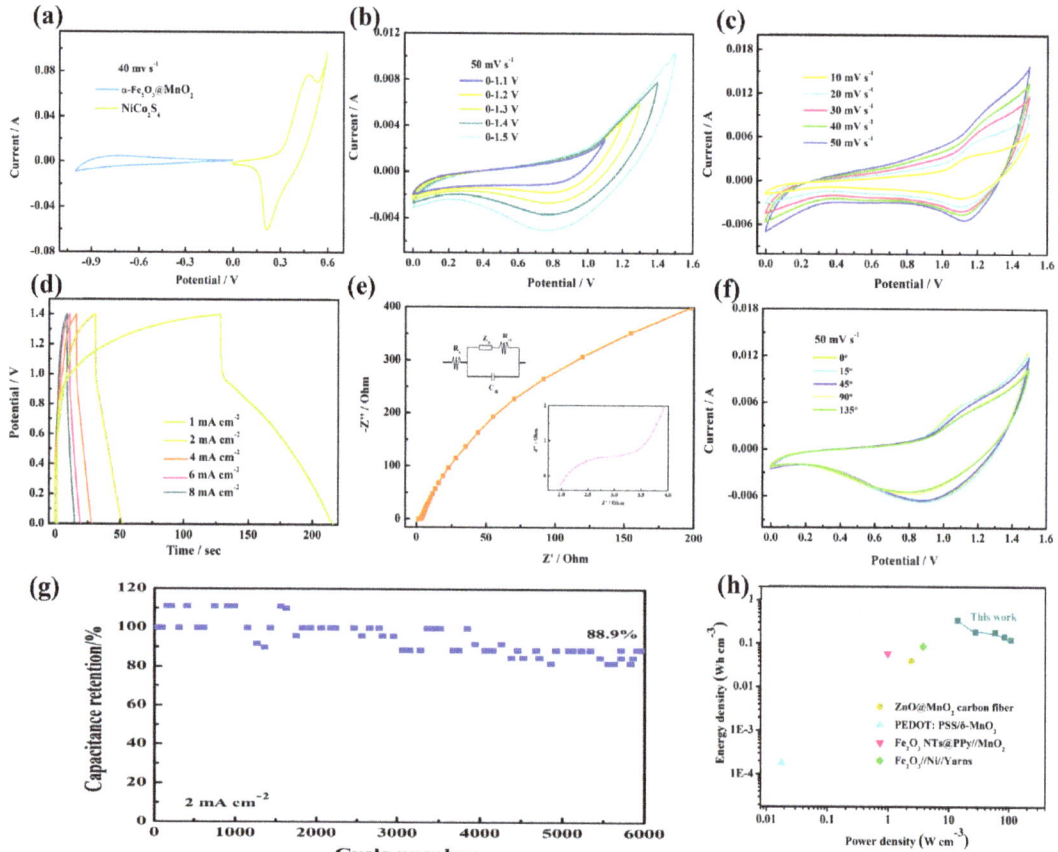

Figure 5. (**a**) CV curves of the α-Fe$_2$O$_3$@MnO$_2$ and NiCo$_2$S$_4$ electrode at 40 mV s^{-1}. (**b**) CV curves in different potential windows at 50 mV s^{-1}. (**c**) CV curves. (**d**) GCD curves. (**e**) EIS. (**f**) CV curves at different bending angles. (**g**) cycling performance at 2 mA cm^{-2}. (**h**) Ragone plot.

At present, electronic devices are developing towards wearable, which puts forward higher requirements for the mechanical flexibility of supercapacitors [37]. We twisted the device and then examined it by cyclic voltammetry (Figure 5f). While device is folded at 15°, 45°, 90° and 135°, the shape sustains virtually unchanged, demonstrating its superior mechanical stability. Figure 5g illustrates that the device maintains 88.9% capacitance retention after 6000 cycles. Figure 5h is the Ragone diagram of α-Fe$_2$O$_3$@MnO$_2$//NiCo$_2$S$_4$. The capacitor values of energy density (E) and power density (P) can be derived based on the Equations (2) and (3):

$$E = 1/2 \times C_a \times V^2 \tag{2}$$

$$P = 3600 \times E/\Delta t \tag{3}$$

where C_a stands for the areal capacitance of the capacitor, V represent the discharge voltage and Δt is the discharge time. At 1 mA cm^{-2}, the energy density of device is 0.102 mWh cm^{-3} at 4.2 W cm^{-2}. This is better than some previously reported materials [38–41] (Table 1).

Table 1. Electrochemical performance of various devices.

Supercapacitor	Capacitance	Energy Density (mWh cm^{-3})	Power Density (W cm^{-2})	Capacitance Retention	Ref.
PEDOT: PSS/δ-MnO$_2$	2.4 F cm^{-3}	0.018	0.018	88%	[38]
Fe$_2$O$_3$NTs@PPy//MnO$_2$	-	0.0594	1	92%	[39]
ZnO@MnO$_2$	26 mF cm^{-2}	0.04	2.44	87.5%	[40]
Fe$_2$O$_3$//Ni/Yarns	0.67 F cm^{-3}	0.086	3.87	87.1%	[41]
α-Fe$_2$O$_3$@MnO$_2$//NiCo$_2$S$_4$	37.8 mF cm^{-2}	0.102	4.2	88.9%	this work

α-Fe$_2$O$_3$@MnO$_2$ delivers excellent performance, which can be explained by the following reasons: (a) Nanostructure uniformly covered on the carbon cloth, which provides outstanding electrical conductivity and flexibility; (b) With α-Fe$_2$O$_3$ as a strong mechanical support and MnO$_2$ as an outer layer, this structure not only protects the morphological structure, but also provides many active sites; (c) The composite utilizes the synergistic effect of α-Fe$_2$O$_3$ and MnO$_2$, so that electrode processes high capacitance and low resistance.

4. Conclusions

In this manuscript, α-Fe$_2$O$_3$@MnO$_2$ nanorods are synthesized through a hydrothermal route and subsequent electrochemical deposition. By combining two oxides of α-Fe$_2$O$_3$ and MnO$_2$, it is favorable to accelerate the electron transport and the oxidation reaction. The synergistic effect between two materials improves electrochemical performance for negative electrode. MnO$_2$ film, after electrodeposition, affects the performance of the electrode material, and the full use of the active area of the film increases, which increases the capacitance of the electrode material. XPS results show that the material processes abundant redox valence states. α-Fe$_2$O$_3$@MnO$_2$ sample presents high specific capacitance and excellent cycling stability. Furthermore, the as-assembled capacitors still show outstanding electrochemical performance and mechanical stability. Therefore, it provides an alternative method for constructing supercapacitor negative materials with higher specific capacitance.

Author Contributions: Conceptualization, Y.N. and D.S.; methodology, Y.N.; software, Z.L.; validation, Y.N., D.S. and Z.L.; formal analysis, D.S.; investigation, Z.L.; resources, Y.N.; data curation, Y.N.; writing—original draft preparation, D.S.; writing—review and editing, Y.N.; visualization, D.S.; supervision, Y.N.; project administration, Y.N.; funding acquisition, Y.N. All authors have read and agreed to the published version of the manuscript.

Funding: This research received no external funding.

Data Availability Statement: Not applicable.

Conflicts of Interest: The authors declare no conflict of interest.

References

1. Ye, S.; Ding, C.M.; Liu, M.Y.; Wang, A.Q.; Huang, Q.G.; Li, C. Water oxidation catalysts for artificial photosynthesis. *Adv. Mater.* **2019**, *31*, 1902069. [CrossRef] [PubMed]
2. Liu, Y.; Wu, X. Hydrogen and sodium ions co-intercalated vanadium dioxide electrode materials with enhanced zinc ion storage capacity. *Nano Energy* **2021**, *86*, 106124. [CrossRef]

3. Cao, X.H.; Tan, C.L.; Zhang, X.; Zhao, W.; Zhang, H. Solution-processed two-dimensional metal dichalcogenide-based nanomaterials for energy storage and conversion. *Adv. Mater.* **2016**, *28*, 6167. [CrossRef] [PubMed]

4. Dai, M.Z.; Zhao, D.P.; Wu, X. Research progress on transition metal oxide based electrode materials for asymmetric hybrid capacitors. *Chin. Chem. Lett.* **2020**, *31*, 2177–2188. [CrossRef]

5. Rakibuddin, M.; Shinde, M.A.; Kim, H. Sol-gel fabrication of NiO and NiO/WO$_3$ based electrochromic device on ITO and flexible substrate. *Ceram. Int.* **2020**, *46*, 8631–8639. [CrossRef]

6. Zheng, X.; Han, Z.C.; Yao, S.Y.; Xiao, H.H.; Chai, F.; Qu, F.Y.; Wu, X. Spinous α-Fe$_2$O$_3$ hierarchical nanostructures anchored on Ni foam for supercapacitors electrodes and visible light driven photocatalysts. *Dalton Trans.* **2016**, *45*, 7094–7103. [CrossRef]

7. Zhao, D.P.; Zhang, R.; Dai, M.Z.; Liu, H.Q.; Jian, W.; Bai, F.Q.; Wu, X. Constructing high efficient CoZnxMn$_2$-xO$_4$ electrocatalyst by regulating the electronic structure and surface reconstruction. *Small* **2022**, *18*, 107268. [CrossRef]

8. Liu, H.Q.; Zhao, D.P.; Hu, P.F.; Chen, K.F.; Wu, X.; Xue, D.F. Design strategies toward achieving high performance Co-MoO4@Co1.62Mo6S8 electrode materials, Mater. *Today Phys.* **2020**, *13*, 100197.

9. Rakibuddin, M.; Shinde, M.A.; Kim, H. Facile solegel fabrication of MoS$_2$ bulk, flake and quantum dot for electrochromic device and their enhanced performance with WO$_3$. *Electrochim. Acta* **2020**, *349*, 136403. [CrossRef]

10. Ahmad, K.; Ahmad, M.A.; Song, G.; Kim, H. Design and fabrication of MoSe$_2$/WO$_3$ thin films for the construction of electrochromic devices on indium tin oxide based glass and flexible substrates. *Ceram. Int.* **2021**, *47*, 34297–34306. [CrossRef]

11. Liu, H.Q.; Zhao, D.P.; Liu, Y.; Tong, Y.L.; Wu, X.; Shen, G.Z. NiMoCo layered double hydroxides for electrocatalyst and supercapacitor electrode. *Sci. China Mater.* **2021**, *64*, 581–591. [CrossRef]

12. Xia, T.; Zhao, D.P.; Xia, Q.; Umar, A.; Wu, X. Realizing high performance flexible supercapacitors by electrode modification. *RSC Adv.* **2021**, *11*, 39045–39050. [CrossRef] [PubMed]

13. Xia, Q.; Xia, T.; Dai, M.Z.; Wu, X.; Zhao, Y.F. A facile synthetic protocol of α-Fe$_2$O$_3$@FeS$_2$ nanocrystals for advanced electrochemical capacitor. *CrystEngComm* **2021**, *23*, 2432–2438. [CrossRef]

14. Xia, Q.; Xia, T.; Wu, X. PPy decorated α-Fe$_2$O$_3$ nanosheets as flexible supercapacitor electrode. *Rare Met.* **2022**, *41*, 1195–1201. [CrossRef]

15. Liu, H.; Zhu, J.; Li, Z.; Shi, Z.; Zhu, J.; Mei, H. Fe$_2$O$_3$/N doped rGO anode hybridized with NiCo LDH/Co(OH)$_2$ cathode for battery-like supercapacitor. *Chem. Eng. J.* **2021**, *403*, 126325. [CrossRef]

16. Yang, J.; Xiao, X.; Chen, P.; Zhu, K.; Cheng, K.; Ye, K.; Wang, G.L.; Cao, D.X.; Yan, J. Creating oxygen-vacancies in MoO$_3$-x nanobelts toward high volumetric energydensity asymmetric supercapacitors with long lifespan. *Nano Energy* **2019**, *58*, 455–465. [CrossRef]

17. Feng, J.X.; Ye, S.H.; Lu, X.F.; Tong, Y.X.; Li, G.R. Asymmetric paper supercapacitor based on amorphous porous Mn$_3$O$_4$ negative electrode and Ni(OH)$_2$ positive Electrode: A novel and high-performance flexible electrochemical energy storage device. *ACS Appl. Mater. Inter.* **2015**, *7*, 11441–11451. [CrossRef]

18. Liu, L.; Lang, J.; Zhang, P.; Hu, B.; Yan, X. Facile synthesis of Fe$_2$O$_3$ nano-dots@nitrogen-doped graphene for supercapacitor electrode with ultralong cycle life in KOH electrolyte. *ACS Appl. Mater. Inter.* **2016**, *8*, 9335–9344. [CrossRef]

19. Ready, D.W. Mass Transport and Sintering in Impure Ionic Solids. *J. Am. Ceram. Soc.* **1966**, *49*, 366. [CrossRef]

20. Liu, T.; Ling, Y.; Yang, Y.; Finn, L.; Collazo, E.; Zhai, T.; Li, Y. Investigation of hematite nanorod–nanoflake morphological transformation and the application of ultrathin nanoflakes for electrochemical devices. *Nano Energy* **2015**, *12*, 169–177. [CrossRef]

21. Racik, M.K.; Manikandan, A.; Mahendiran, M.; Madhavan, J.; Raj, M.V.A.; Mohamed, M.G.; Maiyalagan, T. Hydrothermal synthesis and characterization studies of α-Fe$_2$O$_3$/MnO$_2$ nanocomposites for energy storage supercapacitor application. *Ceram. Int.* **2020**, *46*, 6222–6233. [CrossRef]

22. Sun, Z.P.; Ai, W.; Liu, J.L.; Qi, X.Y.; Wang, Y.L.; Zhu, J.H.; Zhang, H.; Yu, T. Facile fabrication of hierarchical ZnCo$_2$O$_4$/NiO core/shell nanowire arrays with improved lithium-ion battery performance. *Nanoscale* **2014**, *6*, 6563. [CrossRef] [PubMed]

23. Liu, J.P.; Jiang, J.; Cheng, C.W.; Li, H.X.; Zhang, J.X.; Gong, H.; Fan, H.J. Co$_3$O$_4$ nanowire@MnO$_2$ ultrathin nanosheet core/shell arrays: A new class of high-performance pseudocapacitive materials. *Adv. Mater.* **2011**, *23*, 2076. [CrossRef] [PubMed]

24. Sun, X.; Li, Q.; Lv, Y.N.; Mao, Y.B. Three-dimensional ZnO@MnO$_2$ core@shell nanostructures for electrochemical energy storage. *Chem. Commun.* **2013**, *49*, 4456–4458. [CrossRef] [PubMed]

25. Chen, H.; Zhou, M.; Wang, T.; Li, F.; Zhang, Y.X. Construction of unique cupric oxide-manganese dioxide core-shell arrays on a copper grid for high-performance supercapacitors. *J. Mater. Chem. A* **2016**, *4*, 10786. [CrossRef]

26. Liu, Y.; Jiao, Y.; Yin, B.S.; Zhang, S.W.; Qu, F.Y.; Wu, X. Enhanced electrochemical performance of hybrid SnO$_2$@MOx (M = Ni, Co, Mn) core-shell nanostructures grown on flexible carbon fibers as the supercapacitor electrode materials. *J. Mater. Chem. A* **2015**, *3*, 3676. [CrossRef]

27. Seol, M.L.; Nam, I.; Ribeiro, E.L.; Segel, B.; Lee, D.; Palma, T.; Wu, H.L.; Mukherjee, D.; Khomami, B.; Hill, C.; et al. All-Printed in-plane supercapacitors by sequential additive manufacturing process. *ACS Appl. Energy Mater.* **2020**, *3*, 4965–4973. [CrossRef]

28. Sarkar, D.; Khan, G.G.; Singh, A.K.; Mandal, K. High-performance pseudocapacitor electrodes based on α-Fe$_2$O$_3$/MnO$_2$ core-shell nanowire eterostructure arrays. *J. Phys. Chem. C* **2013**, *117*, 15523. [CrossRef]

29. Zhang, Y.; Zeng, T.; Huang, D.X.; Yan, W.; Zhang, Y.Y.; Wan, Q.J.; Yang, N.J. High-energy-density supercapacitors from dual pseudocapacitive nanoelectrodes. *ACS Appl. Energy Mater.* **2021**, *3*, 10685–10694. [CrossRef]

30. Tang, Q.Q.; Wang, W.Q.; Wang, G.C. The perfect matching between the low-cost Fe$_2$O$_3$ nanowire anode and the NiO nanoflake cathode significantly enhances the energy density of asymmetric supercapacitors. *J. Mater. Chem. A* **2015**, *3*, 6662–6670. [CrossRef]

31. Zhao, J.; Li, Z.J.; Yuan, X.C.; Yang, Z.; Zhang, M.; Meng, A.; Li, Q.D. A high-energy density asymmetric supercapacitor based on Fe_2O_3 nanoneedle arrays and $NiCo_2O_4/Ni(OH)_2$ hybrid nanosheet arrays grown on sic nanowire networks as free-standing advanced electrodes. *Adv. Energy Mater.* **2018**, *8*, 1702787. [CrossRef]

32. Zhang, M.; Wu, X.W.; Yang, D.Z.; Qin, L.Y.; Zhang, S.Y.; Xu, T.; Zhao, T.Y.; Yu, Z.Z. Ultraflexible reedlike carbon nanofiber membranes decorated with Ni-Co-S nanosheets and Fe_2O_3-C coreshell nanoneedle arrays as electrodes of flexible quasisolidstate asymmetric supercapacitors. *ACS Appl. Energy Mater.* **2021**, *4*, 1505–1516.

33. Chen, Y.C.; Kang, C.X.; Ma, L.; Fu, L.K.; Li, G.H.; Hu, Q.; Liu, Q.M. MOF-derived Fe_2O_3 decorated with MnO_2 nanosheet arrays as anode for high energy density hybrid supercapacitor. *Chem. Eng. J.* **2021**, *417*, 129243. [CrossRef]

34. Hong, X.Y.; Li, J.H.; Zhu, G.S.; Xu, H.R.; Zhang, X.Y.; Zhao, Y.Y.; Zhang, J.; Yan, D.L.; Yu, A.B. Cobalt-nickel sulfide nanosheets modified by nitrogen-doped porous reduced graphene oxide as high-conductivity cathode materials for supercapacitor. *Electrochim. Acta* **2020**, *362*, 37156. [CrossRef]

35. Ke, Q.; Guan, C.; Zhang, X.; Zheng, M.; Zhang, Y.; Cai, W.; Zhang, H.; Wang, J. Surface charge-mediated formation of H-TiO_2@$Ni(OH)_2$ heterostructures for high-performance supercapacitors. *Adv. Mater.* **2017**, *29*, 1604164. [CrossRef]

36. Kong, W.; Lu, C.C.; Zhang, W.; Pu, J.; Wang, Z.H. Homogeneous core–shell $NiCo_2S_4$ nanostructures supported on nickel foam for supercapacitors. *J. Mater. Chem. A* **2015**, *3*, 12452–12460. [CrossRef]

37. Hu, P.F.; Zhao, D.P.; Liu, H.Q.; Chen, K.F.; Wu, X. Engineering PPy decorated $MnCo_2O_4$ urchins for quasi-solid-state hybrid capacitors. *CrystEngComm* **2019**, *21*, 1600–1606. [CrossRef]

38. Wang, Y.; Zhang, Y.Z.; Dubbink, D.; Elshof, J.E.T. Inkjet Printing of δ-MnO_2 Nanosheets for Flexible Solid-state Microsupercapacitor. *Nano Energy* **2018**, *49*, 481–488. [CrossRef]

39. Wang, Y.; Du, Z.X.; Xiao, J.F.; Cen, W.L.; Yuan, S.J. Polypyrrole-encapsulated Fe_2O_3 nanotube arrays on a carbon cloth support: Achieving synergistic effect for enhanced supercapacitor performance. *Electrochim. Acta* **2021**, *386*, 138486. [CrossRef]

40. Yang, P.H.; Xiao, X.; Li, Y.Z.; Ding, Y.; Qiang, P.F.; Tan, X.H.; Mai, W.J.; Lin, Z.Y.; Wu, W.Z.; Li, T.Q.; et al. Hydrogenated ZnO Core-Shell Nanocables for Flexible Supercapacitors and Self-Powered Systems. *ACS Nano* **2013**, *7*, 2617–2626. [CrossRef]

41. Wu, X.H.; Chen, Y.H.; Liang, K.J.; Yu, X.A.; Zhuang, Q.Q.; Yang, Q.; Liu, S.F.; Liao, S.; Lia, N.; Zhang, H.Y. Fe_2O_3 nanowire arrays on Ni-coated yarns as excellent electrodes for high performance wearable yarn-supercapacitor. *J. Alloys Compd.* **2021**, *886*, 158156. [CrossRef]

Article

Organic Crosslinked Polymer-Derived N/O-Doped Porous Carbons for High-Performance Supercapacitor

Jianhao Lao [1,†], Yao Lu [1,†], Songwen Fang [1], Fen Xu [1,*], Lixian Sun [1,*], Yu Wang [1], Tianhao Zhou [1], Lumin Liao [1,2], Yanxun Guan [1,2], Xueying Wei [3,*], Chenchen Zhang [1], Yukai Yang [1], Yongpeng Xia [1], Yumei Luo [1], Yongjin Zou [1], Hailiang Chu [1], Huanzhi Zhang [1], Yong Luo [1] and Yanling Zhu [1]

1 Guangxi Key Laboratory of Information Materials, Guangxi Collaborative Innovation Center for Structure and Properties for New Energy and Materials, School of Material Science and Engineering, Guilin University of Electronic Technology, Guilin 541004, China; ljh408888@163.com (J.L.); 18677367876@163.com (Y.L.); 1810201010@mails.guet.edu.cn (S.F.); ywang506x@163.com (Y.W.); zhoutianhao233@gmail.com (T.Z.); llm904691049@163.com (L.L.); gyx112405@163.com (Y.G.); Zhang_linba_3760@163.com (C.Z.); yangyukai0530@gmail.com (Y.Y.); ypxia@guet.edu.cn (Y.X.); luoym@guet.edu.cn (Y.L.); zouy@guet.edu.cn (Y.Z.); chuhailiang@guet.edu.cn (H.C.); zhanghuanzhi@guet.edu.cn (H.Z.); 15620323072@139.com (Y.L.); zyl9352021@163.com (Y.Z.)
2 School of Electronic Engineering and Automation, Guilin University of Electronic Technology, Guilin 541004, China
3 School of Architecture and Transportation Engineering, Guilin University of Electronic Technology, Guilin 541004, China
* Correspondence: xufen@guet.edu.cn (F.X.); sunlx@guet.edu.cn (L.S.); http510@guet.edu.cn (X.W.)
† These authors contributed equally to this work.

Citation: Lao, J.; Lu, Y.; Fang, S.; Xu, F.; Sun, L.; Wang, Y.; Zhou, T.; Liao, L.; Guan, Y.; Wei, X.; et al. Organic Crosslinked Polymer-Derived N/O-Doped Porous Carbons for High-Performance Supercapacitor. Nanomaterials 2022, 12, 2186. https://doi.org/10.3390/nano12132186

Academic Editor: Pedro Gómez-Romero

Received: 31 May 2022
Accepted: 23 June 2022
Published: 25 June 2022

Publisher's Note: MDPI stays neutral with regard to jurisdictional claims in published maps and institutional affiliations.

Abstract: Supercapacitors, as a new type of green electrical energy storage device, are a potential solution to environmental problems created by economic development and the excessive use of fossil energy resources. In this work, nitrogen/oxygen (N/O)-doped porous carbon materials for high-performance supercapacitors are fabricated by calcining and activating an organic crosslinked polymer prepared using polyethylene glycol, hydroxypropyl methylcellulose, and 4,4-diphenylmethane diisocyanate. The porous carbon exhibits a large specific surface area (1589 m$^2 \cdot$g^{-1}) and high electrochemical performance, thanks to the network structure and rich N/O content in the organic crosslinked polymer. The optimized porous carbon material (C$_{OCLP-4.5}$), obtained by adjusting the raw material ratio of the organic crosslinked polymer, exhibits a high specific capacitance (522 F$\cdot$g^{-1} at 0.5 A$\cdot$g^{-1}), good rate capability (319 F$\cdot$g^{-1} at 20 A$\cdot$g^{-1}), and outstanding stability (83% retention after 5000 cycles) in a three-electrode system. Furthermore, an energy density of 18.04 Wh$\cdot$kg^{-1} is obtained at a power density of 200.0 W$\cdot$kg^{-1} in a two-electrode system. This study demonstrates that organic crosslinked polymer-derived porous carbon electrode materials have good energy storage potential.

Keywords: supercapacitor; organic crosslinked polymer; porous carbon; electrochemistry

1. Introduction

Solutions to environmental problems, owing to economic development and the excessive use of fossil energy resources, are urgently being sought [1]. Supercapacitors, as a new type of green electrical energy storage device, have drawn increasing attention, owing to their high power density, fast charging/discharging, excellent reversibility, long life cycle, and environmental friendliness [2–4].

Theoretical research on and practical applications of supercapacitors have significantly progressed; however, insufficient energy density and high cost are still challenges requiring resolution [5–7]. Electrode materials, which can be divided into carbon materials [8,9], metal oxides [10,11], and conductive polymers [12,13], play an important role as core components in supercapacitors and are a key step in solving the existing problems. Among them, carbon materials are the most widely used electrode materials because of their high specific surface area, and good electrical conductivity and chemical stability [14–16]. Studies have shown that doping heteroatoms in a carbon-based framework increases the specific capacitance of carbon materials. On the one hand, it can improve the infiltration area between the electrode material and the electrolyte; on the other hand, the heterogeneous atoms can introduce pseudocapacitance during the charging/discharging process, further enhancing the electrochemical performance [17].

Nitrogen doping has been demonstrated to be an effective way to improve the wettability and conductivity of carbon materials and can also provide additional pseudocapacitance for supercapacitors. Generally, nitrogen-doped carbon materials can be prepared using two synthetic strategies, namely by the pyrolysis of nitrogen-containing precursors, such as biomass [18], synthetic polymers [19], small molecules [20], and ionic liquids [21], or by the chemical or thermal modification of premade carbon materials with reagents/gases containing nitrogen atoms [22]. Zhang et al. [23] used urea as a nitrogen-containing precursor and KOH as the activator to prepare a carbon material with an appropriate amount of N doping, which yielded a nitrogen-doped carbon material with a porous structure and large specific surface area. They also found that the capacitance of the carbon material reached up to 446.0 $F \cdot g^{-1}$ at 0.5 $A \cdot g^{-1}$ in a three-electrode system. The symmetrical supercapacitor device assembled with this nitrogen-doped carbon also displayed good performance, with an energy density of 16.3 $Wh \cdot kg^{-1}$ at a power density of 348.3 $W \cdot kg^{-1}$.

Organic crosslinked polymers are mainly composed of elements, such as carbon, nitrogen, oxygen, and hydrogen, which have the characteristics of a network structure. Porous carbon materials prepared using such polymers had a high heteroatom content, specific surface area, and outstanding electrochemical properties [24]. In particular, the structure of organic crosslinked polymers can be adjusted by changing the ratio of raw materials during the synthesis process. Zou et al. [25] prepared a new type of heteroatom-doped porous carbon material with a high specific surface area by carbonizing and activating polyphosphazenes, which exhibited a specific capacitance of 438 $F \cdot g^{-1}$ at a current density of 0.5 $A \cdot g^{-1}$ in a three-electrode system. Chen et al. [26] prepared a porous carbon material by calcining hypercrosslinked polymer (poly (vinylbenzyl chloride-co-divinylbenzene)), which exhibited a specific capacitance of 455 $F \cdot g^{-1}$ at a current density of 0.5 $A \cdot g^{-1}$.

In this work, nitrogen/oxygen(N/O)-doped carbon-based porous materials were fabricated by carbonizing and activating an organic crosslinked polymer with a network structure. The organic crosslinked polymer was synthesized using polyethylene glycol (PEG 6000), hydroxypropyl methylcellulose (HPMC), and 4,4-diphenylmethane diisocyanate (MDI). The carbon material obtained by optimizing the ratio of the raw materials had a large specific surface area (1589 $m^2 \cdot g^{-1}$) and a high specific capacitance of 522 $F \cdot g^{-1}$ at a current density of 0.5 $A \cdot g^{-1}$. Furthermore, its energy density reached 18.04 $Wh \cdot kg^{-1}$ at a power density of 200.0 $W \cdot kg^{-1}$ in a two-electrode system using 1 M Na_2SO_4 as the electrolyte. Mechanistic studies showed that the high electrochemical performance of the obtained carbon was attributed to the network structure and rich N/O content of the crosslinked polymer. Hence, the preparation method for porous carbon materials proposed in this study provides a new approach for the research and development of electrode materials.

2. Materials and Methods

2.1. Materials

Polyethylene glycol (PEG, Mw = 6000), 4,4-diphenylmethane diisocyanate (MDI, analytical grade), hydroxypropyl methylcellulose (HPMC, Mw = 10,000), polytetrafluoroethylene (PTFE), and *N*, *N*-dimethylformamide (DMF) were purchased from Aladdin. Analytical-grade potassium hydroxide (KOH) and acetylene black were obtained from Xilong Science Co., Ltd. (Shantou, China). None of the purchased reagents were purified before use. All aqueous solutions were prepared using ultrapure water (deionized water, resistance 18 $M\Omega \, cm^{-1}$).

2.2. Synthesis of Organic Crosslinked Polymers

The organic crosslinked polymers were prepared by a one-pot method, which is a minor modification based on our previous report [27]. Briefly, PEG 6000 (12.0 g), MDI (1.0 g), and a certain amount of HPMC were stirred in a three-neck flask containing DMF (80 mL) under argon gas and an oil bath with a constant temperature of 75 °C. The organic crosslinked polymer obtained after 30 h of condensation reflux is referred to as OCLP. The mass of HPMC was 3.5, 4.5, and 5.0 g; therefore, the corresponding organic crosslinked polymers were named as $OCLP_{3.5}$, $OCLP_{4.5}$ and $OCLP_{5.0}$, respectively. Figure 1 presents a flowchart of the one-pot method for the preparation of organic crosslinked polymers.

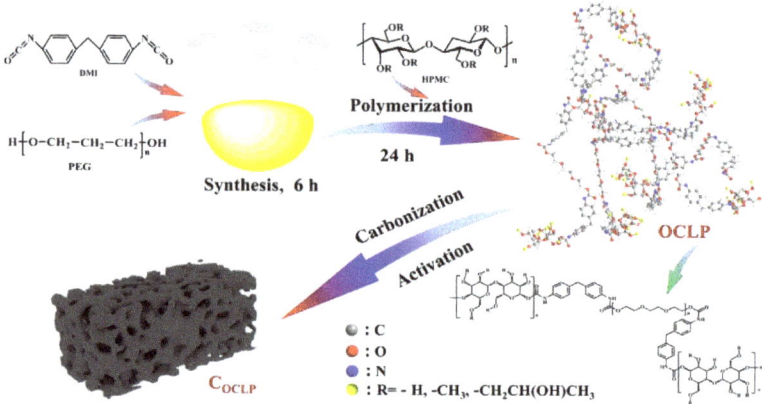

Figure 1. Schematic illustration of the one-pot method for the preparation of organic crosslinked polymer-derived porous carbon.

2.3. Preparation of Porous Carbon Materials

The prepared OCLPs were directly carbonized by heating them in a tube furnace at 500 °C for 2 h under a N_2 atmosphere at a heating rate of 5 °C/min. The resulting carbon precursors were homogeneously ground with KOH in a mass ratio of 1.0:3.0, then calcined in a tube furnace at 600 °C under a N_2 atmosphere for 2 h. The calcined products were stirred with a 1 M hydrochloric acid solution for 2 h, followed by washing with distilled water and anhydrous ethanol sequentially until the filtrate was neutral. The obtained residues were dried in a blast oven at 80 °C for 24 h to obtain porous carbon materials, which were named as $C_{OCLP-3.5}$, $C_{OCLP-4.5}$, and $C_{OCLP-5.0}$, respectively.

2.4. Characterization

Fourier transform infrared (FTIR) spectroscopy was performed on the samples using a Thermo Fisher (Waltham, MA, USA) Nicolet 6700 spectrometer with KBr pellets. A powder X-ray diffractometer (XRD; D8 Advance Bruker, Billerica, MA, USA) operating at 40 kV and 40 mA with Cu Kα radiation (λ = 0.15406 nm) in the 2θ range of 5–90° with 0.01° step increments was used to analyze the microstructure of the materials. The chemical structure and graphitization of the samples were further characterized using Raman spectroscopy (Horiba JY, Palaiseau, France) at an excitation wavelength of 532 nm. The surface micromorphology of the samples was characterized using scanning electron microscopy (SEM; SU8010, HITACHI, Tokyo, Japan) and transmission electron microscopy (TEM; Tecnai G2 F20, FEI Company, Hillsboro, OR, USA), and elemental analysis was performed using energy-dispersive X-ray spectroscopy (EDS). The specific surface area and pore structure characteristics of the samples were characterized using a nitrogen adsorption–desorption analyzer (ASIQM0002-4, Quantachrome, Boynton Beach, Florida,

USA) at −196 °C. Surface element analysis was performed using X-ray photoelectron spectroscopy (XPS; Thermo Scientific Escalab 250Xi, Waltham, MA, USA).

2.5. Electrochemical Measurements

The electrochemical performance of the samples, including galvanostatic charge–discharge (GCD), cyclic voltammetry (CV), and electrochemical impedance (EIS), was measured using a CHI 660E instrument in a three-electrode system. A slurry mixture of carbon material (C_{OCLP}), acetylene black, and PTFE in a weight ratio of 8:1:1 was applied to nickel foam (2 cm × 2 cm) as the working electrode; platinum and Hg/HgO electrodes were used as the counter and reference electrodes, respectively, in the three-electrode system. The voltage was set to −1–0 V and the electrolyte was 6 M KOH. A symmetric supercapacitor was built for a two-electrode system using the C_{OCLP}, a 1 M Na_2SO_4 electrolyte, and a voltage range of 0–1.6 V.

For the three-electrode and two-electrode systems, the weight-specific capacitances ($F \cdot g^{-1}$) of the electrode material were calculated based on the GCD curves using Equations (1) and (2), respectively.

$$C_g = \frac{I \Delta t}{m \Delta V} \tag{1}$$

$$C_g = \frac{2I \Delta t}{m \Delta V} \tag{2}$$

where I (A), Δt (s), ΔV (mV), and m (g) represent the discharge current, discharge time, discharge voltage range, and mass of the active material of a single electrode, respectively.

The energy density (E_{cell}) and power density (P_{cell}) of the symmetrical supercapacitor were calculated using Equations (3) and (4), respectively.

$$E_{cell} = \frac{C_g \Delta V^2}{8 \times 3.6} \tag{3}$$

$$P_{cell} = \frac{3600 \, E_{cell}}{\Delta t} \tag{4}$$

where C_g is obtained from Equation (2), ΔV is the working voltage of the discharge, and Δt is the discharge time.

3. Results and Discussion

3.1. Structural and Morphological Characterization

Figure 2 shows the FTIR spectra of the samples, which indicates that the characteristic absorption peaks for the OCLPs (OCLP$_{3.5}$, OCLP$_{4.5}$, and OCLP$_{5.0}$) are similar. The peaks around 3438, 1639, 1526, and 1106 cm^{-1} correspond to the stretching vibration absorption peaks of the –OH, C=O, C–N, and C–O groups, respectively, which is consistent with the organic crosslinked polymer [27]. The above results illustrate that the OCLPs are a type of organic crosslinked polymer.

The C_{OCLP}s obtained from the OCLPs were characterized using XRD and Raman spectroscopy. Figure 3a summarizes the XRD spectra of the $C_{OCLP-3.5}$, $C_{OCLP-4.5}$, and $C_{OCLP-5.0}$, showing that all the C_{OCLP}s exhibit obvious diffraction peaks at 43°, corresponding to the (100) crystal planes of the graphite structure. The results indicate that $C_{OCLP-3.5}$, $C_{OCLP-4.5}$, and $C_{OCLP-5.0}$ have amorphous graphite structures [28,29]. The diffraction peak intensity of the (100) lattice plane for $C_{OCLP-4.5}$ is the weakest, demonstrating that $C_{OCLP-4.5}$ has the highest structural disorder [30]. Figure 3b shows that there are two characteristic peaks at 1343 and 1594 cm^{-1}, corresponding to the D and G peaks of graphite, respectively. The ratio of the areas of the D peak to the G peak (A_D/A_G) reflects the order degree of the C_{OCLP} structure [31]. The calculated ratios for $C_{OCLP-3.5}$, $C_{OCLP-4.5}$, and $C_{OCLP-5.0}$ are 1.15: 1, 1.18: 1, and 1.13: 1, respectively. This result also illustrates that $C_{OCLP-4.5}$ has more defects because the D peak represents a defect peak caused by the low symmetry or irregularity of the carbon material [32].

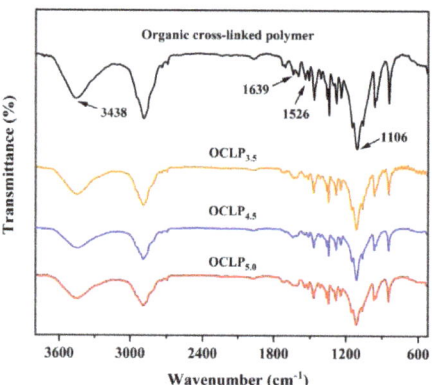

Figure 2. FTIR spectra of the OCLPs and organic crosslinked polymer [27].

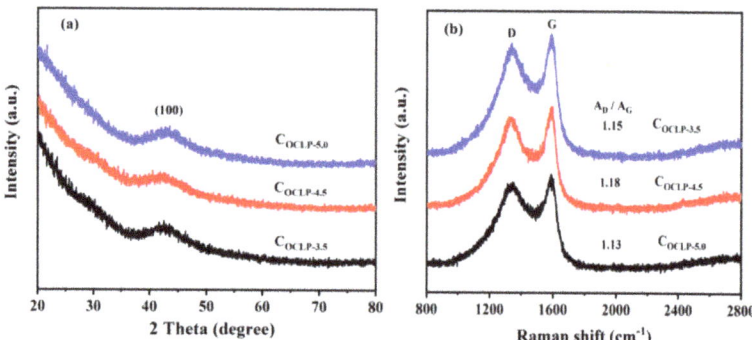

Figure 3. (a) XRD patterns of the C_{OCLPs} and (b) Raman spectra of the C_{OCLPs}.

The surface morphologies of $C_{OCLP-3.5}$, $C_{OCLP-4.5}$, and $C_{OCLP-5.0}$ were characterized using SEM, as shown in Figure 4. Figure 4 indicates that the three C_{OCLPs} are all porous and present a three-dimensional network structure. The number of pores in the C_{OCLP} increases with an increase in the amount of HPMC; however, when the HPMC content is increased to 5.0 g, the pore structure is only partially formed, and the number of pores decreases. The result demonstrates that the pore structure of $C_{OCLP-4.5}$ was excellent. Generally, an abundant number of pores can significantly increase the specific surface area of C_{OCLPs}, thereby providing more storage sites and transport channels for electrolyte ions. This is beneficial for improving the electrochemical performance [33].

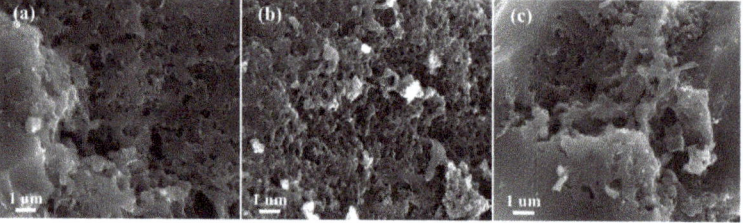

Figure 4. SEM images of samples, (a) $C_{OCLP-3.5}$; (b) $C_{OCLP-4.5}$; (c) $C_{OCLP-5.0}$.

Additionally, Figure 5a further demonstrates that $C_{OCLP-4.5}$ is a porous C_{OCLP}. When $C_{OCLP-4.5}$ is used as the electrode material, these disordered microporous structures can provide sufficient active sites for charge storage [34]. Figure 5b–e are element distribution diagrams obtained from the EDS analysis of $C_{OCLP-4.5}$, showing that carbon, nitrogen, and oxygen were uniformly distributed in the carbon framework. Abundant nitrogen and oxygen can introduce pseudocapacitance and enhance the capacitance performance of the electrode material.

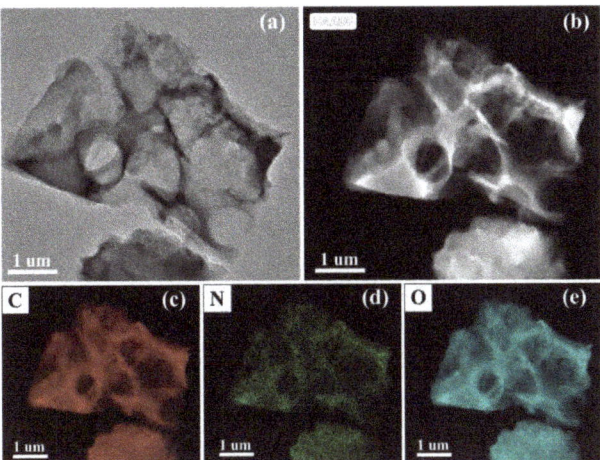

Figure 5. (**a**) TEM and (**b–e**) EDS images of $C_{OCLP-4.5}$.

The C_{OCLPs} were subjected to N_2 adsorption–desorption measurements to explore the pore characteristics. Figure 6 shows that all the C_{OCLPs} exhibit obvious type I isotherm characteristics, indicating that these samples are rich in micropores [35]. Table 1 summarizes the pore structure characteristics of the C_{OCLPs}, showing that the specific surface area and pore volume of these samples are mainly provided by the micropores and mesopores. Among the three samples, $C_{OCLP-4.5}$ has the largest specific surface area (1589 $m^2 \cdot g^{-1}$) and the highest pore volume (0.657 $cm^3 \cdot g^{-1}$), which further confirm that $C_{OCLP-4.5}$ has the best pore structure. Numerous studies have demonstrated that the large specific surface area and rich pore structure of porous carbon material can greatly promote the storage and rapid migration of ions, resulting in the excellent specific capacitance performance of supercapacitors [36,37]. The aqueous electrolytes currently used in supercapacitors are mainly sulfuric acid (H_2SO_4, acidic), KOH (alkaline), and sodium sulfate (Na_2SO_4, neutral). The electrolyte ions in these electrolytes mainly exist as hydrated ions (H^+, K^+, OH^-, Na^+, and SO_4^{2-}). Based on Table 1, it can be found that the C_{OCLPs} obtained can meet the fast migration requirements of these electrolyte ions, thereby significantly improving the conductivity of carbon-based electrodes and enhancing their electrochemical performance.

Table 1. Channel structure parameters of the C_{OCLPs}.

Samples	Specific Surface Area ($m^2 \cdot g^{-1}$)			Pore Volume ($cm^3 \cdot g^{-1}$)		
	Total	Microporous	Mesoporous	Total	Microporous	Mesoporous
$C_{OCLP-3.5}$	942	894	48	0.399	0.353	0.046
$C_{OCLP-4.5}$	1589	1509	80	0.657	0.592	0.065
$C_{OCLP-5.0}$	1102	1040	62	0.482	0.407	0.075

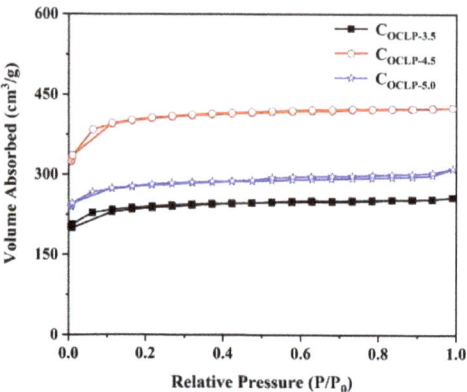

Figure 6. N_2 desorption/adsorption isotherm curves of C_{OCLPs}.

Further analysis of the surface electronic states and elemental compositions of the C_{OCLPs} samples was performed using XPS. Figure 7a shows that there are three peaks in the spectra of all the samples. The binding energies of the three peaks are 285, 400, and 532 eV, corresponding to C 1s, N 1s, and O 1s, respectively. The results also prove that carbon, nitrogen, and oxygen are present in the three samples. Table 2 lists the surface element contents of the three samples. These samples are mainly a carbon-based framework with oxygen and nitrogen. Fine analyses of the C 1s, N 1s, and O 1s spectra of $C_{OCLP-4.5}$ are performed using the peak differentiation fitting method, as shown in Figure 7b–d. The C 1s spectrum (Figure 7b) can be matched by four peaks at 284.8, 285.7, 286.8, and 289.0 eV, corresponding to the C–C, C–N, C–O, and COOR groups, respectively [38]. The N 1s spectrum, shown in Figure 7c, is deconvoluted into four peaks of 398.8, 400.3, 400.8, and 402.4 eV, corresponding to pyridinic-N (N-6) (11.70%), pyrrolic-N (N-5) (52.13%), quaternary-N (N–Q) (29.79%), and oxidized N (N–X) (6.38%), respectively. In particular, the pyridinic-N and pyrrolic-N contents reach 63.83%. A high content of N-6 and N-5 is beneficial for introducing pseudo-capacitance and providing electrochemically active sites and quaternary nitrogen (N–Q) can effectively improve the conductivity of C_{OCLPs} and promote electron transfer in the carbon matrix [35,39]. The deconvoluted O 1s peak displayed four peaks at 531.2, 532.3, 533.3, and 534.2 eV, representing the oxygen atoms in the C=O, C–O/C–OH, COOR, and N–O groups, respectively (shown in Figure 7d) [35,39]. According to a previous report [40], the oxygen groups are evenly distributed in the carbon framework, which can improve the interfacial tension between the carbon-based porous material and electrolyte to reduce the interfacial resistance.

Table 2. Surface element content of the C_{OCLPs}.

Samples	Element Content		
	Carbon (%)	Nitrogen (%)	Oxygen (%)
$C_{OCLP-3.5}$	92.83	1.98	5.19
$C_{OCLP-4.5}$	85.75	1.68	12.75
$C_{OCLP-5.0}$	84.51	2.65	12.84

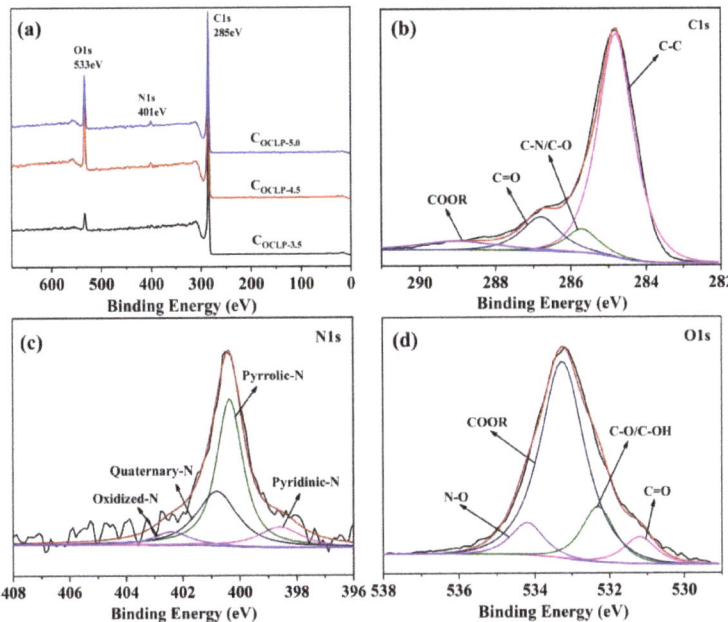

Figure 7. (**a**) XPS spectra of the C_{OCLPs}; (**b–d**) high resolution of C 1s, N 1s and O 1s of $C_{OCLP-4.5}$.

3.2. Electrochemistry Measurements

The electrochemical performances of the electrode materials were evaluated using a three-electrode system. Figure 8a shows the CV plots of the different C_{OCLPs} ($C_{OCLP-3.5}$, $C_{OCLP-4.5}$, and $C_{OCLP-5.0}$) at a sweep rate of 5 mV·s^{-1}. All the samples display a typical rectangular shape, indicating that the capacitive behavior of these materials is mainly electric double-layer capacitance. Concurrently, these curves have a broad peak in the voltage window of −0.8 to −0.3 V, which is caused by the oxidation–reduction reaction of nitrogen and oxygen atoms contained in these samples during the charge and discharge process. Moreover, the pseudo-capacitance introduced by the redox reaction can significantly increase the specific capacitance of carbon electrodes. The $C_{OCLP-4.5}$ sample exhibits the largest encircled area of the CV curve among the three samples, which also illustrates that $C_{OCLP-4.5}$ has the highest specific capacitance. Figure 8b shows the constant GCD curves for the C_{OCLPs} at a current density of 1 A·g^{-1}. The GCD curves for the three samples are all quasi-isosceles triangle shapes, indicating that the capacitance is mainly electric double-layer capacitance (EDLC), and the slight deformation is attributed to the existence of pseudo-capacitance. According to Equation (1), the specific capacitances of $C_{OCLP-3.5}$, $C_{OCLP-4.5}$, and $C_{OCLP-5.0}$ at a current density of 1 A·g^{-1} are 302, 503, and 330 F·g^{-1}, respectively. This result shows that the specific capacitance of $C_{OCLP-4.5}$ is the largest, owing to its large specific surface area (1589 m^2·g^{-1}) and pore volume (0.657 cm^3·g^{-1}). Figure 8c presents the CV curves for $C_{OCLP-4.5}$ at different scanning rates. It reveals that the $C_{OCLP-4.5}$ still maintains a quasi-rectangular shape at scan rates of 5–50 mV·s^{-1}, indicating that the good pore structure of $C_{OCLP-4.5}$ enables the rapid migration of electrolyte ions to result in its good rate capability. Figure 8d presents the GCD curves for $C_{OCLP-4.5}$ at current densities of 0.5–20 A g^{-1}, showing that the GCD curve does not exhibit a significant IR drop at a high current density of 20 A·g^{-1}. Therefore, it demonstrates that the $C_{OCLP-4.5}$ has a high conductivity, good rate capability, and electrochemical reversibility. The specific capacitances are calculated as 522, 503, 432, 396, 363, and 319 F·g^{-1} at current densities of 0.5, 1, 2, 5, 10, and 20 A·g^{-1}, respectively. Comparing the electrochemical performance of $C_{OCLP-4.5}$ with that of the references, the result is listed in Table 3. According to Table 3, the electrochemical performance of $C_{OCLP-4.5}$ is better than that of other electroactive materials reported in the literature. This is attributed to the unique network structure and rich N/O content of the crosslinked polymer fabricated in this study.

Table 3. Comparison of the specific capacitances of the $C_{OCLP-4.5}$ electroactive material to recently reported carbonaceous materials.

Material	Electrolyte	Current Density $(A \cdot g^{-1})$	Capacitance $(F \cdot g^{-1})$	Reference
Grape marc	6 M KOH	0.5	446	[23]
Polyphosphazene	6 M KOH	0.5	438	[25]
Polypyrrole/Polythiophene	KOH	0.5	455	[41]
Cotton stalk	1 M H_2SO_4	0.2	338	[42]
L-tyrosine	KOH	0.3	512	[43]
Coal tar pitch	6 M KOH	0.5	298	[44]
CNTs@Gr-CNF	6 M KOH	0.25	521	[45]
CTAB	6 M KOH	1.0	241	[46]
3-aminophenol-formaldehyde resin	6 M KOH	0.5	381	[47]
Organic crosslinked polymer	6 M KOH	0.5	522	This work

Figure 8e presents the EIS curves for C_{OCLPs} and the equivalent circuit model (the inset of Figure 8e), showing that $C_{OCLP-4.5}$ has the lowest R_{ct} (internal charge transfer resistance) and R_s (contact resistance with the electrolyte) among the three materials. That is, in the high-frequency region, the R_{ct} of $C_{OCLP-4.5}$ is 0.042 Ω, lower than those of $C_{OCLP-3.5}$ (0.152 Ω) and $C_{OCLP-5.0}$ (0.183 Ω). The low R_s demonstrates that the electrolyte ions are readily transferred to the surface of the $C_{OCLP-4.5}$ electrode [48]. Additionally, the linear curve of $C_{OCLP-4.5}$ is almost vertical in the low-frequency region. The EIS results illustrate that the structure of $C_{OCLP-4.5}$ is beneficial for charge transfer and the efficient diffusion of electrolyte ions. For supercapacitors, the cycling stability is a significant parameter to estimate their practical application. Figure 8f shows that $C_{OCLP-4.5}$ retains 83% of its initial specific capacitance value after 5000 cycles at a current density of 5 A g^{-1}. The surface morphology of $C_{OCLP-4.5}$ after cycling was characterized by SEM, as shown in Figure 9. Compared with the $C_{OCLP-4.5}$, before (Figure 4b) shows that the pore structure of $C_{OCLP-4.5}$ has some damage and collapses after 5000 cycles.

A symmetric supercapacitor was constructed using $C_{OCLP-4.5}$ to evaluate its practical application. Figure 10a shows the CV curves for the symmetric supercapacitor at different scan rates. The curves maintained a quasi-rectangular shape at a scan rate of 50 mV·s^{-1}. A slight deformation indicates that the electrochemical behavior of a symmetric supercapacitor is a combination of the EDLC and pseudocapacitance. Figure 10b shows that the GCD curves for the symmetric supercapacitor increased with an increasing current density from 1 to 20 A·g^{-1}. Based on Equation (2), the specific capacitance of $C_{OCLP-4.5}$ is 203 F·g^{-1} at 1 A·g^{-1} and its specific capacitance remains 150 F·g^{-1} at 10 A·g^{-1}, demonstrating a good rate capability even at high current densities for the symmetric supercapacitor. Figure 10c shows the cycle stability curve at a current density of 10 A·g^{-1}. It displays that the capacitance retention of the device is 84.0% after 5000 cycles, reflecting good cycling stability. Figure 10d indicates that the symmetric capacitor obtains an energy density of 18.04 Wh·kg^{-1} at a power density of 200.0 W·kg^{-1} based on Equations (3) and (4), significantly higher than those reported in recent years (13. [25], 10.83 [49], 13.60 [50], 7.00 [51], 13.86 [52], 10.60 [53], and 15.50 Wh·kg^{-1} [54]). Specifically, the symmetric supercapacitor device successfully powers up a light-emitting diode (the inset of Figure 10d). As shown in the video (see Supplementary Materials File S1), the light-emitting diode can last for a while. Obviously, the N/O-doped porous C_{OCLPs} are expected to be used in supercapacitors.

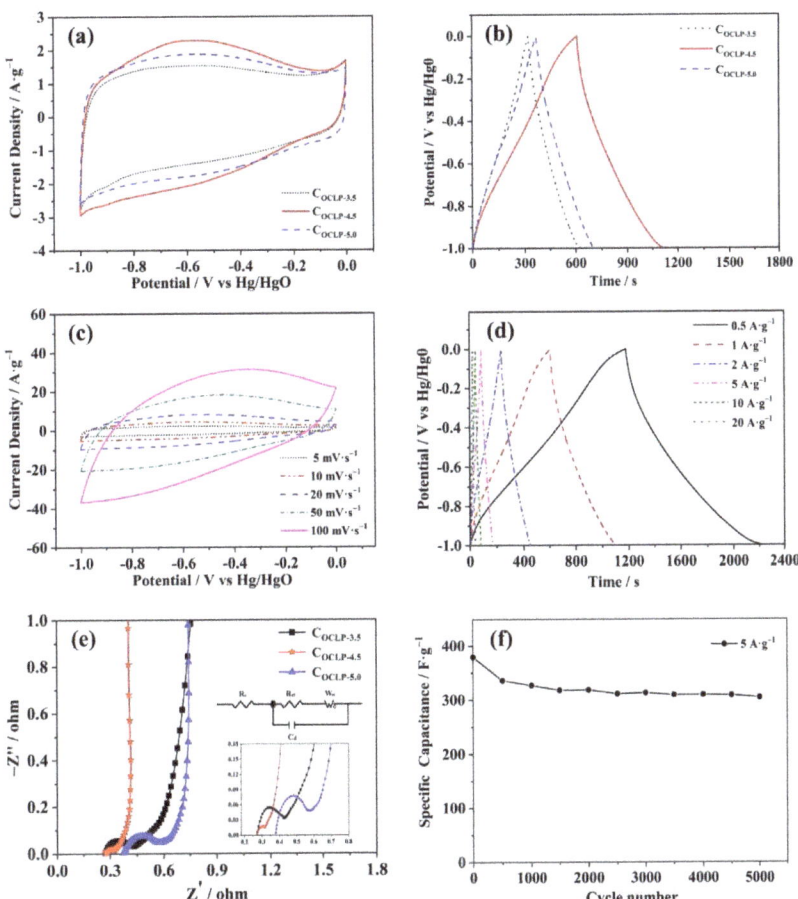

Figure 8. (**a**) CV curves of C$_{OCLPs}$ at a scan rate of 5 mV·s^{-1}; (**b**) GCD curves of C$_{OCLPs}$ at a current density of 1 A·g^{-1}; (**c**) CV curves of C$_{OCLP-4.5}$ at different scan rates; (**d**) GCD curves of C$_{OCLP-4.5}$ at different current densities; (**e**) Nyquist plots of the C$_{OCLPs}$; (**f**) stable cyclic performance of C$_{OCLP-4.5}$.

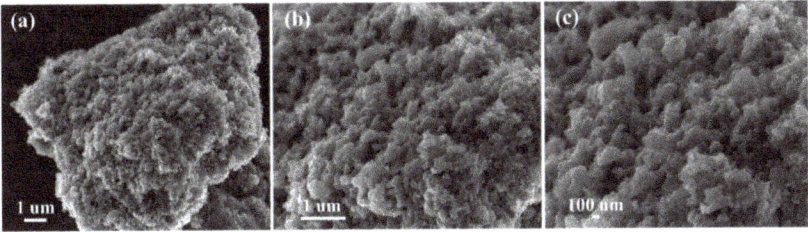

Figure 9. (**a**–**c**) SEM images of C$_{OCLP-4.5}$ with different multiples after 5000 cycles.

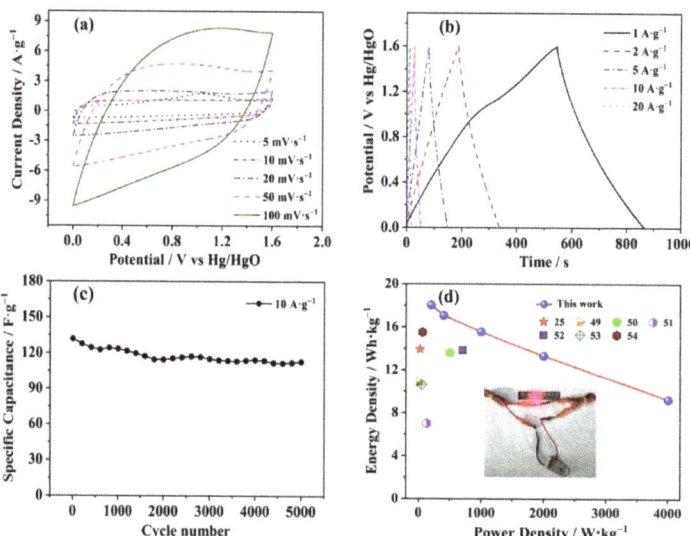

Figure 10. (**a**) CV curves at 5–100 mV·s^{-1}. (**b**) GCD curves at 1–20 A·g^{-1}. (**c**) Stable cyclic performance. (**d**) Ragone plots as compared to other studies and a supercapacitor device lighting up light-emitting diodes.

4. Conclusions

In this study, a network-structured organic crosslinked polymer was used as a carbon source to obtain N/O-doped porous C_{OCLPs}. The results indicated that the $C_{OCLP-4.5}$ obtained by optimizing the raw materials exhibited an excellent electrochemical performance. For instance, the specific capacitance of $C_{OCLP-4.5}$ was as high as 522 F·g^{-1} at a current density of 0.5 A·g^{-1}, and still exhibited 309 F·g^{-1} at 20 A·g^{-1} in a three-electrode system. Furthermore, the symmetric capacitor achieved an energy density of 18.04 Wh·kg^{-1} at a power density of 200.0 W·kg^{-1}. The C_{OCLPs} benefitted from the net structure of organic crosslinked polymers to form hierarchical porous carbon and the pseudocapacitance introduced by heteroatoms. Therefore, the method for fabricating carbon material proposed in this study provides a new strategy for the development of electrode materials with high electrochemical performance.

Supplementary Materials: The following supporting information can be downloaded at: https://www.mdpi.com/article/10.3390/nano12132186/s1, File S1: Lighting up LED lights video.

Author Contributions: Conceptualization, F.X. and Y.L. (Yao Lu); methodology, J.L., Y.L. (Yao Lu) and L.S.; investigation J.L. and Y.L. (Yao Lu); funding acquisition, F.X. and L.S.; writing—original draft preparation, J.L. and S.F.; writing—review and editing, F.X., L.S. and X.W.; data curation, Y.L. (Yong Luo), Y.Z. (Yanling Zhu), Y.W., T.Z., L.L., Y.G., C.Z. and Y.Y.; supervision, Y.X., Y.L. (Yumei Luo), Y.Z. (Yongjin Zou), H.C. and H.Z. All authors have read and agreed to the published version of the manuscript.

Funding: This work was supported by the National Key Research and Development Program of China (2018YFB1502103, 2018YFB1502105), the National Natural Science Foundation of China (51971068, U20A20237 and 51871065), the Scientific Research and Technology Development Program of Guangxi (AA19182014, AD17195073, AA17202030-1), Guangxi Bagui Scholar Foundation, Guangxi Collaborative Innovation Centre of Structure and Property for New Energy and Materials, Guangxi Advanced Functional Materials Foundation and Application Talents Small Highlands, Chinesisch-Deutsche Kooperationsgruppe (GZ1528), Science research and Technology Development project of Guilin (20210216-1), and Innovation Project of GUET Graduate Education (2022YCXS201).

Data Availability Statement: All data are available upon reasonable request.

Conflicts of Interest: The authors declare no conflict of interest.

References

1. Hou, J.; Cao, C.; Idrees, F.; Ma, X. Hierarchical Porous Nitrogen-Doped Carbon Nanosheets Derived from Silk for Ultrahigh-Capacity Battery Anodes and Supercapacitors. *ACS Nano* **2015**, *9*, 2556–2564. [CrossRef] [PubMed]
2. Zou, K.; Deng, Y.; Chen, J.; Qian, Y.; Yang, Y.; Li, Y.; Chen, G. Hierarchically porous nitrogen-doped carbon derived from the activation of agriculture waste by potassium hydroxide and urea for high-performance supercapacitors. *J. Power Sources* **2018**, *378*, 579–588. [CrossRef]
3. Chmiola, J.; Yushin, G.; Dash, R.; Gogotsi, Y. Effect of pore size and surface area of carbide derived carbons on specific capacitance. *J. Power Sources* **2006**, *158*, 765–772. [CrossRef]
4. Wu, Y.; Cao, J.; Zhao, X.; Zhuang, Q.; Zhou, Z.; Huang, Y.; Wei, X. High-performance electrode material for electric double-layer capacitor based on hydrothermal pre-treatment of lignin by ZnCl$_2$. *Appl. Surf. Sci.* **2020**, *508*, 144536. [CrossRef]
5. Li, Z.; Zhang, L.; Amirkhiz, B.S.; Tan, X.; Xu, Z.; Wang, H.; Olsen, B.C.; Holt, C.M.B.; Mitlin, D. Carbonized Chicken Eggshell Membranes with 3D Architectures as High-Performance Electrode Materials for Supercapacitors. *Adv. Energy Mater.* **2012**, *2*, 431–437. [CrossRef]
6. Sun, L.; Tian, C.; Li, M.; Meng, X.; Wang, L.; Wang, R.; Yin, J.; Fu, H. From coconut shell to porous graphene-like nanosheets for high-power supercapacitors. *J. Mater. Chem. A* **2013**, *1*, 6462. [CrossRef]
7. Wang, H.; Xu, Z.; Kohandehghan, A.; Li, Z.; Cui, K.; Tan, X.; Stephenson, T.J.; King ondu, C.K.; Holt, C.M.B.; Olsen, B.C.; et al. Interconnected Carbon Nanosheets Derived from Hemp for Ultrafast Supercapacitors with High Energy. *ACS Nano* **2013**, *7*, 5131–5141. [CrossRef]
8. Chang, Y.; Shi, H.; Yan, X.; Zhang, G.; Chen, L. A ternary B, N, P-Doped carbon material with suppressed water splitting activity for high-energy aqueous supercapacitors. *Carbon* **2020**, *170*, 127–136. [CrossRef]
9. Wang, Y.; Wang, D.; Li, Z.; Su, Q.; Wei, S.; Pang, S.; Zhao, X.; Liang, L.; Kang, L.; Cao, S. Preparation of Boron/Sulfur-Codoped Porous Carbon Derived from Biological Wastes and Its Application in a Supercapacitor. *Nanomaterials* **2022**, *12*, 1182. [CrossRef]
10. Wang, P.; Ding, X.; Zhe, R.; Zhu, T.; Qing, C.; Liu, Y.; Wang, H. Synchronous Defect and Interface Engineering of NiMoO$_4$ Nanowire Arrays for High-Performance Supercapacitors. *Nanomaterials* **2022**, *12*, 1094. [CrossRef]
11. Han, X.; Tao, K.; Wang, D.; Han, L. Design of a porous cobalt sulfide nanosheet array on Ni foam from zeolitic imidazolate frameworks as an advanced electrode for supercapacitors. *Nanoscale* **2018**, *10*, 2735–2741. [CrossRef] [PubMed]
12. Khan, S.; Alkhedher, M.; Raza, R.; Ahmad, M.A.; Majid, A.; Din, E.M.T.E. Electrochemical Investigation of PANI:PPy/AC and PANI:PEDOT/AC Composites as Electrode Materials in Supercapacitors. *Polymers* **2022**, *14*, 1976. [CrossRef] [PubMed]
13. Meng, Q.; Cai, K.; Chen, Y.; Chen, L. Research progress on conducting polymer based supercapacitor electrode materials. *Nano Energy* **2017**, *36*, 268–285. [CrossRef]
14. Shi, L.; Ye, J.; Lu, H.; Wang, G.; Lv, J.; Ning, G. Flexible all-solid-state supercapacitors based on boron and nitrogen-doped carbon network anchored on carbon fiber cloth. *Chem. Eng. J.* **2021**, *410*, 128365. [CrossRef]
15. Wu, D.; Cheng, J.; Wang, T.; Liu, P.; Yang, L.; Jia, D. A Novel Porous N- and S-Self-Doped Carbon Derived from Chinese Rice Wine Lees as High-Performance Electrode Materials in a Supercapacitor. *ACS Sustain. Chem. Eng.* **2019**, *7*, 12138–12147. [CrossRef]
16. Lei, W.; Yang, B.; Sun, Y.; Xiao, L.; Tang, D.; Chen, K.; Sun, J.; Ke, J.; Zhuang, Y. Self-sacrificial template synthesis of heteroatom doped porous biochar for enhanced electrochemical energy storage. *J. Power Sources* **2021**, *488*, 229455. [CrossRef]
17. Jiang, W.; Cai, J.; Pan, J.; Guo, S.; Sun, Y.; Li, L.; Liu, X. Nitrogen-doped hierarchically ellipsoidal porous carbon derived from Al-based metal-organic framework with enhanced specific capacitance and rate capability for high performance supercapacitors. *J. Power Sources* **2019**, *432*, 102–111. [CrossRef]
18. Lima, R.M.A.P.; Dos, R.G.S.; Thyrel, M.; AlcarazEspinoza, J.J.; Larsson, S.H.; de Oliveira, H.P. Facile Synthesis of Sustainable Biomass-Derived Porous Biochars as Promising Electrode Materials for High-Performance Supercapacitor Applications. *Nanomaterials* **2022**, *12*, 866. [CrossRef]
19. Alabadi, A.; Yang, X.; Dong, Z.; Li, Z.; Tan, B. Nitrogen-doped activated carbons derived from a co-polymer for high supercapacitor performance. *J. Mater. Chem. A* **2014**, *2*, 11697–11705. [CrossRef]
20. Park, M.; Ryu, J.; Kim, Y.; Cho, J. Corn protein-derived nitrogen-doped carbon materials with oxygen-rich functional groups: A highly efficient electrocatalyst for all-vanadium redox flow batteries. *Energy Environ. Sci.* **2014**, *7*, 3727–3735. [CrossRef]
21. Paraknowitsch, J.P.; Thomas, A.; Antonietti, M. A detailed view on the polycondensation of ionic liquid monomers towards nitrogen doped carbon materials. *J. Mater. Chem.* **2010**, *20*, 6746. [CrossRef]
22. Inagaki, M.; Toyoda, M.; Soneda, Y.; Morishita, T. Nitrogen-doped carbon materials. *Carbon* **2018**, *132*, 104–140. [CrossRef]
23. Zhang, J.; Chen, H.; Bai, J.; Xu, M.; Luo, C.; Yang, L.; Bai, L.; Wei, D.; Wang, W.; Yang, H. N-doped hierarchically porous carbon derived from grape marcs for high-performance supercapacitors. *J. Alloys Compd.* **2021**, *854*, 157207. [CrossRef]
24. Mohd Abdah, M.A.A.; Azman, N.H.N.; Kulandaivalu, S.; Abdul Rahman, N.; Abdullah, A.H.; Sulaiman, Y. Potentiostatic deposition of poly (3, 4-ethylenedioxythiophene) and manganese oxide on porous functionalised carbon fibers as an advanced electrode for asymmetric supercapacitor. *J. Power Sources* **2019**, *444*, 227324. [CrossRef]
25. Zou, W.; Zhang, S.; Abbas, Y.; Liu, W.; Zhang, Y.; Wu, Z.; Xu, B. Structurally designed heterochain polymer derived porous carbons with high surface area for high-performance supercapacitors. *Appl. Surf. Sci.* **2020**, *530*, 147296. [CrossRef]

26. Chen, Y.; Liu, F.; Qiu, F.; Lu, C.; Kang, J.; Zhao, D.; Han, S.; Zhuang, X. Cobalt-Doped Porous Carbon Nanosheets Derived from 2D Hypercrosslinked Polymer with CoN_4 for High Performance Electrochemical Capacitors. *Polymers* **2018**, *10*, 1339. [CrossRef]
27. Chen, D.; Xu, F.; Sun, L.; Xia, Y.; Wei, S.; Zhang, H. Preparation and thermal property of PEG based composite phase change material. *New Chenical Mater.* **2020**, *48*, 75–79.
28. Chen, C.; Liu, W.; Wang, H.; Peng, K. Synthesis and performances of novel solid–solid phase change materials with hexahydroxy compounds for thermal energy storage. *Appl. Energy* **2015**, *152*, 198–206. [CrossRef]
29. Wang, Y.; Zhang, M.; Dai, Y.; Wang, H.; Zhang, H.; Wang, Q.; Hou, W.; Yan, H.; Li, W.; Zheng, J. Nitrogen and phosphorus co-doped silkworm-cocoon-based self-activated porous carbon for high performance supercapacitors. *J. Power Sources* **2019**, *438*, 227045. [CrossRef]
30. Zhipeng, Q.; Yesheng, W.; Xu, B.; Tong, Z.; Jin, Z.; Jinping, Z.; Zhichao, M.; Weiming, Y.; Peng, F.; Shuping, Z. Biochar-based carbons with hierarchical micro-meso-macro porosity for high rate and long cycle life supercapacitors. *J. Power Sources* **2018**, *376*, 82–90.
31. Kong, L.N.; Yang, W.; Su, L.; Hao, S.G.; Shao, G.J.; Qin, X.J. Nitrogen-doped 3D web-like interconnected porous carbon prepared by a simple method for supercapacitors. *Ionics* **2019**, *25*, 4333–4340. [CrossRef]
32. Boujibar, O.; Ghamouss, F.; Ghosh, A.; Achak, O.; Chafik, T. Activated carbon with exceptionally high surface area and tailored nanoporosity obtained from natural anthracite and its use in supercapacitors. *J. Power Sources* **2019**, *436*, 226882. [CrossRef]
33. Hu, J.; He, W.; Qiu, S.; Xu, W.; Mai, Y.; Guo, F. Nitrogen-doped hierarchical porous carbons prepared via freeze-drying assisted carbonization for high-performance supercapacitors. *Appl. Surf. Sci.* **2019**, *496*, 143643. [CrossRef]
34. Yiju, L.; Guiling, W.; Tong, W.; Zhuangjun, F.; Peng, Y. Nitrogen and sulfur co-doped porous carbon nanosheets derived from willow catkin for supercapacitors. *Nano Energy* **2016**, *19*, 165–175.
35. Jia, H.; Zhang, H.; Wan, S.; Sun, J.; Xie, X.; Sun, L. Preparation of nitrogen-doped porous carbon via adsorption-doping for highly efficient energy storage. *J. Power Sources* **2019**, *433*, 226712. [CrossRef]
36. Javaid, A.; Irfan, M. Multifunctional structural supercapacitors based on graphene nanoplatelets/carbon aerogel composite coated carbon fiber electrodes. *Mater. Res. Express* **2018**, *6*, 16310. [CrossRef]
37. Li, C.; Wu, W.; Wang, P.; Zhou, W.; Wang, J.; Chen, Y.; Fu, L.; Zhu, Y.; Wu, Y.; Huang, W. Fabricating an Aqueous Symmetric Supercapacitor with a Stable High Working Voltage of 2 V by Using an Alkaline-Acidic Electrolyte. *Adv. Sci.* **2019**, *6*, 1801665. [CrossRef]
38. Li, S.; Fan, Z. Nitrogen-doped carbon mesh from pyrolysis of cotton in ammonia as binder-free electrodes of supercapacitors. *Microporous Mesoporous Mater.* **2019**, *274*, 313–317. [CrossRef]
39. Méndez-Morales, T.; Ganfoud, N.; Li, Z.; Haefele, M.; Rotenberg, B.; Salanne, M. Performance of microporous carbon electrodes for supercapacitors: Comparing graphene with disordered materials. *Energy Storage Mater.* **2019**, *17*, 88–92. [CrossRef]
40. Huo, S.; Liu, M.; Wu, L.; Liu, M.; Xu, M.; Ni, W.; Yan, Y. Synthesis of ultrathin and hierarchically porous carbon nanosheets based on interlayer-confined inorganic/organic coordination for high performance supercapacitors. *J. Power Sources* **2019**, *414*, 383–392. [CrossRef]
41. Lu, Y.; Liang, J.; Deng, S.; He, Q.; Deng, S.; Hu, Y.; Wang, D. Hypercrosslinked polymers enabled micropore-dominant N, S Co-Doped porous carbon for ultrafast electron/ion transport supercapacitors. *Nano Energy* **2019**, *65*, 103993. [CrossRef]
42. Cheng, J.; Hu, S.; Sun, G.; Kang, K.; Zhu, M.; Geng, Z. Comparison of activated carbons prepared by one-step and two-step chemical activation process based on cotton stalk for supercapacitors application. *Energy* **2021**, *215*, 119144. [CrossRef]
43. Wang, M.; Han, K.; Qi, J.; Teng, Z.; Zhang, J.; Li, M. Study on performance and charging dynamics of N/O codoped layered porous carbons derived from L-tyrosine for supercapacitors. *Appl. Surf. Sci.* **2022**, *578*, 151888. [CrossRef]
44. Yang, X.; Zhao, S.; Zhang, Z.; Chi, Y.; Yang, C.; Wang, C.; Zhen, Y.; Wang, D.; Fu, F.; Chi, R. Pore structure regulation of hierarchical porous carbon derived from coal tar pitch via pre-oxidation strategy for high-performance supercapacitor. *J. Colloid Interface Sci.* **2022**, *614*, 298–309. [CrossRef] [PubMed]
45. Tolendra, K.; Duy, T.T.; Dinh, C.N.; Nam, H.K.; Kin-tak, L.; Joong, H.L. Ternary graphene-carbon nanofibers-carbon nanotubes structure for hybrid supercapacitor. *Chem. Eng. J.* **2020**, *380*, 122543.
46. Zhang, F.; Zong, S.; Zhang, Y.; Lv, H.; Liu, X.; Du, J.; Chen, A. Preparation of hollow mesoporous carbon spheres by pyrolysis-deposition using surfactant as carbon precursor. *J. Power Sources* **2021**, *484*, 229274. [CrossRef]
47. Juan, D.; Yue, Z.; Haixia, W.; Senlin, H.; Aibing, C. N-doped hollow mesoporous carbon spheres by improved dissolution-capture for supercapacitor. *Carbon* **2020**, *156*, 523–528.
48. Hongmei, H.; Li, M.; Shenna, F.; Mengyu, G.; Liangqing, H.; Huanhuan, Z.; Fei, X.; Minghang, J. Fabrication of 3D ordered honeycomb-like nitrogen-doped carbon/PANI composite for high-performance supercapacitors. *Appl. Surf. Sci.* **2019**, *484*, 1288–1296.
49. Zhou, Z.; Cao, J.; Wu, Y.; Zhuang, Q.; Zhao, X.; Wei, Y.; Bai, H. Waste sugar solution polymer-derived N-doped carbon spheres with an ultrahigh specific surface area for superior performance supercapacitors. *Int. J. Hydrogen Energy* **2021**, *46*, 22735–22746. [CrossRef]
50. He, W.; Haitao, N.; Hongjie, W.; Wenyu, W.; Xin, J.; Hongxia, W.; Hua, Z.; Tong, L. Micro-meso porous structured carbon nanofibers with ultra-high surface area and large supercapacitor electrode capacitance. *J. Power Sources* **2021**, *482*, 228986.
51. Man, W.; Juan, Y.; Siyu, L.; Muzi, L.; Chao, H.; Jieshan, Q. Nitrogen-doped hierarchically porous carbon nanosheets derived from polymer/graphene oxide hydrogels for high-performance supercapacitors. *J. Colloid Interf. Sci.* **2020**, *560*, 69–76.

52. Li, J.; Zou, Y.; Xiang, C.; Xu, F.; Sun, L.; Li, B.; Zhang, J. Osmanthus fragrans-derived N-doped porous carbon for supercapacitor applications. *J. Energy Storage* **2021**, *42*, 103017. [CrossRef]
53. Liu, H.; Song, H.; Hou, W.; Chang, Y.; Zhang, Y.; Li, Y.; Zhao, Y.; Han, G. Coal tar pitch-based hierarchical porous carbons prepared in molten salt for supercapacitors. *Mater. Chem Phys.* **2021**, *265*, 124491. [CrossRef]
54. Zhang, Y.; Wu, C.; Dai, S.; Liu, L.; Zhang, H.; Shen, W.; Sun, W.; Ming, L.C. Rationally tuning ratio of micro- to meso-pores of biomass-derived ultrathin carbon sheets toward supercapacitors with high energy and high power density. *J. Colloid Interf. Sci.* **2022**, *606*, 817–825. [CrossRef] [PubMed]

nanomaterials

Article

First-Principles Study of n*AlN/n*ScN Superlattices with High Dielectric Capacity for Energy Storage

Wei-Chao Zhang [1], Hao Wu [1], Wei-Feng Sun [2,*] and Zhen-Peng Zhang [3]

[1] Key Laboratory of Engineering Dielectrics and Its Application, Ministry of Education, School of Electrical and Electronic Engineering, Harbin University of Science and Technology, Harbin 150080, China; weichaozhang@163.com (W.-C.Z.); 2020310226@stu.hrbust.edu.cn (H.W.)

[2] School of Electrical and Electronic Engineering, Nanyang Technological University, Singapore 639798, Singapore

[3] Power Industry Quality Inspection and Test Center for Electric Equipment, China Electric Power Research Institute, Wuhan 430073, China; zhangzhenpeng@epri.sgcc.com.cn

* Correspondence: weifeng.sun@ntu.edu.sg; Tel.: +65-821-535-62

Abstract: As a paradigm of exploiting electronic-structure engineering on semiconductor superlattices to develop advanced dielectric film materials with high electrical energy storage, the n*AlN/n*ScN superlattices are systematically investigated by first-principles calculations of structural stability, band structure and dielectric polarizability. Electrical energy storage density is evaluated by dielectric permittivity under a high electric field approaching the uppermost critical value determined by a superlattice band gap, which hinges on the constituent layer thickness and crystallographic orientation of superlattices. It is demonstrated that the constituent layer thickness as indicated by larger n and superlattice orientations as in (111) crystallographic plane can be effectively exploited to modify dielectric permittivity and band gap, respectively, and thus promote energy density of electric capacitors. Simultaneously increasing the thicknesses of individual constituent layers maintains adequate band gaps while slightly reducing dielectric polarizability from electronic localization of valence band-edge in ScN constituent layers. The AlN/ScN superlattices oriented in the wurtzite (111) plane acquire higher dielectric energy density due to the significant improvement in electronic band gaps. The present study renders a framework for modifying the band gap and dielectric properties to acquire high energy storage in semiconductor superlattices.

Keywords: semiconductor superlattice; dielectric capacity; energy storage; first-principles calculation

Citation: Zhang, W.-C.; Wu, H.; Sun, W.-F.; Zhang, Z.-P. First-Principles Study of n*AlN/n*ScN Superlattices with High Dielectric Capacity for Energy Storage. *Nanomaterials* **2022**, *12*, 1966. https://doi.org/10.3390/nano12121966

Academic Editor: Jürgen Eckert

Received: 16 April 2022
Accepted: 1 June 2022
Published: 8 June 2022

Publisher's Note: MDPI stays neutral with regard to jurisdictional claims in published maps and institutional affiliations.

1. Introduction

Today renewable sources are urgently developed and expected to dominate the future operation systems of electricity power. However, the inevitable intermittence bearing on renewable sources, such as solar and wind energies, challenges the continuous equilibrium required for temporarily storing electrical energy in adequately prolonged periods of time [1,2]. Even advanced batteries cannot respond sufficiently as fast to complement the promptly fluctuating energy sources [2]. In contrast, high-speed discharging electrostatic capacitors are uniquely preferable to efficiently fulfill the prompt complements in energy support systems [3]. Meanwhile, dielectric capacitors cannot be comprehensively applied to high-power energy storage until now due to the relatively low energy density of dielectric materials in electric discharging work.

In general, it is required for dielectric materials to achieve high energy storage density by increasing the maximum polarization intensity and breakdown field while persisting a low remnant polarization [4–6]. Comprehensive efforts have focused on pursuing antiferroelectric film materials with high energy storage density due to their double hysteresis loops of polarization-field characteristics, which can mostly approach the high-energy-density of 154 J/cm^3, which is comparable with excellent electrochemical supercapacitors [7,8].

However, the energy storage performance of these antiferroelectric films requires a ferroelectric/antiferroelectric coexistence around the morphotropic boundary, which is intensively dependent on chemical composition and thermodynamic temperature [9–11]. It is also unfortunate for nonlinear dielectrics such as antiferroelectrics and relaxors that the inevitable energy dissipation in the charge/discharge cycle from hysteresis leads to low storage efficiency of recoverable energy. Moreover, for ferroelectric materials, it is difficult to approach a high energy density due to their substantial remnant polarization [12]. In comparison, linear dielectrics without remnant polarization and considerable energy loss it is only considered to acquire high energy density by promoting dielectric permittivity and breakdown field strength [13,14].

Recently arising linear dielectrics of III-V semiconductors in forms of solid solutions or superlattices, such as AlScN alloys or AlN/ScN superlattices, have attracted great focus for prospective energy storage due to their nonpolar phase in close proximity with ferroelectric states [15–17]. AlN is the most commonly used barrier material due to its largest band gap in the III-V group semiconductors, which is qualified for applying electric field as high strength as possible, and much promising for energy storage due to its chemical simplicity and low dielectric permittivity under high electric field. Recently observed ferroelectric states appearing in $Al_{1-x}Sc_xN$ films with a substantial remnant polarization in contrast to pure AlN are actually polar or nonpolar but not in the ferroelectric phase [18,19]. The reactive magnetron sputtering method has been successfully applied to prepare $Al_{1-x}Sc_xN$ alloy film, which was expected to be improved for enhancing piezoelectric and ferroelectric responses [20,21]. From these works, it is worthwhile to investigate the AlN/ScN superlattices in a chemical component that is similar to the $Al_{1-x}Sc_xN$ alloys as a representative of newly arising semiconductor film dielectrics with a preferable performance in terms of dielectric energy storage. The electronic band-edge characteristics of the semiconductor superlattices pivot on the quantum well confinement and band alignment of constituent layers, which accounts for the band gap and determines the electrical breakdown field of the electrical capacitor. Previous research lacks the proper consideration of the constituent layer thickness and crystallographic orientation of the AlN/ScN superlattices.

In the present study, we focus on the n*AlN/n*ScN superlattices oriented on the (001) or (111) crystallographic plane of a wurtzite structure, where n denotes the number of AlN or ScN monolayers in constituent layers of superlattices and indicates the constituent layer thickness. Their energy storage characteristics are studied by first-principle calculations of the band-structure and dielectric polarizability dependent on the electrical field and superlattice configurations to explore potential applications in high energy storage. Such artificial layered materials are generally fabricated by the epitaxial growth technology of controlling layer interface in an atom resolution, which provides great flexibility in optimizing electronic states and dielectric polarization by modifying the constituent-layer thickness and crystallographic orientation of superlattices. This also helps us to comprehend the underlying physics of high density and efficiency of energy storage in electrical capacitors.

2. Theoretical Methodology

The pseudopotential plane-wave method is used to carry out first-principle calculations of the crystal structure, electronic structure and polarizability by applying an electric field for (001) and (111) n*AlN/n*ScN superlattices, as implemented by CASTEP of Materials Studio 2020 (Accelrys Inc., Materials Studio version 2020.08, San Diego, CA, USA). The GGA-WC exchange-correlation function was adopted to perform geometry optimization and calculate the dielectric polarizability, while the HSE06 hybrid exchange-correlation function was specified to obtain accurate band structures [22]. The potential field of atomic cores bearing on the electrons is described by on-the-fly generated (OTFG) norm-conserving pseudopotential with the Koelling–Harmon treatment of relativistic effect [23]. Self-consistent field (SCF) iterations are implemented under convergence tolerance of 5×10^{-7} eV/atom in an FFT grid of $72 \times 72 \times 216$, in which the Pulay scheme of charge density mixing in the magnitude of 0.5 is specified to relax the electrons [24,25]. The plane-wave basis-set with cut-off energy of 440.0 eV is modified by the basis-set finiteness correction [26]. Brillouin zone integration is realized by k point sampling on the Monkhorst-Pack $4 \times 4 \times 1$ grid [27]. Crystal structures are geometrically optimized with the BFGS algorithm in delocalized internal coordinates under energy convergence of 5.0×10^{-6} eV/atom with a maximum of 0.02 eV/Å atomic force and 0.001 Å stress [28].

The internal electric fields are theoretically applied to the superlattice crystal structures along layer-plane normal (axis-z) to calculate hysteresis curves of electric polarization versus electric field strength (P-E), in which the geometry optimization for each electric field is performed to represent piezoelectric strain, and linear response formalism based on density-functional perturbation theory is enabled to calculate static dielectric permittivity under the direct-current internal electric fields [29]. According to the calculated band gaps, the intrinsic breakdown electric field is estimated empirically as the universal expression proposed by reference [30]. Cohesive energies in atom average are calculated by $E_{coh} = n[E(Al) + 2E(N) + E(Sc)] - E(sup)$ where $E(Al)$, $E(N)$, $E(Sc)$ and $E(sup)$ represent total energies of Al, N and Sc isolated atoms, and superlattices.

3. Results and Discussion

3.1. Crystal Structure

Atomic configurations in crystal structures of (001) and (111) n*AlN/n*ScN superlattices (n = 1, 2, 3), as shown in Figure 1, have been energetically relaxed by geometry optimization without applying an internal electric field, indicating a diversity of space symmetries alternating with the adjustable superlattice parameters of n and crystallographic orientation. In addition, the space symmetry groups, lattice constants, the thicknesses of individual constituent layers, and cohesive energy per atom obtained are listed in Table 1. For the superlattice configurations, the ScN monolayer (double atomic layer) or the entire constituent layer for constructing superlattice structures is explicitly larger in thickness than the AlN monolayer or constituent layer, indicating that compressive and tensile strains of layer-plane exist in ScN and AlN layers respectively due to a lattice misfit. The thickness of the AlN constituent layer is strictly proportional to n while the ScN constituent layer becomes larger than n times the ScN monolayer thickness, implying that the Sc-N bonding elongation of the relaxing misfit strain along the layer-plane is normal when n increases, which also accounts for the higher cohesive energy per atoms of larger n than that of smaller ones.

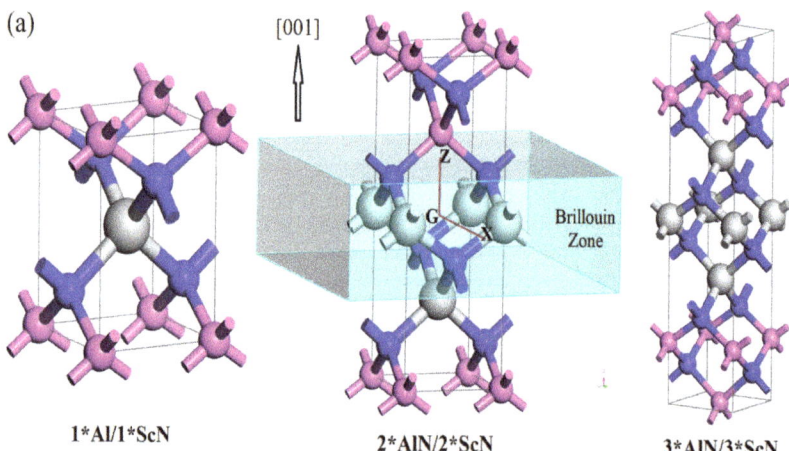

Figure 1. *Cont.*

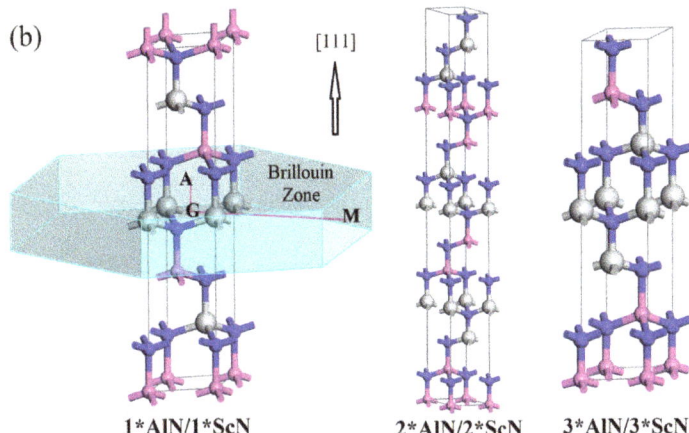

Figure 1. Crystal structures of n*AlN/n*ScN superlattices (*n* = 1, 2, 3) on (**a**) (001) and (**b**) (111) crystallographic faces of wurtzite structure, as indicated by layer-plane normal along (001) and (111) crystallographic orientations respectively, and the dispersion paths of electronic energy band through high symmetry points in the Brillouin zone are also shown. The gray, pink, and blue balls symbolize Sc, Al, and N bonding atoms, respectively.

Table 1. The space symmetry group, lattice constant (*a*/*b*, *c*), thicknesses of AlN and ScN constituent layers (h_{AlN} and h_{ScN}), and cohesive energy per atom (E_{coh}), band gaps E_g and intrinsic breakdown field strength E_b for the AlN/ScN superlattices.

Orientations	Superlattices	Space Groups	*a* = *b*/Å	*c*/Å	h_{AlN}/Å	h_{ScN}/Å	E_{coh}/(eV/atom)	E_g/eV	E_b/(MV·cm⁻¹)
	1*AlN/1*ScN	P-4M2	3.2608	4.6528	2.0997	2.5531	7.9224	3.815	7.75
(001)	2*AlN/2*ScN	PMM2	3.2525	9.3081	4.1715	5.1366	7.9542	3.559	7.23
	3*AlN/3*ScN	P-4M2	3.2507	13.9629	6.2454	7.7175	7.9642	3.535	7.18
	1*AlN/1*ScN	R3M	3.2551	16.1767	2.4945	2.8978	7.9126	4.519	9.18
(111)	2*AlN/2*ScN	R3M	3.2492	32.4172	4.9897	5.81603	7.9324	4.231	8.59
	3*AlN/3*ScN	P3M1	3.2487	16.2102	7.4845	8.7257	7.9380	4.072	8.27

The cohesive energy of the AlN/ScN superlattices approaches the highest and lowest values of 7.96 and 7.91 eV/atom for (001) 3*AlN/3*ScN and (111) 1*AlN/1*ScN super-lattices, respectively, which are all remarkably higher than the III-chalcogenide covalent double-layers and TMD monolayers [31–33]. Bulk AlN and ScN are also calculated by identical first-principle schemes to obtain the cohesive energies of 7.66 and 8.38 eV/atom, which are slightly lower and higher, respectively, than these superlattices. It is an energetic manifestation of high structural stability that both (001) and (111) n*AlN/n*ScN superlattices can be feasibly achieved by matching the AlN and ScN monolayers through Sc-N or Al-N bonding strongly into a periodic layer structure. In comparison to the (001) superlattice orientation, the larger misfit in the (111) layer orientation accounts for the larger extension along the layer-plane normal with increasing constituent layer thickness and results in a lower cohesive energy per atoms.

3.2. Band Structure

Due to the in-layer quantum confinement and large lattice misfit between constituent layers in AlN/ScN superlattices, their band structures are quite different from AlN and ScN bulk materials, as shown in Figure 2. All of these superlattices present large electronic band gaps in the 3.5~4.5 eV range while persisting almost constantly without substantial dependence on constituent layer thickness (n), which is attributed to the simultaneous changing thickness of the individual constituent layer, almost fixing the quantum confinement levels or minibands of superlattices. In particular, for the higher values of n, a smaller dispersion along with the normal layer, as illustrated by the narrower minibands at the valence band-edge in Figure 2, indicates a more localized feature of valence electrons in response to the electric field perpendicular to superlattice layers, which manifests as a lower intensity of dielectric polarization.

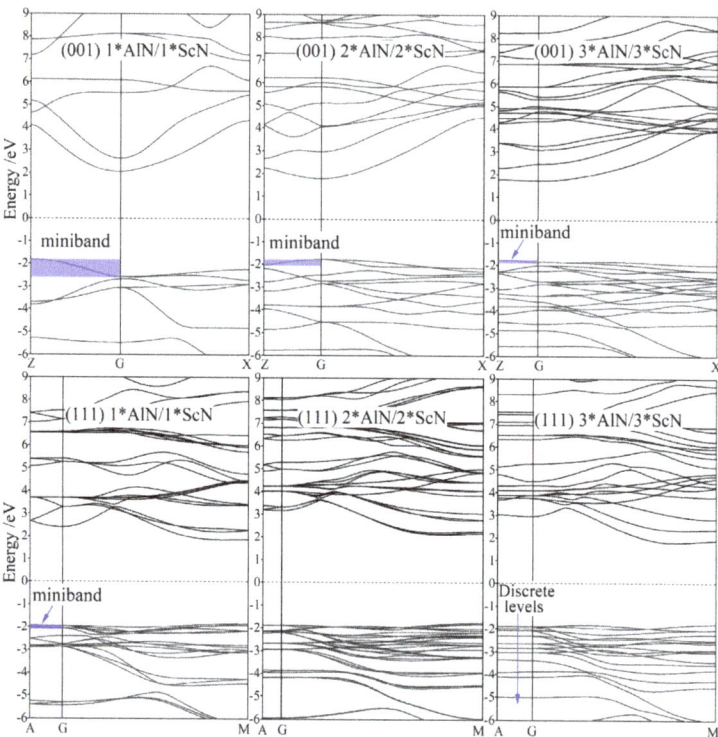

Figure 2. Band structures of the (001) and (111) n*AlN/n*ScN superlattices (*n* = 1,2,3) in the dispersion paths through high symmetry points in the Brillouin zone as indicated in Figure 1; the Fermi level (horizontal dash line) is referenced as energy zero.

The quantum confinement minibands of AlN/ScN superlattices are promptly narrowed down as the superlattice orientation is converted from the low symmetry (001) to high symmetry (111) crystallographic plane of a wurtzite-like structure due to the symmetry-induced degeneration of electronic energy levels. Even when n is raised to 3 for (111) orientation, the multiple electronic minibands with minimal energetic dispersion along with the superlattice's normal layer, as in the *n* < 3 AlN/ScN superlattices, have shrunk into discrete energy levels due to the quantum well confinement of ScN constituent layer sandwiched by sufficiently wider AlN energy barriers. This results in notably larger band gaps of (111)-orientated superlattices than that of the (001) orientation, as shown in Table 1. For a more important consequence of a (111) orientation, the valence electrons are almost completely residing in ScN layers with a considerably lower dielectric polarizability than in the AlN layers, which are dominantly contributed by the uneasily polarized bonding of the valence band-edge electrons derived from the Sc-3*d* and N-2*p* orbitals. To this end, these electronic structure results elucidate why a higher electric polarization can be acquired whilst persisting with a large band gap by simultaneously increasing the AlN and ScN layer thicknesses.

3.3. Dielectric Polarization and Energy Density

Dielectric polarization P under a high electric field has been evidently promoted by increasing the constituent layer thickness as indicated by a larger number of AlN or ScN monolayers in superlattice configurations, as shown in Figure 3a. The $P–E$ relationship obtained from first-principle calculations at diverse points of the electric field intensity is fitted with the analytical functions of $E(P) = aP + bP^3$ based on the Landau free-energy, whereby the energy storage densities are accurately evaluated by an analytical integral of $E(P)$ as $(aP_m^2/2) + (bP_m^4/4)$ where P_m denotes the electric polarization at intrinsic breakdown field, and no polarization arises under a zero external electric field, as shown in the results shown in Figure 3b. In contrast, in the (001) and (111) superlattice orientations, higher energy density can be acquired by the (111)-oriented superlattices, which is attributed to the significant improvement in band gap or intrinsic breakdown field strength for (111) AlN/ScN superlattices as listed in Table 1, whilst without considerable deficiency in dielectric polarization. Meanwhile, the increase of constituent layer thicknesses leads to higher dielectric polarizability under high electric fields for both the (001) and (111) AlN/ScN superlattices. The present first-principles calculations demonstrate that the n*AlN/n*ScN superlattices ($n \leq 3$) are excellent nonlinear dielectrics of energy storage with the highest energy density approaching 304 J/cm^3 by far exceeding the current supercapacitor materials realized in the experiments, as shown in Table 2.

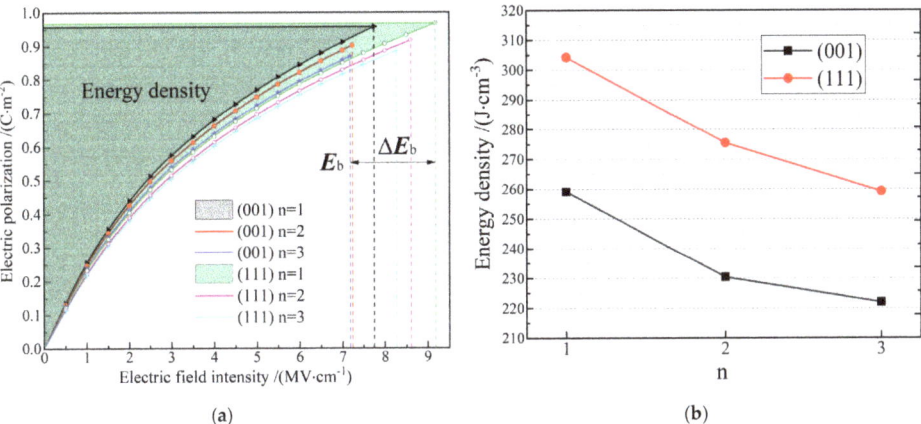

(a)   (b)

Figure 3. (a) $P–E$ hysteresis curves where indicating breakdown field intensity by E_b and energy density areas; (b) energy storage densities as an electric capacitor of (001) and (111) n*AlN/n*ScN superlattices ($n = 1, 2, 3$).

Table 2. Energy densities of the AlN/ScN superlattices in comparison to the recently reported nonlinear dielectrics for energy storage capacitors, where the rGO and EDLC indicate reduced graphene oxide and electrochemical double-layer supercapacitor, respectively.

Material	Energy Density/J·cm^{-3}	Method or Process
(001) 3*AlN/3*ScN superlattice	259	First-principles calculation
(111) 3*AlN/3*ScN superlattice	304	First-principles calculation
Nitrogen-Thiol-rGO Scrolls [34]	215	Nitrogen-doped thiol-functionalization
Pt(111)/Ti/SiO$_2$/Si [35]	99.8	Solid-state reaction
rGO-based EDLC [36]	142	Hydrazine reduction

It is clearly shown in Figure 1 that the majority of Al-N and Sc-N bonds are parallel to the layer-pane in (111) AlN/ScN superlattices which means all these polar bonds cannot contribute to dielectric polarization in response to the electric field perpendicular to layer-pane. In contrast, despite the diversion of an angle from the normal layer-plane in the (001) AlN/ScN superlattices, all of these ionic bonds of intrinsic dipoles are partially devoted to dielectric polarization under a normal electric field, as discriminated by the higher (001) electric polarizabilities than that of the (111) orientation. However, the (111) AlN/ScN superlattices possess remarkably larger band gaps and breakdown fields than the (001) AlN/ScN superlattices. Meanwhile, the larger out-of-plane tensile strain of the ScN constituent layer in the (111) AlN/ScN superlattices, as mentioned in Section 3.2, is another reason accounting for the larger band gaps and higher energy densities than the (001) AlN/ScN superlattices. It is thus flexible and preferable to exploit the superlattice configuration parameters, such as the constituent layer thickness and crystallographic orientation, to engineer band structures and dielectric responses of the AlN/ScN superlattices to suggest a feasible pathway for developing linear dielectrics for energy storage.

4. Conclusions

Employing a first-principle pseudopotential plane-wave method, the n*AlN/n*ScN superlattices with different constituent layer thicknesses and crystallographic orientations have been systematically studied by calculating the atomic structure, band structure and polarizability to elucidate their high energy density of the electrical capacity and predictable experimental feasibility. The large band gaps of the AlN/ScN superlattices can be retained, and the higher dielectric polarizabilities under a high electric field can be acquired by simultaneously increasing the numbers of AlN and ScN monolayers in individual constituent layers of superlattice configurations. The crystallographic orientation in the (111) plane will distinctively promote electronic band gaps while slightly decreasing the static dielectric response to the electric field normal to superlattice layers, which is respectively attributed to the increased absolute misfit of the superlattice layer and the in-plane orientations of major polar bonds. It is preferable to manipulate the superlattice configuration parameters to effectively adjust band structures and dielectric polarization of AlN/ScN superlattices. This study suggests a prospective routine of employing the highly controllable superlattice materials to steer electric polarizability and develop high energy density dielectrics.

Author Contributions: Conceptualization, W.-F.S.; data curation, W.-C.Z. and H.W.; formal analysis, Z.-P.Z.; investigation, W.-F.S.; writing—original draft preparation, W.-F.S.; writing—review and editing, W.-F.S.; project administration, W.-C.Z. All authors have read and agreed to the published version of the manuscript.

Funding: Supported by the Key Laboratory of Engineering Dielectrics and Its Application, Ministry of Education (KFZ202002). This research was funded by the National Natural Science Foundation of China (Grant No. 51337002).

Informed Consent Statement: Informed consent was obtained from all subjects involved in the study.

Data Availability Statement: Theoretical methods and results are available from the authors.

Conflicts of Interest: The authors declare no conflict of interest.

References

1. Gross, R.; Leach, M.; Bauen, A. Progress in renewable energy. *Environ. Int.* **2003**, *29*, 105–122. [CrossRef]
2. Hao, H. A review on the dielectric materials for high energystorage application. *J. Adv. Dielectr.* **2013**, *3*, 1330001. [CrossRef]
3. Chu, B.; Zhou, X.; Ren, K.; Neese, B.; Lin, M.; Wang, Q.; Bauer, F.; Zhang, Q.M. A Dielectric polymer with high electric energy density and fast discharge speed. *Science* **2006**, *313*, 334–336. [CrossRef] [PubMed]
4. Patel, S.; Chauhan, A.; Vaish, R. Enhancing electrical energy storage density in anti-ferroelectric ceramics using ferroelastic domain switching. *Mater. Res. Express* **2014**, *1*, 045502. [CrossRef]
5. Chauhan, A.; Patel, S.; Vaish, R.; Bowen, C.R. Anti-ferroelectric ceramics for high energy density capacitors. *Materials* **2015**, *8*, 8009–8031. [CrossRef]
6. Prateek; Thakur, V.K.; Gupta, R.K. Recent progress on ferroelectric polymer-based nanocomposites for high energy density capacitors: Synthesis, dielectric properties, and future aspects. *Chem. Rev.* **2016**, *116*, 4260. [CrossRef]
7. Peng, B.; Zhang, Q.; Li, X.; Sun, T.; Fan, H.; Ke, S.; Ye, M.; Wang, Y.; Lu, W.; Niu, H.; et al. Giant electric energy density in epitaxial lead-free thin films with coexistence of ferroelectrics and antiferroelectrics. *Adv. Electron. Mater.* **2015**, *1*, 1500052. [CrossRef]

8. Li, Y.Z.; Lin, J.L.; Bai, Y.; Li, Y.X.; Zhang, Z.D.; Wang, Z.J. Ultrahigh-energy storage properties of (PbCa)ZrO$_3$ antiferroelectric thin films via constructing a pyrochlore nanocrystalline structure. *ACS Nano* **2020**, *14*, 6857. [CrossRef]

9. Hao, X.; Wang, Y.; Zhang, L.; Zhang, L.; An, S. Composition-dependent dielectric and energy-storage properties of (Pb,La)(Zr,Sn,Ti)O$_3$ antiferroelectric thick films. *Appl. Phys. Lett.* **2013**, *102*, 163903. [CrossRef]

10. Li, Q.; Chen, L.; Gadinski, M.R.; Zhang, S.; Zhang, G.; Li, H.U.; Iagodkine, E.; Haque, A.; Chen, L.Q.; Jackson, T.N.; et al. Flexible high-temperature dielectric materials from polymer nanocomposites. *Nature* **2015**, *523*, 576. [CrossRef]

11. Emery, S.B.; Cheng, C.J.; Kan, D.; Rueckert, F.J.; Alpay, S.P.; Nagarajan, V.; Takeuchi, I.; Wells, B.O. Phase coexistence near a morphotropic phase boundary in Sm-doped BiFeO$_3$ films. *Appl. Phys. Lett.* **2010**, *97*, 152902. [CrossRef]

12. Xu, B.; Wang, D.; Iniguez, J.; Bellaiche, L. Finite-temperature properties of rare-earth-substituted BiFeO$_3$ multiferroic solid solutions. *Adv. Funct. Mater.* **2015**, *25*, 552–558. [CrossRef]

13. Peddigari, M.; Palneedi, H.; Hwang, G.T.; Ryu, J. Linear and nonlinear dielectric ceramics for high-power energy storage capacitor applications. *J. Korean Ceram. Soc.* **2019**, *56*, 1. [CrossRef]

14. Zhou, H.Y.; Liu, X.Q.; Zhu, X.L.; Chen, X.M. CaTiO$_3$ linear dielectric ceramics with greatly enhanced dielectric strength and energy storage density. *J. Am. Ceram. Soc.* **2018**, *101*, 1999. [CrossRef]

15. Tasnádi, F.; Alling, B.; Häglund, C.; Wingqvist, G.; Birch, J.; Hultman, L.; Abrikosov, I.A. Origin of the Anomalous Piezoelectric Response in Wurtzite Sc$_x$Al$_{1-x}$N Alloys. *Phys. Rev. Lett.* **2010**, *104*, 137601. [CrossRef]

16. Yazawa, K.; Drury, D.; Zakutayev, A.; Brennecka, G.L. Reduced coercive field in epitaxial thin film of ferroelectric wurtzite Al$_{0.7}$Sc$_{0.3}$N. *Appl. Phys. Lett.* **2021**, *118*, 162903. [CrossRef]

17. Jiang, Z.; Paillard, C.; Vanderbilt, D.; Xiang, H.; Bellaiche, L. Designing Multifunctionality via Assembling Dissimilar Materials: Epitaxial AlN/ScN Superlattices. *Phys. Rev. Lett.* **2019**, *123*, 096801. [CrossRef]

18. Xu, B.; Wang, D.; Zhao, H.J.; Íñiguez, J.; Chen, X.M.; Bellaiche, L. Hybrid Improper ferroelectricity in multiferroic superlattices: Finite-temperature properties and electric-field-driven switching of polarization and magnetization. *Adv. Funct. Mater.* **2015**, *25*, 3626. [CrossRef]

19. Fichtner, S.; Wolff, N.; Lofink, F.; Kienle, L.; Wagner, B. AlScN: A III-V semiconductor based ferroelectric. *J. Appl. Phys.* **2019**, *125*, 114103. [CrossRef]

20. Akiyama, M.; Kamohara, T.; Kano, K.; Teshigahara, A.; Takeuchi, Y.; Kawahara, N. Enhancement of piezoelectric response in scandium aluminum nitride alloy thin films prepared by dual reactive cosputtering. *Adv. Mater.* **2009**, *21*, 593. [CrossRef]

21. Yasuoka, S.; Shimizu, T.; Tateyama, A.; Uehara, M.; Yamada, H.; Akiyama, M.; Hiranaga, Y.; Cho, Y.; Funakubo, H. Effects of deposition conditions on the ferroelectric properties of (Al$_{1-x}$Sc$_x$)N thin films. *J. Appl. Phys.* **2020**, *128*, 114103. [CrossRef]

22. Wu, Z.; Cohen, R.E. More accurate generalized gradient approximation for solids. *Phys. Rev. B* **2006**, *73*, 235116. [CrossRef]

23. Lejaeghere, K.; Van Speybroeck, V.; van Oost, G.; Cottenier, S. Error estimates for solid-state density-functional theory predictions: An overview by means of the ground-state elemental crystals. *Crit. Rev. Solid State Mater. Sci.* **2014**, *39*, 1–24. [CrossRef]

24. Payne, M.C.; Teter, M.P.; Allan, D.C.; Arias, T.A.; Joannopoulos, J.D. Iterative minimization techniques for ab initio total energy calculations: Molecular dynamics and conjugate gradients. *Rev. Mod. Phys.* **1992**, *64*, 1045–1097. [CrossRef]

25. Kresse, G.; Furthmuller, J. Efficient iterative schemes for *ab initio* total-energy calculations using a plane-wave basis set. *Phys. Rev. B* **1996**, *54*, 11169–11186. [CrossRef]

26. Monkhorst, H.J.; Pack, J.D. Special points for Brillouin-zone integrations-A reply. *Phys. Rev. B* **1977**, *16*, 1748.

27. Milman, V.; Lee, M.H.; Payne, M.C. Ground-state properties of CoSi$_2$ determined by a total-energy pseudopotential method. *Phys. Rev. B* **1994**, *49*, 16300. [CrossRef]

28. Pfrommer, B.G.; Cote, M.; Louie, S.G.; Cohen, M.L. Relaxation of crystals with the quasi-newton method. *J. Comput. Phys.* **1997**, *131*, 133–140. [CrossRef]

29. Baroni, S.; de Gironcoli, S.; dal Corso, A.; Giannozzi, P. Phonons and related crystal properties from density-functional perturbation theory. *Rev. Mod. Phys.* **2001**, *73*, 515–562. [CrossRef]

30. Wang, L.M. Relationship between intrinsic breakdown field and bandgap of materials. In Proceedings of the 25th International Conference on Microelectronics, Belgrade, Serbia, 14–17 May 2006; pp. 576–579.

31. Wang, N.; Cao, D.; Wang, J.; Liang, P.; Chen, X.; Shu, H. Semiconducting edges and flake-shape evolution of monolayer GaSe: Role of edge reconstructions. *Nanoscale* **2018**, *10*, 12133–12140. [CrossRef]

32. Xu, X.D.; Sun, W.F. First-principles investigation of GaInSe$_2$ monolayer as a janus material. *J. Chin. Ceram. Soc.* **2020**, *48*, 507–513.

33. Lv, M.H.; Li, C.M.; Sun, W.F. Spin-orbit coupling and spin-polarized electronic structures of janus vanadium-dichalcogenide monolayers: First-principles calculations. *Nanomaterials* **2022**, *12*, 382. [CrossRef] [PubMed]

34. Rani, J.R.; Thangavel, R.; Oh, S.I.; Lee, Y.S.; Jang, J.H. An ultra-high-energy density supercapacitor; fabrication based on Thiol-functionalized graphene oxide scrolls. *Nanomaterials* **2019**, *9*, 148. [CrossRef]

35. Silva, J.P.; Silva, J.M.; Oliveira, M.J.; Weingärtner, T.; Sekhar, K.C.; Pereira, M.; Gomes, M.J. High-performance ferroelectric-dielectric multilayered thin films for energy storage capacitors. *Adv. Funct. Mater.* **2019**, *29*, 1807196. [CrossRef]

36. Marcano, D.C.; Kosynkin, D.V.; Berlin, J.M.; Sinitskii, A.; Sun, Z.; Slesarev, A.; Alemany, L.B.; Lu, W.; Tour, J.M. Improved Synthesis of Graphene Oxide. *ACS Nano* **2010**, *4*, 4806–4814. [CrossRef]

nanomaterials

Article

Sophisticated Structural Tuning of NiMoO₄@MnCo₂O₄ Nanomaterials for High Performance Hybrid Capacitors

Yifei Di [1], Jun Xiang [1,*], Nan Bu [1], Sroeurb Loy [1], Wenduo Yang [1], Rongda Zhao [1,*], Fufa Wu [1,*], Xiaobang Sun [1] and Zhihui Wu [2]

[1] School of Materials Science and Engineering, Liaoning University of Technology, Jinzhou 121001, China; diyifei97@126.com (Y.D.); bn1376980216@163.com (N.B.); loysroeurb999@gmail.com (S.L.); 18342841189@163.com (W.Y.); lilac_rain@126.com (X.S.)
[2] Liaoning Brother Electronics Technology Co., Chaoyang 122000, China; fyzl86@126.com
* Correspondence: xiangj@lnut.edu.cn (J.X.); rongdazhaoln@126.com (R.Z.); ffwu@lnut.edu.cn (F.W.); Tel./Fax: +86-416-4199650 (R.Z.)

Abstract: NiMoO₄ is an excellent candidate for supercapacitor electrodes, but poor cycle life, low electrical conductivity, and small practical capacitance limit its further development. Therefore, in this paper, we fabricate NiMoO₄@MnCo₂O₄ composites based on a two-step hydrothermal method. As a supercapacitor electrode, the sample can reach 3000 mF/cm² at 1 mA/cm². The asymmetric supercapacitor (ASC), NiMoO₄@MnCo₂O₄//AC, can be constructed with activated carbon (AC) as the negative electrode, the device can reach a maximum energy density of 90.89 mWh/cm³ at a power density of 3726.7 mW/cm³ and the capacitance retention can achieve 78.4% after 10,000 cycles.

Keywords: supercapacitors; NiMoO₄@MnCo₂O₄; microstructure; electrochemical performance; cycling stability

Citation: Di, Y.; Xiang, J.; Bu, N.; Loy, S.; Yang, W.; Zhao, R.; Wu, F.; Sun, X.; Wu, Z. Sophisticated Structural Tuning of NiMoO₄@MnCo₂O₄ Nanomaterials for High Performance Hybrid Capacitors. *Nanomaterials* **2022**, *12*, 1674. https://doi.org/10.3390/nano12101674

Academic Editor: Diego Cazorla-Amorós

Received: 14 April 2022
Accepted: 9 May 2022
Published: 14 May 2022

Publisher's Note: MDPI stays neutral with regard to jurisdictional claims in published maps and institutional affiliations.

1. Introduction

With the development of the world economy, environmental pollution is caused by the excessive burning of traditional fossil fuels, which poses a serious threat to the goal of human sustainable development [1]. Supercapacitors (SCs), as a new environmentally friendly electrochemical energy storage device, have attracted extensive attention from researchers. The selection of electrode material is an important factor for energy storage performance. Developing an electrode material with excellent electrochemical performance has become key to the future development of SCs [2–5]. Transition metal oxides possess high specific capacitance, superior cycling performance and abundant valence states, such as NiMoO₄, MnCo₂O₄, NiCo₂O₄ and ZnCo₂O₄. They have been widely reported due to their large theoretical capacitance, excellent redox performance and environmental friendliness [6–10].

NiMoO₄ is a very suitable electrode material for SCs because of its advantages of better electrochemical performance and low price [11–13]. However, there are still many problems such as low theoretical utilization value, poor cycle life and low conversion performance at a higher rate [14]. Xuan [15] et al. prepared a NiMoO₄@Co₃O₄ composite nanoarray electrode. The pseudocapacitance performance of the prepared NiMoO₄@Co₃O₄-5H composite was 1722.3 F/g at the current density of 1 A/g, and the capacitance retention rate of 91% was realized by the 6000 cycles test. Feng [16] et al. prepared hierarchical flower-like NiMoO₄@Ni₃S₂ composite material on a 3D nickel foam matrix by the hydrothermal method. The specific capacity was 870 C/g at 0.6 A/g, and the capacity retention rate was 81.2% after 8000 cycles. Transition metal oxide MnCo₂O₄ with excellent electrochemical performance is very suitable for the electrode material of SCs, because its Mn ion can offer high electron conductivity and excellent rate performance, and cobalt ion has high oxidation potential. However, they can also demonstrate poor application, such as poor cycling performance, poor electrical conductivity and so on, which greatly affect the practical application of SCs [17,18]. Cheng [19] et al. prepared porous $MnCo_2O_4$@NiO nanosheets by hydrothermal synthesis and calcination. The specific capacitance of the electrode material was 508.3 F/g at 2 A/g current density. The 2000 cycles test was applied at 10 A/g current density, and it presented the capacitance retention performance of 89.7%. Liu [20] et al. prepared $MnCo_2O_4$@MnO_2 nanosheet arrays with core–shell structure on nickel foam by two-step hydrothermal treatment. The surface capacitance of the electrode was 3.39 F/cm² at a current density of 3 mA/cm². Furthermore, the capacity retention rate was 92.5% by 3000 cycles test at a current density of 15 mA/cm². It could be seen that the composites exhibited excellent electrochemical properties due to their excellent conductivity [21–24]. It was also confirmed that $NiMoO_4$ and $MnCo_2O_4$ have great potential as electrode materials for SCs [25]. The composite electrodes constructed from these two materials can effectively improve the conductivity, specific surface area, and number of reaction sites, thereby improving the overall electrochemical performance. [26–28].

In this work, $NiMoO_4$@$MnCo_2O_4$ composite electrode material is obtained by the two-step hydrothermal synthesis method. The results show that the $NiMoO_4$@$MnCo_2O_4$ electrode has better electrochemical performance than single $NiMoO_4$ or $MnCo_2O_4$ electrode, and its electrochemical performance is greatly improved after the composite. At the current density of 1 mA/cm², the specific capacitance of single $NiMoO_4$ electrode material is 1656 mF/cm², and the specific capacitance of the single $MnCo_2O_4$ electrode material is 224 mF/cm². Finally, the $NiMoO_4$@$MnCo_2O_4$ electrode material is 3000 mF/cm². After 10,000 cycles, the capacity retention rate of $NiMoO_4$@$MnCo_2O_4$ electrode material is 96%. $NiMoO_4$@$MnCo_2O_4$//AC devices show high electrochemical performance with a maximum energy density of 90.89 mWh/cm³ and a power density of 3726.7 mW/cm³.

2. Experimental Section

2.1. Preparation of $NiMoO_4$ Nano Pompon-Like Structure Electrode Material

In a typical process, 6 mmol $Na_2MoO_4 \cdot 2H_2O$, 6 mmol $Ni(NO_3)_2 \cdot 6H_2O$, 1 mmol NH_4F, and 1 mmol $CO(NH_2)_2$ was added to 50 mL deionized water. After magnetic stirring, the nickel foam was put into the solution and reacted at 120 °C for 12 h, and then it was cleaned by deionized water and anhydrous ethanol to remove surface impurities. The $NiMoO_4$ precursor was obtained by drying for 6 h in a drying oven at 60 °C and annealing for 2 h in air at 350 °C.

2.2. Preparation of $NiMoO_4$@$MnCo_2O_4$ Urchin-like Core-Shell Structure Electrode Material

In a similar process to above, 6 mmol $Mn(CH_3COO)_2 \cdot 4H_2O$, 6 mmol $Co(NO_3)_3 \cdot 6H_2O$, 5 mmol NH_4F and 5 mmol $CO(NH_2)_2$ were dissolved in 50 mL deionized water to obtain a homogeneous solution. The nickel foam with $NiMoO_4$ was put into this solution, and it kept 140 °C for 8 h. After cooling down to room temperature, the samples were washed, dried, and annealed for 2 h at 350 °C. The mass loading of $NiMoO_4$, $MnCo_2O_4$, and $NiMoO_4$@$MnCo_2O_4$ is 1.27, 1.02, and 1.91 mg/cm², respectively.

2.3. Materials Characterizations

The elemental composition and valence of the samples were characterized by X-ray powder diffraction (XRD, D/max-2500/PC, Rigaku Corporation, Tokyo, Japan) with Cu Kα (λ = 1.5406 Å) and X-ray photo-electron spectroscopy (XPS, ESCALAB250, FEI Company, Waltham, MA, USA). The structure and morphology were investigated by emission scanning electron microscopy (SEM, Sigma500, Zeiss, Jena, Germany), and high-resolution transmission electron microscopy (HRTEM, Tecnai G2 S-Twin F20, FEI Company, Waltham, MA, USA).

2.4. Electrochemical Measurements

The electrochemical characteristics of the products were tested by Shanghai CHI660E electrochemical workstation. The sample material was applied as the working electrode, the platinum electrode was utilized as the auxiliary electrode, and Hg/HgO electrode was employed as the reference electrode. The working electrode was processed as a circle with a diameter of 1 cm. Moreover, 3 M KOH solution was used as the electrolyte and the ultrasonic-treated nickel foam was served as the collector. Through cyclic voltammetry (CV), galvanostatic charging–discharging (GCD), electrochemical impedance spectroscopy (EIS) and cycling performance measurements, the electrochemical properties of electrode materials and their application value were analyzed.

Energy density (E) can be obtained from the integral area of discharging curves. Specific capacitance (Cs), power density (P), and coulombic efficiency (η) can be calculated by the following equations:

$$C_s = I\Delta t_d / S\Delta V \tag{1}$$

$$P = 3600E / \Delta t_d \tag{2}$$

$$\eta = \Delta t_d / \Delta t_c \tag{3}$$

where I is the current value, Δt_d and Δt_c represent the discharging time and charging time, S is the geometrical area of the electrode, and ΔV denotes the voltage window.

2.5. Fabrication of Asymmetric Supercapacitors

Asymmetric supercapacitors were constructed with $NiMoO_4@MnCo_2O_4$ as the positive electrode and active carbon as the negative one. The active carbon electrode was made of active carbon, acetylene black, and polyvinylidene fluoride with N-methylpyrrolidone as the solvent in a mass ratio of 7:2:1. The slurry was evenly coated on the nickel foam. The active carbon electrode was vacuum dried for 24 h at 60 °C. The electrolyte of ASCs was PVA-KOH. The preparation process was as follows: 3 g PVA and 3 g KOH were mixed in 30 mL deionized water, and the mixture was heated in an 80 °C water bath for 1 h and stirred continuously until clear.

3. Results and Discussion

The $NiMoO_4@MnCo_2O_4$ composite electrode was synthesized by a two-step hydrothermal method, as shown in Figure 1. Firstly, $NiMoO_4$ precursor is grown on nickel foam. Secondly, $NiMoO_4$ can be obtained by calcination. Thirdly, the nano needle-like $MnCo_2O_4$ precursor was coated on $NiMoO_4$ by the second hydrothermal preparation. Finally, the samples were calcined to obtain $NiMoO_4@MnCo_2O_4$ on nickel foam.

As seen from the XRD results of $NiMoO_4$, $MnCo_2O_4$ and $NiMoO_4@MnCo_2O_4$ electrode materials, it can be observed that the three strong peaks are diffraction peaks of the foamed nickel substrate in Figure 2. When 2θ values are 26.57°, 29.14°, 33.73° and 60.01°, the crystal planes correspond to (220), (310), (22$\bar{2}$) and (060). The crystal structure is consistent with that of $NiMoO_4$ (JCPDS No. 45-0142). Meanwhile, the values of 2θ are 30.53°, 35.99°, 57.90° and 63.62° and the diffraction peaks correspond to (220), (311), (511) and (440) crystal planes, which is consistent with the crystal structure of $MnCo_2O_4$ (JCPDS No. 23–1237). Therefore, the diffraction peaks of $NiMoO_4@MnCo_2O_4$ electrode material prepared under the condition of the best ratio correspond to the diffraction peaks of a single compound.

Figure 3 shows the morphologies of $NiMoO_4$, $MnCo_2O_4$ and $NiMoO_4@MnCo_2O_4$ electrode materials. As seen from Figure 3a,b, the $NiMoO_4$ electrode material is nano pompon-like, and there are many intersecting nano needle-like structures densely growing on the nickel foam substrate. As shown in Figure 3c,d, $MnCo_2O_4$ electrode material possesses a nano needle-like structure and uniformly grows on the nickel foam substrate. Figure 3e,f show the micromorphology of $NiMoO_4@MnCo_2O_4$ electrode material. It can be observed that a large number of uniformly distributed nano needle-like $MnCo_2O_4$ and nano pompon-like $NiMoO_4$ grow together to form a uniform and orderly arrangement of nano urchin-like morphology, which increases the specific surface area of $NiMoO_4$ electrode and presents a great deal of active sites for rapid transfer between ions and active substances. The gap between the nano needle-like structures allows sufficient Faraday chemical reactions between the active substance and electrolyte, which enhances the electrochemical storage performance. Figure 3g,h show TEM images of NiMoO$_4$@MnCo$_2$O$_4$ electrode material. Figure 3g exhibits the morphology after the composite of NiMoO$_4$ and MnCo$_2$O$_4$. It can be seen from Figure 3h that NiMoO$_4$@MnCo$_2$O$_4$ composite material shows two kinds of lattice fringes; the lattice fringes with the spacing of 0.154 nm correspond to the (060) crystal plane of NiMoO$_4$, and the lattice fringes with the spacing of 0.146 nm correspond to the (440) crystal plane of MnCo$_2$O$_4$. From the stable microstructure of NiMoO$_4$@MnCo$_2$O$_4$, it can be inferred that the composite has multiple ion and electron transport channels and a larger specific surface area, therefore it is beneficial to shorten the ion diffusion path, which makes it advantageous for high storage capacity and rate capacity.

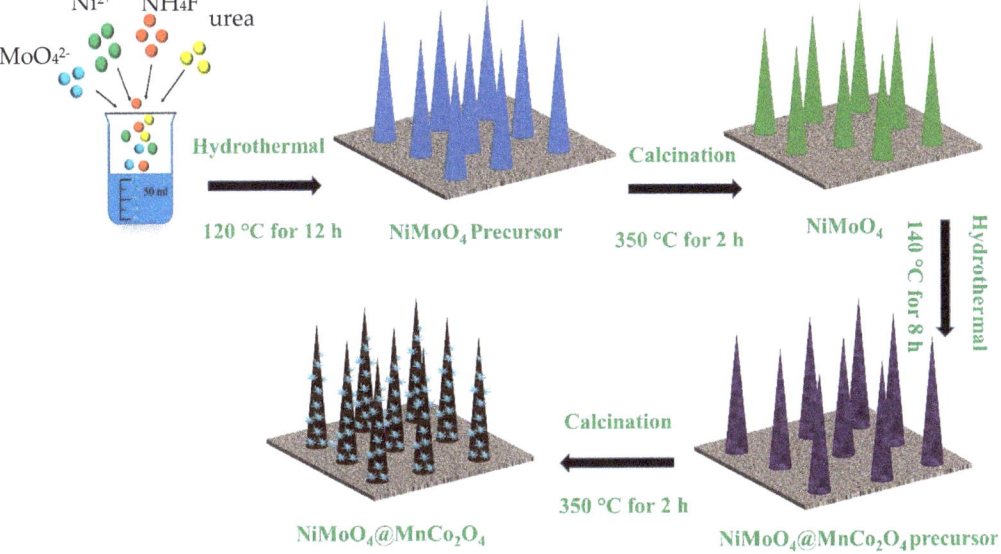

Figure 1. Synthesis schematic of NiMoO$_4$@MnCo$_2$O$_4$ composite electrode.

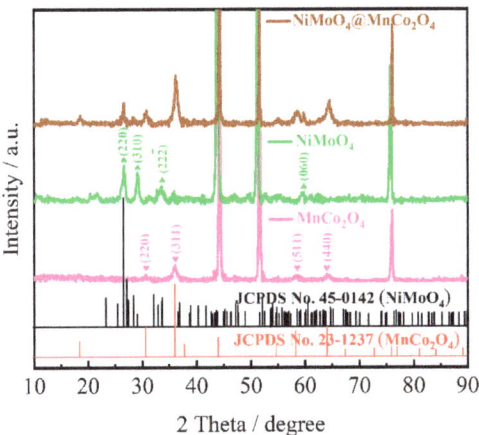

Figure 2. XRD patterns of NiMoO$_4$, MnCo$_2$O$_4$ and NiMoO$_4$@MnCo$_2$O$_4$ electrode materials.

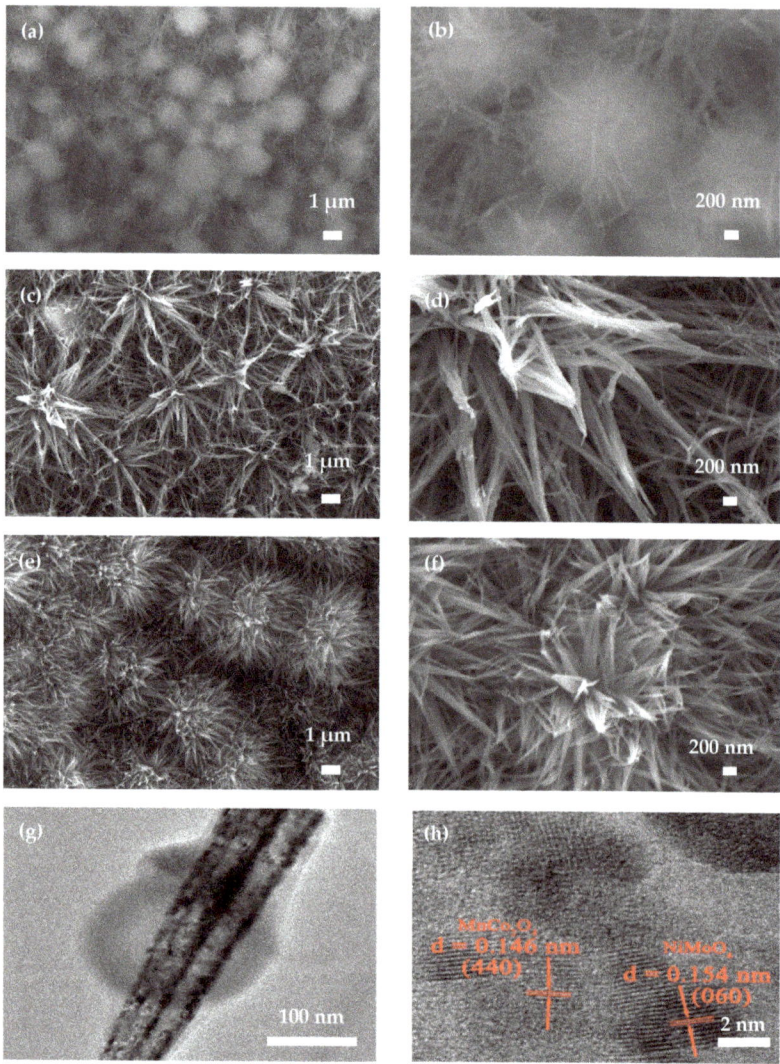

Figure 3. (**a–f**) Microstructure of NiMoO$_4$, MnCo$_2$O$_4$ and NiMoO$_4$@MnCo$_2$O$_4$ electrode materials at different multiples; (**g,h**) TEM of NiMoO$_4$@MnCo$_2$O$_4$ electrode material.

In order to further investigate the elemental component and different valence states of the prepared NiMoO$_4$@MnCo$_2$O$_4$ composite, XPS tests were carried out on the samples. Figure 4a presents the full measurement scanning spectrum showing the presence of Mn 2p, Co 2p, Mo 3d, Ni 2p, O 1s and C 1s, among which O 1s and C 1s elements are mixed impurities in the test process. In order to identify the detailed valence states of Mn, the high resolution XPS spectrum is present in Figure 4b. The Mn 2p$_{3/2}$ and Mn 2p$_{1/2}$ are found in the two main peaks, respectively, which can be divided into four peaks after fine fitting. The two peaks with a binding energy of 641.4 eV and 652.9 eV can be ascribed to the presence of Mn^{2+}. The peaks corresponding to Mn^{3+} are distributed with a binding energy of 644.6 eV and 654.2 eV, respectively. Meanwhile, there is a satellite peak (defined as "Sat.") at a position with a binding energy of 644.6 eV. According to the Co 2p spectrum of Figure 4c, it was found that two peaks appear at 780 eV and 795.3 eV, corresponding to the two excitation spectra of Co $2p_{3/2}$ and Co $2p_{1/2}$. The diffraction peaks corresponding to Co^{2+} have a binding energy of 781.5 eV and 797.3 eV, respectively. The diffraction peaks corresponding to Co^{3+} have a binding energy of 779.9 eV and 795.2 eV, respectively. In Figure 4d, the peaks of Mo 3d spectrum at 231.6 eV and 234.8 eV belong to Mo $3d_{5/2}$ and Mo $3d_{3/2}$, respectively. In Figure 4e, Ni 2p spectra can be well fitted into two main peaks, characterized by Ni^{2+} and Ni^{3+} oxidation states. Each peak has its own satellite peak (defined as "Sat.") at 861.6 eV and 879.9 eV, respectively. Two fitting peaks at 855.1 eV (Ni $2p_{3/2}$) and 872.9 eV (Ni $2p_{1/2}$) belong to Ni^{2+}, and two fitting peaks at 855.9 eV (Ni $2p_{3/2}$) and 873.8 eV (Ni $2p_{1/2}$) belong to Ni^{3+}. Figure 4f shows the O 1s region, which can be divided into two peaks (529.8 eV and 531.8 eV). For the binding energy of 529.8 eV, it is attributed to the formation of M-O bond (M=Co, Mn). Therefore, XPS data confirm that the synthesis of $NiMoO_4@MnCo_2O_4$ is successful [29–31].

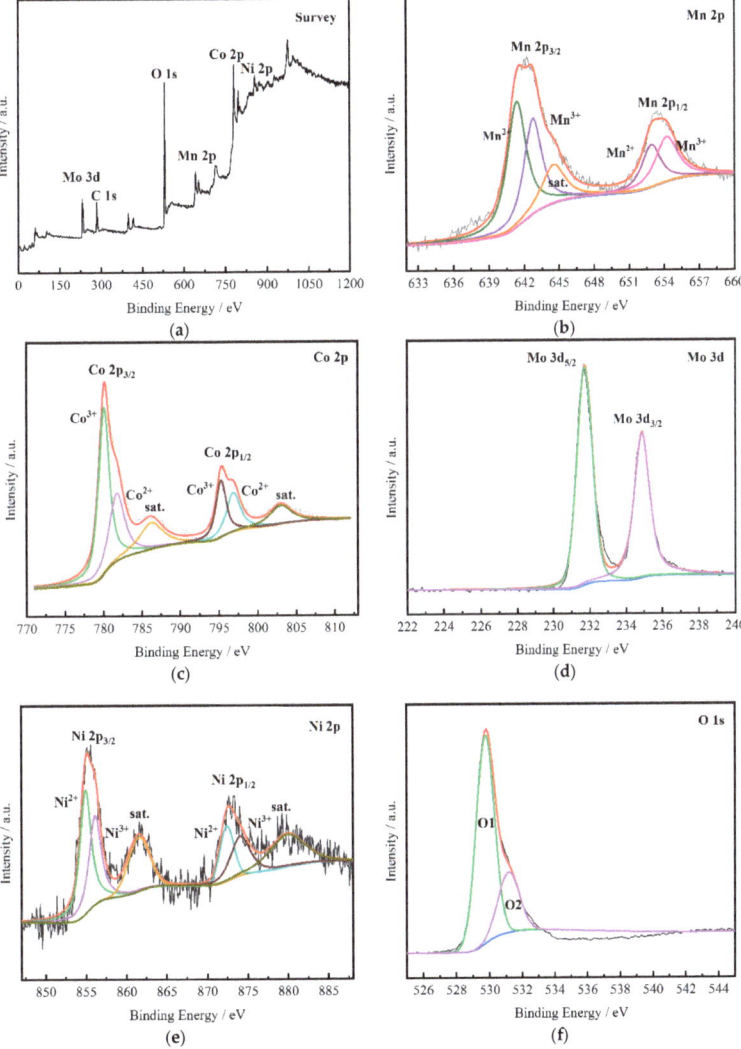

Figure 4. XPS diagram of $NiMoO_4@MnCo_2O_4$ electrode material: (**a**) Full measurement spectrum; (**b**) Mn 2p; (**c**) Co 2p; (**d**) Mo 3d; (**e**) Ni 2p; (**f**) O 1s.

Figure 5a shows the cyclic voltammetry (CV) curves of NiMoO$_4$@MnCo$_2$O$_4$ electrode material, which is measured by a scanning rate of 10–100 mV/s and a voltage window of 0–0.5 V, showing excellent rate performance. The visible redox peaks are seen from the curves, indicating that redox reaction occurs in the process of energy storage. Figure 5b presents the galvanostatic charge–discharge (GCD) curves with current density of 1, 2, 4, 8, and 10 mA/cm^2, the areal capacitance is 3000, 1076, 964, 696, and 580 mF/cm^2, respectively. The high electrochemical performance is mainly attributed to the nano urchin-like morphology of the material. The nano needle-like structure densely and uniformly distributed on the urchin-like surface provides a larger surface area for electrolyte contact, thus improving the electrochemical performance of the composite.

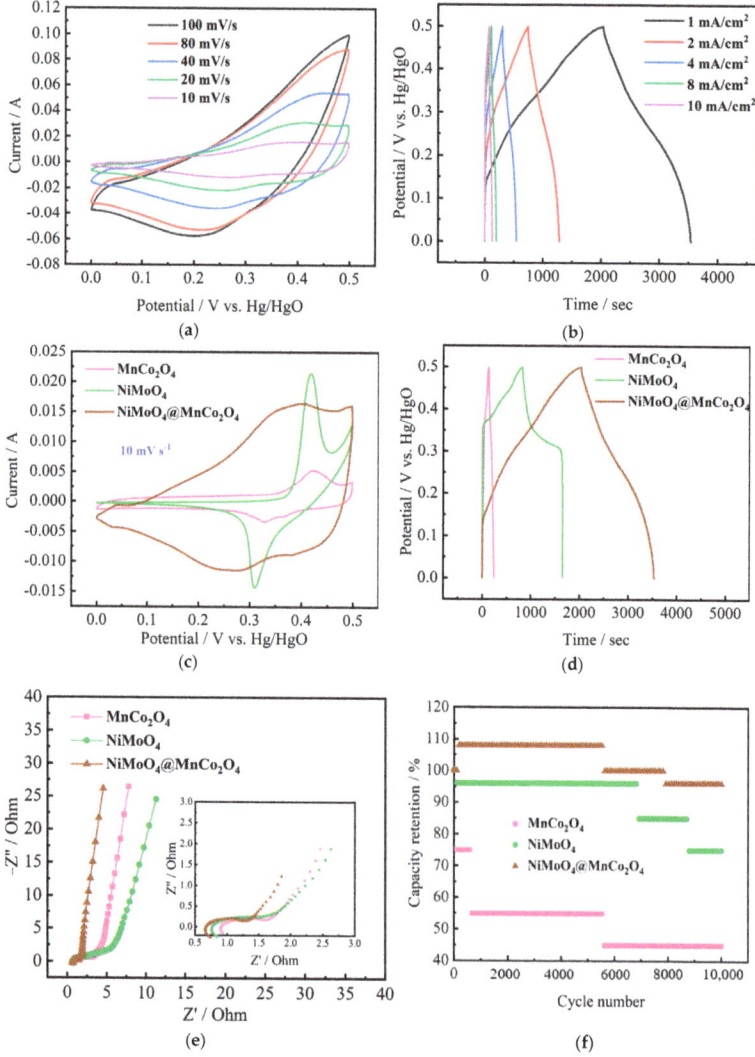

Figure 5. Electrochemical tests of three electrode materials: (**a**) CV curves of NiMoO$_4$@MnCo$_2$O$_4$; (**b**) GCD curves of NiMoO$_4$@MnCo$_2$O$_4$; (**c**) CV curves of three electrode materials; (**d**) GCD curves of the three electrode materials; (**e**) EIS curves of the three electrode materials (inset is the high-frequency region); (**f**) Long cycle curves of the three electrode materials.

In order to show the advantages of the composite electrode, $NiMoO_4$, $MnCo_2O_4$ and $NiMoO_4@MnCo_2O_4$ electrode materials are used as working electrodes, respectively, and necessary tests are carried out in a three-system with 3 M KOH solution. Studies have shown that the capacitance of $NiMoO_4$ in an alkaline environment is mainly attributed to the reversible redox reaction between the valence states of Ni element, while Mo element does not participate in any reaction, but it helps to improve the conductivity of molybdate. Figure 5c reveals the CV curves of $NiMoO_4$, $MnCo_2O_4$ and $NiMoO_4@MnCo_2O_4$ electrodes at 10 mV/s. Visible redox peaks can be seen from the curves. By comparing the three CV curves, it is obviously observed that the $NiMoO_4@MnCo_2O_4$ electrode has a larger integral area than $NiMoO_4$ and $MnCo_2O_4$ electrode, so it has a larger specific capacitance. These excellent electrochemical properties can be credited to the singular nano urchin-like structure and a series of redox reactions, which not only involve Co^{2+} and Mn^{2+}, but also come from Ni^{2+}, thus increasing the redox peak. The specific redox reaction mechanism is as follows:

$$NiMoO_4: NiMoO_4 \rightarrow Ni^{2+} + MoO_4{}^{2-} \tag{4}$$

$$Ni^{2+} + 2OH^- \rightarrow Ni(OH)_2 \tag{5}$$

$$Ni(OH)_2 + OH^- \rightarrow NiOOH + H_2O + e^- \tag{6}$$

$$MnCo_2O_4: MnCo_2O_4 + H_2O + OH^- \rightarrow MnOOH + 2CoOOH + e^- \tag{7}$$

$$MnOOH + OH^- \rightarrow MnO_2 + H_2O + e^- \tag{8}$$

$$CoOOH + OH^- \rightarrow CoO_2 + H_2O + e^- \tag{9}$$

Figure 5d shows the GCD curves of $NiMoO_4$, $MnCo_2O_4$ and $NiMoO_4@MnCo_2O_4$ composite electrode material measured at the current density of 1 mA/cm^2. It is observed that the charge and discharge time of $NiMoO_4@MnCo_2O_4$ composite electrode material is the longest, which corresponds to the maximum CV curve area of $NiMoO_4@MnCo_2O_4$ in Figure 5c. By calculation, the specific capacitances of the three electrodes can reach 1656, 224 and 3000 mF/cm^2. The specific capacitance of $NiMoO_4@MnCo_2O_4$ is compared, as shown in Table 1, which is higher than that of some previous literatures [32–36]. The charging–discharging time of $NiMoO_4@MnCo_2O_4$ composite electrode material is the longest, and the symmetry of the charging and discharging cycle indicates that the electrode has excellent reversibility. The capacitance performance is attributed to the nano urchin-like morphology of the material, which provides a larger electrolyte contact area. Therefore, the electrochemical properties of composite electrode material are improved. To further explore the charge transfer ability of the prepared electrodes, EIS measurements were carried out, as shown in Figure 5e. The inset exhibits that compared with two single electrodes, in the high frequency region, the $NiMoO_4@MnCo_2O_4$ sample has a smaller semicircle arc and x-axis intercept, which represents the charge transfer resistance (R_{ct}) and solution resistance (R_s), indicating that the composite has a faster ion-electron transfer rate at the electrode and electrolyte interface, and smaller intrinsic resistance. The corresponding R_s values of $NiMoO_4$, $MnCo_2O_4$, and $NiMoO_4@MnCo_2O_4$ are 0.91, 0.77 and 0.67 Ω, respectively. In the low frequency region, the composite material shows the higher straight-line slope, which accounts for faster electrolyte ion mobility. Cycling performance (10 mA cm^{-2}) of the as-prepared electrodes is displayed in Figure 5f. Compared with $NiMoO_4$ (75%) and $MnCo_2O_4$ (45%), $NiMoO_4@MnCo_2O_4$ (96%) shows a better cycling lifespan after undergoing the charging–discharging process 10,000 times.

In order to study the application of $NiMoO_4@MnCo_2O_4$ in SCs, the positive electrode and negative electrode of ASCs are $NiMoO_4@MnCo_2O_4$ electrode and active carbon (AC) electrode, respectively. Figure 6 shows the electrochemical curves of the assembled device. Figure 5a shows the CV curves at the scanning rate of 100 mV/s. The voltage windows of the device are 1.1 V, 1.2 V, 1.3 V, 1.4 V, 1.5 V and 1.6 V, respectively. The shapes of all curves are nearly the same, indicating that the device can operate at 1.1 V–1.6 V and the maximum voltage window can reach 1.6 V at the same time. Figure 6b shows the CV curves of $NiMoO_4@MnCo_2O_4//AC$ at scanning rates of 5–100 mV/s. With the increase in scanning rate, the shapes of the CV curves increase, which is mainly attributed to the synergy between materials. These curves have obvious redox peaks, indicating that the asymmetric SCs have pseudocapacitance characteristics. Meanwhile, with increasing scanning rate, the integral area of the curves is enhanced. The GCD curves with different current densities are shown in Figure 6c, which indicates that the linear trend of the curve is obvious at high current densities. The voltage window is 1.5 V, and the surface capacitance of the device can be calculated according to the formula. When the current densities are 1, 2, 4, 8 and 10 mA/cm^2, the surface capacitances are 58.53, 22.73, 12.13, 1.9 and 1.13 mF/cm^2, respectively. Figure 6d shows the charge transfer characteristics of the prepared electrode studied by EIS test. The slope is larger in the low frequency region, indicating that the diffusion resistance of the assembled asymmetric SC is lower. The inset shows the R_s value is only 1 Ω. Figure 6e shows the long cycling test with 10,000 times at 10 mA cm^{-2} and coulombic efficiency. The capacity retention rate of the assembled asymmetric SC is 78.4%. The decrease in capacity may be due to the morphology damage caused by long-term redox reaction of electrode materials, which reduces the potential activity of the surface of the material. The coulombic efficiency of ASCs keeps nearly 100% during 10,000 charging–discharging tests. From Figure 6f, the Ragone plot offers an expression of the trend of the energy density with the corresponding power density. Importantly, the maximum energy density of the $NiMoO_4@MnCo_2O_4//AC$ device reaches 90.89 mWh/cm^3 at the power density of 3726.7 mW/cm^3, which is better than some reported devices [37–41].

Table 1. Electrochemical performance comparison of $NiMoO_4@MnCo_2O_4$ with previous literatures.

Materials	Capacity	Current Density	Electrolyte	Capacitance Retention	Ref.
$NiCo_2O_4/rGO/NiO$	2.644 F cm^{-2}	1 mA cm^{-2}	3 M KOH	97.5% (3000 cycles)	[32]
Fe_2O_3/Fe dendrite	2.166 F cm^{-2}	1 mA cm^{-2}	1 M KOH	90% (1000 cycles)	[33]
$NiCo_2O_4/C$	2.057 F cm^{-2}	1 mA cm^{-2}	2 M KOH	81% (10,000 cycles)	[34]
rGO/PPy	0.807 F cm^{-2}	1 mA cm^{-2}	1 M H_2SO_4	78% (2000 cycles)	[35]
$C@MnNiCo-OH/Ni_3S_2$	2.332 F cm^{-2}	1 mA cm^{-2}	3 M KOH	89.45% (5000 cycles)	[36]
$NiMoO_4@MnCo_2O_4$	3 F cm^{-2}	1 mA cm^{-2}	3 M KOH	96% (10,000 cycles)	This work

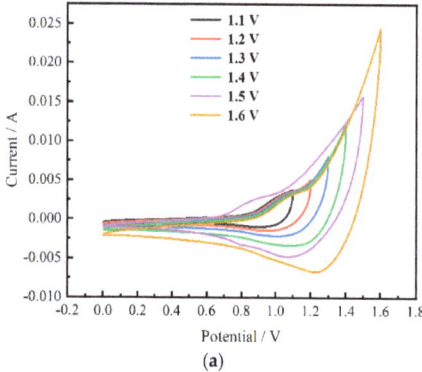

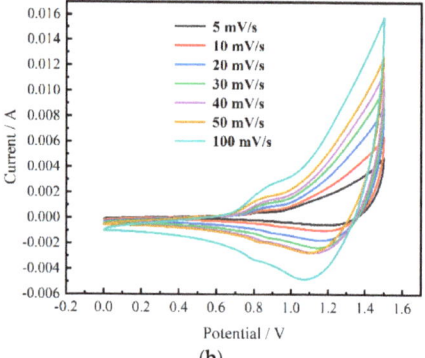

Figure 6. *Cont.*

Nanomaterials **2022**, *12*, 1674

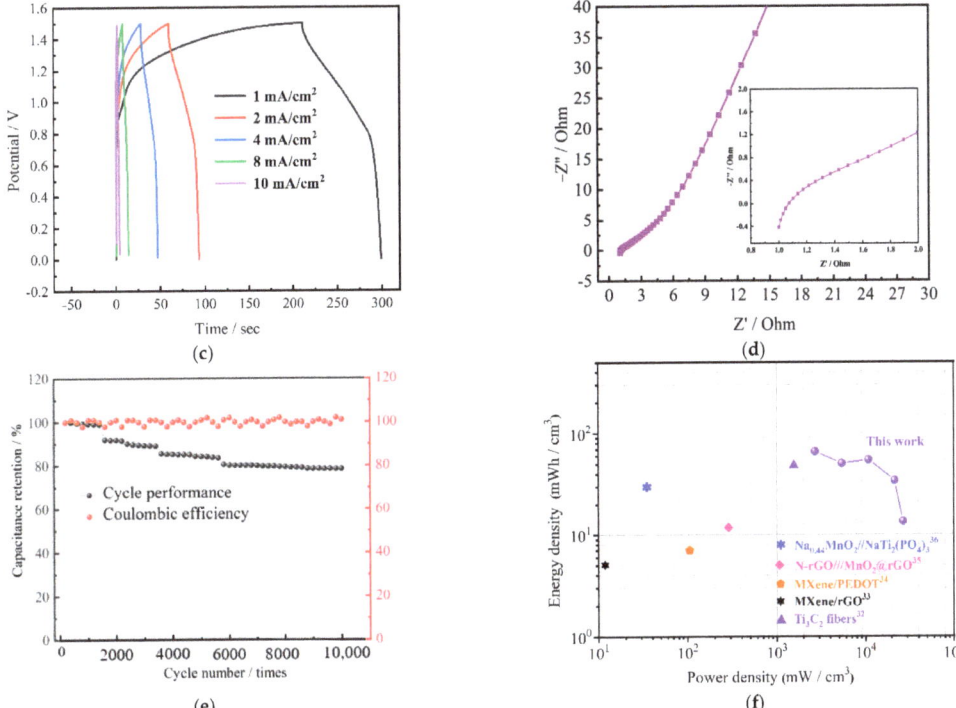

Figure 6. Electrochemical testing of NiMoO₄@MnCo₂O₄ composite assembled devices: (**a**) Cyclic voltammetry curves under different voltage windows; (**b**) Cyclic voltammetry curves of different scanning speeds; (**c**) GCD curves with different current densities; (**d**) Impedance diagram (inset is the high-frequency region); (**e**) Cycling stability and coulombic efficiency; (**f**) Ragone plots.

4. Conclusions

A new type of NiMoO₄@MnCo₂O₄ composite electrode material has been successfully prepared on nickel foam by the two-step hydrothermal method, and its phase structures, micromorphology and electrochemical properties are characterized and analyzed. Due to the synergistic effect between the NiMoO₄ nano pompon-like structure and MnCo₂O₄ nano needle-like structure, the prepared nano urchin-like NiMoO₄@MnCo₂O₄ core–shell nanostructure presents good pseudocapacitance properties. NiMoO₄@MnCo₂O₄ samples show better electrochemical performance than single NiMoO₄ or MnCo₂O₄ electrode materials, which exhibit a high specific capacitance of 3000 mF/cm². After 10,000 cycles, the capacity retention rate is 96%. In addition, the NiMoO₄@MnCo₂O₄//AC assembled device delivers a high energy density of 90.89 mWh/cm³ at a power density of 3726.7 mW/cm³.

Author Contributions: Conceptualization, Y.D. and N.B.; methodology, W.Y.; software, N.B. and S.L.; validation, W.Y., S.L. and J.X.; formal analysis, F.W.; investigation, R.Z.; resources, R.Z.; data curation, N.B.; writing—original draft preparation, Y.D.; writing—review and editing, W.Y.; visualization, X.S. and Z.W.; supervision, F.W.; project administration, J.X.; funding acquisition, J.X. All authors have read and agreed to the published version of the manuscript.

Funding: This research was funded by [Jun Xiang] grant number [51702143], and the APC was funded by [the National Natural Science Foundation of China]. This research was funded by [Fufa Wu] grant number [51971106], and the APC was funded by [the National Natural Science Foundation of China]. This research was funded by [Jun Xiang] grant number [2021-MS-320], and the APC was funded by [the Natural Science Foundation of Liaoning Province]. This research was funded by [Rongda Zhao] grant number [2019-MS-171], and the APC was funded by [the Natural Science Foundation of

Liaoning Province]. This research was funded by [Jun Xiang] grant number [LJKQZ2021145], and the APC was funded by [the Youth Project of Education Department of Liaoning Province].

Acknowledgments: This work was funded by the National Natural Science Foundation of China (NSFC, Grant Nos. 51702143 and 51971106), the Natural Science Foundation of Liaoning Province (Grant Nos. 2021-MS-320 and 2019-MS-171), the Youth Project of Education Department of Liaoning Province (No. LJKQZ2021145).

Conflicts of Interest: The authors declare no conflict of interest.

References

1. Simon, P.; Gogotsi, Y.; Dunn, B. Where do batteries end and supercapacitors begin. *Science* **2014**, *343*, 1210–1211. [CrossRef] [PubMed]
2. Borysiewicz, M.A.; Ekielski, M.; Ogorzałek, Z.; Wzorek, M.; Kaczmarsk, J.; Wojciechowski, T. Highly transparent supercapacitors based on ZnO/MnO$_2$ nanostructures. *Nanoscale* **2017**, *9*, 7577–7589. [CrossRef] [PubMed]
3. Jin, E.M.; Lim, J.G.; Jeong, S.M. Facile synthesis of graphene-wrapped CNT-MnO$_2$, nanocomposites for asymmetric electrochemical capacitors. *J. Ind. Eng. Chem.* **2017**, *54*, 421–427. [CrossRef]
4. Li, Y.; Xia, Z.B.; Gong, Q.; Liu, X.H.; Yang, Y.; Chen, C.; Qian, C.H. Green synthesis of free standing cellulose/graphene oxide/polyaniline aerogel electrode for high-performance flexible all-solid-state supercapacitors. *Nanomaterials* **2020**, *10*, 1546. [CrossRef]
5. Zhao, D.P.; Dai, M.Z.; Liu, H.Q.; Duan, Z.X.; Tan, X.J.; Wu, X. Bifunctional ZnCo$_2$S$_4$@CoZn$_{13}$ hybrid electrocatalysts for high efficient overall water splitting. *J. Energy Chem.* **2022**, *69*, 292–300. [CrossRef]
6. Kannan, V.; Kim, H.J.; Park, H.C.; Kim, H.S. Single-step direct hydrothermal growth of NiMoO$_4$ nanostructured thin film on stainless steel for supercapacitor electrodes. *Nanomaterials* **2018**, *8*, 563. [CrossRef]
7. Zhao, D.P.; Dai, M.Z.; Liu, H.Q.; Zhu, X.F.; Wu, X. PPy film anchored on ZnCo$_2$O$_4$ nanowires facilitating efficient bifunctional electrocatalysis. *Mater. Today Energy* **2021**, *20*, 100637. [CrossRef]
8. Chen, S.; Yang, G.; Jia, Y.; Zheng, H.J. Three-dimensional NiCo$_2$O$_4$@ NiWO$_4$ core–shell nanowire arrays for high performance supercapacitors. *J. Mater. Chem. A* **2017**, *5*, 1028–1034. [CrossRef]
9. Cai, D.P.; Wang, D.D.; Liu, B.; Wang, Y.R.; Liu, Y.; Wang, L.L.; Li, H.; Huang, H.; Li, Q.H.; Wang, T.H. Comparison of the electrochemical performance of NiMoO$_4$, nanorods and hierarchical nanospheres for super capacitor applications. *ACS Appl. Mater. Interfaces* **2013**, *5*, 12905–12910. [CrossRef]
10. Zhao, D.P.; Zhang, R.; Dai, M.Z.; Liu, H.Q.; Jian, W.; Bai, F.Q.; Wu, X. Constructing High Efficiency CoZn$_x$Mn$_{2−x}$O$_4$ Electrocatalyst by Regulating the Electronic Structure and Surface Reconstruction. *Small* **2022**, *18*, 2107268. [CrossRef]
11. Xiao, K.; Xia, L.; Liu, G.X.; Wang, S.Q.; Ding, L.X.; Wang, H.H. Honeycomb-like NiMoO$_4$, ultrathin nanosheet arrays for high-performance electrochemical energy storage. *J. Mater. Chem. A* **2015**, *3*, 6128–6135. [CrossRef]
12. Liu, M.C.; Kang, L.; Kong, L.B.; Lu, C.; Ma, X.J.; Li, X.M.; Luo, Y.C. Facile synthesis of NiMoO$_4$·xH$_2$O nanorods as a positive electrode material for supercapacitors. *RSC Adv.* **2013**, *3*, 6472–6478. [CrossRef]
13. Wang, Z.J.; Wei, G.J.; Du, K.; Zhao, X.X.; Liu, M.; Wang, S.T.; Zhou, Y.; An, C.H.; Zhang, J. Ni foam-supported carbon-sheathed NiMoO$_4$, nanowires as integrated electrode for high-performance hybrid supercapacitors. *ACS Sustain. Chem. Eng.* **2017**, *5*, 5964–5971. [CrossRef]
14. Zhang, Z.Q.; Zhang, H.D.; Zhang, X.Y.; Yu, D.Y.; Ji, Y.; Sun, Q.S.; Wang, Y.; Liu, X.Y. Facile synthesis of hierarchical CoMoO$_4$@NiMoO$_4$, core-shell nanosheet arrays on nickel foam as an advanced electrode for asymmetric supercapacitors. *J. Mater. Chem. A* **2016**, *4*, 18578–18584. [CrossRef]
15. Xuan, H.C.; Wang, R.; Yang, J.; Zhang, G.H.; Liang, X.H.; Li, Y.P.; Xie, Z.G.; Han, P.D. Synthesis of NiMoO$_4$@Co$_3$O$_4$ hierarchical nanostructure arrays on reduced graphene oxide/Ni foam as binder-free electrode for asymmetric supercapacitor. *J. Mater. Sci.* **2021**, *56*, 9419–9433. [CrossRef]
16. Thiagarajan, K.; Bavani, T.; Arunachalam, P.; Lee, S.J.; Theerthagiri, J.; Madhavan, J.; Pollet, B.G.; Choi, M.Y. Nanofiber NiMoO$_4$/g-C$_3$N$_4$ composite electrode materials for redox supercapacitor applications. *Nanomaterials* **2020**, *10*, 392. [CrossRef]
17. Zhao, Y.Y.; Zhang, P.; Fu, W.B.; Ma, X.W.; Zhou, J.Y.; Zhang, X.J.; Li, J.; Xie, E.Q.; Pan, X.J. Understanding the role of Co$_3$O$_4$ on stability between active hierarchies and scaffolds: An insight into NiMoO$_4$ composites for supercapacitors. *Appl. Surf. Sci.* **2017**, *416*, 160–167. [CrossRef]
18. Veerasubramani, G.K.; Krishnamoorthy, K.; Sivaprakasam, R.; Kim, S.J. Sonochemical synthesis, characterization, and electrochemical properties of MnMoO$_4$ nanorods for supercapacitor applications. *Mater. Chem. Phys.* **2014**, *147*, 836–842. [CrossRef]
19. Cheng, B.B.; Zhang, W.; Yang, M.; Zhang, Y.J.; Meng, F.B. Preparation and study of porous MnCo$_2$O$_4$@NiO nanosheets for high-performance supercapacitor. *Ceram. Int.* **2019**, *45*, 20451–20457. [CrossRef]
20. Liu, H.Y.; Guo, Z.X.; Wang, S.B.; Xun, X.C.; Chen, D.; Lian, J.S. Reduced core-shell structured MnCo$_2$O$_4$@ MnO$_2$ nanosheet arrays with oxygen vacancies grown on Ni foam for enhanced-performance supercapacitors. *J. Alloy. Compd.* **2020**, *846*, 156504. [CrossRef]
21. Jiang, H.; Ma, J.; Li, C.Z. Mesoporous carbon incorporated metal oxide nanomaterials as supercapacitor electrodes. *Adv. Mater.* **2012**, *24*, 4197–4202. [CrossRef]

22. Yan, J.; Fan, Z.J.; Wei, T.; Cheng, J.; Shao, B.; Wang, K.; Song, L.P.; Zhang, M.L. Carbon nanotube/MnO_2 composites synthesized by microwave-assisted method for supercapacitors with high power and energy densities. *J. Power Sources* **2009**, *194*, 1202–1207. [CrossRef]

23. Pasero, D.; Reeves, N.; West, A.R. Co-doped Mn_3O_4: A possible anode material for lithium batteries. *J. Power Sources* **2005**, *141*, 156–158. [CrossRef]

24. Lv, J.L.; Hideo, M.; Yang, M. A novel mesoporous $NiMoO_4$@rGO nanostructure for supercapacitor applications. *Mater. Lett.* **2017**, *194*, 94–97.

25. Senthilkumar, B.; Sankar, K.V.; Selvan, R.K.; Danielle, M.; Manickam, M. Nano α-$NiMoO_4$ as a new electrode for electrochemical supercapacitors. *RSC Adv.* **2013**, *3*, 352–357. [CrossRef]

26. Moosavifard, S.E.; Shamsi, J.; Ayazpour, M. 2D high-ordered nanoporous $NiMoO_4$ for high-performance supercapacitors. *Ceram. Int.* **2015**, *41*, 1831–1837. [CrossRef]

27. Chen, H.; Yu, L.; Zhang, J.M.; Liu, C.P. Construction of hierarchical $NiMoO_4$@MnO_2 nanosheet arrays on titanium mesh for supercapacitor electrodes. *Ceram. Int.* **2016**, *42*, 18058–18063. [CrossRef]

28. Chen, T.T.; Wang, G.N.; Ning, Q.Y. Rationally designed three-dimensional $NiMoO_4$/polypyrrole core-shell nanostructures for high-performance supercapacitors. *Nano* **2017**, *12*, 1750061. [CrossRef]

29. Dai, M.Z.; Zhao, D.P.; Liu, H.Q.; Zhu, X.F.; Wang, X.; Wang, B. Nanohybridization of Ni–Co–S nanosheets with $ZnCo_2O_4$ nanowires as supercapacitor electrodes with long cycling stabilities. *ACS Appl. Energy Mater.* **2021**, *4*, 2637–2643. [CrossRef]

30. Liu, H.Q.; Zhao, D.P.; Dai, M.Z.; Zhu, X.F.; Qu, F.Y.; Umar, A.; Wu, X. PEDOT decorated $CoNi_2S_4$ nanosheets electrode as bifunctional electrocatalyst for enhanced electrocatalysis. *Chem. Eng. J.* **2022**, *428*, 131183. [CrossRef]

31. Liu, H.Q.; Zhao, D.P.; Liu, Y.; Tong, Y.L.; Wu, X.; Shen, G.Z. NiMoCo layered double hydroxides for electrocatalyst and supercapacitor electrode. *Sci. China Mater.* **2021**, *64*, 581–591. [CrossRef]

32. Li, D.L.; Gong, Y.N.; Wang, M.S.; Pan, C.X. Preparation of sandwich-like $NiCo_2O_4$/rGO/NiO heterostructure on nickel foam for high-performance supercapacitor electrodes. *Nano-Micro. Lett.* **2017**, *9*, 16. [CrossRef] [PubMed]

33. Zhang, X.L.; Liu, R.L. Construction of coral shaped $Ni(OH)_2$/Ni dendrite and Fe_2O_3/Fe dendrite electrodes for cable-shaped energy storage devices. *J. Electrochemical. Soc.* **2019**, *166*, A2636–A2642. [CrossRef]

34. Li, Y.Y.; Wang, Q.Y.; Shao, J.; Li, K.; Zhao, W.W. $NiCo_2O_4$/C Core-shell nanoneedles on Ni Foam for all-solid state asymmetric supercapacitors. *ChemistrySelect* **2020**, *5*, 5501–5506. [CrossRef]

35. Yang, C.; Zhang, L.L.; Hu, N.T.; Yang, Z.; Wei, H.; Zhang, Y.F. Reduced graphene oxide/polypyrrole nanotube papers for flexible all-solid-state supercapacitors with excellent rate capability and high energy density. *J. Power Sources* **2016**, *302*, 39–45. [CrossRef]

36. Zhao, H.Q.; Wang, J.M.; Sui, Y.W.; Wei, F.X.; Qi, J.Q.; Meng, Q.K.; Ren, Y.J.; He, Y.Z. Construction of layered C@MnNiCo-OH/Ni_3S_2 core-shell heterostructure with enhanced electrochemical performance for asymmetric supercapacitor. *J. Mater. Sci. Mater. Electron.* **2021**, *32*, 11145–11157. [CrossRef]

37. Zhu, C.; Geng, F.X. Macroscopic MXene ribbon with oriented sheet stacking for high-performance flexible supercapacitors. *Carbon Energy* **2020**, *3*, 142–152. [CrossRef]

38. Seyedin, S.; Yanza, E.R.S.; Razal, J.M. Knittable energy storing fiber with high volumetric performance made from predominantly MXene nanosheets. *J. Mater. Chem. A* **2017**, *5*, 24076–24082. [CrossRef]

39. Zhang, J.Z.; Seyedin, S.; Qin, S.; Wang, Z.Y.; Moradi, S.; Yang, F.L.; Lynch, P.A.; Yang, W.R.; Liu, J.Q.; Wang, X.G.; et al. Highly conductive $Ti_3C_2T_x$ MXene hybrid fibers for flexible and elastic fiber-shaped supercapacitors. *Small* **2019**, *15*, 1804732. [CrossRef]

40. Yu, D.S.; Goh, K.L.; Zhang, Q.; Li, W.; Wang, H.; Jiang, W.C.; Chen, Y. Controlled functionalization of carbonaceous fibers for asymmetric solid-state micro-supercapacitors with high volumetric energy density. *Adv. Mater.* **2014**, *26*, 6790–6797. [CrossRef]

41. Guo, Z.W.; Zhao, Y.; Ding, Y.X.; Dong, X.L.; Chen, L.; Cao, J.Y.; Wang, C.C.; Xia, Y.Y.; Peng, H.S.; Wang, Y.G. Multi-functional flexible aqueous sodium-ion batteries with high safety. *Chem* **2017**, *3*, 348–362. [CrossRef]

 *nanomaterials*

Article

A Flexible TENG Based on Micro-Structure Film for Speed Skating Techniques Monitoring and Biomechanical Energy Harvesting

Zhuo Lu [1], Changjun Jia [2,*], Xu Yang [3], Yongsheng Zhu [2], Fengxin Sun [2], Tianming Zhao [4], Shouwei Zhang [1,*] and Yupeng Mao [1,2,*]

[1] School of Physical Education, Northeast Normal University, Changchun 130024, China; luz560@nenu.edu.cn
[2] Physical Education Department, Northeastern University, Shenyang 110819, China; 2001276@stu.neu.edu.cn (Y.Z.); 2171435@stu.neu.edu.cn (F.S.)
[3] Changchun Polytechnic Tourism School, Changchun 130022, China; yangxu0316@126.com
[4] College of Sciences, Northeastern University, Shenyang 110819, China; zhaotm@stumail.neu.edu.cn
* Correspondence: 2071367@stu.neu.edu.cn (C.J.); zhangsw178@nenu.edu.cn (S.Z.); maoyupeng@pe.neu.edu.cn (Y.M.)

Abstract: Wearable motion-monitoring systems have been widely used in recent years. However, the battery energy storage problem of traditional wearable devices limits the development of human sports training applications. In this paper, a self-powered and portable micro-structure triboelectric nanogenerator (MS-TENG) has been made. It consists of micro-structure polydimethylsiloxane (PDMS) film, fluorinated ethylene propylene (FEP) film, and lithium chloride polyacrylamide (LiCl-PAAM) hydrogel. Through the micro-structure, the voltage of the MS-TENG can be improved by 7 times. The MS-TENG provides outstanding sensing properties: maximum output voltage of 74 V, angular sensitivity of 1.016 V/degree, high signal-to-noise ratio, and excellent long-term service stability. We used it to monitor the running skills of speed skaters. It can also store the biomechanical energy which is generated in the process of speed skating through capacitors. It demonstrates capability of sensor to power electronic calculator and electronic watch. In addition, as a flexible electrode hydrogel, it can readily stretch over 1300%, which can help improve the service life and work stability of MS-TENG. Therefore, MS-TENG has great application potential in human sports training monitoring and big data analysis.

Keywords: self-powered; wearable flexible sensor; energy harvesting; human motion monitoring; triboelectric nanogenerator

 check for updates

Citation: Lu, Z.; Jia, C.; Yang, X.; Zhu, Y.; Sun, F.; Zhao, T.; Zhang, S.; Mao, Y. A Flexible TENG Based on Micro-Structure Film for Speed Skating Techniques Monitoring and Biomechanical Energy Harvesting. *Nanomaterials* **2022**, *12*, 1576. https://doi.org/10.3390/nano12091576

Academic Editor: Christian M. Julien

Received: 17 March 2022
Accepted: 5 May 2022
Published: 6 May 2022

Publisher's Note: MDPI stays neutral with regard to jurisdictional claims in published maps and institutional affiliations.

1. Introduction

In the Beijing Winter Olympic Games, a total of 14 gold medals were won in speed skating, including 10 Olympic records and 1 world record. The good results of athletes are inseparable from scientific training. Among this, the data support provided by sports training monitoring for speed skating training and competition is the key link. The speed skating competition is fierce. Speed skaters wear 2 mm-wide skates to complete high-speed skating on the ice. High-quality athletes' technical motions are the fundamental guarantee of this high-speed movement. The athletes' physical agility is the main influencing factor of their competitive ability and excellent performance. Moreover, the physical agility decline caused by technical instability exists in most athletes' training and competition [1–6]. High speed cameras and inertial sensors have been used to monitor the changes in athletes' real-time technical motions during taxiing [7–10]. However, the accuracy of motion monitoring is limited by the large space demand, complex circuit, and large battery volume of cameras, inertial sensors, and portable sensors. Therefore, it is an urgent problem to develop a portable, economical, self-powered, and real-time motion monitoring sensor to assist the development of speed skating.

In recent years, the Zhong Lin Wang team invented the triboelectric nanogenerator (TENG) based on Maxwell displacement current theory [11–13], which has been widely discussed and developed rapidly. This technology has a great application potential in the fields of blue energy, self-powered systems and portable sensors [14–19]. TENG mainly consists of two different materials [20–23]. It can convert low-frequency mechanical energy from surroundings into electrical energy, such as human motion mechanical energy [24–29]. Due to the electrical signal being closely related to the surroundings, TENG seems to be an ideal candidate for motion monitoring. Unfortunately, common TENG with a metal electrode can be easily destroyed and uncomfortable to wear, so it cannot be further applied to biological systems [30,31]. Chen et al. propose a kind of hydrogel with high conductivity, transports and flexibility [32,33]. Combing with conductive hydrogel and surface modification with micro-structure to fabricate TENG, proposed by Zhao et al. [34–36], the sensitivity and response of the self-powered sensor would be dramatically enhanced.

In this work, we develop a micro-structure triboelectric nanogenerator (MS-TENG). It consists of micro-structure polydimethylsiloxane (PDMS) film, fluorinated ethylene propylene (FEP) film, and lithium chloride polyacrylamide (LiCl-PAAM) hydrogel (Figure 1). Through introduction of micro-structure on dielectric surface, the output voltage of the MS-TENG can be improved by 7 times. In our experiment, MS-TENG can be attached to the athlete's body surface easily and it can collect the technical motion information accurately (movement structure, bending angle and frequency). The triboelectric signal can not only be used as biosensor signal, but also can power microelectronics. In addition, by replacing metal electrodes with hydrogel, the response, stability, lifetime and comfort level have been improved. Therefore, MS-TENG can be applied to sports training monitoring and big data analysis of speed skating or other sports. As a new generation of motion-monitoring equipment, it has great application potential.

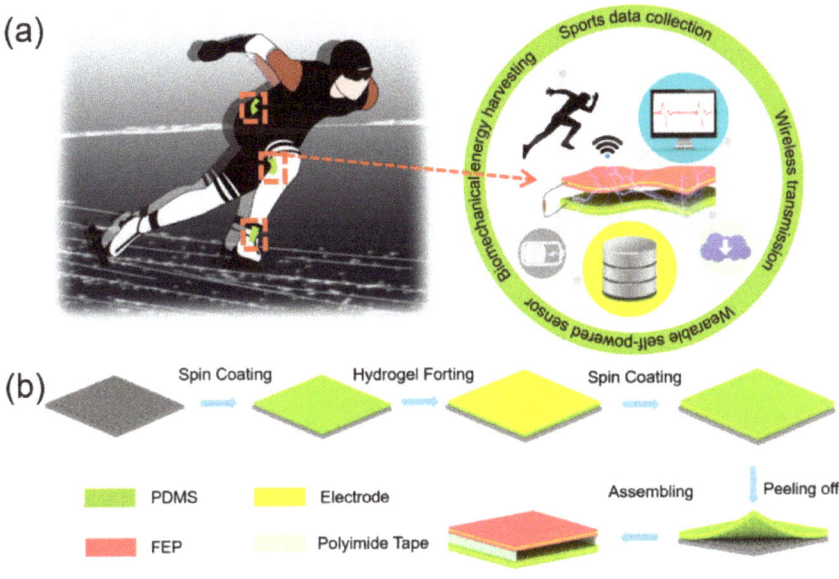

Figure 1. *Cont.*

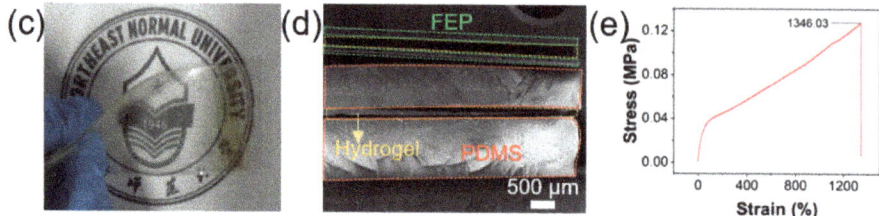

Figure 1. (a) The application scenarios of MS-TENG; (b) the fabricating process of the MS-TENG; (c) the optical image of MS-TENG at bending state; (d) the SEM image of the MS-PDMS; (e) tensile strength testing of hydrogel.

2. Materials and Methods

2.1. Materials

Fluorinated ethylene propylene and Polyimide tape were purchased from Zeyou plastic Co., Ltd. (Suzhou, China). DOW CORNING Sylgard 184 was purchased from Xinheng Trading Co., Ltd. (Tianjin, China). N,N-dimethylformamide (DMF), Acrylamide (AM), Lithium chloride (LiCl), N,N'-methylene diacrylamide (MBA), Ammonium persulphate (APS), and N,N,N',N'-tetramethylethylenediamine (TMEDA) were purchased from Jintong letai chemical industry products Co., Ltd. (Beijing, China).

2.2. Methods

Synthesis of lithium chloride polyacrylamide hydrogel pre-solution: AM was used as the monomer, MBA was used as the crosslinking agent, and APS was used as the initiator. The whole reaction was processed under room temperature. The specific steps were as follows: AM powder and LiCl particles were dissolved in 50 mL deionized water at a speed of 500 rpm, wherein the concentrations of AM and LiCl were 3 mol/L and 5 mol/L, respectively. After continuous magnetic stirring for 10 min, MBA and APS were added to the solution, and the molar ratios of MBA and APS to AM monomer were 0.02 and 0.03 mol% respectively. Then, the particles were stirred until they were dissolved completely, and then it was kept for 1 h to obtain the pre-solution.

Preparation of micro-structure PDMS triboelectric layer: The PDMS mixture of base and crosslinker (the weight ratio of base to cross linker was 10:1) was stirred for at least 20 min and degassed in vacuum for 10 min to remove air bubbles at room temperature. PDMS mixture was spin-coated (900 rpm, 20 s) on a silicon mold with microstructure, and cured at 80 °C for 1 h. A few drops of TMEDA was added in the uniform hydrogel solution which is used as an accelerator. Subsequently, the hydrogel solution was spin-coated on PDMS. After the hydrogel was solidified, the PDMS mixture was spin-coated on the above hydrogel membrane again. After the PDMS was solidified, PDMS films with micro-structure were obtained by peeling off the sandwich PDMS from the Si mold surface carefully.

Manufacture of triboelectric nanogenerator: The FEP triboelectric layer with sandwich structure was composed of FEP film and LiCl-PAAM hydrogel. Finally, the double-electrode TENG with microstructure (MS-TENG) consisted of PDMS triboelectric layer, FEP triboelectric layer, and polyimide (50 μm). Polyimide was used as spacer layer, which provided space for two triboelectric layers.

2.3. Characterization and Measurement

The MS-TENG was fixed on the stepping motor to simulate joint movement. The different amplitudes and frequencies were used to hit sensors repeatedly and periodically and triboelectric single was generated by the MS-TNEG. Signals were collected by oscilloscopes (sto 1102 c, Shenzhen, China). The morphology and structure of the sensor were carried out by an optical microscope (Sunshine Instrument Co., Ltd., SDPTOP-CX 40m, Ningbo, China).

3. Results and Discussion

To achieve accurate, reliable, and convenient assessment of the technical movements of speed skaters, conformal and real-time measurement of athlete's joint and articular chains is necessary. We proposed a self-powered and flexible sensor (MS-TENG) which consists of PDMS elastomer, FEP film, and ionic conductive hydrogel. According to MS-TENG output signals, a coach can adjust a skater's technical movements and develop a suitable plan for an athlete, so that they can scientifically and systematically enhance athlete's performance. As shown in Figure 1a, MS-TENG can attach to the joints of the skater flexibly. Based on the triboelectric effect, the output signals of MS-TENG are sensing signals. It can collect the information of athletes' joint bending angle, movement frequency and movement structure, and it provides the basis for big data analysis. We have made a comparison between the existing articles in the field of manufacturing sensors that we have referred to in this research. The results of this comparison are shown in Table S1 [37–44]. Compared with other works, the MS-TENG has the advantages of self-powered, soft, and high-outputting properties. Hydrogel has been used as flexible electrode, which improves the service life and working stability of MS-TENG. The application value of the sensor has been verified. The manufacture process of MS-TENG is shown in Figure 1b. In brief, the PDMS mixture is spin-coated on a silicon mold. After curing, the complementary structure of the epidermis pattern is uniformly transferred from the silicon mold to the PDMS. Later, the hydrogel pre-solution with TMEDA is spin-coated on the bottom PDMS layer, and then the PDMS mixture is spin-coated above the hydrogel again. After curing, a microstructure PDMS triboelectric layer can be obtained. Finally, the PDMS triboelectric layer and FEP triboelectric layer are assembled together by polyimide tape. Figure 1c is an optical image of the MS-TENG at bending state. It shows the flexible, soft, and thin characteristics of MS-TENG. Figure 1d shows the cross-sectional scanning electron microscope (SEM) image of MS-TENG which clearly shows the structure of MS-TENG. Figure S1 shows the SEM images of PDMS, FEP, and hydrogel, respectively. The microstructure of PDMS surface is shown in Figure S1a. Since MS-PDMS has a large effective contact area and higher surface energy. It leads to more charge accumulation on the contact surface, higher potential, and better electrical performance. Figure S2 shows the Fourier-transform infrared (FTIR) spectrum of the FEP, PDMS, and hydrogel. Then we investigated the mechanical properties of hydrogel, which is an important parameter for practical applications. The tensile strength measurement process of hydrogel is shown in Figure S3. As shown in Figure 1e, hydrogel can stretch over 1300%, which compared with traditional metal electrode, hydrogel electrode has excellent flexibility [45,46]. In addition, this work expands the application of hydrogels in other fields [47–49].

The working mechanism of the MS-TENG is schematically illustrated in Figure 2a. In the original state (Figure 2a(I)), charge transfer does not take place before the triboelectric materials contacts. When a pressure force is applied to MS-TENG (Figure 2a(II)) charge transfer takes places at the interface between PDMS and FEP, due to the differences of the electronegativity [50]. Since the surface electron affinity of PDMS is higher than FEP, electrons transfer from the FEP surface to PDMS surface, leaving equal positive triboelectric charges on the FEP surface. When a pressure force disappears, PDMS and PEF begin to separate and the electrons transfer from the top electrode to the bottom electrode via external circuit due to the electrostatic force (Figure 2a(III)). As shown in Figure 2a(IV), the MS-TENG reaches the equilibrium state, and electrons do not transfer from the top electrode to the bottom electrode anymore. Finally, when the pressure force appears again, the electrons transfer from bottom electrode to top electrode via external circuit due to the electrostatic force (Figure 2a(V)), outputting a reversing electrical signal. Therefore, the AC electricity can be continuously generated by periodical contact-separation between PDMS film and FEP film. In order to understand the working mechanism of the MS-TENG, the visualized simulation via COMSOL software is shown in Figure 2b, and the corresponding simulated output electric potential is depicted by color variation. We measure the peak voltage of MS-TENG and without microstructure TENG under variable applied force from

0 to 50 N (as shown in Figure 2c). It shows that the two peak voltages increase with the increase of force, but the MS-TENG is more sensitive to response of force. The output performance of MS-TENG under different load resistances is shown in Figure 2d. The output voltage increases with the load resistance increasing. Instantaneous electric power is the 11 µW at 9 MΩ. Meanwhile, we tested the resistance of the hydrogel. When the hydrogel is stretched from 1 cm to 15 cm, the resistance increases rapidly and then stabilizes gradually, the maximum resistance reaches 564.2 kΩ (Figure S4). Even if the resistance of hydrogel changes with the stretching, the inherent resistance of MS-TENG is much larger than that of the hydrogel. Therefore, the change of hydrogel resistance does not affect the output voltage. In order to explore the characteristics of the TENG generator, the energy conversion efficiency of the MS-TENG is investigated. The efficiency of the MS-TENG is defined as the ratio between the input mechanical energy and the generated electrical energy delivered to the load. The formula of energy conversion efficiency are as follows:

$$\eta = \frac{E_{ele}}{W_{total}} = \frac{\int I^2 R dt}{W_G} \tag{1}$$

where E_{ele} and W_{total} stand for the electric energy and the total work done by the ambient, W_G represents the work done by gravity. As shown in Figure S5, E_{ele} shows an energy output that is measured under the best matched load (9 MΩ). According to calculation, the energy conversion efficiency of TENG is 0.08%. Figure 2e shows the voltage wave of the MS-TENG, when the pressing/releasing speed is from 5 to 20 cm/s. It can be observed that the peak voltage decreases with the pressing/releasing speed decreasing also, and the pulse width increases gradually with the pressing/releasing speed decreasing.

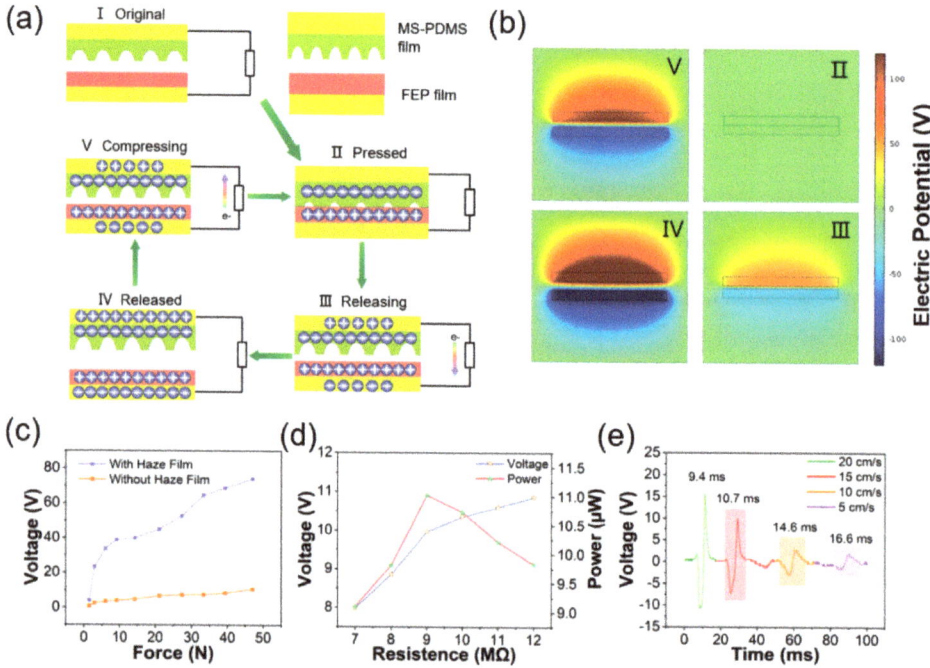

Figure 2. (**a**) Working principle of the MS-TENG; (**b**) FEA simulation of the MS-TENG electric potential distribution; (**c**) peak voltage of MS-TENG and without microstructure TENG under variable applied force from 0 to 50 N; (**d**) the output performance of MS-TENG under different load resistances; (**e**) the voltage wave of MS-TENG under speed of 5, 10, 15, and 20 cm/s.

The physiological structure of the human body determines the working mode of the upper and lower limbs. Combined with torso coordination and cooperation, various forms of human movement are formed. With the body movement forms changing, many movements are formed such as push, pull, stretch, swing, among others. Further, many motion of human can drive TENG to work. Before practical application, it is necessary to study the effects of different mechanical stimuli on the output of MS-TENG to prove its practicability. Figure 3a is a system of MS-TENG monitoring body joint movements. We manufactured a MS-TENG which size is 8×3 cm^2. The stepping motor with programmable system and slide rail simulates joint movement to apply different deformations to the sensor. All the measurements are carried out at room temperature (22 °C) and 25% relative humidity. The output triboelectric voltage of the MS-TENG at the same frequency (1 Hz) and different bending angles (as shown in Figure 3b). When the angles are 168, 166, 164, and 162°, the output triboelectric voltage is 13.8, 15, 17.56, and 19.27 V, respectively, and the output voltage increases with the bending angle increasing. In order to show the sensitivity of MS-TENG, the linear relationship between bending angles and output voltages is shown in Figure S6. The red line is a linear fit. The linear fitting of Formula (2) is as follows:

$$y = 184.16 - 1.016x \tag{2}$$

where y represents the triboelectric voltage (V) and x represents the bending angle (degree). The linearity is up to 0.99. Figure 3c shows the relationship between output triboelectric voltage and frequency. When the bending angle is 160°, the frequencies are 1, 1.5, 2, and 2.5 Hz, and the output triboelectric voltages are 1.85, 1.88, 1.86, and 1.92 V respectively. Figure 3d shows the response of MS-TENG at different bending angles and frequencies. The response of MS-TENG can be calculated from the following equation:

$$\% = \left| \frac{V_0 - V_i}{V_i} \right| \times 100\%, \tag{3}$$

where V_0 and V_i are the outputting voltage of 168° (first data) and other voltages. The response of MS-TENG is 0, 8, 21.4, and 30% when it is in different bending angles, and when the frequency is 1, 1.5, 2, and 2.5 Hz, the response of MS-TENG is 0, 0, 0, and 0%. These data indicate that MS-TENG can monitor the joint change of angle and frequency accurately, and these data are used be big data analysis to enhance athlete's sports technology. The durability of MS-TENG is shown in Figure 3e. After many tests, the output is almost constant (~22 V). The excellent durability and high output power of the MS-TENG shows the potential of practical application in the future.

(a)

Figure 3. *Cont.*

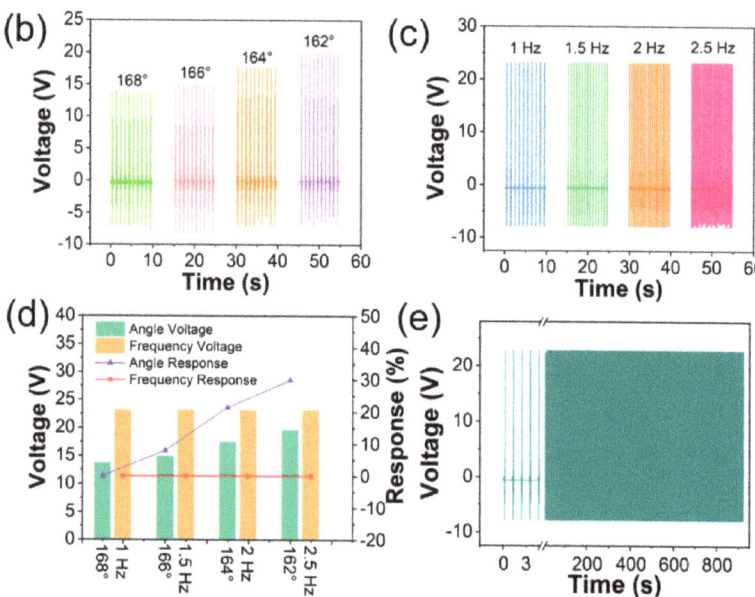

Figure 3. (**a**) MS-TENG monitors the body joint motion system; (**b**) outputting triboelectric voltage of MS-TENG at different bend angles; (**c**) outputting triboelectric voltage of MS-TENG at different frequencies; (**d**) response of output triboelectric voltage of MS-TENG at different bending angles and frequencies; (**e**) durability property of MS-TENG.

On the basis of the superior performance of electrical output and splendid sensing property to force, MS-TENG can be used to monitor skaters' motion techniques. The human movement system consists of bones, joints, and muscles. By using the flexibility of MS-TENG, it can be attached to the joints of the skater flexibly. With flexion and extension of the targeted joints, MS-TENG would then be compressed and released, converting mechanical signals into voltage signals simultaneously (Figure 4a). The oscilloscope synchronously collects the voltage signals of MS-TENG which is attached to the ankle, knee, and coxa of athlete 1 (as shown in Figure 4b). All the above sensors are 8×3 cm^2 in size. The detailed collection process is shown in Movies S1–S3. In addition, athlete 2 also performed the same motion test, and the voltage signal is as shown in Figure S7. The results are summarized in Table 1. At the same joint motion, the output voltage of athlete 1 is higher, but his variance is large. To sum up, it shows that athlete 1 is a strength-type player, and his technical stability needs to be improved. Athlete 2 is a technique-type player, and his strength needs to be enhanced. Speed skating is a periodic event, and it is a fitness and technique sport. Speed skaters possess great physical strength and excellent technique. According to the monitoring results, the coach can arrange technical training for athlete 1 appropriately, so that athlete 1 can form the correct motion concept and maintain the good stability, thus he can improve his competitive ability. The coach can arrange strength training load for athlete 2 to ensure that he can adapt to the load requirements of the competition. To avoid being thrown off the track, the athlete can adjust his barycenter at curve-skating and push off the ice with the outside edge of his left skate and the inside edge of his right skate, which can keep his body tilted toward the center of the circle. In this state, the athlete uses the centripetal force that is formed by the supporting reaction force of the body's barycenter and the resultant force of body gravity to counter the centrifugal force which is generated by circular motion. Figure 4c is the output voltage of MS-TENG which is attached to left/right coxa when athlete simulates curve-skating at different inclination angles. The incline angle formed by the body, and the ice is closely related to the athlete's speed at curve-skating. The lower the incline angle, the smaller the skating radius, and the faster the speed. At high, moderate, and low inclination angles, the voltage of left coxa is 1.16, 1.54, and 1.78 V, and voltage of the right coxa is 1.01, 1.29, and 2 V. The result shows that the lower the incline angle, the higher the voltage. Figure 4d,f shows the output voltage of MS-TENG attached to ankle when two athletes simulate straight-skating. Detailed drawings of voltage curves are shown in Figure 4e,g. It shows that two athletes do leg extension-ankle extension-leg retraction movements. Because the sole of the Clap skate has a special hinge device (Figure S8), the heel of the skate can be separated from the skate. Therefore, athletes can take an action to extend their ankles in the process of pushing off the ice. This allows the edge to stay in contact with the ice longer; thus, it is important to improve the athlete pushing-off effect. As shown in Figure 4g, three signal waves correspond to the three movements of athlete 2's leg-extension–ankle-extension–leg-retraction movements. However, only two signal waves of athlete 1's leg-extension–leg-retraction can be observed in Figure 4e, and the above results show that the technical action of athlete 1 needs to be improved. With the rapid progress of science, the mobile phone has become a necessary tool in people's lives. People collect a lot of information through mobile phones. If mobile phones can collect the motion monitoring information, it would be more convenient to monitor the motion. Therefore, a wireless sensor system consisting of a flexible MS-TENG, a digital multimeter with a Bluetooth module and a mobile phone was established to verify the feasibility of human motion monitoring (Movie S4). Through digital multimeter transmitting the flexible-sensor-collected signals, an app in the mobile phone can monitor the voltage change in real time. Through analysis of these data, the technical information of athletes can be learned, which could provide quantifiable, objective, accurate, and reliable support in sports training.

Table 1. Comparison of athletes' data.

	Athlete 1	Athlete 2
Average voltage of ankle	3.28 V	1.78 V
Variance of ankle voltage	4.817	0.203
Average voltage of knee	4.17 V	0.62 V
Variance of knee voltage	0.555	0.0135
Average voltage of coxa	0.9 V	0.45 V
Variance of coxa voltage	0.168	0.002

The mechanical energy which is generated by human motion belongs to good-quality renewable forms of energy, because it is not limited by time, place, or other objective factors. meanwhile it is sustainable and easily accessible. The MS-TENG can be used to harvest biomechanical energy which is generated by human motions. Figure 5a shows the equivalent circuit of the self-charge system. The electrical energy output is from MS-TENG which can be stored in an energy storage device (such as a capacitor) to power electronic devices. Figure 3c shows the voltage–time curve that MS-TNEG charges different capacitors. It can charge 1, 3.3, 4.7, and 10 μF capacitors to 3.7, 2.6, 1.8, and 1 V when the frequency is 5 Hz for 35 s. MS-TENG charging a 4.7 μF capacitor is shown in Movie S5. As shown in Figure 5c,d, an electronic calculator and an electronic watch can work about 15 s after hitting the sensor (Movies S6 and S7). These demonstrations indicate that the MS-TENG has great potential in a fully self-powered and sustainable electronic system.

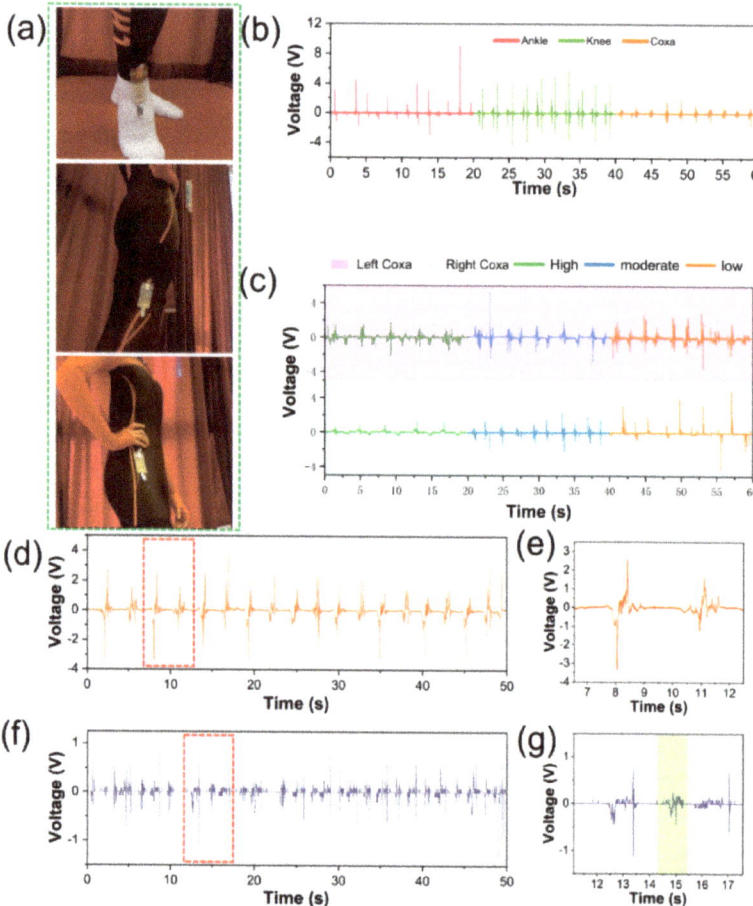

Figure 4. (**a**) Images of the MS-TENG attached to the ankle, knee, and coxa; (**b**) the output triboelectric voltage of MS-TENG attached to athlete 1's ankle, knee, and coxa; (**c**) the output triboelectric voltage of MS-TENG attached to left/right coxa when athlete simulates curve-skating at different inclination angles; (**d**) the output triboelectric voltage of MS-TENG attached to ankle when athlete 1 simulates straight-skating; (**e**) athlete 1—detailed drawings of voltage curves simulating straight-skating; (**f**) the output triboelectric voltage of MS-TENG attached to ankle when athlete 2 simulates straight-skating; (**g**) athlete 2—detailed drawings of voltage curves simulating straight-skating.

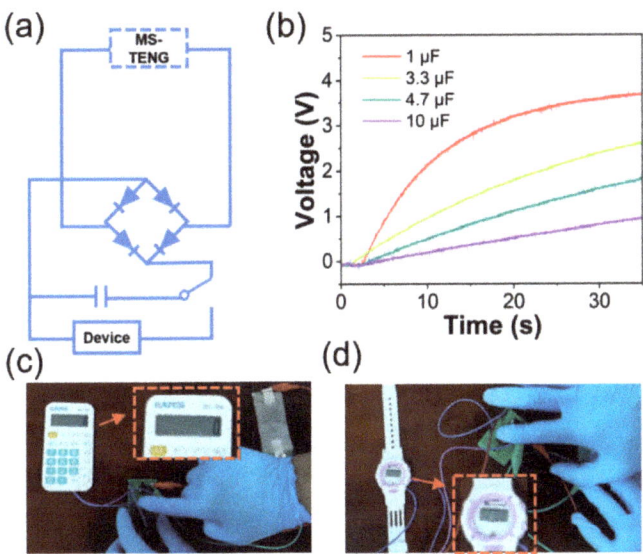

Figure 5. (**a**) The equivalent circuit of a self-powered system; (**b**) charging voltage of different capacitor which is charged by MS-TENG; (**c**) powering for an electronic calculator; (**d**) powering for an electronic watch.

4. Conclusions

In summary, a flexible TENG based on a micro-structure (MS-TENG) is fabricated with a facile and low-cost fabrication method. Moreover, the fabrication method can be used as a universal strategy for improving the output of TENG. Through the micro-structure, the voltage of the MS-TENG can be im-proved by 7 times. We prepared a hydrogel to replace a traditional electrode to overcome the vulnerability of traditional metallic electrode in TENG during long-term service with large deformation. The MS-TENG provides outstanding sensing properties: maximum output voltage of 74 V, angular sensitivity of 1.016 V/degree, high signal-to-noise ratio and excellent long-term service stability. We used it to monitor the running skills of speed skaters. Based on the triboelectric effect, it can accurately convert the technical action information (such as motion, bending angle, and frequency) of athletes in training into triboelectric signals for outputting. Moreover, there is no external power supply for the whole process. In addition, MS-TENG can collect energy from human mechanical motion to drive small electronic devices (such as electronic calculators and electronic watches). The self-powered sensing and sustainable energy conversion realized by MS-TENG show its potential as a new generation of motion-monitoring equipment.

Supplementary Materials: The following supporting information can be downloaded at: https://www.mdpi.com/article/10.3390/nano12091576/s1, Figure S1: Scanning electron microscope (SEM) images of the PDMS (a), FEP (b), and hydrogel (c), respectively; Figure S2: Fourier-transform infrared (FTIR) spectrum of the PDMS, FEP, and hydrogel; Figure S3: The tensile strength measurement process of hydrogel; Figure S4: The electrical conductivity of hydrogel; Figure S5: Output current of the MS-TENG at a load resistance of 9 MΩ; Figure S6: The linear relationship of angles and voltages; Figure S7: The output triboelectric voltage of MS-TENG attached to athlete 2's ankle, knee, and coxa; Figure S8: The Clap skate with a hinge device; Table S1: The self-powered flexible sensor comparison with other works; Movie S1: The output signal of the ankle is collected; Movie S2: The output signal of the knee is collected; Movie S3: The output signal of the coxa is collected; Movie S4: The wireless monitoring system consisting of MS-TENG, Bluetooth multimeter, and mobile phone; Movie S5: MS-TENG charges a 4.7 μF capacitor; Movie S6: Powering for an electronic calculator; Movie S7: Powering for an electronic watch.

Author Contributions: The manuscript was written through the contributions of all authors. Z.L., C.J., S.Z. and Y.M. designed, fabricated, and tested the MS-TENG. The data was collected, sorted out and analyzed by Y.Z. and F.S. X.Y. and T.Z. made the verification and visualization. Z.L. and C.J. wrote the manuscript. Z.L., S.Z. and Y.M. finished the writing (comments and editors). All authors have read and agreed to the published version of the manuscript.

Funding: Research and demonstration on key technologies of National Scientific Training Base Construction. 2018YFF0300806; Winter Sports Training monitoring technical services. 2020021300002.

Institutional Review Board Statement: Not applicable.

Informed Consent Statement: Not applicable.

Data Availability Statement: The data presented in this study are available in Supplementary Materials.

Acknowledgments: The authors would like to thank the collaboration of all volunteers who participated in data collection.

Conflicts of Interest: The authors declare no conflict of interest. The funders had no role in the design of the study; in the collection, analyses, or interpretation of data; in the writing of the manuscript, or in the decision to publish the results.

References

1. Spiteri, T.; McIntyre, F.; Specos, C.; Myszka, S. Cognitive Training for Agility: The Integration Between Perception and Action. *Strength Cond. J.* **2018**, *40*, 39–46. [CrossRef]
2. Guimaraes, E.; Baxter-Jones, A.D.G.; Williams, A.M.; Tavares, F.; Janeira, M.A.; Maia, J. Tracking Technical Skill Development in Young Basketball Players: The INEX Study. *Int. J. Environ. Res. Public Health* **2021**, *18*, 4094. [CrossRef] [PubMed]
3. Alesi, M.; Bianco, A.; Luppina, G.; Palma, A.; Pepi, A. Improving Children's Coordinative Skills and Executive Functions: The Effects of a Football Exercise Program. *Percept. Mot. Ski.* **2016**, *122*, 27–46. [CrossRef] [PubMed]
4. Liang, Y.P.; Kuo, Y.L.; Hsu, H.C.; Hsia, Y.Y.; Hsu, Y.W.; Tsai, Y.J. Collegiate baseball players with more optimal functional movement patterns demonstrate better athletic performance in speed and agility. *J. Sport Sci* **2019**, *37*, 544–552. [CrossRef] [PubMed]
5. Schwesig, R.; Laudner, K.G.; Delank, K.S.; Brill, R.; Schulze, S. Relationship between Ice Hockey-Specific Complex Test (IHCT) and Match Performance. *Appl. Sci.* **2021**, *11*, 3080. [CrossRef]
6. Slater, L.V.; Vriner, M.; Zapalo, P.; Arbour, K.; Hart, J.M. Difference in Agility, Strength, and Flexibility in Competitive Figure Skaters Based on Level of Expertise and Skating Discipline. *J. Strength Cond. Res.* **2016**, *30*, 3321–3328. [CrossRef]
7. Oguchi, T.; Ae, M.; Schwameder, H. Step characteristics of international-level skeleton athletes in the starting phase of official races. *Sports Biomech.* **2021**, 1–14, *online ahead of print.* [CrossRef]
8. Kim, K.; Kim, J.S.; Purevsuren, T.; Khuyagbaatar, B.; Lee, S.; Kim, Y.H. New method to evaluate three-dimensional push-off angle during short-track speed skating using wearable inertial measurement unit sensors. *Proc. Inst. Mech. Eng. Part H J. Eng. Med.* **2019**, *233*, 476–480. [CrossRef]
9. Tomita, Y.; Iizuka, T.; Irisawa, K.; Imura, S. Detection of Movement Events of Long-Track Speed Skating Using Wearable Inertial Sensors. *Sensors* **2021**, *21*, 3649. [CrossRef]
10. Purevsuren, T.; Khuyagbaatar, B.; Kim, K.; Kim, Y.H. Investigation of Knee Joint Forces and Moments during Short-Track Speed Skating Using Wearable Motion Analysis System. *Int. J. Precis Eng. Man* **2018**, *19*, 1055–1060.
11. Wang, Z.L. On Maxwell's displacement current for energy and sensors: The origin of nanogenerators. *Mater. Today* **2017**, *20*, 74–82. [CrossRef]
12. Niu, S.M.; Wang, S.H.; Lin, L.; Liu, Y.; Zhou, Y.S.; Hu, Y.F.; Wang, Z.L. Theoretical study of contact-mode triboelectric nanogenerators as an effective power source. *Energy Environ. Sci.* **2013**, *6*, 3576–3583. [CrossRef]
13. Liu, Y.; Niu, S.M.; Wang, Z.L. Theory of Tribotronics. *Adv. Electron. Mater.* **2015**, *1*, 1500124. [CrossRef]
14. Li, X.J.; Luo, J.J.; Han, K.; Shi, X.; Ren, Z.W.; Xi, Y.; Ying, Y.B.; Ping, J.F.; Wang, Z.L. Stimulation of ambient energy generated electric field on crop plant growth. *Nat. Food* **2022**, *3*, 133–142. [CrossRef]
15. Wu, C.S.; Wang, A.C.; Ding, W.B.; Guo, H.Y.; Wang, Z.L. Triboelectric Nanogenerator: A Foundation of the Energy for the New Era. *Adv. Energy Mater.* **2019**, *9*, 1802906. [CrossRef]
16. Wang, Z.L. Triboelectric Nanogenerator (TENG)-Sparking an Energy and Sensor Revolution. *Adv. Energy Mater.* **2020**, *10*, 2000137. [CrossRef]
17. Zhao, T.M.; Fu, Y.M.; Sun, C.X.; Zhao, X.S.; Jiao, C.X.; Du, A.; Wang, Q.; Mao, Y.P.; Liu, B.D. Wearable biosensors for real-time sweat analysis and body motion capture based on stretchable fiber-based triboelectric nanogenerators. *Biosens. Bioelectron.* **2022**, *205*, 114115. [CrossRef]

18. Lu, Z.; Zhu, Y.S.; Jia, C.J.; Zhao, T.M.; Bian, M.Y.; Jia, C.F.; Zhang, Y.Q.; Mao, Y.P. A Self-Powered Portable Flexible Sensor of Monitoring Speed Skating Techniques. *Biosensors* **2021**, *11*, 108. [CrossRef]
19. Mao, Y.P.; Yue, W.; Zhao, T.M.; Shen, M.L.; Liu, B.; Chen, S. A Self-Powered Biosensor for Monitoring Maximal Lactate Steady State in Sport Training. *Biosensors* **2020**, *10*, 75. [CrossRef]
20. Yoo, D.; Go, E.Y.; Choi, D.; Lee, J.W.; Song, I.; Sim, J.Y.; Hwang, W.; Kim, D.S. Increased Interfacial Area between Dielectric Layer and Electrode of Triboelectric Nanogenerator toward Robustness and Boosted Energy Output. *Nanomaterials* **2019**, *9*, 71. [CrossRef]
21. Ding, Z.Y.; Zou, M.; Yao, P.; Zhu, Z.Y.; Fan, L. A Triboelectric Nanogenerator Based on Sodium Chloride Powder for Self-Powered Humidity Sensor. *Nanomaterials* **2021**, *11*, 2657. [CrossRef] [PubMed]
22. Chen, H.M.; Xu, Y.; Zhang, J.S.; Wu, W.T.; Song, G.F. Self-Powered Flexible Blood Oxygen Monitoring System Based on a Triboelectric Nanogenerator. *Nanomaterials* **2019**, *9*, 778. [CrossRef] [PubMed]
23. Tofel, P.; Castkova, K.; Riha, D.; Sobola, D.; Papez, N.; Kastyl, J.; Talu, S.; Hadas, Z. Triboelectric Response of Electrospun Stratified PVDF and PA Structures. *Nanomaterials* **2022**, *12*, 349. [CrossRef]
24. Zi, Y.L.; Guo, H.Y.; Wen, Z.; Yeh, M.H.; Hu, C.G.; Wang, Z.L. Harvesting Low-Frequency (<5 Hz) Irregular Mechanical Energy: A Possible Killer Application of Triboelectric Nanogenerator. *ACS Publ.* **2016**, *10*, 4797–4805.
25. Pu, X.; Liu, M.M.; Chen, X.Y.; Sun, J.M.; Du, C.H.; Zhang, Y.; Zhai, J.Y.; Hu, W.G.; Wang, Z.L. Ultrastretchable, transparent triboelectric nanogenerator as electronic skin for biomechanical energy harvesting and tactile sensing. *Sci. Adv.* **2017**, *3*, e1700015. [CrossRef] [PubMed]
26. Lu, X.; Zheng, L.; Zhang, H.D.; Wang, W.H.; Wang, Z.L.; Sun, C.W. Stretchable, transparent triboelectric nanogenerator as a highly sensitive self-powered sensor for driver fatigue and distraction monitoring. *Nano Energy* **2020**, *78*, 105359. [CrossRef]
27. Jing, X.; Li, H.; Mi, H.Y.; Feng, P.Y.; Tao, X.M.; Liu, Y.J.; Liu, C.T.; Shen, C.Y. Enhancing the Performance of a Stretchable and Transparent Triboelectric Nanogenerator by Optimizing the Hydrogel Ionic Electrode Property. *ACS Appl. Mater. Inter.* **2020**, *12*, 23474–23483. [CrossRef]
28. Wang, C.; Qu, X.C.; Zheng, Q.; Liu, Y.; Tan, P.C.A.; Shi, B.J.; Ouyang, H.; Chao, S.Y.; Zou, Y.; Zhao, C.C.; et al. Stretchable, Self-Healing, and Skin-Mounted Active Sensor for Multipoint Muscle Function Assessment. *ACS Nano* **2021**, *15*, 10130–10140. [CrossRef]
29. Zhu, Y.S.; Sun, F.X.; Jia, C.J.; Zhao, T.M.; Mao, Y.P. A Stretchable and Self-Healing Hybrid Nano-Generator for Human Motion Monitoring. *Nanomaterials* **2022**, *12*, 104. [CrossRef]
30. Wang, C.H.; Li, X.S.; Hu, H.J.; Zhang, L.; Huang, Z.L.; Lin, M.Y.; Zhang, Z.R.; Yin, Z.N.; Huang, B.; Gong, H.; et al. Monitoring of the central blood pressure waveform via a conformal ultrasonic device. *Nat. Biomed. Eng.* **2018**, *2*, 687–695. [CrossRef]
31. Shin, J.H.; Yan, Y.; Bai, W.B.; Xue, Y.G.; Gamble, P.; Tian, L.M.; Kandela, I.; Haney, C.R.; Spees, W.; Lee, Y.; et al. Bioresorbable pressure sensors protected with thermally grown silicon dioxide for the monitoring of chronic diseases and healing processes. *Nat. Biomed. Eng.* **2019**, *3*, 37–46. [CrossRef] [PubMed]
32. Chen, Z.S.; Yu, J.H.; Xu, M.F.; Zeng, H.Z.; Tao, K.; Wu, Z.X.; Wu, J.; Miao, J.M.; Chang, H.L.; Yuan, W.Z. Highly Deformable and Transparent Triboelectric Physiological Sensor Based on Anti-Freezing and Anti-Drying Ionic Conductive Hydrogel. In Proceedings of the 34th IEEE International Conference on Micro Electro Mechanical Systems (MEMS), Electr network, 25–29 January 2021.
33. Han, X.; Jiang, D.J.; Qu, X.C.; Bai, Y.; Cao, Y.; Luo, R.Z.; Li, Z. A Stretchable, Self-Healable Triboelectric Nanogenerator as Electronic Skin for Energy Harvesting and Tactile Sensing. *Materials* **2021**, *14*, 1689. [CrossRef] [PubMed]
34. Jiang, B.; Long, Y.; Pu, X.; Hu, W.G.; Wang, Z.L. A stretchable, harsh condition-resistant and ambient-stable hydrogel and its applications in triboelectric nanogenerator. *Nano Energy* **2021**, *86*, 106086. [CrossRef]
35. Zhao, G.R.; Zhang, Y.W.; Shi, N.; Liu, Z.R.; Zhang, X.D.; Wu, M.Q.; Pan, C.F.; Liu, H.L.; Li, L.L.; Wang, Z.L. Transparent and stretchable triboelectric nanogenerator for self-powered tactile sensing. *Nano Energy* **2019**, *59*, 302–310. [CrossRef]
36. Yu, J.B.; Hou, X.J.; He, J.; Cui, M.; Wang, C.; Geng, W.P.; Mu, J.L.; Han, B.; Chou, X.J. Ultra-flexible and high-sensitive triboelectric nanogenerator as electronic skin for self-powered human physiological signal monitoring. *Nano Energy* **2020**, *69*, 104437. [CrossRef]
37. Su, Y.J.; Chen, C.X.; Pan, H.; Yang, Y.; Chen, G.R.; Zhao, X.; Li, W.X.; Gong, Q.C.; Xie, G.Z.; Zhou, Y.H.; et al. Muscle Fibers Inspired High-Performance Piezoelectric Textiles for Wearable Physiological Monitoring. *Adv. Funct. Mater.* **2021**, *31*, 2010962. [CrossRef]
38. Cho, D.; Li, R.; Jeong, H.; Li, S.P.; Wu, C.S.; Tzavelis, A.; Yoo, S.; Kwak, S.S.; Huang, Y.G.; Rogers, J.A. Bitter Flavored, Soft Composites for Wearables Designed to Reduce Risks of Choking in Infants. *Adv. Mater.* **2021**, *33*, 2103857. [CrossRef]
39. Sriphan, S.; Vittayakorn, N. Facile roughness fabrications and their roughness effects on electrical outputs of the triboelectric nanogenerator. *Smart Mater. Struct.* **2018**, *27*, 105026. [CrossRef]
40. Kim, K.N.; Lee, J.P.; Lee, S.H.; Lee, S.C.; Baik, J.M. Ergonomically designed replaceable and multifunctional triboelectric nanogenerator for a uniform contact. *RSC Adv.* **2016**, *6*, 88526–88530. [CrossRef]
41. Park, H.J.; Jeong, J.M.; Son, S.G.; Kim, S.G.; Lee, M.; Kim, S.J.; Jeong, J.; Hwang, S.Y.; Park, J.; Eom, Y.; et al. Fluid-Dynamics-Processed Highly Stretchable, Conductive, and Printable Graphene Inks for Real-Time Monitoring Sweat during Stretching Exercise. *Adv. Funct. Mater.* **2021**, *31*, 2011059. [CrossRef]

42. Deng, W.; Yang, T.; Jin, L.; Yan, C.; Huang, H.; Chu, X.; Wang, Z.; Xiong, D.; Tian, G.; Gao, Y.; et al. Cowpea-structured PVDF/ZnO nanofibers based flexible self-powered piezoelectric bending motion sensor towards remote control of gestures. *Nano Energy* **2019**, *55*, 516–525. [CrossRef]

43. He, M.; Du, W.; Feng, W.; Li, S.; Wang, W.; Zhang, X.; Yu, A.; Wan, L.; Zhai, J. Flexible and stretchable triboelectric nanogenerator fabric for biomechanical energy harvesting and self-powered dual-mode human motion monitoring. *Nano Energy* **2021**, *86*, 106058. [CrossRef]

44. Wang, L.; Liu, W.; Yan, Z.; Wang, F.; Wang, X. Stretchable and Shape-Adaptable Triboelectric Nanogenerator Based on Biocompatible Liquid Electrolyte for Biomechanical Energy Harvesting and Wearable Human–Machine Interaction. *Adv. Funct. Mater.* **2021**, *31*, 2007221. [CrossRef]

45. Li, G.; Deng, Z.H.; Cai, M.K.; Huang, K.X.; Guo, M.X.; Zhang, P.; Hou, X.Y.; Zhang, Y.; Wang, Y.J.; Wang, Y.; et al. A stretchable and adhesive ionic conductor based on polyacrylic acid and deep eutectic solvents. *Npj Flex. Electron.* **2021**, *5*, 23. [CrossRef]

46. Norioka, C.; Inamoto, Y.; Hajime, C.; Kawamura, A.; Miyata, T. A universal method to easily design tough and stretchable hydrogels. *Npg Asia Mater.* **2021**, *13*, 34. [CrossRef]

47. Schulz, V.; Zschoche, S.; Zhang, H.P.; Voit, B.; Gerlach, G. Macroporous Smart Hydrogels for Fast-responsive Piezoresistive Chemical Microsensors. In Proceedings of the 25th Eurosensors Conference, Athens, Greece, 4–7 September 2011.

48. Erfkamp, J.; Guenther, M.; Gerlach, G. Piezoresistive Hydrogel-Based Sensors for the Detection of Ammonia. *Sensors* **2019**, *19*, 971. [CrossRef]

49. Erfkamp, J.; Guenther, M.; Gerlach, G. Enzyme-Functionalized Piezoresistive Hydrogel Biosensors for the Detection of Urea. *Sensors* **2019**, *19*, 2858. [CrossRef]

50. Zhao, T.M.; Wang, Q.; Du, A. Self-Powered Flexible Sour Sensor for Detecting Ascorbic Acid Concentration Based on Triboelectrification/Enzymatic-Reaction Coupling Effect. *Sensors* **2021**, *21*, 373. [CrossRef]

Review

Recent Advancements of Polyaniline/Metal Organic Framework (PANI/MOF) Composite Electrodes for Supercapacitor Applications: A Critical Review

Rajangam Vinodh [1,†], Rajendran Suresh Babu [2,†], Sangaraju Sambasivam [3,†], Chandu V. V. Muralee Gopi [4], Salem Alzahmi [5,6,*], Hee-Je Kim [7,*], Ana Lucia Ferreira de Barros [2] and Ihab M. Obaidat [3,6,*]

[1] Department of Electronics Engineering, Pusan National University, Busan 46241, Korea; vinoth6482@gmail.com
[2] Laboratory of Experimental and Applied Physics, Centro Federal de Educação Tecnológica Celso suckow da Fonesca, Av. Maracanã Campus 229, Rio de Janeiro 20271-110, Brazil; ryesbabu@gmail.com (R.S.B.); ana.barros@cefet-rj.br (A.L.F.d.B.)
[3] Department of Physics, United Arab Emirates University, Al Ain P.O. Box 15551, United Arab Emirates; sambaphy@gmail.com
[4] Department of Electrical Engineering, University of Sharjah, Sharjah P.O. Box 27272, United Arab Emirates; naga5673@gmail.com
[5] Department of Chemical & Petroleum Engineering, United Arab Emirates University, Al Ain P.O. Box 15551, United Arab Emirates
[6] National Water and Energy Center, United Arab Emirates University, Al Ain P.O. Box 15551, United Arab Emirates
[7] Department of Electrical and Computer Engineering, Pusan National University, Busan 46241, Korea
* Correspondence: s.alzahmi@uaeu.ac.ae (S.A.); heeje@pusan.ac.kr (H.-J.K.); iobaidat@uaeu.ac.ae (I.M.O.)
† These authors contributed equally to this work.

Citation: Vinodh, R.; Babu, R.S.; Sambasivam, S.; Gopi, C.V.V.M.; Alzahmi, S.; Kim, H.-J.; de Barros, A.L.F.; Obaidat, I.M. Recent Advancements of Polyaniline/Metal Organic Framework (PANI/MOF) Composite Electrodes for Supercapacitor Applications: A Critical Review. *Nanomaterials* 2022, 12, 1511. https://doi.org/10.3390/nano12091511

Academic Editor: Xiang Wu

Received: 14 March 2022
Accepted: 26 April 2022
Published: 29 April 2022

Publisher's Note: MDPI stays neutral with regard to jurisdictional claims in published maps and institutional affiliations.

Abstract: Supercapacitors (SCs), also known as ultracapacitors, should be one of the most promising contenders for meeting the needs of human viable growth owing to their advantages: for example, excellent capacitance and rate efficiency, extended durability, and cheap materials price. Supercapacitor research on electrode materials is significant because it plays a vital part in the performance of SCs. Polyaniline (PANI) is an exceptional candidate for energy-storage applications owing to its tunable structure, multiple oxidation/reduction reactions, cheap price, environmental stability, and ease of handling. With their exceptional morphology, suitable functional linkers, metal sites, and high specific surface area, metal–organic frameworks (MOFs) are outstanding materials for electrodes fabrication in electrochemical energy storage systems. The combination of PANI and MOF (PANI/MOF composites) as electrode materials demonstrates additional benefits, which are worthy of exploration. The positive impacts of the two various electrode materials can improve the resultant electrochemical performances. Recently, these kinds of conducting polymers with MOFs composites are predicted to become the next-generation electrode materials for the development of efficient and well-organized SCs. The recent achievements in the use of PANI/MOFs-based electrode materials for supercapacitor applications are critically reviewed in this paper. Furthermore, we discuss the existing issues with PANI/MOF composites and their analogues in the field of supercapacitor electrodes in addition to potential future improvements.

Keywords: polyaniline; metal–organic framework; supercapacitors; energy density; specific capacitance; stability

1. Introduction

In recent times, the energy crisis has resurfaced as a severe social issue that is stifling growth and eventually endangering human survival [1]. Due to the economic surge, global consumption for sustainable and alternative energy resources is growing relentlessly alongside a vigorous worldwide upsurge in concern regarding ecological issues such as global warming, inappropriate climate change (including wildfire, melting glaciers, floods, drought, increasing in ocean level), and most important, the sustainability of oil reserves. Energy storage and conversion technologies that are renewable, safe, clean, and long-lasting have become a hot research topic [2–5]. Advances in the development of clean, renewable, safe, and practical energy storage systems such as batteries and supercapacitors and fuel cells [6–10] have attracted widespread interest from the scientific community. In recent times, electrochemical energy storage devices have gained considerable attention due to their higher energy efficiency and ecological power systems [11–13]. SCs are presently found in consumer electronics, tools, power supply, voltage stabilization, microgrid, renewable energy storage, energy harvesting, streetlights, medical applications, military, and automotive applications [14–19]. Recently, a commercial corporation offered a 48 V ultra-capacitor module with 1,000,000 duty cycles or a ten-year DC life and 48 V DC working voltage [20]. The modules were engineered explicitly for hybrid bus and construction equipment to provide cost-effective solutions. Furthermore, Maxwell Technologies and LS Mtron Corporations offered different voltage module SCs with a high cycle life and 48 V DC working voltage [21].

Figure 1 shows a Ragone plot of the specific energy (Wh kg^{-1}) versus the specific power (W kg^{-1}), which is used to evaluate the performance of various energy storage technologies. The logarithmic scale of both vertical and horizontal axes and the performances of different systems can be accessibly evaluated. The first version of this type of graph was used to compare the performance of batteries. However, it is appropriate for comparing any kind of energy storage systems. The fuel cells are high-energy-density devices, while SCs are high-power-density devices, as shown in this diagram. Batteries have intermediate power (P_d) and energy (E_d) densities. Furthermore, no electrochemical device can compete with an internal combustion engine, as shown in Figure 1. Hence, to compete with the combustion engine, the E_d and P_d values of electrochemical systems must increase [22]. Batteries can deliver specific energy between 150 and 500 Wh kg^{-1} [23–28] but are limited to their poor specific power because of sluggish electron and ion transport at high rates. To sustain a higher energy output, their discharge time is usually more than 600 s or even 60 min. In contrast, electric double-layer capacitors (EDLCs) which are characterized by high specific power can completely release their energy within less than 10 s, providing a power output between 10- and 20- kW kg^{-1} [29–32]. The specific energy and specific power based on the recently reported work with respect to the supercapacitors has been presented in Table S1 (please refer to the supporting information section).

Unlike fuel cells and batteries, SCs are electrochemical capacitors that store electric charges in electric double layers that form at the electrode–electrolyte interface. SCs are presently found in consumer electronics, memory storage devices, and industrial power/energy organization systems. The SC is composed of high surface-area electrodes (such as anode and cathode), an electrolyte (for example, aqueous medium/organic medium), and a separator (which avoids short circuits among anode and cathode). The electrode is an important element that controls the performance of the SC. The construction of ultrahigh performance SC electrodes includes various serious characteristics such as high specific surface area, extraordinary conductivity, stability based on temperature, optimizing the distribution of pore size, appropriate processing, adequate corrosive resistance, and cost efficiency [33–38]. Hence, the selection of appropriate materials and optimizing the electrode design are vital approaches to convert SCs into more energy-efficient energy storage devices than secondary ion batteries [39–45].

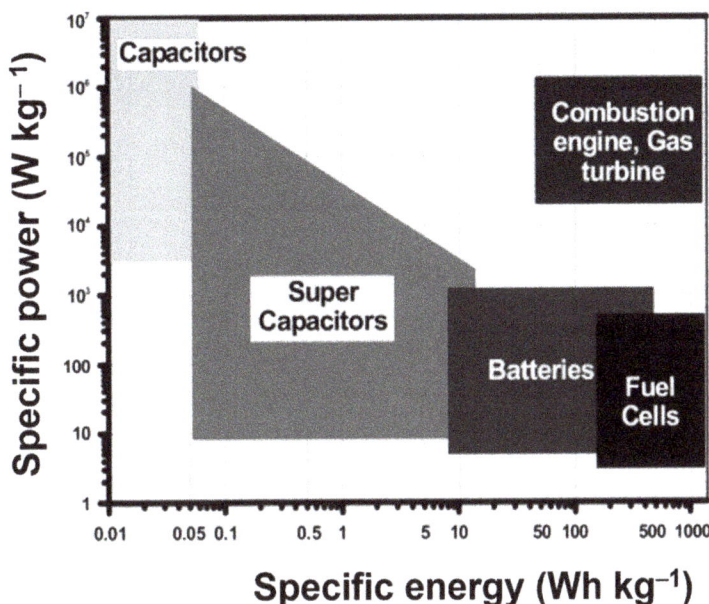

Figure 1. Ragone plot for the various electrochemical energy storage devices. Reproduced with permission from [22]. Copyright 2004 American Chemical Society.

1.1. Classification of Supercapacitors

Supercapacitors are divided into three kinds, namely an electric double layer capacitor (EDLC), pseudocapacitor (PC), and battery hybrid supercapacitor (BHS) based on the mechanism of energy storage, as illustrated in Figure 2.

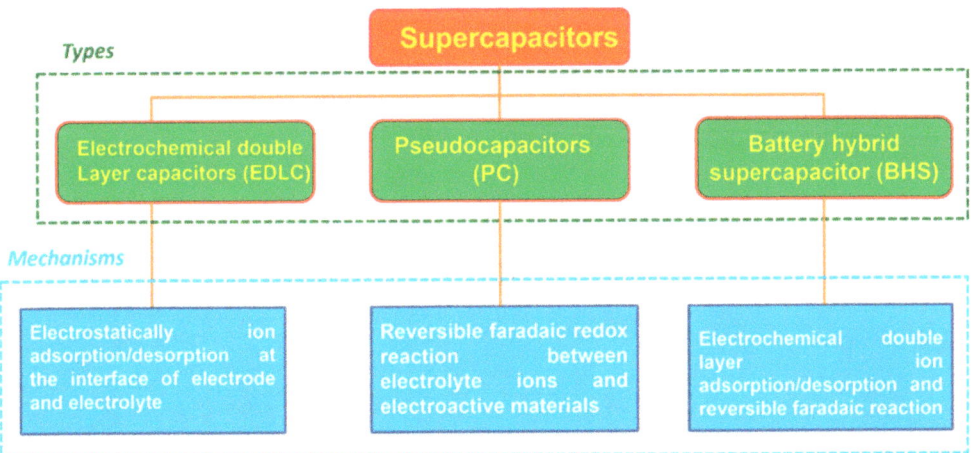

Figure 2. Types of supercapacitors and its mechanism.

1.1.1. EDLCs

EDLCs include two separate carbon-based materials employed as electrodes: an electrolyte as well as a separator. EDLC can store the charges electrostatically, which is a non-Faradaic process that does not require the transfer of charges between the electrode/electrolyte interfaces [46]. EDLCs use the electric double-layer model for energy storage mechanism. The electrons migrate from the anode to the cathode via the external loop during the charging process, with anions moving toward the cathode and cations moving toward the anode in the electrolyte. The electrons and ions flow in opposite directions during the discharging process. The energy storage process is non-Faradaic, and there are no redox reactions, since no charges flow across the electrode–electrolyte contact. Because of the non-Faradaic charge storage mechanism, the volume and morphology of the electrode materials hardly changes, resulting in EDLCs' extended cycle-life [47,48]. Furthermore, the mechanism of charge storage in EDLCs allows quick energy uptake, delivery, and exceptional power output. EDLCs have the potential to withstand millions of cycles compared to batteries with maximum capacity. In lithium-ion batteries (LIBs), when high potential positive electrodes or graphite negative electrodes are employed, the charging process does not need an electrolyte; this leads to a solid electrolyte intermediate [49]. In general, EDLCs employ carbon electrode materials, such as graphene, activated carbon, nano-architectured carbon, and carbon aerogels, for the accumulation of charge via the reversible adsorption/desorption of ions at the electrode/electrolyte interface [50–52]. EDLC materials have been studied extensively owing to their high SSA [53], good electrical conductivity, and excellent mechanical stability [54], but they suffer from a low specific capacitance [55].

1.1.2. PCs

The Faradaic charge-storage mechanism, such as redox reactions, involves the transfer of charge between the electrolyte and electrode. In PCs, when a potential is applied to the electrode material, a redox reaction occurs at the surface of the electrode and electrolyte, which cause the charges to pass through the double layer and results in Faradaic current via the SC cell. When compared to EDLCs, the Faradaic mechanism used in PCs allows for higher specific capacitance and energy density [47,56]. Suitable materials for PCs are thoroughly being explored such as transition metal oxides (TMOs), which provide a relatively high specific capacitance and greater specific energy with a good intrinsic conductivity [57,58], making them exceptional candidates for high-performance SCs. Unfortunately, PCs suffer from inferior cyclic stability performance due to the frequent swelling and shrinking of the polymer chains during the doping/de-doping procedure [59–63]. For TMOs, the major drawback is the low conductivity, which hinders them from reaching the high theoretical specific capacitance value.

1.1.3. Hybrid Supercapacitors

A hybrid supercapacitor is a supercapacitor with asymmetric electrodes, one with electrostatic capacitance, and the other with electrochemical capacitance. The hybrid supercapacitors reached previously unachievable performance characteristics. Furthermore, they combine the great features associated with PCs and EDLCs into one integrated supercapacitor. Although hybrid supercapacitors are less studied compared to EDLCs and pseudocapacitors, efforts are increasing in terms of developing improved hybrid supercapacitors and creating accurate quantifiable models. Developing the high energy density and long-term cycling stability of hybrid supercapacitors has overtaken EDLCs as a class of core SCs [64]. Hybrid supercapacitors are divided into three groups, which differ by their arrangement of electrodes: asymmetric, composite, and battery type.

Composite

Carbonaceous materials are mixed with conducting polymers and (or) metal oxides to fabricate composite electrodes, demonstrating that a single electrode may store energy in both chemical and physical modes. There are two types of composites: (i) binary composites—the electrode material is combination of two materials, and (ii) ternary composites—the electrode material comprises three different materials [47].

Asymmetric

Asymmetric-type supercapacitors are combining the process of Faradaic and non-Faradaic by connecting the electrodes of the pseudocapacitor with EDLCs. In this manner, the conducting polymer or metal oxide is employed as the cathode and the carbon-based material is employed as the anode [47].

Battery Type

The battery-type supercapacitors are a one-of-a-kind integration of a battery and SC electrode materials. This design demonstrates the requirements for greater power density batteries and greater energy density capacitors by integrating SC and battery characteristics in a single cell to achieve both battery and SC properties. Battery-type materials have been widely developed and studied for hybrid supercapacitors because of their richer Faradaic reactions and higher energy density. However, the redox reactions that emerged in bulk materials and the phase transformation process may result in sluggish kinetics and poor rate capability, which need to be improved further [65]. There is a typical feature in electrochemical tests for the battery-type electrode materials: they possess obvious redox peaks and a nonlinear potential platform, while those from capacitive and pseudo-capacitive materials are quite different [66]. Therefore, the specific capacity with a unit of C/g (mAh/g) instead of F/g for specific capacitance is employed to express the capability of charge storage for the battery-grade materials. Binary transition metal oxides (BTMOs) such as $NiCo_2O_4$ [67], $MgCo_2O_4$ [68], $CuCo_2O_4$ [69], and $ZnCo_2O_4$ [70] have been reported as battery-grade electrode materials. In their crystal structure, some metals can provide variable oxide states for plenty of redox reactions, and thus, the specific capacity is expected to be enhanced [71].

2. Conducting Polymers (CPs)

Owing to their unique features, CPs have been regarded, to date, as reliable and excellent electrode materials for pseudocapacitors. Several CPs, for example PANI, polypyrrole (PPy), and polythiophene (PTh), are important for energy-storage applications. These materials have variety of advantages, including excellent conductivity, flexibility, low cost, and ease of preparation [72]. Furthermore, many scientists have investigated the CPs electrodes for their electrochemical performances and attempted to enhance their properties in several ways. In this section, we evaluate the current state of research on pure PANI-based electrode materials for supercapacitor applications.

PANI

PANI is an excellent CP, which can be polymerized with monomer of aniline by various techniques, and it has many advantages because of its facile preparation, easy acid/base chemistry (insertion/desertion), and ecological sustainability [73]. PANI has turned into one of the most efficient materials for PC electrodes. The morphology of PANI nanostructures has a significant impact on their electrochemical performances; therefore, it is very important to employ a suitable and high-efficacy preparation technique to produce PANI with the appropriate nanostructure. Indeed, the chemical or electrochemical polymerization of PANI is rather simple. PANI prefers to form nanofibers in an aqueous solution during chemical oxidative polymerization [74], and there are several polymerization methods for obtaining PANI nanostructures [75–77]. Interfacial polymerization is quite simple and one of the least expensive and general methods to prepared PANI.

Sivakkumar et al. [78] used an interfacial polymerization process to make PANI nanofibers. Their electrochemical characteristics were evaluated in a two-electrode cell configuration with an aqueous electrolyte where the device was reported to exhibit extraordinary specific capacitance of 554 F g^{-1} at 1 A g^{-1}. However, it showed very poor cyclic stability where the initial value of capacitance declined sharply. The theoretical and experimental capacitances of PANI in sulfuric acid medium were reported by Li et al. [79]. Because the specific capacitance of PANI depends on both the conductivity of PANI and the diffusion of counter-anions, the PANI theoretical capacitance value is approximately 2000 F g^{-1}, whereas the experimental values calculated by various methods are less than the theoretical value.

In conclusion, a large quantity of bare PANI electrode material has been investigated for the use in supercapacitors, but its electrochemical performance, notably cycle stability, did not meet commercial application criteria. The poor cyclability of the supercapacitors results in a rapid decrease in its specific capacitance and thus a shorter cycle-life. Hence, for improving the performance of supercapacitors, the scientific community has attempted to mix PANI with carbonaceous materials, metal oxides, metal hexacyanoferrates and/or MOFs to produce various PANI-based composites, particularly electrochemical characteristics [80–84].

3. MOFs

Over the last decade, MOFs, also called coordination polymers, have attracted the attention of materials research. They are constructed as a "node-spacer" of nanosized materials. MOFs contain metal centers (cluster/ions), which are coupled through organic linkers (groups comprising imidazole/carboxyl) to synthesize crystalline, durable, and often very fine porous structures. MOFs exhibit a variety of improvements over the traditional porous materials: for example, rationally designed and highly desirable crystal structures of achievable crystal engineering. Furthermore, the high synthetic flexibility of MOFs with the ease of combining different chemical functionalization leads to engineering MOFs with lightweight organic linkers that result in a high specific surface area and excellent porosity that are inaccessible to traditional materials such as zeolites and porous carbon [85–87].

For instance, Vinodh et al. reported on the effect of Co/Zn ratio on the synthesis of zeolitic imidazole frameworks (ZIFs), where it displayed remarkable ability on the specific surface area, crystal structure, pore size, and electrochemical performances [88]. The maximum BET surface area of ZIF with Co/Zn = 0.5 was found to be 1043.65 m^2 g^{-1}. The ZIF with Co/Zn = 0.5 electrode exhibited a specific capacitance maximum of 30 F g^{-1} at a current density of 0.2 A g^{-1}. Furthermore, ZIF with a Co/Zn = 0.5 electrode retained 91.7% of its initial capacitance over 2000 GCD cycles.

Although their weak conductivity does not ensure higher specific capacitance, pristine MOFs and their derived structures possess an enhanced quantity of pores, leading to higher specific surface areas, as previously noted [89]. Furthermore, the energy density and power density values are not at the preferred levels. To mitigate such deficiencies, different techniques have been introduced: for example, the MOFs intercalation with CPs such as PANI, PPy, and polyethylene dioxythiophene (PEDOT) [90,91]. The CPs have been developed to synthesize, delivering high pseudocapacitance and excellent stability on the long term. In supercapacitors, the charge storage mechanism of their CP electrodes is Faradaic [91]. Combining CPs and MOFs produced a supercapacitor electrode material with remarkable electrochemical properties. PANI is one of the most extensively utilized CPs for such applications due to its simplistic synthesis, excellent conductivity, and high pseudocapacitance behavior [92].

4. PANI/MOF Composite Electrode Material for Supercapacitor Applications

Wang et al. reported the reduction in MOFs bulk resistance with efficient methodology using interweaving MOF crystals into PANI chains that are electrically coated on MOFs [93]. Briefly, cobalt-based MOF crystals (ZIF-67) were deposited on carbon cloth (CC), and then, PANI was electrochemically deposited to provide a flexible porous electrode (PANI-ZIF-67-CC) without changing the MOF primary structure. From the electrochemical examination, the prepared PANI-ZIF-67-CC showed an outstanding areal capacitance of 2146 mF cm^{-2} at the sweep rate of 10 mV s^{-1}. Furthermore, a symmetric flexible solid-state supercapacitor (SFSS) was constructed and evaluated.

Xu et al. synthesized a simple stirring method of ZIF-67 and PANI composites (ZIF-67/PANI) [94]. Additionally, sulfur was incorporated into ZIF-67/PANI using sulfurization (Co_3S_4/PANI). The electron transfer process was enhanced by introducing sulfur for its lower electronegativity. The specific capacitance of Co_3S_4/PANI achieved was 1106 F g^{-1} at 1 A g^{-1}, which is approximately 11 times higher than that of ZIF-67. The constructed asymmetric supercapacitors (ASC) device showed a high energy density of 40.75 Wh kg^{-1} at a specific power of 800 W kg^{-1} and displayed excellent cyclic life. In addition, the fabricated ASC retained 88% of its initial capacitance over 20,000 charge/discharge cycles at a higher current density (5 A g^{-1}). Furthermore, the authors stated that the outstanding electrochemical performances suggested that the fabricated electrode could possess virtuous market prospects and could be an appropriate candidate in energy storage fields.

Iqbal et al. reported cobalt intercalated in a composite of MOF/PANI for the supercapattery device applications [95]. The ASC supercapattery device (AC//MOF/PANI) was fabricated using the activated carbon (AC) and MOF/PANI as the anode and cathode, respectively (Figure 3a). The working voltage window of the constructed ASC was the combination of voltage windows of both of the electrodes.

Figure 3b shows the cyclic voltammetry (CV) of both MOF/PANI, and AC electrodes that were recorded individually, in a three-electrode compartment to examine the plausible wide voltage window. Furthermore, Figure 3c shows the galvanostatic charge/discharge (GCD) plateaus for the fabricated ASC device. The GCD curves are neither triangular nor humped shapes but have a combination of both shapes, which are in good arrangement with the CV traces. The GCD profiles for the ASC device at various current densities ranging from 1 to 3 A g^{-1} are depicted in Figure 3d between the cut-off window of 0 and 1.6 V. The GCD plateaus at different densities of current are nearly linear (symmetrical) with the minimal ohmic drop indicating a reduction in the internal resistance and excellent rate capability confirming the high columbic efficiencies of the fabricated device. The ASC device showed a specific capacity maximum of 104.5 C g^{-1} at 1 A g^{-1}.

Figure 3e exhibits the electrochemical impedance spectroscopy (EIS) examinations which display the finest performance and exceptional electrical conductivity of the supercapattery device. Furthermore, the constructed ASC device delivered outstanding performance with the energy density of 23.2 Wh kg^{-1} with higher power density of 1600 W kg^{-1} at 1 A g^{-1} along with outstanding stability (3000 GCD cycles and endure specific capacity of 146%).

Yao et al. have prepared porous carbon frameworks derived from MOFs (PC-MOFs) as the substrate and deposited PANI via in situ polymerization [96]. The structurally stable porous carbon frameworks derived from MOFs and the homogeneously immobilized conducting PANI nanowires resulted in a PC-MOFs/PANI hybrid electrode with a supreme capacitance of 534.16 F g^{-1} at 0.2 A g^{-1} and an extreme capacitance maintenance of 211% at 2 A g^{-1} after 20,000 GCD curves. In addition, the constructed symmetrical supercapacitors (SSC) resulted in excellent electrochemical performance (specific power of 9.72 µWh cm^{-2}) and outstanding cyclability (94.4% at 10,000 cycles), which can be powered with commercial LED.

In another prominent work, Salunkhe et al. fabricated SSC based on a core–shell 3D structure consisting of MOF derived nanoporous carbon-PANI composite electrodes [97]. A pictorial representation of the preparation methodology for the achievement of a core–shell structure of nanoporous carbon-PANI nanocomposites is revealed in Figure 4A. This configuration has the advantage of improving the mechanical strength of the polymer without blocking the carbon core's electronic conductivity as well as providing a direct diffusion path to the core. The unique multifaced nanoarchitecture avoids the general issue of stacking caused by one-dimensional CNTs or two-dimensional graphene, and thus, it allows ions to penetrate deeper into the material more easily.

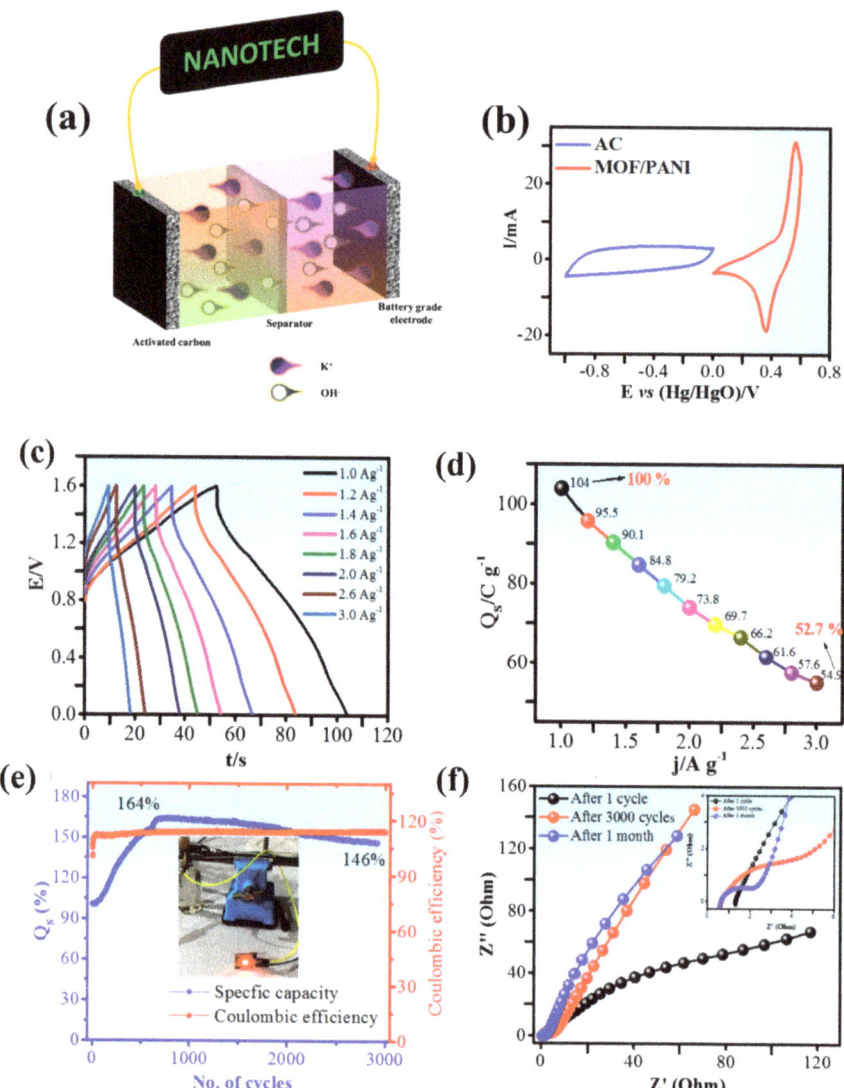

Figure 3. Pictorial illustration of the ASC device, AC//MOF/PANI (**a**); separate CV profiles of AC anode and MOF/PANI cathode (**b**); GCD curves at various densities of current (**c**); specific capacity vs. current density plot (**d**); specific capacity and columbic efficiency plot (**e**); and EIS of before stability test, after 3000 GCD cycles and one month later (**f**). Reproduced with the permission from [95]. Copyright 2020 Elsevier.

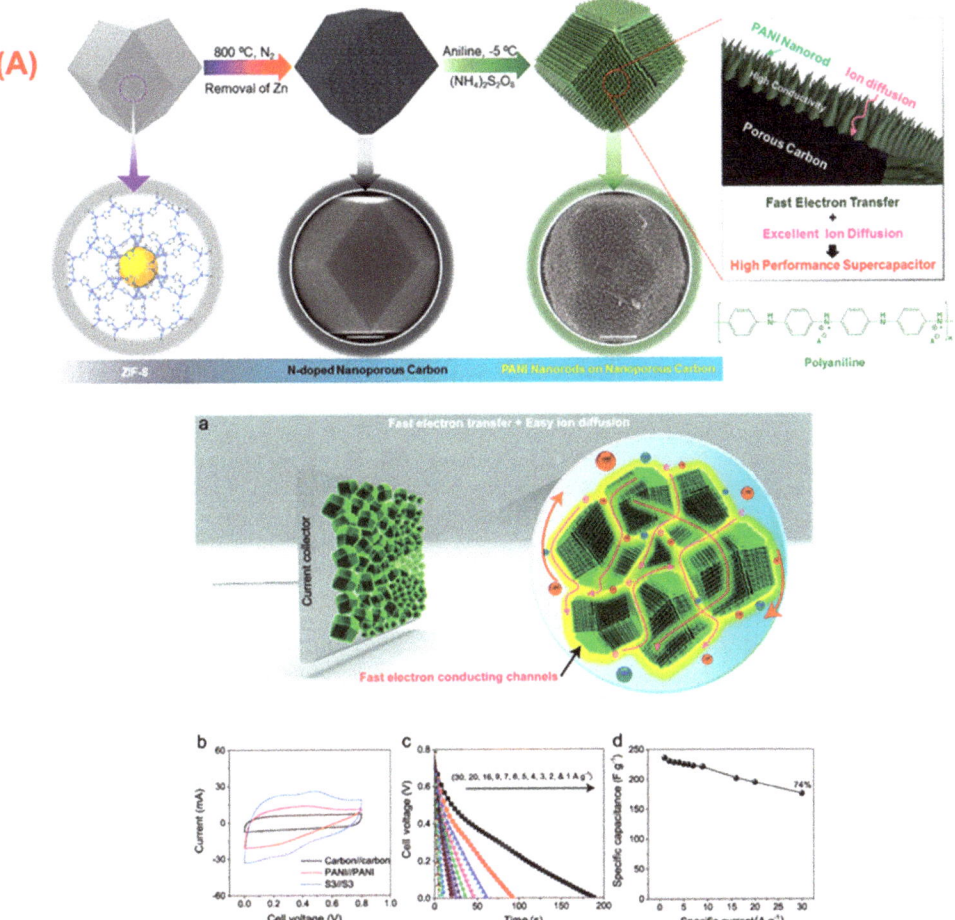

Figure 4. (**A**) Schematic illustration and preparation methodology of the nanoporous carbon/PANI core–shell nanocomposites from the ZIF-8; electrochemical performances, (**a**) distinctive multifaced nanoarchitecture avoids the general issue of stacking caused by one-dimensional CNTs or two-dimensional graphene, which allows ions to penetrate deeper into the material more easily; (**b**) CVs of carbon//carbon, PANI//PANI and carbon–PANI//carbon–PANI (S3), capacitors in 1 M H_2SO_4 electrolyte. (**c**) Discharge profile for the S3 capacitor at various current densities; (**d**) plot of specific capacitance vs. current density. Reproduced from [97].

In addition, the PANI nanorod arrays deliver the ions with simple contact to the carbon core, which lead to the improved interaction of these nanocomposites. In addition, the PANI nanorods provide electrons with rapid conducting routes (electron highways) to attain the current collector surface (Figure 4a). The synthesized composites allowed well-organized electrochemical entry to the electrolyte ions. The comparative CVs of the three different materials are revealed in Figure 4b. The GCD studies were examined at different densities of current ranges from 1 to 30 A g^{-1}. As seen in Figure 4c, the GCD plots are linear and without ohmic drop up to 30 A g^{-1}. Consequently, a higher value of capacitances (between 300 and 1100 F g^{-1}) was attained (Figure 4d). The SSC assembled with this composite material displayed a supreme specific energy of 21 Wh kg^{-1} at a specific power of 12 kW kg^{-1}. Approximately 86% of its original specific capacitance was maintained over 20,000 GCD profiles.

Milakin et al. prepared a composite of PANI/Fe-BTC by the in situ polymerization of aniline monomer in the presence of Fe-BTC [98]. The increasing ratio of aniline and Fe-BTC was found to enhance the gravimetric capacitances value of the composite electrode materials, achieving superior capacitance of 346 F g^{-1} at a sweep rate of 20 mV s^{-1}. In addition, the enhanced pseudocapacitance behavior and the significantly better reversibility throughout the electrochemical techniques displayed by the prepared composite electrode (PANI/Fe-BTC) compared to virgin PANI could be beneficial for supercapacitor applications.

Wang et al. studied a novel flexible solid-state micro supercapacitor (MSCs) with good specific power, outstanding cyclic stability, and excellent mechanical flexibility [99]. The MSCs were constructed by layer-by-layer electrodeposition of microporous PANI and the MOFs crystals on the substrate of laser-induced graphene. Due to the combined effects of MOFs with higher pore structure and the outstanding conductivity of PANI chains, the resultant MSCs showed layer-dependent capacitance performances, resulting in a very high areal specific capacitance of 719.2 mF cm^{-2} at 0.5 mA cm^{-2}. The obtained specific capacitance value was approximately 370 folds better than that of MSCs made by the virgin LIG. Furthermore, the fabricated MSCs retain almost 87.6% of its initial specific capacitance over 6000 GCD curves, illustrating their remarkable cycling stability. In addition, the usage of MSCs for light-emitting diode and their constant mechanical flexibility demonstrate their outstanding potential as electricity for the small and wearable electronics.

Guo et al. developed a high-performance carbonized composite electrode material (Zn-MOF/PANI) from aniline monomer, 8-hydroxyquinoline, and zinc acetate by a facile process for supercapacitor applications [100]. The electrochemical characteristics of the carbonized composite electrode were explored by GCD and CV techniques. The maximum capacitance of 477 F g^{-1} at 1 A g^{-1} was achieved for MOF/PANI composite material.

Shao et al. employed a stable interpenetration polymer network (IPN) structure using extremely stable microscopic MOFs with various synergistic effects to improve the conductivity and electrochemical characteristics, using an efficient approach to grow the molecular chains of PANI in the pores of UiO-66 (PANI/UiO-66) [101]. Furthermore, the prepared composite electrode, PANI/UiO-66, displayed a specific capacitance maximum of 1015 F g^{-1} at 1 A g^{-1}. The assembled supercapacitor displayed a promising capacitance of 647 F g^{-1} at a current density of 1 A g^{-1} and an extraordinary cyclic stability (retains almost 91% of its original specific capacitance over 5000 GCD curves). The bending angle test designates that the attained SC was bendable, and only 10% of its original value declined over 800 twisting cycles with 180° (bending angle). Therefore, the authors suggested that the flexible solid-state supercapacitor (FSSC) could be a potential contender in energy storage device.

Liu et al. established a facile and efficient approach to prepare MOFs derived SC by an in situ network of ZIF-67 particles covered by conducting polyaniline [102]. The attained ZIF-67/PANI electrode material possesses an extraordinarily huge porous surface area and excellent electrical conductivity, ensuring an astonishingly superior specific capacity of 1123.65 C g^{-1} (2497 F g^{-1}) at 1 A g^{-1} in a three-electrode configuration and a remarkable cycling performance (capacitance retention of 92.3% over 9000 cycles at 5 A g^{-1}) for ZIF-67@PANI-2. Furthermore, ZIF-67@PANI-2 displayed a high specific power of 504.72 Wkg^{-1} at a high specific energy of 71.1 Wh kg^{-1} at 1 A g^{-1}.

Xu et al. grew leaflike ZIF nanosheets (ZIF-L) into carbon fiber paper (CFP) by a simple single-step immersing technique with the absence of binders and conductive additives [103]. In contrast, three-dimensional ZIF-67 nanoparticles were also employed as electrode materials. The meager intrinsic conductivity and poor capacitance of ZIFs were enhanced by interlacing with polyaniline. The composite CFP/ZIF-L/PANI showed an area capacitance of 730 mF cm^{-2} at 10 mV s^{-1}, which is higher than that of CFP/ZIF-

67/PANI (608 mF cm^{-2}). In addition, the CFP/ZIF-L/PANI electrode maintained 82.6% of its initial specific capacitance over 3000 GCD cycles.

Udayan et al. employed a facile approach to alter ZIF-8 with polyaniline through a precise interfacial polymerization technique to synthesize ZIF-8/PANI nanocomposites [104]. The present methodology evades the accumulation of ZIF-8/PANI, lifts the consumption of active materials, and disclosures additional active sites, thus making it advantageous for simple electron transfer. Owing to its unique multiporous architecture, ZIF-8/PANI had a large specific surface area of 610.8 m^2 g^{-1}, and the ZIF-8/PANI electrode showed a supreme specific capacitance of 395.4 F g^{-1} at a current density of 0.2 A g^{-1}. A solid-state ASC constructed with ZIF-8/PANI displayed an excellent performance over a wide operating voltage window from 0 to 2.5 V without non-aqueous electrolytes. It showed a specific areal capacitance of 28.1 mF cm^{-2} at 0.1 mA cm^{-2}. The solid-state ASC also displayed a high specific energy (3.2 μW h cm^{-2}) and specific power (1.1 mW cm^{-2}), remarkable cycling stability, and flexibility.

Neisi et al. fabricated a nanocomposite, PANI/Cu-MOF, by a two-step procedure comprising the chemical polymerization of aniline monomer and Cu-MOFs at ambient temperature [105]. The composite electrode illustrates better capacitive characteristics compared with the bare Cu-MOF. In addition, the CV outcomes demonstrated that the PANI/Cu-MOF electrode possesses a superior specific capacitance (734 F g^{-1} at 5 mV s^{-1}) with decent electrochemical cyclic stability.

Ternary MOF composite materials have attracted more attention compared to binary MOF-derived composite electrodes profiting from the synergetic effect of three different constituents [106]. Further inclusive properties are assembled by several components.

For example, Gong et al. prepared a multiporous (micro, meso, and macropores) architecture electrode material with three-dimensional porous carbon nanotubes sponges (porous CNTS) as a base surface for the successive incorporation of PANI and MOF [107]. The different pores-enriched architecture of the sponge favored the penetration of precursors as well as the uniform dispersion of PANI and MOF in the nanotubes. The multiporous architecture of CNTS not only offers a communication pathway for electrons but also provides networks for the rapid distribution of ions. The layered MOF provides an additional ion storage reservoir, while the MOFs are connected to the insulating PANI wires. In addition, the composite structure requires no mechanical binders or conductive additives and has excellent capacity combined with compressive, flexible, and moderately extraordinary specific capacitance.

The specific capacitance characteristic of CNTS was synergistically enhanced by the incorporation of ZIF-8, ZIF-67, and PANI. The specific capacitance value increased from 89 to 746 F g^{-1}, and a highest specific energy of 28.9 Wh kg^{-1} was achieved. Furthermore, the prepared composite electrode was compressive, flexible, and has outstanding specific capacitance. Therefore, it could open new avenue for flexible energy storage devices.

He et al. prepared a multi-component hybrid of copper MOF-derived copper oxide@mesoporous carbon (CuO$_x$@mC) entrenched with PANI and reduced graphene oxide (rGO) by in situ polymerization (CuO$_x$@mC@PANI@rGO) [108]. The sequence of as-synthesized CuO$_x$@mC@PANI@rGO composites was investigated for supercapacitor applications, and the schematic representation of the reaction protocol is depicted in Figure 5A.

Due to the ordered octahedral structure of CuO$_x$@mC composites, a uniform and extremely well-organized interface layer of PANI with rGO nanosheets was formed on the surface of the CuO$_x$@mC architecture. This effective conductive network could increase ion transport and redox behavior at the electrode/electrolyte interface, resulting in enhanced electrical conductivity and supercapacitor performances. TEM and HR-TEM images of CuO$_x$@mC$_{700}$, CuO$_x$@mC$_{700}$@PANI, and CuO$_x$@mC$_{700}$@PANI@rGO are presented in Figure 5a–d. From Figure 5a, we can see that the polyhedron crystals are approximately 500 nm in size, in which CuO$_x$ particles are highly distributed in amorphous carbon. In the meantime, the HR-TEM image (Figure 5b) illustrated the distance of 0.20 and 0.21 nm, which can be indexed for the interplanar spacing between the cubic phase of Cu (200) and (111) planes, respectively. Furthermore, the $CuO_x@mC_{700}@PANI$ exhibited a huge and uneven surface together with many nanowires on the external surface, as shown in Figure 5c.

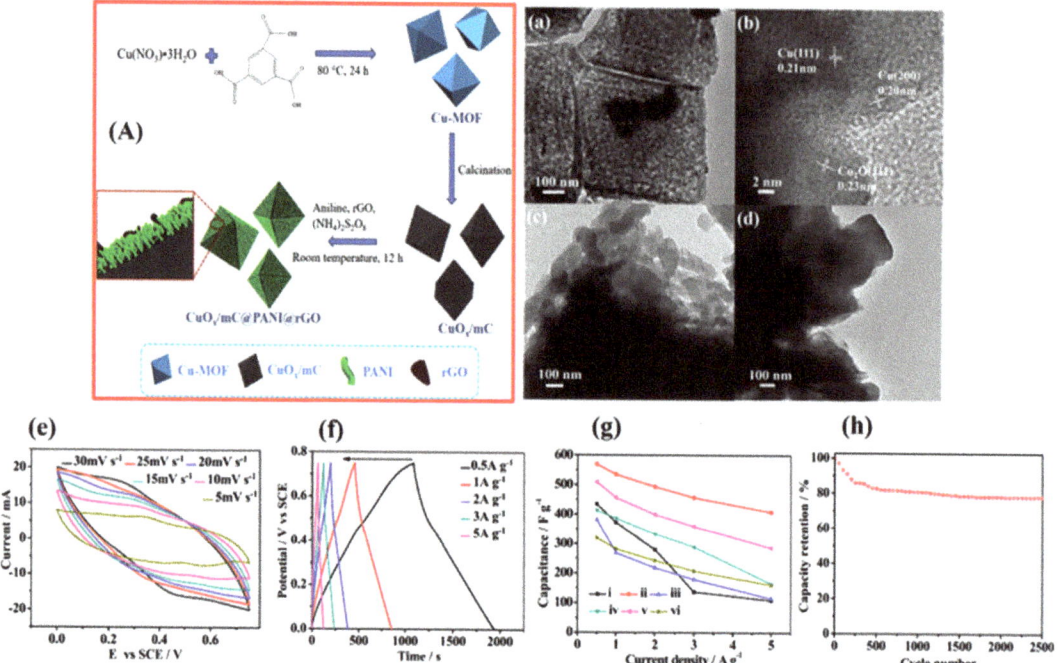

Figure 5. (**A**) Pictorial representation of the synthesis of $CuO_x@mC_{700}@PANI@rGO$ composites; (**a,b**) represents the TEM and HR-TEM pictures of $CuO_x@mC_{700}$; (**c,d**) HR-TEM images of $CuO_x@mC_{700}@PANI$ and $CuO_x@mC_{700}@PANI@rGO$, respectively; (**e**) CV curves of $CuO_x@mC_{700}@PANI@rGO$ at different sweep rates; (**f**) GCD profiles of $CuO_x@mC_{700}@PANI@rGO$ with various current densities; (**g**) comparative plot of the specific capacitance of different electrode materials versus current density variation; (**h**) cyclability of $CuO_x@mC_{700}@PANI@rGO$ electrode at 2.0 A g^{-1}. Reproduced with the permission from [108]. Copyright 2018 Elsevier.

Figure 5d demonstrates the stretchy and crumpled landscapes of rGO sheets which were examined after the incorporation of rGO nanosheets. The specific capacitance characteristics of the prepared composite electrode, $CuO_x@mC_{700}@PANI@rGO$, were further explored by continuous electrochemical measurements. Figure 5e demonstrates the cyclic voltammogram plots of $CuO_x@mC_{700}@PANI@rGO$ at different sweep rates ranging between 5 and 30 mV s^{-1}. With increasing the sweep rate, the current densities of the CV plots also increased. Nevertheless, the redox peaks shifted negatively and positively due to the electrode resistance [109].

Figure 5f displays the galvanostatic charge–discharge (GCD) plateaus in which typical triangular shapes were obtained at different densities of current, signifying excellent capacitance characteristic and reversibility. In addition, the specific capacitance decreases with raising the current density. By varying the pyrolysis temperature of Cu-MOF, the ternary $CuO_x@mC_{700}@PANI@rGO$, attained at 700 °C, displayed a superior specific capacitance of 534.5 F g^{-1} and extraordinary cyclability (Figure 5g). In contrast, the resulted $CuO_x@mC@PANI$ displayed a specific capacitance only of 456.0 F g^{-1} at 1 A g^{-1}. Furthermore, it retains 70% of its initial capacitance over 2500 GCD curves (Figure 5h). This research has led to new insights into the study of metal oxide–carbon hybrids with morphologically controlled microstructures, where the beneficial function in the PANI is thought to be a hidden approach to improve the performance of these composites in supercapacitors.

Liu et al. prepared an electrode material, for supercapacitor applications, through the in situ formation of ZIF-8 onto the surface of ZnO followed by the deposition of thin PANI film (PANI/ZnO/ZIF-8/G/PC) [110]. The exceptional electrode architecture efficiently improved the performance of the supercapacitors. The assembled electrode, PANI/ZnO/ZIF-8/G/PC, exhibited a superior areal capacitance value of 1.378 F cm^{-2} at 1 mA cm^{-1} compared with the existing textile-based electrode materials (WO_3/polyester/graphene and cotton/graphene). Furthermore, the authors constructed PANI/ZnO/ZIF-8/G/PC electrode in a flexible supercapacitor, where it delivered a good specific energy of 235 µWh cm^{-3} at a specific power of 1542 µW cm^{-3}.

The composites of CPs with MOFs helped assemble highly efficient electrode materials, especially for PANI-based SCs. Nevertheless, such composite electrodes operate with Faradic redox reactions, which eventually decompose the electrolyte and shorten the lifetime of the supercapacitor device. To date, very few reports have investigated PANI/MOF-based electrode materials for supercapacitor applications [111]. This might be due to its comparatively inferior water stability and the wide distribution of most MOFs, which can create difficulties in identifying the appropriate preparation methodology for perceiving PANI/MOF composites. In addition to PANI, forthcoming inquiries on this topic may be driven by the process of various substitutes, such as PPy, PEDOT, and P3HT. The combination of such CPs with MOFs could lead to the arrangement of highly well-organized and flexible electrodes for high-performance supercapacitors [112,113]. Table 1 comprises various PANI/MOF-derived electrode materials for supercapacitor applications.

Table 1. PANI/MOF composite–based electrode materials for supercapacitor applications.

S. No.	Electrode Materials	Specific Capacitance		Electrolyte	Specific Energy	Specific Power	Cyclability/ Capacitance Retention	Ref.
		3-ES	2-ES					
1	PANI-ZIF-67/CC	2146 mF cm^{-2} @ 10 mV s^{-1}	SSC: 35 mF cm^{-2}	3-ES: 3 M KCl; SSC: H$_2$SO$_4$/PVA	0.0161 mWh cm^{-3}	0.833 W cm^{-3}	2000 GCD cycles; 80%	[93]
2	Co$_3$S$_4$/PANI	106 F g^{-1} @ 1 A g^{-1}	ASC: 114.6 F g^{-1} @ 1 A g^{-1}	3-ES and ASC: 6 M KOH	40.75 Wh kg^{-1}	800 W kg^{-1}	20,000 GCD cycles; 88%	[94]
3	MOF/PANI	162.5 C g^{-1} @ 0.4 A g^{-1}	ASC: 104.5 C g^{-1} @ 1 A g^{-1}	3-ES and ASC: 1 M KOH	23.2 Wh kg^{-1}	1600 W kg^{-1}	3000 GCD cycles; 146%	[95]
4	PC-MOFs/PANI	534.16 F g^{-1} @ 0.2 A g^{-1}	SSC: 140 F g^{-1} @ 0.2 A g^{-1}	SSC: H$_2$SO$_4$/PVA	9.72 μWh cm^{-2}	199.99 μW cm^{-2}	10,000 cycles; 94.4%	[96]
5	MOF/PANI	1100 F g^{-1} @ 1 mV s^{-1}	SSC: 236 140 F g^{-1} @ 1 A g^{-1}	1 M H$_2$SO$_4$	21 Wh kg^{-1}	400 W kg^{-1}	20,000 GCD cycles; 86%	[97]
6	PANI/Fe-BTC	346 F g^{-1} @ 20 mV s^{-1}	—	0.5 M H$_2$SO$_4$	—	—	—	[98]
7	MOF/PANI	719.2 mF cm^{-2} @ 0.5 mA cm^{-2}	MSCs: 528.5 mF cm^{-2} @ 10 mA cm^{-2}	MSCs: H$_2$SO$_4$/PVA	443.7 mW cm^{-3}	3218.4 μW cm^{-2}	6000 GCD cycles; 87.6%	[99]
8	Zn-MOF/PANI	477 F g^{-1} @ 1 A g^{-1}	—	1 M H$_2$SO$_4$	—	—	—	[100]
9	PANI/UiO-66	1015 F g^{-1} @ 1 A g^{-1}	SSC: 647 F g^{-1} @ 1 A g^{-1}	H$_2$SO$_4$/PVA	78.8 Wh kg^{-1}	200 W kg^{-1}	5000 GCD cycles; 91%	[101]
10	ZIF-67@PANI	2497 F g^{-1} @ 1 A g^{-1}	SSC: 512 F g^{-1} @ 1 A g^{-1}	KOH	71.1 Wh kg^{-1}	504.72 W kg^{-1}	9000 GCD cycles; 92.3%	[102]
11	CFP/ZIF-L/PANI	681 mF cm^{-2} @ 1 mA cm^{-2}	—	3 M KCl	—	—	3000 GCD cycles; 82.6%	[103]
12	ZIF-8/PANI	395.4 F g^{-1} @ 0.2 A g^{-1}	ASC: 28.1 mF cm^{-2} @ 0.1 mA cm^{-2}	1 M H$_2$SO$_4$	3.2 μWh cm^{-2}	1.11 mW cm^{-2}	1000 GCD cycles; 78.4%	[104]
13	PANI/Cu-MOF	734 F g^{-1} @ 5 mV s^{-1}	—	6 M KOH	—	—	4000 GCD cycles; 98%	[105]
14	CNT/MOF/PANI	342.5 F g^{-1} @ 1 A g^{-1}	—	—	28.9 Wh kg^{-1}	~800 W kg^{-1}	—	[107]
15	CuO$_x$@mC@PANI@rGO	534.5 F g^{-1} @ 1 A g^{-1}	—	1 M H$_2$SO$_4$	—	—	2500 GCD cycles; 80%	[108]
16	PANI/ZnO/ZIF-8/G/PC	1.378 F cm^{-2} @ 1 mA cm^{-2}	SSC: —	SSC: H$_2$SO$_4$/PVA	235 μWh cm^{-3}	1542 μW cm^{-3}		[110]

Note: 3-ES: Three-electrode system; 2-ES: Two-electrode system; KCl: Potassium chloride; CNT: Carbon nanotube; PC: Porous carbon; CFP: Carbon fiber paper; SSC: Symmetric supercapacitor; ASC: Asymmetric supercapacitor; MSCs: Micro-supercapacitors; KOH: Potassium hydroxide; H$_2$SO$_4$: Sulfuric acid; PVA: Polyvinyl alcohol.

5. Conclusions and Future Perspectives

The demand for alternative energy resources and storage systems is increasing as conventional fossil fuels are gradually decreasing. Fossil fuels are sources of conventional energy production but have been gradually transitioned to the existing advanced technologies with a prominence of renewable resources such as solar, tidal, and wind. Despite consistent increases in energy prices, the customers' needs are mounting rapidly due to an increase in populations, economic growth, per capita consumption, supply at remote places, and stationary forms for machines and portable electronics. The energy storage may allow the flexible generation and delivery of stable energy for meeting the demands of end users. The requirements for energy storage will triple the current values by 2050 where unique devices and systems are required. Protecting the ecology is an important effort related to the requirement of new technologies. Electrode material plays a major role in defining the practical viability of any energy storage device. For example, supercapacitors that can be used in practice should attain the technical needs of excellent specific capacity, specific energy, and specific power as well as long-term cyclability.

Briefly, the present review article describes the recent developments in electrode materials with their design, synthesis, and use of supercapacitors. PANI shows high specific capacitance value but displays a shorter life-cycle, whereas MOFs exhibit poor conductivity and specific capacitance. To overcome the shortcoming of PANI and enhance the conductivity and specific capacitance, PANI and MOF were composited.

There is no doubt that a wide range of PANI/MOFs and their derivatives have been well-initiated to catch enhancements in electrochemical behavior in recent years. Nevertheless, there are still numerous disputes and prospects for researchers/scientific communities to further investigate this interesting topic. The future of the PANI supercapacitor mainly relies on the adequate structure of the associated nanocomposites. However, a commercial supercapacitor is not based on a simple nanocomposite that mixes two composites; instead, a delicate structure is needed to place the MOF with the interacting surfaces between the polymer chains of PANI or vice versa. This design can take advantage of the flexibility of PANI for the development of flexible supercapacitors, which are in high demand. There are relatively minimal precursors or templates involving various MOFs (for example ZIF-8, ZIF-67, MOF-5, MOF-74, MIL-101) available to create an MOF and its derivative materials for supercapacitors. Even better starting materials and templates need to be created to attain new functional materials with unique architectures.

A wide cut-off voltage is frequently considered to be one of the crucial factors to enhance the supercapacitor performance. Still, the main barrier of the use of aqueous electrolyte is the dissociation of water that occurs when the voltage surpasses 1.23 V. PANI/MOFs as supercapacitor electrodes are still concerned with aqueous medium and the bench scale level. Therefore, in-depth research of solid-state supercapacitors (SSC) with a wide voltage window is required. Ecological, inexpensive, and high-yielding PANI/MOFs-based energy storage devices are expected to exist soon due to the development of PANI/MOFs technology.

In conclusion, the commercial usage of PANI/MOFs supercapacitors are still at the laboratory stage. Additional inputs in this vast research field will promote the expansion of PANI/MOF research into the next-generation environment-friendly energy storage systems. As this research area has seen tremendous growth, we can assume the development of highly efficient energy storage devices with the forthcoming developments in pilot-scale machineries of PANI/MOFs.

Supplementary Materials: The following supporting information can be downloaded at: https://www.mdpi.com/article/10.3390/nano12091511/s1, Table S1: Specific energy and specific power of recently reported articles in supercapacitors. References [6,114–122] are cited in the supplementary materials.

Author Contributions: Investigation and writing—original draft, R.V.; Investigation and visualization, R.S.B.; Conceptualization, S.S. and C.V.V.M.G.; Validation and Project administration, S.A.;

Supervision, H.-J.K.; Supervision and validation, A.L.F.d.B. and I.M.O. All authors have read and agreed to the published version of the manuscript.

Funding: This work was financially supported by the UAEU-Strategic research program under Grant no. 12R128.

Institutional Review Board Statement: Not applicable.

Informed Consent Statement: Not applicable.

Data Availability Statement: No new data were created or analyzed in this study. Data sharing is not applicable to this article.

Acknowledgments: The authors gratefully acknowledge the financial support from BK21 Plus Creative Human Resource Education and Research Programs for ICT Convergence in the 4th Industrial Revolution, Pusan National University, Busan, South Korea. R. Suresh Babu would like to thank FAPERJ for the financial support of post-doctoral senior fellowship (E-26/202.333/2021).

Conflicts of Interest: The authors declare no conflict of interest.

References

1. Goodenough, J.B. Electrochemical energy storage in a sustainable modern society. *Energy Environ. Sci.* **2014**, *7*, 14–18. [CrossRef]
2. Vinodh, R.; Sasikumar, Y.; Kim, H.-J.; Atchudan, R.; Yi, M. Chitin and chitosan based biopolymer derived electrode materials for supercapacitor applications: A critical review. *J. Ind. Eng. Chem.* **2021**, *104*, 155–171. [CrossRef]
3. Vinodh, R.; Gopi, C.V.V.M.; Kummara, V.G.R.; Atchudan, R.; Ahamad, T.; Sambasivam, S.; Yi, M.; Obaidat, I.M.; Kim, H.-J. A review on porous carbon electrode material derived from hypercross-linked polymers for supercapacitor applications. *J. Energy Storage* **2020**, *32*, 101831. [CrossRef]
4. Gopi, C.V.V.M.; Vinodh, R.; Sambasivam, S.; Obaidat, I.M.; Kim, H.-J. Recent progress of advanced energy storage materials for flexible and wearable supercapacitor: From design and development to applications. *J. Energy Storage* **2020**, *27*, 101035.
5. Kim, H.-J.; Krishna, T.; Zeb, K.; Vinodh, R.; Gopi, C.V.V.M.; Sambasivam, S.; Raghavendra, K.V.G.; Obaidat, I.M. A comprehensive review of Li-ion battery materials and their recycling techniques. *Electronics* **2020**, *9*, 1161. [CrossRef]
6. Kim, I.; Vinodh, R.; Gopi, C.V.V.M.; Kim, H.-J.; Babu, R.S.; Deviprasath, C.; Devendiran, M.; Kim, S. Novel porous carbon electrode derived from hypercross-linked polymer of poly(divinylbenzene-co-vinyl benzyl chloride) for supercapacitor applications. *J. Energy Storage* **2021**, *43*, 103287. [CrossRef]
7. Atchudan, R.; Samikannu, K.; Suguna, P.; Edison, T.N.J.I.; Vinodh, R.; Lee, Y.R. *Aesculus turbinata* biomass-originated nanoporous carbon for energy storage applications. *Mater. Lett.* **2022**, *309*, 131445. [CrossRef]
8. Atchudan, R.; Edison, T.N.J.I.; Suguna, P.; Vinodh, R.; Babu, R.S.; Sundramoorthy, A.K.; Renita, A.A.; Lee, Y.R. Facile synthesis of nitrogen-doped porous carbon materials using waste biomass for energy storage applications. *Chemosphere* **2022**, *289*, 133225. [CrossRef]
9. Vinodh, R.; Rana, P.J.S.; Gopi, C.V.V.M.; Yang, Z.; Atchudan, R.; Venkatachalam, K.; Kim, H.-J. Polyaniline-13X zeolite composite-supported platinum electrocatalysts for direct methanol fuel cell applications. *Polym. Int.* **2019**, *68*, 929–935. [CrossRef]
10. Vinodh, R.; Deviprasath, C.; Gopi, C.V.V.M.; Kummara, V.G.R.; Atchudan, R.; Ahamad, T.; Kim, H.-J.; Yi, M. Novel 13X Zeolite/PANI electrocatalyst for hydrogen and oxygen evolution reaction. *Int. J. Hydrogen Energy* **2020**, *45*, 28337–28349. [CrossRef]
11. Liu, J.; Zhang, J.G.; Yang, Z.; Lemmon, J.P.; Imhoff, C.; Graff, G.L.; Li, L.; Hu, J.; Wang, C.; Xiao, J.; et al. Materials science and materials chemistry for large scale electrochemical energy storage: From transportation to electrical grid. *Adv. Funct. Mater.* **2012**, *23*, 929–946. [CrossRef]
12. Lin, M.-C.; Gong, M.; Lu, B.; Wu, Y.; Wang, D.-Y.; Guan, M.; Angel, M.; Chen, C.; Yang, J.; Hwang, B.-J.; et al. An ultrafast rechargeable aluminium-ion battery. *Nature* **2015**, *520*, 324–328. [CrossRef]
13. Merlet, C.; Rotenberg, B.; Madden, P.A.; Taberna, P.-L.; Simon, P.; Gogotsi, Y.; Salanne, M. On the molecular origin of supercapacitance in nanoporous carbon electrodes. *Nat. Mater.* **2012**, *11*, 306. [CrossRef]
14. Farhadi, M.; Mohammed, O. Real-time operation and harmonic analysis of isolated and non-isolated hybrid DC microgrid. *IEEE Trans. Ind. Appl.* **2014**, *50*, 2900–2909. [CrossRef]
15. Inthamoussou, F.A.; Pegueroles-Queralt, J.; Bianchi, F.D. Control of a Supercapacitor Energy Storage System for Microgrid Applications. *IEEE Trans. Energy Convers.* **2013**, *28*, 690–697. [CrossRef]
16. Nippon Chemi-Con. *Stanley Electric and Tamura Announce the Development of "Super CaLeCS," an Environment-Friendly EDLC-Powered LED Street Lamp*; Press Release Nippon Chemi-Con Corp.: Tokyo, Japan, 2010.
17. Kotz, R.; Carlen, M. Principles and applications of electrochemical capacitors. *Electrochim. Acta* **2000**, *45*, 2483–2498. [CrossRef]
18. Jaafar, A.; Sareni, B.; Roboam, X.; Thiounn-Guermeur, M. Sizing of a hybrid locomotive based on accumulators and ultracapacitors. In Proceedings of the IEEE Vehicle Power and Propulsion Conference, Lille, France, 1–3 September 2010; pp. 1–6.
19. Fröhlich, M.; Klohr, M.; Pagiela, S. Energy storage system with ultracaps on board of railway vehicles. In Proceedings of the 8th World Congress on Railway Research, Seoul, Korea, 18–22 May 2008.

20. Bonaccorso, F.; Colombo, L.; Yu, G.; Stoller, M.; Tozzini, V.; Ferrari, A.; Pellegrini, V. Graphene, related two-dimensional crystals, and hybrid systems for energy conversion and storage. *Science* **2015**, *347*, 1246501. [CrossRef]
21. Şahin, M.E.; Blaabjerg, F.; Sangwongwanich, A. A Comprehensive Review on Supercapacitor Applications and Developments. *Energies* **2022**, *15*, 674. [CrossRef]
22. Winter, M.; Brodd, R.J. What are batteries, fuel cells and supercapacitors? *Chem. Rev.* **2004**, *104*, 4245–4269. [CrossRef]
23. Zhang, Y.; Ma, Q.; Wang, S.; Liu, X.; Li, L. Poly (vinyl alcohol)-assisted fabrication of hollow carbon spheres/reduced graphene oxide nanocomposites for high-performance lithium-ion battery anodes. *ACS Nano* **2018**, *12*, 4824–4834. [CrossRef]
24. Wu, F.; Yang, H.; Bai, Y.; Wu, C. Paving the path toward reliable cathode materials for aluminum-ion batteries. *Adv. Mater.* **2019**, *31*, 1806510. [CrossRef]
25. Gummow, R.J.; Vamvounis, G.; Kannan, M.B.; He, Y. Calcium-ion batteries: Current state-of-the-art and future perspectives. *Adv. Mater.* **2018**, *30*, 1801702. [CrossRef]
26. Konarov, A.; Voronina, N.; Jo, J.H.; Bakenov, Z.; Sun, Y.-K.; Myung, S.-T. Present and future perspective on electrode materials for rechargeable zinc-ion batteries. *ACS Energy Lett.* **2018**, *3*, 2620–2640. [CrossRef]
27. Palomares, V.; Casas-Cabanas, M.; Castillo-Martínez, E.; Han, M.H.; Rojo, T. Update on Na-based battery materials. A growing research path. *Energy Environ. Sci.* **2013**, *6*, 2312–2337. [CrossRef]
28. Xiong, T.; Yu, Z.G.; Wu, H.; Du, Y.; Xie, Q.; Chen, J.; Zhang, Y.W.; Pennycook, S.J.; Lee, W.S.V.; Xue, J. Defect engineering of oxygen-deficient manganese oxide to achieve high-performing aqueous zinc ion battery. *Adv. Energy Mater.* **2019**, *9*, 1803815. [CrossRef]
29. Yang, S.; Wang, S.; Liu, X.; Li, L. Biomass derived interconnected hierarchical micro-meso-macro-porous carbon with ultrahigh capacitance for supercapacitor. *Carbon* **2019**, *147*, 540–549. [CrossRef]
30. Zhang, Y.; Yang, S.; Wang, S.; Liu, X.; Li, L. Microwave/freeze casting assisted fabrication of carbon frameworks derived from embedded upholder in tremella for superior performance supercapacitors. *Energy Storage Mater.* **2019**, *18*, 447–455. [CrossRef]
31. Zhang, Y.; Liu, X.; Wang, S.; Li, L.; Dou, S. Bio-nanotechnology in high-performance supercapacitors. *Adv. Energy Mater.* **2017**, *7*, 1700592. [CrossRef]
32. Beguin, F.; Presser, V.; Balducci, A.; Frackowiak, E. Carbons and electrolytes for advanced supercapacitors. *Adv. Mater.* **2014**, *26*, 2219–2251. [CrossRef]
33. Miller, J.R.; Burke, A.F. Electrochemical capacitors: Challenges and opportunities for real-world applications. *Electrochem. Soc. Interface* **2008**, *17*, 53–57. [CrossRef]
34. Park, S.; Jayaraman, S. Smart textiles: Wearable electronic systems. *MRS Bull.* **2003**, *28*, 585–591. [CrossRef]
35. Shim, B.S.; Chen, W.; Doty, C.; Xu, C.; Kotov, N.A. Smart electronic yarns and wearable fabrics for human biomonitoring made by carbon nanotube coating with polyelectrolytes. *Nano Lett.* **2008**, *8*, 4151–4157. [CrossRef]
36. Selvaraj, A.R.; Muthusamy, A.; Kim, H.-J.; Senthil, K.; Prabakar, K. Ultrahigh surface area biomass derived 3D hierarchical porous carbon nanosheet electrodes for high energy density supercapacitors. *Carbon* **2021**, *174*, 463–474. [CrossRef]
37. Chmiola, J.; Largeot, C.; Taberna, P.-L.; Simon, P.; Gogotsi, Y. Monolithic Carbide-Derived Carbon Films for Micro-Supercapacitors. *Science* **2010**, *328*, 480–483. [CrossRef]
38. Chmiola, J.; Yushin, G.; Gogotsi, Y.; Portet, C.; Simon, P.; Taberna, P.-L. Anomalous increase in carbon capacitance at pore sizes less than 1 nanometer. *Science* **2006**, *313*, 1760–1763. [CrossRef]
39. Wang, D.W.; Li, F.; Liu, M.; Lu, G.Q.; Cheng, H.M. 3D Aperiodic hierarchical porous graphitic carbon material for high-rate electrochemical capacitive energy storage. *Angew. Chem. Int. Ed.* **2008**, *47*, 373–376. [CrossRef]
40. Wang, Y.; Foo, C.Y.; Hoo, T.K.; Ng, M.; Lin, J. Designed smart system of the sandwiched and concentric architecture of $RuO_2/C/RuO_2$ for high performance in electrochemical energy storage. *Chem. Eur. J.* **2010**, *16*, 3598–3603. [CrossRef]
41. Ramya, R.; Sivasubramanian, R.; Sangaranarayanan, M. Conducting polymers-based electrochemical supercapacitors—Progress and prospects. *Electrochim. Acta* **2013**, *101*, 109–129. [CrossRef]
42. Snook, G.A.; Kao, P.; Best, A.S. Conducting-polymer-based supercapacitor devices and electrodes. *J. Power Sources* **2011**, *196*, 1–12. [CrossRef]
43. Dubal, D.P.; Holze, R. Self-assembly of stacked layers of Mn_3O_4 nanosheets using a scalable chemical strategy for enhanced, flexible, electrochemical energy storage. *J. Power Sources* **2013**, *238*, 274–282. [CrossRef]
44. Dubal, D.P.; Kim, J.G.; Kim, Y.; Holze, R.; Kim, W.B. Demonstrating the highest supercapacitive performance of branched MnO_2 nanorods grown directly on flexible substrates using controlled chemistry at ambient temperature. *Energy Technol.* **2013**, *1*, 125–130. [CrossRef]
45. Gund, G.S.; Dubal, D.P.; Patil, B.H.; Shinde, S.S.; Lokhande, C.D. Enhanced activity of chemically synthesized hybrid graphene oxide/Mn_3O_4 composite for high performance supercapacitors. *Electrochim. Acta* **2013**, *92*, 205–215. [CrossRef]
46. Kiamahalleh, M.V.; Zein, S.H.S.; Najafpour, G.; SATA, S.A.; Buniran, S. Multiwalled carbon nanotubes based nanocomposites for supercapacitors: A review of electrode materials. *Nano* **2012**, *7*, 1230002. [CrossRef]
47. Halper, M.S.; Ellenbogen, J.C. *Supercapacitors: A Brief Overview*; The MITRE Corporation: McLean, VA, USA, 2006; pp. 1–34.
48. Choi, H.; Yoon, H. Nanostructured electrode materials for electrochemical capacitor applications. *Nanomaterials* **2015**, *5*, 906–936. [CrossRef]
49. Simon, P.; Gogotsi, Y. Materials for electrochemical capacitors. *Nanosci. Technol.* **2009**, 320–329.

50. Kulandaivalu, S.; Sulaiman, Y. Recent advances in layer-by-layer assembled conducting polymer based composites for supercapacitors. *Energies* **2019**, *12*, 2107. [CrossRef]

51. Wong, S.I.; Sunarso, J.; Wong, B.T.; Lin, H.; Yu, A.; Jia, B. Towards enhanced energy density of graphene-based supercapacitors: Current status, approaches, and future directions. *J. Power Sources* **2018**, *396*, 182–206. [CrossRef]

52. Afif, A.; Rahman, S.M.H.; Tasfiah Azad, A.; Zaini, J.; Islan, M.A.; Azad, A.K. Advanced materials and technologies for hybrid supercapacitors for energy storage—A review. *J. Energy Storage* **2019**, *25*, 100852. [CrossRef]

53. Wu, Y.; Ran, F. Vanadium nitride quantum dot/nitrogen-doped microporous carbon nanofibres electrode for high-performance supercapacitors. *J. Power Sources* **2017**, *344*, 1–10. [CrossRef]

54. Xianwen, M.; Hatton, T.A.; Gregory, C.R. A reviewof electrospun carbon fibres as electrode materials for energy storage. *Curr. Org. Chem.* **2013**, *17*, 1390–1401.

55. Tong, L.; Liu, J.; Boyer, S.M.; Sonnenberg, L.A.; Fox, M.T.; Ji, D.; Feng, J.; Bernier, W.E.; Jones Jr, W.E. Vapor-phase polymerized poly(3,4-ethylenedioxythiophene) (PEDOT)/TiO_2 composite fibres as electrode materials for supercapacitors. *Electrochim. Acta* **2017**, *224*, 133–141. [CrossRef]

56. Chen, S.M.; Ramachandran, R.; Mani, V.; Saraswathi, R. Recent advancements in electrode materials for the high-performance electrochemical supercapacitors: A review. *Int. J. Electrochem. Sci.* **2014**, *9*, 4072–4085.

57. Gan, J.K.; Lim, Y.S.; Pandikumar, A.; Huang, N.M.; Lim, H.N. Graphene/polypyrrole coated carbon nanofibre core–shell architecture electrode for electrochemical capacitors. *RSC Adv.* **2015**, *5*, 12692–12699. [CrossRef]

58. Chang, W.-M.; Wang, C.-C.; Chen, C.-Y. Plasma-induced polyaniline grafted on carbon nanotube-embedded carbon nanofibres for high-performance supercapacitors. *Electrochim. Acta* **2016**, *212*, 130–140. [CrossRef]

59. Tian, Y.; Yang, C.; Song, X.; Liu, J.; Zhao, L.; Zhang, P.; Gao, L. Engineering the volumetric effect of Polypyrrole for auto-deformable supercapacitor. *Chem. Eng. J.* **2019**, *374*, 59–67. [CrossRef]

60. Zhuo, H.; Hu, Y.; Chen, Z.; Zhong, L. Cellulose carbon aerogel/PPy composites for highperformance supercapacitor. *Carbohydr. Polym.* **2019**, *215*, 322–329. [CrossRef]

61. Yang, Z.; Ma, J.; Bai, B.; Qiu, A.; Losic, D.; Shi, D.; Chen, M. Free-standing PEDOT/polyaniline conductive polymer hydrogel for flexible solid-state supercapacitors. *Electrochim. Acta* **2019**, *322*, 134769. [CrossRef]

62. Wang, H.; Gao, M.; Zhu, Y.; Zhou, H.; Liu, H.; Gao, L.; Wu, M. A flexible 3-D structured carbon molecular sieve@PEDOT composite electrode for supercapacitor. *J. Electroanal. Chem.* **2018**, *826*, 191–197. [CrossRef]

63. Ravit, R.; Abdullah, J.; Ahmad, I.; Sulaiman, Y. Electrochemical performance of poly(3, 4-ethylenedioxythipohene)/nanocrystalline cellulose (PEDOT/NCC) film for supercapacitor. *Carbohydr. Polym.* **2019**, *203*, 128–138. [CrossRef]

64. Sharma, P.; Kumar, V. Current technology of supercapacitors: A review. *J. Electron. Mater.* **2020**, *49*, 3520–3532. [CrossRef]

65. Wang, T.; Chen, H.C.; Yu, F.; Zhao, X.S.; Wang, H. Boosting the cycling stability of transition metal compounds-based supercapacitors. *Energy Storage Mater.* **2019**, *16*, 545–573. [CrossRef]

66. Sun, J.; Xu, C.; Chen, H. A review on the synthesis of $CuCo_2O_4$-based electrode materials and their applications in supercapacitors. *J. Mater.* **2021**, *7*, 98–126. [CrossRef]

67. Jiang, W.; Hu, F.; Yan, Q.; Wu, X. Investigation on electrochemical behaviors of $NiCo_2O_4$ battery-type supercapacitor electrodes: The role of an aqueous electrolyte. *Inorg. Chem. Front.* **2017**, *4*, 1642–1648. [CrossRef]

68. Wang, Y.; Li, S.; Sun, J.; Zhang, Y.; Chen, H.; Xu, C. Simple solvothermal synthesis of magnesium cobaltite microflowers as a battery grade material with high electrochemical performances. *Ceram. Int.* **2019**, *45*, 14642–14651. [CrossRef]

69. Vijayakumar, S.; Nagamuthu, S.; Ryu, K.S. $CuCo_2O_4$ flowers/Ni-foam architecture as a battery type positive electrode for high performance hybrid supercapacitor applications. *Electrochim. Acta* **2017**, *238*, 99–106. [CrossRef]

70. Sun, J.; Li, S.; Han, X.; Liao, F.; Zhang, Y.; Gao, L.; Chen, H.; Xu, C. Rapid hydrothermal synthesis of snowflake-like $ZnCo_2O_4$/ZnO mesoporous microstructures with excellent electrochemical performances. *Ceram. Int.* **2019**, *45*, 12243–12250. [CrossRef]

71. Guo, X.; Chen, C.; Zhang, Y.; Xu, Y.; Pang, H. The application of transition metal cobaltites in electrochemistry. *Energy Storage Mater.* **2019**, *23*, 439–465. [CrossRef]

72. Ryu, K.S.; Kim, K.M.; Park, N.G.; Park, Y.J.; Chang, S.H. Symmetric redox supercapacitor with conducting polyaniline electrodes. *J. Power Sources* **2002**, *103*, 305–309. [CrossRef]

73. Li, D.; Huang, J.X.; Kaner, R.B. Polyaniline nanofibers: A unique polymer nanostructure for versatile applications. *Acc. Chem. Res.* **2008**, *42*, 135–145. [CrossRef]

74. Huang, J.X.; Kaner, R.B. Nanofiber formation in the chemical polymerization of aniline: A mechanistic study. *Angew. Chem.* **2004**, *116*, 5941–5945. [CrossRef]

75. Chiou, N.R.; Epstein, A.J. Polyaniline nanofibers prepared by dilute polymerization. *Adv. Mater.* **2005**, *17*, 1679–1683. [CrossRef]

76. Huang, J.X.; Virji, S.; Weiller, B.H.; Kaner, R.B. Polyaniline nanofibers: Facile synthesis and chemical sensors. *J. Am. Chem. Soc.* **2003**, *125*, 314–315. [CrossRef] [PubMed]

77. Huang, J.X.; Kaner, R.B. A general chemical route to polyaniline nanofibers. *J. Am. Chem. Soc.* **2004**, *126*, 851–855. [CrossRef] [PubMed]

78. Sivakkumar, S.R.; Kim, W.J.; Choi, J.A.; Macfarlane, D.R.; Forsyth, M.; Kim, D.W. Electrochemical performance of polyaniline nanofibres and polyaniline/multi-walled carbon nanotube composite as an electrode material for aqueous redox supercapacitors. *J. Power Sources* **2007**, *171*, 1062–1068. [CrossRef]

79. Li, H.L.; Wang, J.X.; Chu, Q.X.; Wang, Z.; Zhang, F.B.; Wang, S.C. Theoretical and experimental specific capacitance of polyaniline in sulfuric acid. *J. Power Sources* **2009**, *190*, 578–586. [CrossRef]
80. Xu, J.J.; Wang, K.; Zu, S.Z.; Han, B.H.; Wei, Z.X. Hierarchical nanocomposites of polyaniline nanowire arrays on graphene oxide sheets with synergistic effect for energy storage. *ACS Nano* **2010**, *4*, 5019–5026. [CrossRef]
81. Meng, C.Z.; Liu, C.H.; Chen, L.Z.; Hu, C.H.; Fan, S.S. Highly flexible and all-solid-state paperlike polymer supercapacitors. *Nano Lett.* **2010**, *10*, 4025–4031. [CrossRef]
82. Srinivasan, R.; Elaiyappillai, E.; Nixon, E.J.; Lydia, I.S.; Johnson, P.M. Enhanced electrochemical behaviour of Co-MOF/PANI composite electrode for supercapacitors. *Inorg. Chim. Acta* **2020**, *502*, 119393. [CrossRef]
83. Maier, M.A.; Babu, R.S.; Sampaio, D.M.; de Barros, A.L.F. Binder-free polyaniline interconnected metal hexacyanoferrates nanocomposites (Metal = Ni, Co) on carbon fibers for flexible supercapacitors. *J. Mater. Sci. Mater. Electron.* **2017**, *28*, 17405–17413. [CrossRef]
84. Babu, R.S.; de Barros, A.L.F.; Maier, M.A.; Sampaio, D.M.; Balamurugan, J.; Lee, J.H. Novel polyaniline/manganese hexacyanoferrate nanoparticles on carbon fiber as binder-free electrode for flexible supercapacitors. *Compos. B Eng.* **2018**, *143*, 141–147. [CrossRef]
85. Furukawa, H.; Cordova, K.E.; O'Keeffe, M.; Yaghi, O.M. The chemistry and applications of metal-organic frameworks. *Science* **2013**, *341*, 1230444. [CrossRef] [PubMed]
86. Zhou, H.-C.J.; Kitagawa, S. Metal-organic frameworks (MOFs). *Chem. Soc. Rev.* **2014**, *43*, 5415–5418. [CrossRef] [PubMed]
87. Miao, Y.-R.; Suslick, K.S. Chapter nine—Mechanochemical reactions of metal-organic frameworks. *Adv. Inorg Chem.* **2018**, *71*, 403–434.
88. Chul, H.D.; Vinodh, R.; Gopi, C.V.V.M.; Deviprasath, C.; Kim, H.-J.; Yi, M. Effect of the cobalt and zinc ratio on the preparation of zeolitic imidazole frameworks (ZIFs): Synthesis, characterization and supercapacitor applications. *Dalton Trans.* **2019**, *48*, 14808–14819. [CrossRef] [PubMed]
89. Kumar, P.; Kumar, P.; Bharadwaj, L.M.; Paul, A.; Deep, A. Luminescent nanocrystal metal organic framework based biosensor for molecular recognition. *Inorg. Chem. Commun.* **2014**, *43*, 114–117. [CrossRef]
90. Shown, I.; Ganguly, A.; Chen, L.C.; Chen, K.H. Conducting polymer-based flexible supercapacitor. *Energy Sci. Eng.* **2015**, *3*, 2–26. [CrossRef]
91. Kim, J.; Lee, J.; You, J.; Park, M.-S.; Al Hossain, M.S.; Yamauchi, Y.; Kim, J.H. Conductive polymers for next-generation energy storage systems: Recent progress and new functions. *Mater. Horiz.* **2016**, *3*, 517–535. [CrossRef]
92. Wang, Y.; Zhang, W.; Wu, X.; Luo, C.; Wang, Q.; Li, J.; Hu, L. Conducting polymer coated metal-organic framework nanoparticles: Facile synthesis and enhanced electromagnetic absorption properties. *Synth. Met.* **2017**, *228*, 18–24. [CrossRef]
93. Wang, L.; Feng, X.; Ren, L.; Piao, Q.; Zhong, J.; Wang, Y.; Li, H.; Chen, Y.; Wang, B. Flexible solid-state supercapacitor based on a metal-organic framework interwoven by electrochemically-deposited PANI. *J. Am. Chem. Soc.* **2015**, *137*, 4920–4923. [CrossRef]
94. Xu, M.; Guo, H.; Zhang, T.; Zhang, J.; Wang, X.; Yang, W. High-performance zeolitic imidazolate frameworks derived three-dimensional Co₃S₄/polyaniline nanocomposite for supercapacitors. *J. Energy Storage* **2021**, *35*, 102303. [CrossRef]
95. Iqbal, M.Z.; Faisal, M.M.; Ali, S.R.; Farid, S.; Afzal, A.M. Co-MOF/polyaniline-based electrode material for high performance supercapattery devices. *Electrochim. Acta* **2020**, *346*, 136039. [CrossRef]
96. Yao, M.; Zhao, X.; Zhang, Q.; Zhang, Y.; Wang, Y. Polyaniline nanowires aligned on MOFs-derived nanoporous carbon as high-performance electrodes for supercapacitor. *Electrochim. Acta* **2021**, *390*, 138804. [CrossRef]
97. Salunkhe, R.R.; Tang, J.; Kobayashi, N.; Kim, J.; Ide, Y.; Tominaka, S.; Kim, J.H.; Yamauchi, Y. Ultrahigh performance supercapacitors utilizing core–shell nanoarchitectures from a metal–organic framework-derived nanoporous carbon and a conducting polymer. *Chem. Sci.* **2016**, *7*, 5704–5713. [CrossRef] [PubMed]
98. Konstantin, A.M.; Gavrilov, N.; Pasti, I.A.; Moravkova, Z.; Acharya, U.; Unterweger, C.; Breitenbach, S.; Zhigunov, A.; Bober, P. Polyaniline-metal organic framework (Fe-BTC) composite for electrochemical applications. *Polymer* **2020**, *208*, 122945.
99. Wang, M.; Ma, Y.; Ye, J. Controllable layer-by-layer assembly of metal-organic frameworks/polyaniline membranes for flexible solid-state microsupercapacitors. *J. Power Sources* **2020**, *474*, 228681. [CrossRef]
100. Guo, S.N.; Zhu, Y.; Yan, Y.Y.; Min, Y.L.; Fan, J.C.; Xu, Q.J.; Yun, H. (Metal-Organic Framework)-Polyaniline sandwich structure composites as novel hybrid electrode materials for high-performance supercapacitor. *J. Power Sources* **2016**, *316*, 176–182. [CrossRef]
101. Shao, L.; Wang, Q.; Ma, Z.; Ji, Z.; Wang, X.; Song, D.; Liu, Y.; Wang, N. A high-capacitance flexible solid-state supercapacitor based on polyaniline and metal-organic framework (UiO-66) composites. *J. Power Sources* **2018**, *379*, 350–361. [CrossRef]
102. Liu, P.-Y.; Zhao, J.-J.; Dong, Z.-P.; Liu, Z.-L.; Wang, Y.-Q. Interwoving polyaniline and a metal-organic framework grown in situ for enhanced supercapacitor behavior. *J. Alloys Compd.* **2021**, *854*, 157181. [CrossRef]
103. Xu, M.; Wang, X.; Ouyang, X.; Xu, Z. Two-Dimensional Metal-organic framework nanosheets grown on carbon fiber paper interwoven with polyaniline as an electrode for supercapacitors. *Energy Fuels* **2021**, *35*, 19818–19826. [CrossRef]
104. Udayan, A.P.M.; Sadak, O.; Gunasekaran, S. Metal-organic framework/polyaniline nanocomposites for lightweight energy storage. *ACS Appl. Energy Mater.* **2020**, *3*, 12368–12377. [CrossRef]
105. Neisi, Z.; Ansari-Asl, Z.; Dezfuli, A.S. Polyaniline/Cu(II) metal-organic frameworks composite for high performance supercapacitor electrode. *J. Inorg. Organomet. Polym. Mater.* **2019**, *29*, 1838–1847. [CrossRef]

106. Tian, D.; Wang, C.; Lu, X. Metal—Organic frameworks and their derived functional materials for supercapacitor electrode application. *Adv. Energy Sustain. Res.* **2021**, *2*, 2100024. [CrossRef]
107. Gong, J.; Xu, Z.; Tang, Z.; Zhong, J.; Zhang, L. Highly compressible 3-D hierarchical porous carbon nanotube/metal organic framework/polyaniline hybrid sponges supercapacitors. *AIP Adv.* **2019**, *9*, 055032. [CrossRef]
108. He, L.; Liu, J.; Yang, L.; Song, Y.; Wang, M.; Peng, D.; Zhang, Z.; Fang, S. Copper metal-organic framework-derived CuO_x-coated three-dimensional reduced graphene oxide and polyaniline composite: Excellent candidate free-standing electrodes for high-performance supercapacitors. *Electrochim. Acta* **2018**, *275*, 133–144. [CrossRef]
109. Wang, Y.G.; Li, H.Q.; Xia, Y.Y. Ordered whisker like polyaniline grown on the surface of mesoporous carbon and its electrochemical capacitance performance. *Adv. Mater.* **2006**, *18*, 2619–2623. [CrossRef]
110. Liu, Y.-N.; Jin, L.-N.; Wang, H.-T.; Kang, X.-H.; Bian, S.-W. Fabrication of three-dimensional composite textile electrodes by metal-organic framework, zinc oxide, graphene and polyaniline for all-solid-state supercapacitors. *J. Colloid Interface Sci.* **2018**, *530*, 29–36. [CrossRef]
111. Sundriyal, S.; Kaur, H.; Bhardwaj, S.K.; Mishra, S.; Kim, K.-H.; Deep, A. Metal-organic frameworks and their composites as efficient electrodes for supercapacitor applications. *Coord. Chem. Rev.* **2018**, *369*, 15–38. [CrossRef]
112. Fu, D.; Li, H.; Zhang, X.-M.; Han, G.; Zhou, H.; Chang, Y. Flexible solid-state supercapacitor fabricated by metal-organic framework/graphene oxide hybrid interconnected with PEDOT. *Mater. Chem. Phys.* **2016**, *179*, 166–173. [CrossRef]
113. Mulzer, C.R.; Shen, L.; Bisbey, R.P.; McKone, J.R.; Zhang, N.; Abruña, H.D.; Dichtel, W.R. Superior Charge storage and power density of a conducting polymer-modified covalent organic framework. *ACS Cent. Sci.* **2016**, *2*, 667–673. [CrossRef]
114. Zhang, L.; Yao, H.; Li, Z.; Sun, P.; Liu, F.; Dong, C.; Wang, J.; Li, Z.; Wu, M.; Zhang, C.; et al. Synthesis of delaminated layered double hydroxides and their assembly with graphene oxide for supercapacitor application. *J. Alloys Compd.* **2017**, *711*, 31–41. [CrossRef]
115. Yu, L.; Shi, N.; Liu, Q.; Wang, J.; Yang, B.; Wang, B.; Yan, H.; Sun, Y.; Jing, X. Facile synthesis of exfoliated Co–Al LDH–carbon nanotube composites with high performance as supercapacitor electrodes. *Phys. Chem. Chem. Phys.* **2014**, *16*, 17936–17942. [CrossRef] [PubMed]
116. Vinodh, R.; Babu, R.S.; Atchudan, R.; Kim, H.-J.; Yi, M.; Samyn, L.M.; de Barros, A.L.F. Fabrication of High-Performance Asymmetric Supercapacitor Consists of Nickel Oxide and Activated Carbon (NiO//AC). *Catalysts* **2022**, *12*, 375. [CrossRef]
117. Atchudan, R.; Edison, T.N.J.I.; Perumal, S.; Thirukumaran, P.; Vinodh, R.; Lee, Y.R. Green synthesis of nitrogen-doped carbon nanograss for supercapacitors. *J. Taiwan Inst. Chem. Eng.* **2019**, *102*, 475–486. [CrossRef]
118. Xing, T.; Ouyang, Y.; Chen, Y.; Zheng, L.; Wu, C.; Wang, X. P-doped ternary transition metal oxide as electrode material of asymmetric supercapacitor. *J. Energy Storage* **2020**, *28*, 101248. [CrossRef]
119. Masikhwa, T.M.; Barzegar, F.; Dangbegnon, J.K.; Bello, A.; Madito, M.J.; Momodu, D.; Manyala, N. Asymmetric supercapacitor based on VS_2 nanosheets and activated carbon materials. *RSC Adv.* **2016**, *6*, 38990–39000. [CrossRef]
120. Neeraj, N.S.; Mordina, B.; Srivastava, A.K.; Mukhopadhyay, K.; Prasad, N.E. Impact of process condi-tions on the electrochemical performances of NiMoO_4 nanorods and activated carbon based asymmetric supercapacitor. *Appl. Surf. Sci.* **2019**, *473*, 807–819. [CrossRef]
121. Gao, L.; Xiong, L.; Xu, D.; Cai, J.; Huang, L.; Zhou, J.; Zhang, L. Distinctive Construction of Chitin-Derived Hierarchically Porous Carbon Microspheres/Polyaniline for High-Rate Supercapacitors. *ACS Appl. Mater. Interfaces* **2018**, *10*, 28918–28927. [CrossRef]
122. Salleh, N.A.; Kheawhom, S.; Mohamad, A.A. Chitosan as biopolymer binder for graphene in superca-pacitor electrode. *Results Phys.* **2021**, *25*, 104244. [CrossRef]

Article

Zinc-Ion Storage Mechanism of Polyaniline for Rechargeable Aqueous Zinc-Ion Batteries

Jiangfeng Gong [1,*], Hao Li [1], Kaixiao Zhang [1], Zhupeng Zhang [1], Jie Cao [1], Zhibin Shao [1], Chunmei Tang [1,*], Shaojie Fu [2], Qianjin Wang [2] and Xiang Wu [3,*]

[1] College of Science, Hohai University, Nanjing 210098, China; lihao5799@163.com (H.L.); kxzhang@126.com (K.Z.); zzp18752006001@163.com (Z.Z.); caojie@hhu.edu.cn (J.C.); zbshao@hhu.edu.cn (Z.S.)
[2] National Laboratory of Microstructures, Nanjing University, Nanjing 210093, China; fushaojie@nju.edu.cn (S.F.); qjwang@nju.edu.cn (Q.W.)
[3] School of Materials Science and Engineering, Shenyang University of Technology, Shenyang 110870, China
* Correspondence: jfgong@hhu.edu.cn (J.G.); cmtang@hhu.edu.cn (C.T.); wuxiang05@sut.edu.cn (X.W.)

Abstract: Aqueous multivalent ion batteries, especially aqueous zinc-ion batteries (ZIBs), have promising energy storage application due to their unique merits of safety, high ionic conductivity, and high gravimetric energy density. To improve their electrochemical performance, polyaniline (PANI) is often chosen to suppress cathode dissolution. Herein, this work focuses on the zinc ion storage behavior of a PANI cathode. The energy storage mechanism of PANI is associated with four types of protonated/non-protonated amine or imine. The PANI cathode achieves a high capacity of 74 mAh g^{-1} at 0.3 A g^{-1} and maintains 48.4% of its initial discharge capacity after 1000 cycles. It also demonstrates an ultrahigh diffusion coefficient of $6.25 \times 10^{-9} \sim 7.82 \times 10^{-8}$ cm^{-2} s^{-1} during discharging and $7.69 \times 10^{-10} \sim 1.81 \times 10^{-7}$ cm^{-2} s^{-1} during charging processes, which is one or two orders of magnitude higher than other reported studies. This work sheds a light on developing PANI-composited cathodes in rechargeable aqueous ZIBs energy storage devices.

Keywords: zinc-ion batteries; conducting polymers; polyaniline; zinc-ion diffusion

Citation: Gong, J.; Li, H.; Zhang, K.; Zhang, Z.; Cao, J.; Shao, Z.; Tang, C.; Fu, S.; Wang, Q.; Wu, X. Zinc-Ion Storage Mechanism of Polyaniline for Rechargeable Aqueous Zinc-Ion Batteries. *Nanomaterials* 2022, 12, 1438. https://doi.org/10.3390/nano12091438

Academic Editors: Christian M. Julien and Ullrich Scherf

Received: 18 February 2022
Accepted: 14 April 2022
Published: 23 April 2022

Publisher's Note: MDPI stays neutral with regard to jurisdictional claims in published maps and institutional affiliations.

1. Introduction

To build a low-carbon society, green energy sources such as solar energy and wind energy were developed rapidly. A challenge exists in terms of how we can adapt these intermittency renewables to the electricity grid. Thus, it is essential to develop large-scale electrochemical energy storage technologies. In recent years, much effort has been focused on aqueous multivalent ion batteries (zinc-ion batteries (ZIBs) [1,2], magnesium-ion batteries [3], calcium-ion batteries [4], and aluminum-ion batteries [5]) according to the following reasons: (1) The aqueous electrolytes are much safer than flammable organic electrolytes. (2) The ionic conductivity of aqueous electrolytes ($\sim$1 S cm^{-1}) is much higher than that of organic electrolytes ($\sim$1$-$10 mS cm^{-1}), which enable a fast intercalation/de-intercalation rate. (3) During charge/discharge processes, multivalent ions enable more than one electron transfer, which imply that multivalent ion batteries can offer high gravimetric energy densities.

Exploring high-performance electrode active materials is a critical factor to construct advanced energy storage batteries. Until now, a variety of active materials has been developed and assembled as rechargeable batteries [6–9]. During long cyclic usage, the electrochemical performance of the assembled rechargeable batteries inevitably shows degradation, and the reliability of the batteries is seriously limited. J.W. Wang et al. studied the lithiation/delithiation of micro-sized Sn particles using the in situ transmission electron microscopy technique; the results demonstrated that degradation is attributed to particle pulverization generated by the lithiation-induced, large and inhomogeneous volume changes [10]. Similar conclusions were demonstrated by Y. Sun in pulverized V_2O_5 powder [11]. To overcome the capacity fade originating from irreversible phase conversion and structure dissolution, a conductive polymer was often chosen by researchers to suppress cathode dissolution. Among the family of conductive polymers, polyaniline (PANI) was the most popular media because of its high conductivity and reversible electrochemical response during anodic oxidation and cathodic reduction. For example, J. H. Huang et al. designed a polyaniline-intercalated-layered MnO_2, in which the PANI polymer eliminated phase change and alleviated volume change upon cation insertion/extraction [12]. W.J. Li et al. designed a vanadium oxide (V_2O_{5-x})/PANI superlattice to strengthen the alternative layered structure, where the PANI layer restrains the dissolution of V_2O_{5-x} active materials in aqueous electrolytes, which worked as structural stabilizer, enabling a high-rate capability and a long-term cycling life [13]. The PANI-GO/CNT cathode and PANI-intercalated VOH were also widely reported [14,15]. In such energy storage systems, PANI jumbles with host materials, which show the synergistic energy storage effect. Despite reports showing high specific capacity, the reasons for the improved performance after adding PANI remain ambiguous and need to be further explored; in particular, there is a lack of comprehensive studies on the charge storage mechanism and ion transport kinetics of PANI.

In view of the attractive properties of aqueous multivalent ion batteries, we investigated the electrochemical performance and ion transport kinetics of PANI cathode to further understand Zn^{2+} storage mechanisms. It was found that the charge/discharge processes of PANI can be controlled by protonation, and it is associated with four types of nitrogen, including non-protonated amine $-NH-$, protonated amine $-NH^+-$, non-protonated imine $-N=$, and protonated imine $-NH^+=$. The assembled PANI/Zn cell achieves a high capacity of 74 mAh g^{-1} at 0.3 A g^{-1} and maintains 48.4% of its initial discharge capacity after 1000 cycles. Importantly, the Zn^{2+} diffusion coefficient in the PANI cathode is within the range of 6.25×10^{-9} to 7.82×10^{-8} cm^{-2} s^{-1} for discharge processes and 7.69×10^{-10} to 1.81×10^{-7} cm^{-2} s^{-1} for charge processes, which is one or two orders of magnitude higher than any other reported cathode materials for ZIBs [11–13,16–18]. Our findings herein will inspire the modification of PANI-intercalated cathode materials for high performance ZIBs.

2. Materials and Methods

2.1. Chemical Reagents

All chemical reagents were of analytical grade and were used as received without further purification. Sulphuric acid (H_2SO_4, 98%) and aniline (99.5%) were purchased from Chengdu Kelong Chemical Reagent Co. (Chengdu, China). Stainless steel films and zinc foil were purchased from Guangdong Canrd New Energy Technology Co., Ltd. (Dongguan, China). All solutions were prepared with deionized water.

2.2. Materials Preparation

PANI films were anodically electrodeposited by cyclic voltametric (CV) methods on an electrochemical workstation (CHI660E, Chenhua, Shanghai, China). Saturated calomel electrode (SCE, the potential vs. SHE is 199 mV) and platinum sheets were used as the reference electrode and counter electrode, respectively. After cleaning by plasma bombardment to optimize hydrophilicity, the stainless-steel substrates were carefully coated with a thick film Polyvinyl chloride (PVC) with an exposed surface area of 1.54 cm^2. The electrolyte was prepared by dropping 2.72 mL H_2SO_4 into 200 mL 2 M aniline solution with vigorous stirring until obtaining a clear brown solution. The PANI film was electroplated at a scan rate of 25 mV s^{-1} for 30 cycles ranging from -0.1 to 0.9 V. After deposition, the as-prepared PANI film was carefully washed with distilled water to remove unreacted aniline and dried out in a drying cabinet. The mass loading of active material was around 1.0 mg cm^{-2}

2.3. Physicochemical Characterizations

An X-ray diffractometer (XRD, D8 ADVANCE, Bruker, Karlsruhe, Germany) using Cu Kα radiation (λ = 1.5418 Å) was used to analyze the phases and structures of the deposited films. A scanning electron microscope (SEM, Quanta 200, FEI, Hillsborough, OR, USA) was used to study the morphologies and microstructures of the samples. Transmission electron microscopy (TEM, Tecnai F20, FEI, Hillsborough, OR, USA) and high-resolution TEM images were taken to confirm the size as well as the crystalline structure of the PANI film. Integrated elemental compositions over an area was collected using energy dispersive X-ray spectroscopy (EDS GENESIS Apex, EDAX Inc. Mahwah, NJ, USA) equipped with TEM. X-ray photoelectron spectroscopy (XPS ESCALAB 250 Xi, Thermo Fisher Scientific, Waltham, MA, USA) measurements were performed by using a monochromatic Al Kα X-ray beam (1486.6 eV), The binding energies were calibrated using C 1s peak (BE = 284.6 eV) as a standard.

2.4. Electrochemical Measurements

The PANI/Zn batteries were assembled using PANI film with stainless-steel substrates as the cathode, Zn foil (diameter: 15.6 mm, thickness: 50 μm) as the anode, and Whatman glass fiber as the separator in CR2032 coin cells. A 2 M quantity of Zn $(CF_3SO_3)_2$ was used as the aqueous electrolyte. All cells were assembled in the ambient environment. The electrochemical performance measurements were performed by a multichannel battery testing system (CT-4008, Neware, Shenzhen, China) with a voltage window of 0.3–1.8 V (vs. Zn^{2+}/Zn) at 20 °C. The specific capacity was calculated based on the mass of PANI in cathode. CV curves were collected on an electrochemical workstation (CHI660, Chenhua, Shanghai, China) within the same voltage window at different scan rates from 0.1 to 1 mV s^{-1}. The electrochemical impedance spectra (EIS) were performed in a frequency range of 10^{-2}~10^5 Hz with an AC voltage amplitude of 5 mV (CHI660, Chenhua, Shanghai, China).

3. Results

The PANI electrode was prepared on stainless steel through a facile electrodeposition method. During the electrochemical polymerization process, aniline monomers polymerized and formed long-chain PANI. The typical microstructures of PANI are presented in Figure 1a, which shows a continuous three-dimensional network. The pure PANI film shows short rods clusters with diameters of ~50 nm and lengths of 150–200 nm (Figure 1b), which can provide enough electrochemical active sites for adsorbing ions. The high-resolution TEM image in Figure 1c shows their short worm-like characterization, and the selected area electron diffraction (SAED) of PANI (inset of Figure 1c) presents dispersed diffraction rings, which illustrates the amorphous character of the sample. The XRD pattern (Supplementary Figure S1) shows the amorphous nature of PANI. However, some signals are launched at 2θ = 6.3°. The signal is assigned as the periodicity distance between the dopant and N atom on adjacent main chains [19,20]. Figure 1d–f shows the EDS mapping of the PANI film, and the dashed line shows the outline of Figure 1b. The elements of N, O, and S are distributed uniformly, implying the homogenous doping of SO_4^{2-} in the polyaniline's long chain.

XPS was also carried out to characterize the valence states and chemical composition of PANI film. The survey XPS scans of the PANI films indicate the presence of sulfur (S 2p, 168.85 eV), carbon (C 1s, 286.32 eV), nitrogen (N 1s, 401.07 eV), and oxygen (O 1s, 533.21 eV), as shown in Supplementary Figure S2. C and N are expected to originate from PANI film, while S may derive from H_2SO_4 in electrochemical solutions. The N 1s core level XPS spectrum can be deconvoluted into four peaks, as shown in Figure 2a. The peak at 398.48 eV corresponds to −N= (quinoid imine), the main peak at 399.38 eV is ascribed to −NH− (benzenoid amine), and the two remaining peaks located at 400.53 and 401.79 eV may be attributed to protonated nitrogen −NH$^+$− and −NH$^+$= [21]. Moreover, the XPS analysis of S 2p can be deconvoluted into S 2p$_{1/2}$ (169.6eV) and S 2p$_{3/2}$ (168.6eV) in Figure 2b, and the S 2p peak is fitted with the spin-orbit doublets of sulfate groups. The doped $SO_4{}^{2-}$ remaining in PANI's long chain could play the role of rapid balance charges during redox reactions [22]. The morphological and structural advantages of the PANI cathode described above are favorable for ion diffusion and Zn ion storage during charge/discharge.

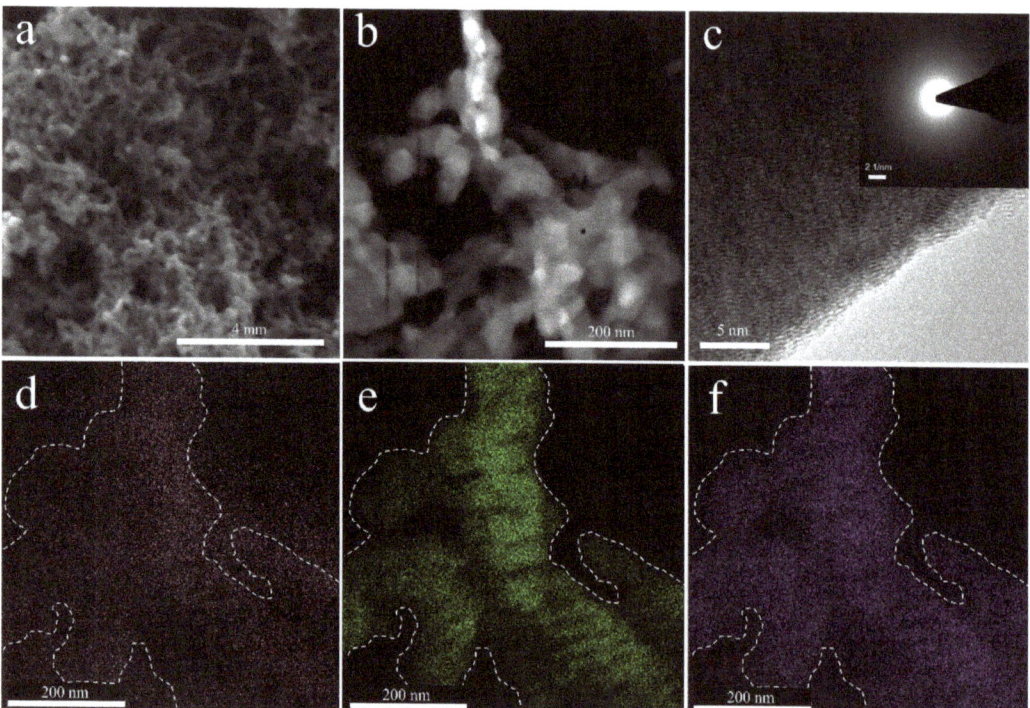

Figure 1. (**a**) SEM, (**b**) TEM, and (**c**) high-resolution TEM images of PANI film. The inset of (**c**) shows the corresponding SAED image. (**d**–**f**) Corresponding EDS mapping of the nitrogen (N), oxygen (O), and sulfur (S) in PANI film.

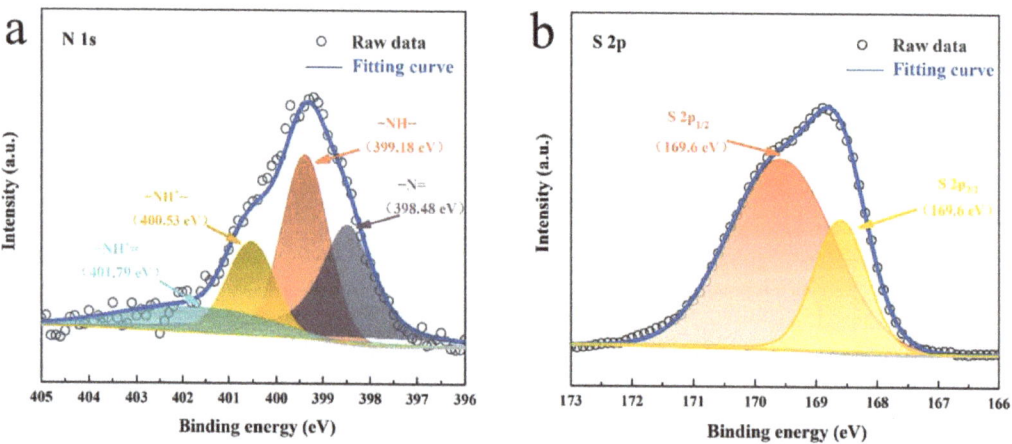

Figure 2. Core level XPS of (**a**) N 1s and (**b**) S 2p.

Nanomaterials **2022**, *12*, 1438

The electrochemical profile is characterized in the typical 2032 cell. Figure 3a shows CV curve tested at 0.1 mV s^{-1} with the potential window of 0.3~1.8 V. There is one pair of cathodic peaks (R and O$_2$ marked in Figure 3a) and one small shoulder (O$_1$) next to the O$_2$ peak. These represent the reduction/oxidation process during adsorption/desorption of Zn^{2+}. The galvanostatic discharge–charge curves of PANI in Figure 3b show a steeper slope, especially at large current densities and a high discharge capacity of 74 mA h g^{-1} at 0.3 A g^{-1}. The rapid charge–discharge speed corresponds to the fast ion absorb–desorption and redox reaction. In rate capability tests (Figure 3c), the PANI electrode with a mass loading of 1 mg cm^{-2} delivers a relatively stable capacity. With an increase in current density from 0.3 to 0.5, 0.7, 1, and 2 A g^{-1}, the cell delivers specific capacities of 68, 68, 58, and 40 mA h g^{-1}, respectively. When the current densities decrease back to 0.3 A g^{-1} from 2 A g^{-1}, the capacities recover to the initial values, suggesting a stable structure and great electrochemical reversibility. The EIS spectra and the equivalent circuit model of PANI are presented in Figure 3d. The impedance measurements are taken after discharging at the 1st cycle and 50th cycle. Both spectra comprised a semicircle in the high frequency region, which originated from the solid/electrolyte interfacial resistance. The interception between semicircle and the real axis corresponds to the migrating resistance of Zn^{2+} ions through the surface layer (R$_s$), and the semicircle represents the charge transfer resistance (Rct) [23]. The Rct value transformed from 398.2 Ω to 195.5 Ω after cycling, which might be attributed to the activation of materials. While at the low frequency region, the inclined line is caused by the Zn^{2+} ions' chemical diffusion impedance (Warburg impedance). The result of EIS further demonstrates that the PANI film with amorphous nature effectively enhances the electrochemical kinetics by decreasing the impedances. As shown in Figure 3e, the 3D conductive network PANI ZIBs could maintain a discharge capacity of 30 mAh g^{-1} (48.4% of its initial discharge capacity) after 1000 cycles with high Coulombic efficiency close to 100%. The polymer retains the 3D network's morphology without being peeled off from the substrate. Such high stability of the electrode guarantees excellent capacity retention.

To comprehensively understand the energy storage kinetics of the Zn/PANI batteries, CV curves at various scan rates are shown in Figure 4a. With the increased scan rates from 0.1 mV s^{-1} to 1 mV s^{-1}, the CV curves keep similar shapes with subtle shifts in redox peaks, indicating a fast and stable Zn^{2+} adsorption/desorption process even at the high scan rates. Their peak currents (*i*) and scan rates (*v*) have a relationship [6,13,24,25]: $i = av^b$, where *a* and *b* are adjustable parameters. When the value of *b* is close to 0.5, the reaction process relies on the control of ionic diffusion processes. When the value of *b* reaches 1, the corresponding electrochemical behavior is controlled by capacitance. According to the slopes of the log(*i*) vs. log(*v*) plots of all peaks in Figure 4b, the calculated *b* values for peaks O$_1$, O$_2$, and R are 0.57, 0.85, and 0.97, respectively. The value of peak O$_1$ is very close to 0.5, which is mostly dominated by diffusion-controlled capacitance. While peak O$_2$ and R imply that the surface-dominated pseudocapacitance contribution plays a major role in the following charge storage stage. Along with the increase in scan rate, the capacitive contribution increases and finally reaches to about 96.62% at a scan rate of 1 mV s^{-1} (Figure 4c,d). The large capacitive contribution at low sweep rates suggests a unique pseudocapacitive effect, which can be attributed to the unique surface-dominated reaction. This kind of capacitive-dominated behavior further indicates that fast electrochemical kinetics can match fast surface reactions. Galvanostatic intermittent titration technique (GITT) measurements were further carried out to reveal the kinetics of Zn^{2+} diffusion in PANI electrodes during the cycles. The discharge/charge curves and corresponding diffusion coefficient of Zn^{2+} (D) in GITT measurement for PANI electrodes during the cycles are shown in Figure 4e. The details of the diffusion coefficient calculation are shown in Supplementary Materials. The calculated D values of PANI cathode ranges from 6.25 × 10^{-9} to 7.82 × 10^{-8} cm^{-2} s^{-1} during the two discharge processes and 7.69 × 10^{-10} to 1.81 × 10^{-7} cm^{-2} s^{-1} during the charge processes, which is one or two orders of magnitude higher than other reported manganese oxide and vanadium oxide cathode materials for ZIBs (Table 1) [9,11,13,15,17,25–32]. This result demonstrates that the diffusion kinetics of Zn^{2+} through PANI is quicker and easier, which may be attributed to lower number of electrostatic interactions between Zn^{2+} and host sites reduced by the 3D conductive network's morphology. PANI also boosts the electrical conductivity of the electrode, allowing for sufficient electrical charge transfers to accommodate the rapid diffusion of Zn^{2+} in the electrode.

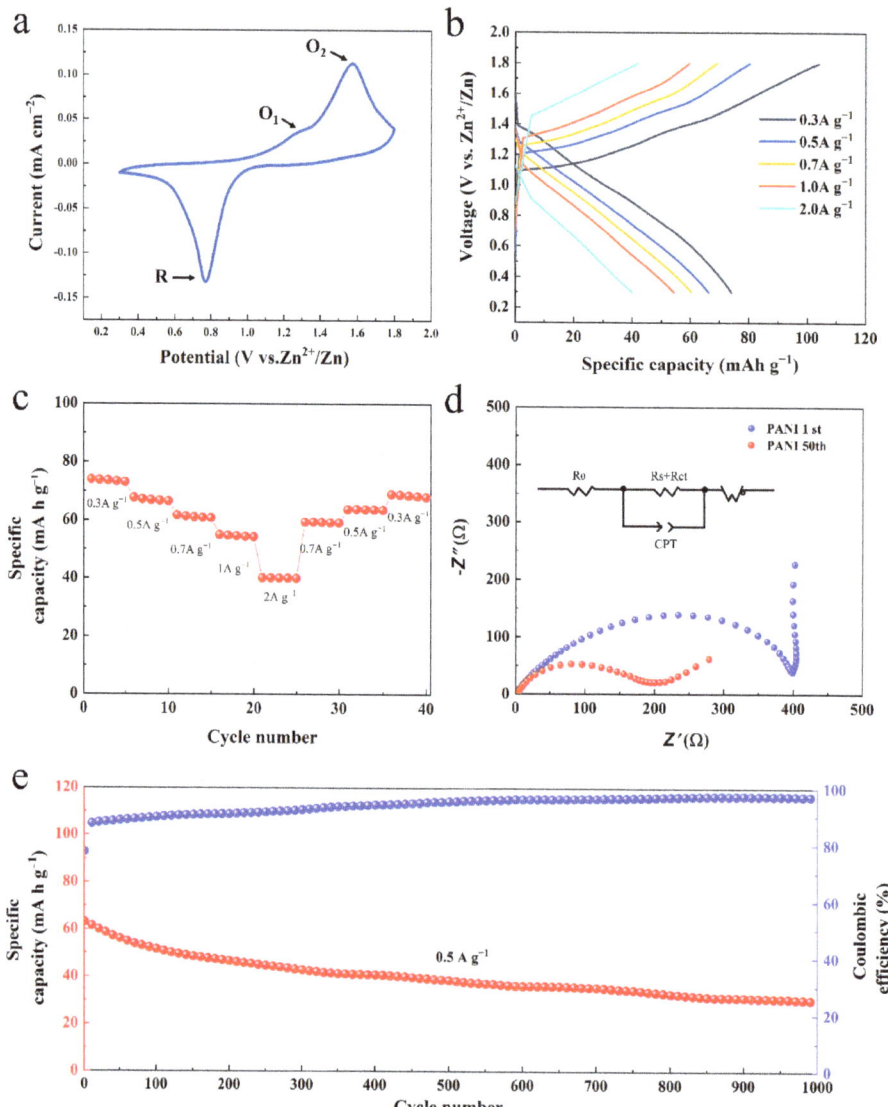

Figure 3. Electrochemical performance of PANI electrode. (**a**) CV curve measured at 0.1 mV s^{-1}. (**b**) Typical galvanostatic charge/discharge curves scanned from 0.3 A g^{-1} to 2 A g^{-1}. (**c**) Rate performance with the charge/discharge current densities varying from 300 to 2000 mA g^{-1}. (**d**) EIS Nyquist plots and (**e**) cycling performance of the cell.

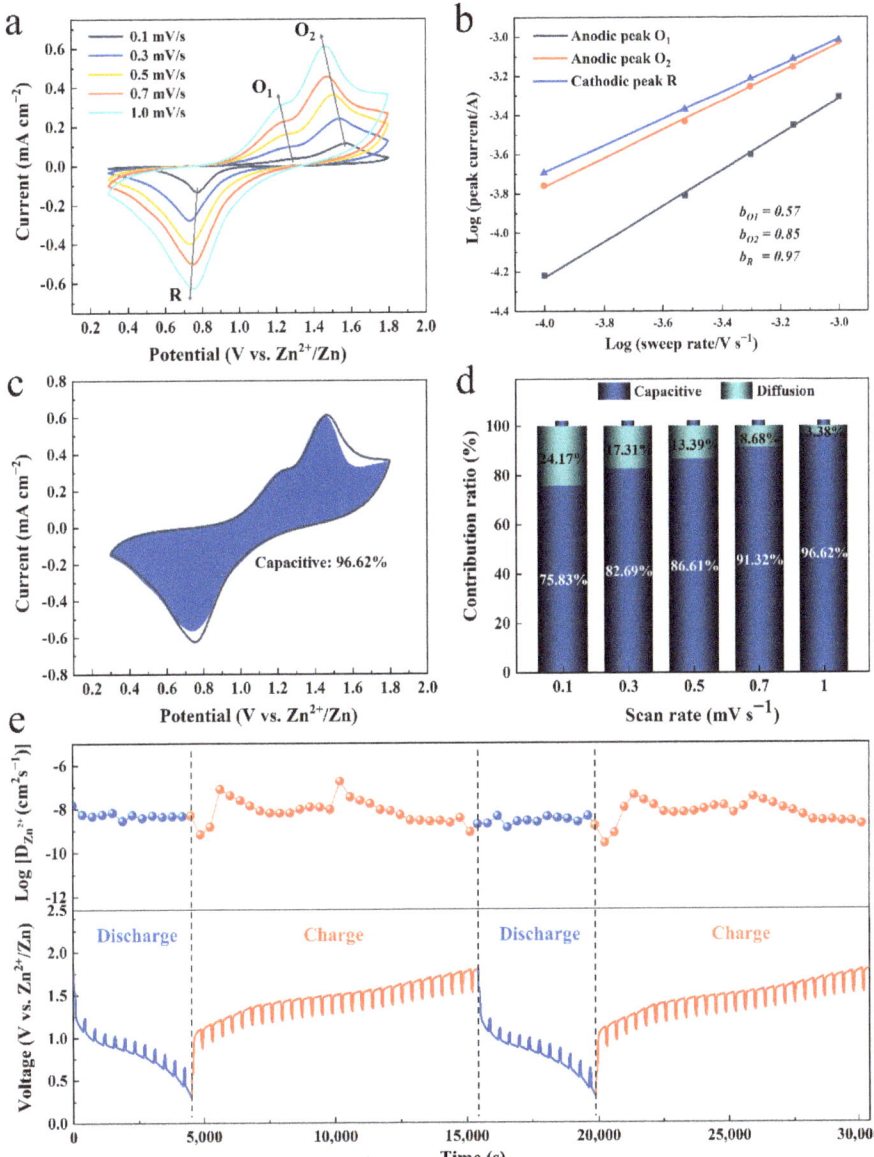

Figure 4. Diffusion kinetics characterization of PANI cathode. (**a**) CV curves of PANI electrode at different scan rates. (**b**) log(*v*) vs. log(*i*) plots at the peak current. (**c**) Contribution ratio of diffusion-controlled vs. capacitive-controlled capacities obtained at 0.1 mV s^{-1}. (**d**) Contribution ratio of diffusion-controlled vs. capacitive-controlled capacities at different scan rates. (**e**) GITT analysis results for PANI electrode and corresponding Zn^{2+} diffusion coefficient.

Table 1. Diffusion coefficient of Zn^{2+} in referenced cathode materials.

Active Materials	Electrolyte	Diffusion Coefficient $(cm^{-2}\ s^{-1})$	Reference
V_2O_5@CNTs	1 M $ZnSO_4$	$10^{-10} \sim 10^{-8}$ (Discharging) $10^{-12} \sim 10^{-8}$ (Charging)	[9]
V_2O_5	2 M $ZnSO_4$	1.32×10^{-12} (Discharging) 3.82×10^{-11} (Charging)	[11]
$V_2O_5 \cdot nH_2O$	2 M $ZnSO_4$	2.4×10^{-9} (Discharging)	[13]
PANI$-$VOH	3 M $Zn(CF_3SO_3)_2$	$10^{-16} \sim 10^{-13}$ (Discharging) $10^{-14} \sim 10^{-13}$ (Charging)	[15]
V_2O_5	$ZnSO_4$	$10^{-11} \sim 10^{-9}$ (Discharging)	[17]
MnVO/VOH	3 M $Zn(CF_3SO_3)_2$	$3.22 \times 10^{-12} \sim$ (Discharging) $1.46 \times 10^{-12} \sim$ (Charging)	[25]
$Mn_{0.15}V_2O_5 \cdot nH_2O$	1 M $Zn(ClO_4)_2$	$10^{-12} \sim 10^{-10}$ (Discharging)	[26]
Graphene Scroll Coated α-MnO_2	2 M $ZnSO_4$ 0.2 M $MnSO_4$	$10^{-17} \sim 10^{-12}$ (Discharging)	[27]
MnO_2 nanospheres	2 M $ZnSO_4$ 0.2 M $MnSO_4$	$10^{-15} \sim 10^{-12}$ (Discharging)	[28]
δ-MnO_2	3 M $ZnSO_4$ 0.15 M $MnSO_4$	$10^{-13} \sim 10^{-9}$ (Discharging) $10^{-11} \sim 10^{-9}$ (Charging)	[29]
$(NH_4)_2V_{10}O_{25} \cdot 8H_2O$	3 M $Zn(CF_3SO_3)_2$	$10^{-10} \sim 10^{-9}$ (Discharging)	[30]
$V_5O_{12} \cdot 6H_2O$ (VOH)	3 M $Zn(CF_3SO_3)_2$	$10^{-11} \sim 10^{-10}$ (Discharging)	[31]
$K_2V_8O_{21}$	2 M $ZnSO_4$	$1.99 \times 10^{-11} \sim 2.23 \times 10^{-10}$ (Discharging)	[32]
PANI	2 M $Zn(CF_3SO_3)_2$	$6.25 \times 10^{-9} \sim 7.82 \times 10^{-8}$ (Discharging) $7.69 \times 10^{-10} \sim 1.81 \times 10^{-7}$ (Charging)	This work

To further understand the charge storage mechanism of PANI during the reversible redox reaction, ex situ XPS of N 1s analyses (Figure 5a) were performed on a PANI cathode at different charge/discharge voltages. The N 1s XPS spectra of fully charged PANI cathode were fitted with four peaks related to non-protonated amine $-NH-$, protonated amine $-NH^+-$, non-protonated imine $-N=$, and protonated imine $-NH^+=$, located at 399.38 eV, 400.53 eV, 398.48 eV, and 401.79 eV, respectively. The XPS of S 1s (Supplementary Figure S3) can be divided into two pairs of characteristic peaks, SO_4^{2-} and SO_3^{-}. During the charge process (from I to III), the peak intensities of $-NH^+=$ at 401.79 eV and $-N=$ at 398.48 eV are strengthened gradually, whereas in the following discharge process (from III to V), the peak intensities are weakened and ultimately recovered to the original state, which is arising from the reversible reactions between protonated and non-protonated PANI. After full discharge, the XPS analysis' results in point I show only two components of $-NH-$ and $-NH^+-$ with the proportion of 59% and 41%, respectively (Figure 5b). When the battery charges from the initial 0.3 to 1.8 V (from I to III), the intensity of $-NH-$ and $-NH^+-$ decreases while the intensity of $-NH^+=$ and $-N=$ increases, as a result of the protonation process. The N 1s XPS spectrum of fully charged PANI cathode is fitted with four peaks related to $-NH-$ (38%), $-NH^+-$ (16%), $-N=$ (22%), and $-NH^+=$ (24%), respectively. Generally, the former one N signal is referred to the reduced state, while the last three N signals correspond to the oxidized state. While SO_4^{2-} increases to balance the charge and the $-SO_3^-H^+$ external dopant PANI cathode is charged to 1.8 V (state III), the amount of oxidized state increases and reduced state decreases. In the oxidation process, the oxidation of the non-protonated components is more facile than that of the protonated $-NH^+-$ due to an easier loss of electrons for the former.

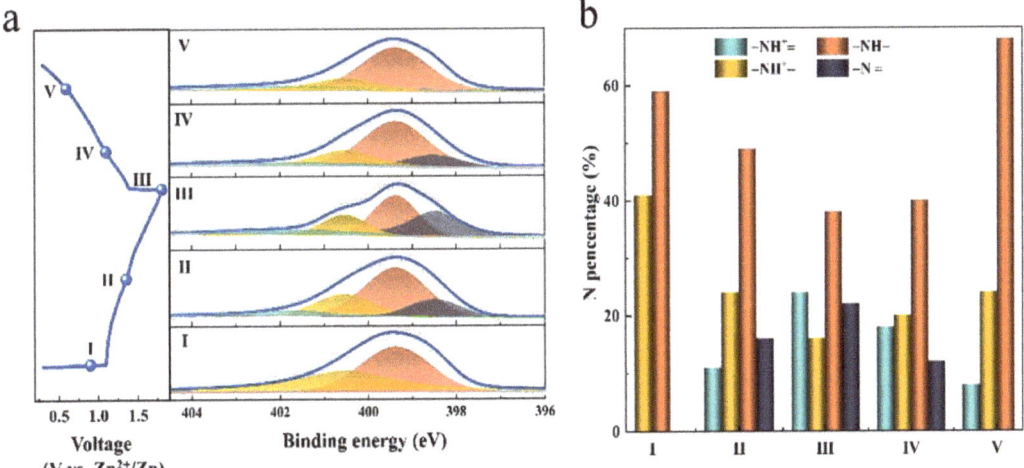

Figure 5. Structure evolution of PANI electrode during cycling. (**a**) Evolution of ex situ N 1s XPS spectra during the charge/discharge process labeled as I−V in the left panel. (**b**) The calculated N contents from N 1s XPS.

Based on the above analysis, we propose a transformation process of PANI electrodes and adsorption/desorption mechanism of Zn^{2+} (Figure 6). In the charging stage, with increases in oxidation voltage, the non-protonated −NH− becomes oxidized to −NH+−. Then, −NH+− further oxidized with respect to −NH+= and −N=. This phenomenon is consistent with the connected double oxidation peaks in the CV curve (Figure 3a). The first oxidation step may be the main contribution to the first small oxidation peak, and the latter oxidation reaction is described by the second oxidation peak. As for discharge processes (from III to V), the peaks of N 1s spectra will go through opposite changes in their intensities compared to the charge process due to the reduction of −NH+−, −N=, and −NH+=. During the reduction reaction, H+ can be consumed to encourage −N= to transform into −NH+−, together with the external doping of SO_4^{2-} to balance the charge. The leftover OH^- results in the formation of basic zinc sulfate by a similar manner observed in the previous demonstration [12]. In this process, protonated amine and protonated imine supply abundant active sites for Zn^{2+} ions adsorption and desorption. This sequential transformation can help illuminate the fast diffusion kinetics and energy storage capacities of PANI electrodes.

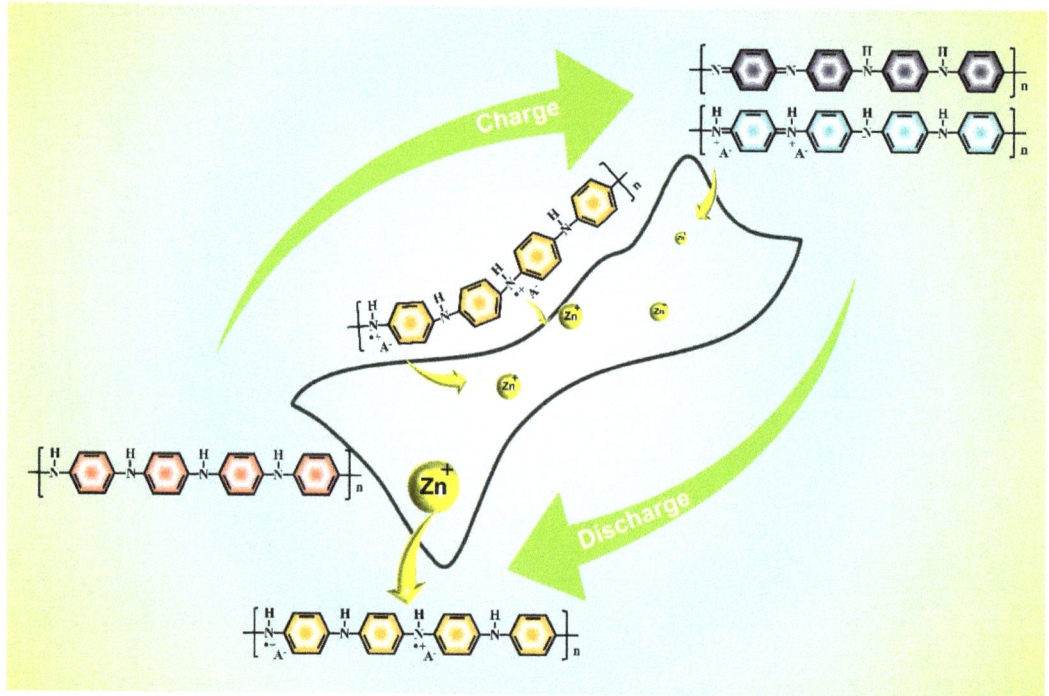

Figure 6. Diagram showing the sequential transformation of protonated and non-protonated PANI. Zn^{2+} adsorption and desorption at the active sites supplied by protonated amine and imine during the process.

4. Conclusions

We have comprehensively investigated zinc ion storage behaviors in three-dimensional conductive-network-structured PANI. The energy storage mechanism of PANI exhibits four types of N changes during protonated and non-protonated process. The PANI cathode ZIBs shows a high capacity of 74.25 mAh g^{-1} at 300 mA g^{-1} and maintains 48.4% of its initial discharge capacity after 1000 cycles. A corresponding kinetic analysis demonstrated that the diffusion coefficients of PANI are within the range of 6.25×10^{-9} to 7.82×10^{-8} cm^{-2} s^{-1} for discharge processes and 7.69×10^{-10} to 1.81×10^{-7} cm^{-2} s^{-1} for charge processes, and the values are one or two orders of magnitude higher than other reported cathode materials. The analysis revealed in this work provides new ideas for understanding the roles of PANI intercalated in host materials. It shows high reference values for active materials using conductive polymers as intercalators to develop energy storage devices with high energy density and stable performance.

Supplementary Materials: The following supporting information can be downloaded at: https://www.mdpi.com/article/10.3390/nano12091438/s1, Figure S1: X-ray diffraction pattern of PANI film electrode; Figure S2: XPS survey scans of the PANI films in the binding energy range of 0–1300 eV. Figure S3: S 1s XPS spectra of the PANI films during the charge/discharge process Supporting Notes: Calculated details of the GITT.

Author Contributions: Conceptualization, J.G.; methodology, J.G.; validation, J.G., C.T. and X.W.; formal analysis, J.G., K.Z. and H.L.; investigation, J.G., K.Z. and H.L.; resources, S.F. and Q.W.; data curation, H.L., J.C. and Z.Z.; writing—original draft preparation, H.L.; writing—review and editing, J.G.; supervision, J.G., C.T. and X.W.; project administration, J.G.; funding acquisition, J.G. and Z.S. All authors have read and agreed to the published version of the manuscript.

Funding: This work was financially assisted by the National Natural Science Foundation of China (Nos 22075068, and 62074051).

Institutional Review Board Statement: Not applicable.

Informed Consent Statement: Not applicable.

Data Availability Statement: Not applicable.

Acknowledgments: The authors are very thankful to Zhiqiang Wang form the University of Western Ontario for his work hard in English writing proficiency.

Conflicts of Interest: The manuscript was written through the contributions of all authors. All authors have given approval to the final version of the manuscript. The authors declare no competing financial interests.

References

1. Ming, J.; Guo, J.; Xia, C.; Wang, W.; Alshareef, H.N. Zinc-Ion Batteries: Materials, Mechanisms, and Applications. *Mater. Sci. Eng. R. Rep.* **2019**, *135*, 58–84. [CrossRef]
2. Jia, X.; Liu, C.; Neale, Z.G.; Yang, J.; Cao, G. Active Materials for Aqueous Zinc Ion Batteries: Synthesis, Crystal Structure, Morphology, and Electrochemistry. *Chem. Rev.* **2020**, *120*, 7795–7866. [CrossRef] [PubMed]
3. Wang, F.; Fan, X.; Gao, T.; Sun, W.; Ma, Z.; Yang, C.; Han, F.; Xu, K.; Wang, C. High-Voltage Aqueous Magnesium Ion Batteries. *ACS Central Sci.* **2017**, *3*, 1121–1128. [CrossRef]
4. Arroyo-de Dompablo, M.E.; Ponrouch, A.; Johansson, P.; Palacín, M.R. Achievements, Challenges, and Prospects of Calcium Batteries. *Chem. Rev.* **2020**, *120*, 6331–6357. [CrossRef] [PubMed]
5. Chen, H.; Xu, H.; Wang, S.; Huang, T.; Xi, J.; Cai, S.; Guo, F.; Xu, Z.; Gao, W.; Gao, C. Ultrafast All-Climate Aluminum-Graphene Battery with Quarter-Million Cycle Life. *Sci. Adv.* **2017**, *3*, eaao7233. [CrossRef] [PubMed]
6. Li, J.-C.; Gong, J.; Zhang, X.; Lu, L.; Liu, F.; Dai, Z.; Wang, Q.; Hong, X.; Pang, H.; Han, M. Alternate Integration of Vertically Oriented CuSe@FeOOH and CuSe@MnOOH Hybrid Nanosheets Frameworks for Flexible In-Plane Asymmetric Microsupercapacitors. *ACS Appl. Energy Mater.* **2020**, *3*, 3692–3703. [CrossRef]
7. Liu, Y.; Wu, X. Review of Vanadium-Based Electrode Materials for Rechargeable Aqueous Zinc Ion Batteries. *J. Energy Chem.* **2021**, *56*, 223–237. [CrossRef]
8. Sun, W.; Xiao, L.; Wu, X. Facile Synthesis of NiO Nanocubes for Photocatalysts and Supercapacitor Electrodes. *J. Alloy. Compd.* **2019**, *772*, 465–471. [CrossRef]
9. Chen, H.; Qin, H.; Chen, L.; Wu, J.; Yang, Z. V$_2$O$_5$@CNTs as Cathode of Aqueous Zinc Ion Battery with High Rate and High Stability. *J. Alloy. Compd.* **2020**, *842*, 155912. [CrossRef]
10. Wang, J.; Fan, F.; Liu, Y.; Jungjohann, K.L.; Lee, S.W.; Mao, S.X.; Liu, X.; Zhu, T. Structural Evolution and Pulverization of Tin Nanoparticles during Lithiation-Delithiation Cycling. *J. Electrochem. Soc.* **2014**, *161*, F3019–F3024. [CrossRef]
11. Li, Y.; Huang, Z.; Kalambate, P.K.; Zhong, Y.; Huang, Z.; Xie, M.; Shen, Y.; Huang, Y. V$_2$O$_5$ Nanopaper as a Cathode Material with High Capacity and Long Cycle Life for Rechargeable Aqueous Zinc-Ion Battery. *Nano Energy* **2019**, *60*, 752–759. [CrossRef]
12. Huang, J.; Wang, Z.; Hou, M.; Dong, X.; Liu, Y.; Wang, Y.; Xia, Y. Polyaniline-Intercalated Manganese Dioxide Nanolayers as a High-Performance Cathode Material for an Aqueous Zinc-Ion Battery. *Nat. Commun.* **2018**, *9*, 2906. [CrossRef] [PubMed]
13. Li, W.; Han, C.; Gu, Q.; Chou, S.; Wang, J.; Liu, H.; Dou, S. Electron Delocalization and Dissolution-Restraint in Vanadium Oxide Superlattices to Boost Electrochemical Performance of Aqueous Zinc-Ion Batteries. *Adv. Energy Mater.* **2020**, *10*, 2001852. [CrossRef]
14. Du, W.; Xiao, J.; Geng, H.; Yang, Y.; Zhang, Y.; Ang, E.H.; Ye, M.; Li, C.C. Rational-Design of Polyaniline Cathode Using Proton Doping Strategy by Graphene Oxide for Enhanced Aqueous Zinc-Ion Batteries. *J. Power Sources* **2020**, *450*, 227716. [CrossRef]
15. Wang, M.; Zhang, J.; Zhang, L.; Li, J.; Wang, W.; Yang, Z.; Zhang, L.; Wang, Y.; Chen, J.; Huang, Y.; et al. Graphene-like Vanadium Oxygen Hydrate (VOH) Nanosheets Intercalated and Exfoliated by Polyaniline (PANI) for Aqueous Zinc-Ion Batteries (ZIBs). *ACS Appl. Mater. Interfaces* **2020**, *12*, 31564–31574. [CrossRef]
16. Kataoka, F.; Ishida, T.; Nagita, K.; Kumbhar, V.S.; Yamabuki, K.; Nakayama, M. Cobalt-Doped Layered MnO$_2$ Thin Film Electrochemically Grown on Nitrogen-Doped Carbon Cloth for Aqueous Zinc-Ion Batteries. *ACS Appl. Energy Mater.* **2020**, *3*, 4720–4726. [CrossRef]
17. Qin, H.; Chen, L.; Wang, L.; Chen, X.; Yang, Z. V2O5 Hollow Spheres as High Rate and Long Life Cathode for Aqueous Rechargeable Zinc Ion Batteries. *Electrochim. Acta* **2019**, *306*, 307–316. [CrossRef]
18. Alfaruqi, M.H.; Mathew, V.; Gim, J.; Kim, S.; Song, J.; Baboo, J.P.; Choi, S.H.; Kim, J. Electrochemically Induced Structural Transformation in a γ-MnO2 Cathode of a High Capacity Zinc-Ion Battery System. *Chem. Mater.* **2015**, *27*, 3609–3620. [CrossRef]
19. Pan, L.; Pu, L.; Shi, Y.; Sun, T.; Zhang, R.; Zheng, Y.O. Hydrothermal Synthesis of Polyaniline Mesostructures. *Adv. Funct. Mater.* **2006**, *16*, 1279–1288. [CrossRef]
20. Yang, Y.; Wan, M. Chiral Nanotubes of Polyaniline Synthesized by a Template-Free Method. *J. Mater. Chem.* **2002**, *12*, 897–901. [CrossRef]

21. Liu, Y.; Xie, L.; Zhang, W.; Dai, Z.; Wei, W.; Luo, S.; Chen, X.; Chen, W.; Rao, F.; Wang, L.; et al. Conjugated System of PEDOT:PSS-Induced Self-Doped PANI for Flexible Zinc-Ion Batteries with Enhanced Capacity and Cyclability. *ACS Appl. Mater. Interfaces* **2019**, *11*, 30943–30952. [CrossRef] [PubMed]

22. Shi, H.-Y.; Ye, Y.-J.; Liu, K.; Song, Y.; Sun, X. A Long-Cycle-Life Self-Doped Polyaniline Cathode for Rechargeable Aqueous Zinc Batteries. *Angew. Chem. Int. Ed.* **2018**, *57*, 16359–16363. [CrossRef] [PubMed]

23. Alruwashid, F.S.; Dar, M.A.; Alharthi, N.H.; Abdo, H.S. Effect of Graphene Concentration on the Electrochemical Properties of Cobalt Ferrite Nanocomposite Materials. *Nanomaterials* **2021**, *11*, 2523. [CrossRef] [PubMed]

24. Li, J.-C.; Gong, J.; Yang, Z.; Tian, Y.; Zhang, X.; Wang, Q.; Hong, X. Design of 2D Self-Supported Hybrid CuSe@PANI Core/Shell Nanosheet Arrays for High-Performance Flexible Microsupercapacitors. *J. Phys. Chem. C* **2019**, *123*, 29133–29143. [CrossRef]

25. Liu, C.; Neale, Z.; Zheng, J.; Jia, X.; Huang, J.; Yan, M.; Tian, M.; Wang, M.; Yang, J.; Cao, G. Expanded Hydrated Vanadate for High-Performance Aqueous Zinc-Ion Batteries. *Energy Environ. Sci.* **2019**, *12*, 2273–2285. [CrossRef]

26. Geng, H.; Cheng, M.; Wang, B.; Yang, Y.; Zhang, Y.; Li, C.C. Electronic Structure Regulation of Layered Vanadium Oxide via Interlayer Doping Strategy toward Superior High-Rate and Low-Temperature Zinc-Ion Batteries. *Adv. Funct. Mater.* **2020**, *30*, 1907684. [CrossRef]

27. Wu, B.; Zhang, G.; Yan, M.; Xiong, T.; He, P.; He, L.; Xu, X.; Mai, L. Graphene Scroll-Coated α-MnO$_2$ Nanowires as High-Performance Cathode Materials for Aqueous Zn-Ion Battery. *Small* **2018**, *14*, 1703850. [CrossRef]

28. Wang, J.; Wang, J.-G.; Liu, H.; Wei, C.; Kang, F. Zinc Ion Stabilized MnO$_2$ Nanospheres for High Capacity and Long Lifespan Aqueous Zinc-Ion Batteries. *J. Mater. Chem. A* **2019**, *7*, 13727–13735. [CrossRef]

29. Chen, L.; Yang, Z.; Cui, F.; Meng, J.; Jiang, Y.; Long, J.; Zeng, X. Ultrathin MnO$_2$ Nanoflakes Grown on N-Doped Hollow Carbon Spheres for High-Performance Aqueous Zinc Ion Batteries. *Mater. Chem. Front.* **2020**, *4*, 213–221. [CrossRef]

30. Wei, T.; Li, Q.; Yang, G.; Wang, C. Highly Reversible and Long-Life Cycling Aqueous Zinc-Ion Battery Based on Ultrathin (NH$_4$)$_2$V$_{10}$O$_{25}$·8H$_2$O Nanobelts. *J. Mater. Chem. A* **2018**, *6*, 20402–20410. [CrossRef]

31. Zhang, N.; Jia, M.; Dong, Y.; Wang, Y.; Xu, J.; Liu, Y.; Jiao, L.; Cheng, F. Hydrated Layered Vanadium Oxide as a Highly Reversible Cathode for Rechargeable Aqueous Zinc Batteries. *Adv. Funct. Mater.* **2019**, *29*, 1807331. [CrossRef]

32. Tang, B.; Fang, G.; Zhou, J.; Wang, L.; Lei, Y.; Wang, C.; Lin, T.; Tang, Y.; Liang, S. Potassium Vanadates with Stable Structure and Fast Ion Diffusion Channel as Cathode for Rechargeable Aqueous Zinc-Ion Batteries. *Nano Energy* **2018**, *51*, 579–587. [CrossRef]

Article

Temperature-Dependent Fractional Dynamics in Pseudo-Capacitors with Carbon Nanotube Array/Polyaniline Electrodes

Igor O. Yavtushenko [1], Marat Yu. Makhmud-Akhunov [1], Renat T. Sibatov [2,*], Evgeny P. Kitsyuk [2] and Vyacheslav V. Svetukhin [2]

1   Laboratory of Diffusion Processes, Ulyanovsk State University, 432017 Ulyanovsk, Russia; yavigor@mail.ru (I.O.Y.); maratmau@mail.ru (M.Y.M.-A.)
2   Scientific-Manufacturing Complex "Technological Centre", 124498 Moscow, Russia; kitsyuk.e@gmail.com (E.P.K.); v.svetukhin@tcen.ru (V.V.S.)
*   Correspondence: ren_sib@bk.ru

Abstract: Pseudo-capacitors with electrodes based on polyaniline and vertically aligned multiwalled carbon nanotubes (PANI/VA-MWCNT) composite are studied. Fractional differential models of supercapacitors are briefly discussed. The appropriate fractional circuit model for PANI/MWCNT pseudo-capacitors is found to be a linearized version of the recently proposed phase-field diffusion model based on the fractional Cahn–Hilliard equation. The temperature dependencies of the model parameters are determined by means of impedance spectroscopy. The fractional-order α is weakly sensitive to temperature, and the fractional dynamic behavior is related to the pore morphology rather than to thermally activated ion-hopping in PANI/MWCNT composite.

Keywords: pseudocapacitor; carbon nanotube; polyaniline; fractional-order circuit model; memory effect; fractional derivative

check for
updates

Citation: Yavtushenko, I.O.; Makhmud-Akhunov, M.Y.; Sibatov, R.T.; Kitsyuk, E.P.; Svetukhin, V.V. Temperature-Dependent Fractional Dynamics in Pseudo-Capacitors with Carbon Nanotube Array/Polyaniline Electrodes. *Nanomaterials* **2022**, *12*, 739. https://doi.org/10.3390/nano12050739

Academic Editors: Xiang Wu and Jung Woo Lee

Received: 7 January 2022
Accepted: 17 February 2022
Published: 22 February 2022

1. Introduction

Conductive polymers such as polythiophene, polyaniline (PANI), polyacetylene, etc. have redox properties and can be used as electrode materials for electrochemical power sources. Among these polymers, PANI has proven to be one of the most promising due to its high capacitance characteristics, ease of processing and environmental friendliness [1–4]. Redox centers in the polymer backbone are not sufficiently stable during many cyclic redox processes [2]. Special additives in composites (activated carbon, nanotubes, graphene, transition metal oxides) are used to eliminate disadvantages of pure PANI such as rapid degradation during cycling and slow ion transfer kinetics. The presence of nanotubes in such composites promotes efficient charge transfer that reduces the internal resistance of the electrodes. Carbon nanotubes (CNTs) increase the electrical conductivity of material regardless of polymer redox state; in addition, a structure with optimal porosity can be created.

For several applications, the stacking of individual CNTs during growth is of great importance. The complex morphology of the entangled nanotube agglomerates leads to a slowdown in the transport of charge carriers [5], including subdiffusive anomalous transport [6]. To eliminate this disadvantageous feature, electrodes based on an array of vertically aligned (oriented) multiwalled carbon nanotubes (VA-MWCNTs) are used [5,7]. This geometry contributes to an increase in the electronic and ionic conductivity in the composite, and the specific active surface area can be larger than in the case of an entangled CNT network.

In this paper, PANI/VA-MWCNT pseudo-capacitors are prepared and the temperature-dependent charging–discharging dynamics of these devices is studied. As is known, fractional-order technique using fractional calculus and fractional equivalent circuits [3] is effective to describe the dynamics of supercapacitors [6,8–15]. Recently, the time-fractional phase-field model has been applied to describe PANI/MWCNT pseudo-capacitors [16]. Here, the simplified representation of a fractional circuit model is used to describe the impedance spectra, cyclic voltammograms, and charging–discharging in potentiostatic mode. The proposed model is a linearized version of the nonlinear model based on the fractional Cahn–Hilliard equation of phase-field diffusion and suitable for the analysis of temperature-dependent fractional dynamics.

Recently, Kopka [17] studied the effect of temperature on the derivative order in the fractional model of supercapacitor. The rate of electrochemical reactions is related to the temperature of the supercapacitor, so the order of fractional derivative model should be temperature-dependent. Here, we determine the temperature dependencies of the fractional model parameters for PANI/VA-MWCNT pseudo-capacitors.

2. Materials and Experimental Methods

Pseudo-capacitors with electrodes based on the PANI/VA-MWCNT nanocomposite were prepared. The nanotubes are presented in the form of a vertically aligned array (VA-MWCNT) grown on a 0.5 cm^2 titanium substrate. The fabrication process starts with wet cleaning and thermal oxidation of a bare silicon base plate to isolate the substrate from the electrodes. Then, Ti and Ni layers were evaporated onto the substrate with the magnetron sputtering system. Ti serves as the current collector material, and Ni particles are the catalyst for CNT forest growth. After the preparation of the VA-MWCNT array shown in Figure 1, the CNT forest was covered with a thin layer of PANI (emeraldine form), obtained by the chemical method of aniline solution oxidation. SEM images (top plan view) of the MWCNT array and the array covered by PANI layers are presented in Figure 2. A two-stage method of coating by polyaniline was used. After drying the first layer, the second layer was applied. The thickness of each layer is approximately equal to 150 nm, and was determined by the method of atomic force microscopy. It is known [1] that the formation of a PANI layer on the CNT surface begins with the adsorption of aniline molecules, which then form oligomers during oxidative polymerization. When such a mechanism is implemented, the properties of the resulting PANI are significantly affected by the nature of the surface groups of the initial CNTs.

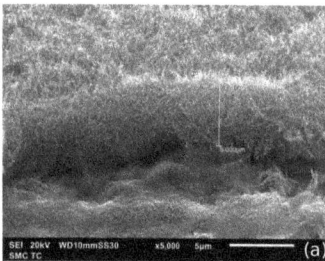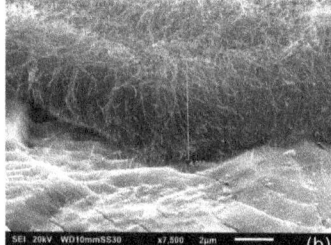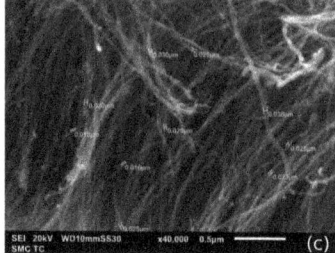

Figure 1. SEM images of a grown MWCNT array on a titanium plate used by us for preparation of PANI/VA-MWCNT pseudocapacitor. Scale bar: 5 µm (**a**), 2 µm (**b**) and 0.5 µm (**c**).

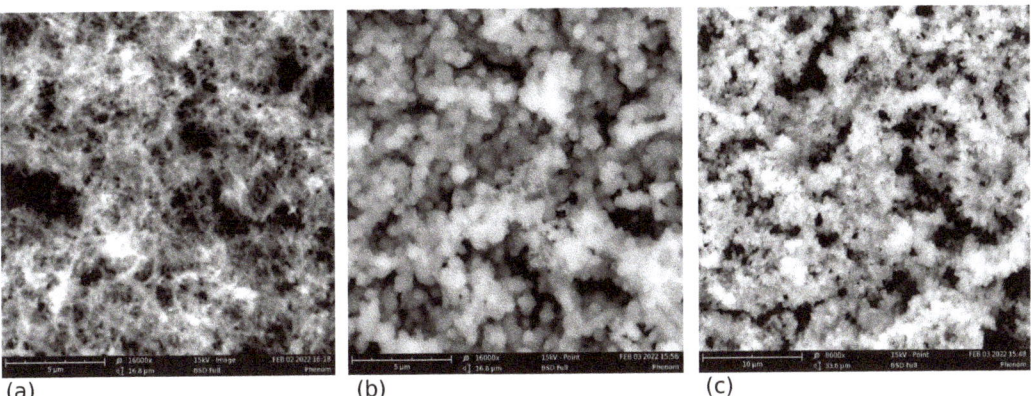

Figure 2. SEM images (top plan view) of the MWCNT array (**a**) and the array covered by single PANI layer (**b**,**c**). Scale bar: 5 μm (**a**), 5 μm (**b**) and 10 μm (**c**).

The electrolyte in our system is a solution of phosphoric acid H_3PO_4 and polyvinyl alcohol (PVA). A schematic representation of the PANI/VA-MWCNT pseudocapacitor is shown in Figure 3a.

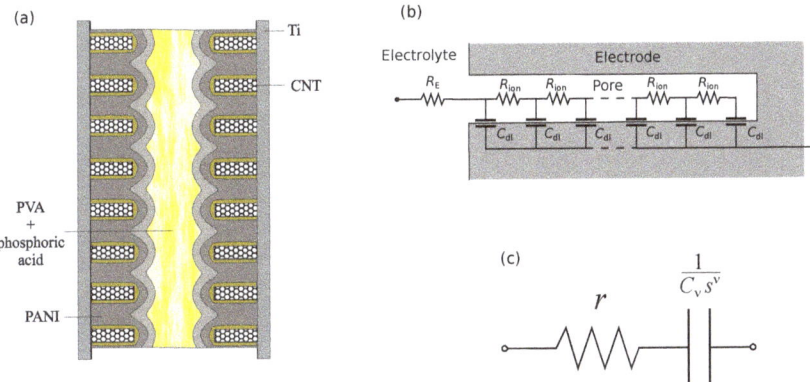

Figure 3. Schematic representation of a cell with electrodes based on vertically aligned MWCNT array/PANI composite (**a**), RC transmission line (De Levie model) for a single pore (**b**), and the simplest equivalent circuit model of a supercapacitor (**c**).

We determine the model parameters of pseudo-capacitors by fitting impedance spectra, cyclic voltammograms and charging–discharging curves. These data were obtained by measurements with a P-45X potentiostat–galvanostat (Electrochemical Instruments company). For cyclic voltammetry, the voltage ranges from −0.5 to 0.5 V, potential scan rates are 20, 50 and 100 mV/s. For the impedance spectroscopy measurement, the frequency ranges from 0.1 Hz to 50 kHz, and the voltage amplitude is 50 mV.

3. Fractional Differential Models of Supercapacitors

It is common to analyze impedance results using a physical model that is expressed by a system of mathematical equations. If this system is linear, it can be usually represented by an equivalent circuit. The parameters of circuit electrical components are related to the physical and chemical properties of electrolyte, electrode and their interface. The charging–discharging kinetics of supercapacitors and pseudo-capacitors is largely determined by the diffusion of ions in electrodes and electrolyte.

The de Levie model [18] successfully describes the impedance of a porous electrode containing oblong pores (see, e.g., [19,20]). The impedance of a single pore is modeled by a transmission line (Figure 3b) with the assumption that specific resistances of solution and local impedance do not depend on the depth inside pore, and the solid phase is assumed to be perfectly conducting [18]. The half-integer impedance $Z = \sqrt{R/j\omega C}$ characterizes this transmission line. A more general form is given by a constant phase element (CPE),

$$Z(\omega) = \frac{1}{C_\nu (j\omega)^\nu}. \tag{1}$$

CPE coupled in series with resistor r (Figure 3c) represents a simplified supercapacitor model considered in references [14,21],

$$Z(s) = R + \frac{1}{C_\nu s^\nu}, \quad s = j\omega, \tag{2}$$

and used in [22,23] to characterize electric double layer (EDL) supercapacitor impedance. Here, s can be associated with the Laplace variable. In [24], this impedance model is used to predict the transient response of a supercapacitor to a voltage-step signal. The de Levie model successfully describes electrodes with pores of similar geometric parameters, particularly nanocrystalline TiO_2 films [25], and other metal oxides electrodes. Transition metal oxides such as TiO_2, NiO_x, RuO_2, MnO_x are widely used in the development of supercapacitor electrodes [26–28]. It is noteworthy that the hierarchical structure of the electrode surface also leads to a fractional impedance, which can be derived within the recursive fractal ladder model [29].

In review [24], the authors discuss three equivalent circuit models for EDL supercapacitors. The first of them is shown in Figure 3c. Another model is a combination of a resistor and three CPEs. In this case, the impedance is

$$Z(s) = R + \frac{1}{C_\alpha s^\alpha} + \frac{1}{C_\beta s^\beta} + \frac{1}{C_3 s^{\alpha+\beta}}. \tag{3}$$

This model was used in [10] to describe the dynamics of supercapacitor HE0120C-0027A 120 F in the frequency range 1 mHz–1 kHz. The third circuit model is given by impedance

$$Z(s) = R + k\frac{(1+s/\omega_0)^\alpha}{s^\beta}. \tag{4}$$

This was proposed in [23] and successfully applied to the description of supercapacitor EPCOS 5 F. The fitted parameters are as follows $\alpha = 0.5190$, $\beta = 0.9765$, $k = 0.3440\ \Omega/s^\beta$.

For the above-listed impedance models, the corresponding charging–discharging equations for current and voltage contain fractional derivatives. On the other hand, due to heterogeneity and complexity of porous electrodes, anomalous diffusive kinetics of ions can take place [13–15]. Anomalous diffusion is characterized by power law expansion of the diffusion packet, $\Delta(t) \propto t^{\alpha/2}$, with $\alpha \neq 1$. The case $0 < \alpha < 1$ is classified as subdiffusion, and the case $\alpha > 1$ as superdiffusion. Mathematical treatment of self-similar anomalous diffusion is usually based on diffusion equations with fractional derivatives.

The simplest fractional diffusion equation has the form

$$\frac{\partial c(x,t)}{\partial t} = K\ _0D_t^{1-\nu}\frac{\partial^2 c(x,t)}{\partial x^2}, \tag{5}$$

where

$$_0D_t^{1-\nu}c(x,t) = \frac{1}{\Gamma(\nu)}\frac{\partial}{\partial t}\int_0^t \frac{c(x,\tau)}{(t-\tau)^{1-\nu}}d\tau, \quad 0 < \nu \leq 1,$$

is the fractional Riemann–Liouville derivative of order $1 - \nu$ [30].

Using anomalous diffusion equations with fractional derivatives, one could generalize impedances for different geometries and boundary conditions (see [25,31] and references therein). The simplest example is subdiffusive generalization of Warburg's impedance for a semi-infinite medium [31]

$$Z(j\omega) = B(i\omega)^{-(1-\nu/2)}, \tag{6}$$

where B is a frequency-independent constant.

3.1. Havriliak–Negami Response

To evaluate electrolyte diffusion parameters in porous media, electrochemical impedance spectroscopy is often used. In [18,32,33], the relationship between pore size and electrochemical properties of electrodes has been studied. One of the approaches to assessing the properties of a supercapacitor from impedance spectra is based on the formal representation of a supercapacitor as a dielectric liquid in which the molecular relaxation of the system is assessed in a wide range of frequencies and associated with its structure. Such a view is convenient for using widely known models of dielectric relaxation, such as the Debye, Cole–Cole, and Havriliak–Negami (HN) models. Unlike capacitance, resistance and leakage current, dielectric permeability is an intensive rather than extensive characteristic of the system. Studies [34,35] have shown that structural confinement has a significant effect on molecular relaxation, and it is possible to estimate the contribution of surface morphology by the electrolytic molecular component [34].

The circuit element based on the HN function describes the asymmetric and broad nature of dielectric dispersion [36]. In the case of a linear response, the relationship between current and voltage can be represented as follows

$$i(t) = K\frac{d}{dt}\int_0^\infty \phi(t')\,u(t-t')dt'.$$

Turning to the Fourier transforms, we obtain

$$\frac{\tilde{i}(j\omega)}{\tilde{\phi}(j\omega)} = K \cdot j\omega \cdot \tilde{u}(j\omega).$$

In the case of a system with the HN response, we have

$$[1 + (j\omega\tau)^\alpha]^\beta\,\tilde{i}(j\omega) = K \cdot j\omega \cdot \tilde{u}(j\omega).$$

The inverse Fourier transformation leads to a fractional differential relationship

$$[1 + \tau^\alpha\,{}_{-\infty}D_t^\alpha]^\beta\,i(t) = g(t), \quad g(t) = K\dot{V}(t)$$

In the case of a step input $V(t) = V_0 l(t)$, we have

$$[1 + \tau^\alpha\,{}_{-\infty}D_t^\alpha]^\beta\,i(t) = KV_0\delta(t).$$

Here, $l(t)$ is the Heaviside step function.

The solution of this equation is expressed through the generalized Mittag–Leffler function proposed by Prabhakar [37],

$$E_{\alpha,\gamma}^\beta(z) = \sum_{n=0}^\infty \frac{\Gamma(\beta+n)}{\Gamma(\beta)\Gamma(\alpha n+\gamma)n!}z^n.$$

Using the Laplace transform

$$\int_0^\infty e^{-st}\,t^{\gamma-1}E_{\alpha,\gamma}^\beta(at^\alpha)dt = \frac{s^{\alpha\beta-\gamma}}{(s^\alpha-a)^\beta},$$

one can express the relaxation function in the form

$$f(t) = i(t)/KV_0 = \tau^{-\alpha\beta} t^{\alpha\beta-1} E_{\alpha,\alpha\beta}^{\beta} \left(-(t/\tau)^{\alpha} \right).$$

The asymptotic behavior of the solution at large and small times is given by power laws

$$f(t) \sim \Gamma(1 - \alpha\beta) \, \tau^{-\alpha\beta} t^{-1+\alpha\beta}, \quad t \to 0,$$
$$f(t) \sim \alpha\beta\tau^{\alpha} [\Gamma(1 - \alpha)]^{-1} \, t^{-1-\alpha}, \quad t \to \infty.$$

If $\beta = 0$, response $f(t)$ is expressed through the two-parameter Mittag-Leffler function.

The HN response is often considered to be a general expression for the universal relaxation law [38]. This universality implies the similarity of relaxation laws in different materials. This universality holds for dielectric relaxation in dipolar and nonpolar materials, for hopping transport in semiconductors, conduction in ionic materials, delayed luminescence decay, surface conduction on insulators, kinetics of chemical reactions, mechanical relaxation, magnetic relaxation. Despite the completely different internal mechanisms, the processes show striking similarity [38]. This universality stimulates the search for an appropriate stochastic model for the universal relaxation law. Investigations of such kind have been carried out in many works (see e.g., [39–42]). Based on the solution of fractional relaxation equation [43] and HN response [44], the memory recovery effect was demonstrated. Corresponding relaxation curves are described by the exponential law at initial stage and power law for long-time asymptotics. Charging–discharging curves in PANI/VA-MWCNT demonstrate the similar behavior (see Section 3.3).

3.2. Phase-Field Model

In [16], a generalized diffusion impedance model for materials with a subdiffusion phase transition is proposed. The model is based on the fractional Cahn–Hilliard equation with fractional time derivatives. A one-dimensional cell with reflecting and absorbing boundaries is considered. Phase-field generalizations of anomalous diffusion models AD-Ib and AD-Ia presented in [31] are described by the following time-fractional equations [16] with Caputo and Riemann–Liouville derivatives:

$$_0^C D_t^\alpha c = M \, \nabla (c \, \nabla\mu_a), \quad _0^{RL} D_t^\alpha c = M \, \nabla (c \, \nabla\mu_a).$$

Here, M is the ambipolar mobility, and μ_a is the effective (ambipolar) chemical potential (see details in [16]).

The corresponding impedance models were denoted as Z^{C-CH} and Z^{RL-CH}, respectively. Letters denote the type of used fractional time derivative (Caputo or Riemann–Liouville):

$$Z^{C-CH} \propto \frac{\sqrt{\Lambda(\omega)}}{i\omega} F(\omega), \qquad Z^{RL-CH} \propto \frac{1}{\sqrt{\Lambda(\omega)}} F(\omega) \qquad (7)$$

with

$$\Lambda(\omega) = (i\omega)^{\alpha}.$$

The form of $F(\omega)$ depends on boundary conditions. For a cell with reflecting boundary, in [16], it was obtained

$$F_{\text{refl}}(\omega) = \frac{\left(1 + \sqrt{1 - 4\chi\Lambda(\omega)}\right)^{3/2} \coth\left[\frac{1}{\sqrt{2\chi}}\sqrt{1 - \sqrt{1 - 4\chi\Lambda(\omega)}}\right] - \left(1 - \sqrt{1 - 4\chi\Lambda(\omega)}\right)^{3/2} \coth\left[\frac{1}{\sqrt{2\chi}}\sqrt{1 + \sqrt{1 - 4\chi\Lambda(\omega)}}\right]}{\sqrt{1 - 4\chi\Lambda(\omega)}}.$$

The frequency dependencies of the PANI/VA-MWCNT pseudocapacitor impedance were described by an equivalent circuit (Figure 4a) containing generalized fractional elements Z_{refl}^{C-CH} or Z_{refl}^{RL-CH} defined by (7).

The proposed equivalent circuit was substantiated by the following arguments [16]. The ion transport is interpreted in terms of the one-dimensional diffusion model. The schematic representation of the pseudocapacitor is given in Figure 3a. According to the de Levie model [18,19], CPE describes the EDL capacity formed around the MWCNTs. Diffusion of ions in the interelectrode space is described by the open Warburg impedance. Generalized fractional element Z_{refl}^{RL-CH} corresponds to phase-field ion diffusion in PANI filling the VA-MWCNT array. Reflecting boundary condition is assumed for the base of nanotube array. The RL-CH model (Riemann–Liouville type) implies non-conserving ion density, and it is related to the EDL formation by fraction of ions during phase-field diffusion in PANI filling the MWCNT forest. The series resistor corresponds to the summarized resistance of MWCNTs, polymer and electrolyte.

3.3. Linearized Model

The model described in the previous section implies phase-field diffusion of ions in PANI filling the MWCNT forest. The Cahn–Hilliard equation and the corresponding circuit model are nonlinear [16]. The expressions for impedance are obtained after linearization (for details, see [16]). The equivalent scheme is dependent on state of charge. For simplicity, under small voltage perturbations phase-field diffusion can be replaced by ordinary diffusion (Figure 4). Such a replacement implies the dependence of diffusion coefficient on the reference values of ion concentration. Below, we will show that this simplified model describes the observed impedance spectra of PANI/VA-MWNT pseudo-capacitors quite well.

To study the effect of PANI layer thickness on the characteristics of PANI/VA-MWCNT pseudocapacitor, samples with one and two PANI layers were studied. The fitted parameters for impedance spectra are provided in Table 1. The used equivalent circuit is shown in Figure 4b. A comparison of the model impedance spectra with the measured ones is presented in Figure 5. Cyclic voltammograms of PANI/VA-MWNT pseudo-capacitors with single and double PANI layers demonstrated in Figure 6 indicate that the sample with double PANI layer is characterized by higher capacity (0.05 F) than the single layer pseudo-capacitor (0.025 F).

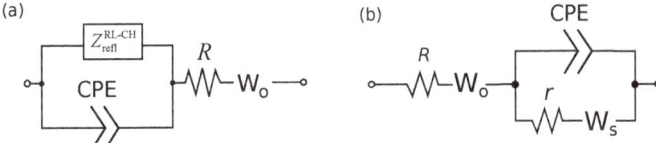

Figure 4. Equivalent circuit model of PANI-MWCNT pseudocapacitor with fractional phase-field element (**a**), and simplified model (**b**).

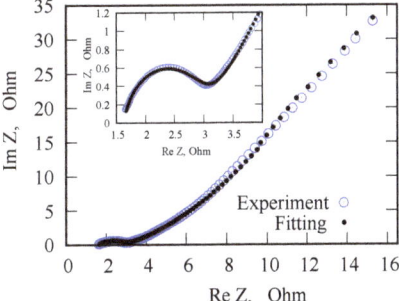

Figure 5. Nyquist plot for PANI/VA-MWCNT pseudo-capacitor.

Table 1. Parameters of the equivalent circuit model.

Parameter	PANI/VA-MWCNT 1	PANI/VA-MWCNT 2
R, Ohm	4.124	1.623
r, Ohm	0.704	1.360
C_α, 10^{-4} $s^\alpha \cdot$Ohm^{-1}	5.558	1.270
α	0.8068	0.8654
W_s, Ohm$\cdot$s$^{-1/2}$	6.497	10.513
b_s, s$^{1/2}$	2.408	2.272
W_o, Ohm$\cdot$s$^{-1/2}$	8.081	0.348
b_o, s$^{1/2}$	0.238	0.0263

CPE is characterized by impedance $Z_{CPE} = C_\alpha^{-1}(j\omega)^{-\alpha}$. Two Warburg elements ($W_s$ and W_o) are included into the circuit. Element W_s corresponds to the one-dimensional diffusion in a finite cell with an absorbing boundary, the W_o element is the same with a reflecting boundary. The corresponding impedances are

$$Z_{W_s} = \frac{W_s}{\sqrt{j\omega}} \tanh\left(b_s\sqrt{j\omega}\right), \quad Z_{W_o} = \frac{W_o}{\sqrt{j\omega}} \coth\left(b_o\sqrt{j\omega}\right),$$

where W_s and W_o are Warburg coefficients, $b_{s,o} = d/\sqrt{D}$, where d is thickness of the Nernst diffusion layer, D is the diffusion coefficient.

The EIS Spectrum Analyzer software is used to fit the impedance spectroscopy data. The Levenberg–Marquard algorithm with amplitude minimization has been chosen.

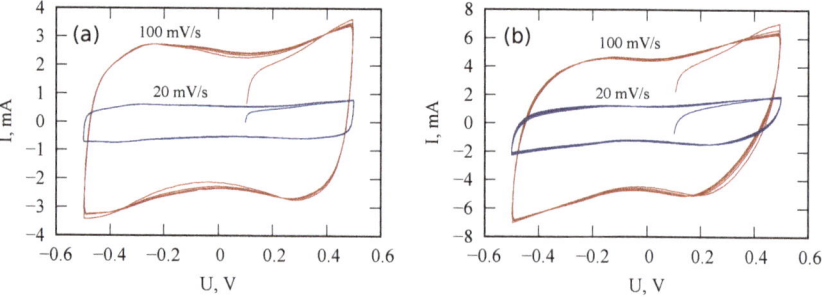

Figure 6. Cyclic voltammograms of PANI/VA-MWNT supercapacitors with one (**a**) and two (**b**) PANI layers. Scan rates are 20 and 100 mV/s.

Figure 7 demonstrates the charging current curve and discharging curves for different charging times ($\theta = 30, 60, 120, 240$ s). A slight jump noticeable on the curves is associated with a change in the measuring range of the device. The initial stage is successfully approximated by an exponential function with a relaxation time $\tau = 12$ s. Long-term relaxation is dependent on prehistory of charging process. This is a sign of nonlocality in time behavior that is consistent with the fractional circuit model discussed in this work.

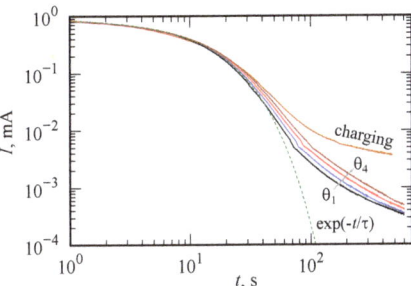

Figure 7. Charging and discharging curves $|I(t)|$ in the log–log scale. Charging time θ is varied: $\theta = 30, 60, 120, 240$ s. The initial stage is successfully approximated by an exponential function with a relaxation time $\tau = 12$ s.

4. Temperature-Dependent Fractional Dynamics in PANI/VA-MWCNT Pseudo-Capacitors

The rate of electrochemical reactions is related to the temperature of the supercapacitor, so the order of fractional derivative model should be temperature-dependent [17]. Here, we determine the temperature dependencies of the fractional model parameters for PANI/MWCNT pseudo-capacitors.

The proposed fractional circuit is consistent with the results of measurements by cyclic voltammetry, impedance spectroscopy and charge–discharge in potentiostatic mode. We made sure that the proposed model works satisfactorily for different temperatures. The temperature dependencies of the fractional model parameters are studied. Equivalent circuit model parameters for $T = 25\,^\circ$C, $T = 35\,^\circ$C, $T = 45\,^\circ$C, and $T = 55\,^\circ$C are listed in Table 2. Cyclic voltammograms of PANI/VA-MWCNT pseudocapacitor for different temperatures are demonstrated in Figure 8.

Table 2. Equivalent circuit model parameters for different temperatures.

Parameter	$T = 25\,^\circ$C	$T = 35\,^\circ$C	$T = 45\,^\circ$C	$T = 55\,^\circ$C
R, Ohm	3.8431	3.1989	2.8983	2.6658
r, Ohm	2.301	2.085	1.737	1.345
C_α, 10^{-5} $s^\alpha \cdot$Ohm^{-1}	4.47	4.34	4.23	4.13
α	0.88363	0.88906	0.89299	0.89952
W_s, Ohm$\cdot$s$^{-1/2}$	40.478	39.609	38.988	35.463
b_s, s$^{1/2}$	6.7649	18.647	25.284	7.0558
W_o, Ohm$\cdot$s$^{-1/2}$	3.1208	2.3203	0.54176	1.656
b_o, s$^{1/2}$	0.0907	0.085134	0.020342	0.066649

The resistances R and r decrease with increasing temperature (Figure 9). Apparently, both parameters are associated with the transfer of ions in the PANI/MWCNT structure. The α parameter is weakly sensitive to temperature, which means that it is related to the pore morphology. It is expected that in the case of the dominant role of ion-hopping transport in PANI, a significant dependence of α on temperature would be observed. The C_α parameter decreases, even though voltammograms indicate a slight increase in the supercapacitor's capacity with increasing temperature. Changes in the parameters of Warburg elements can be traced from the data in Table 2.

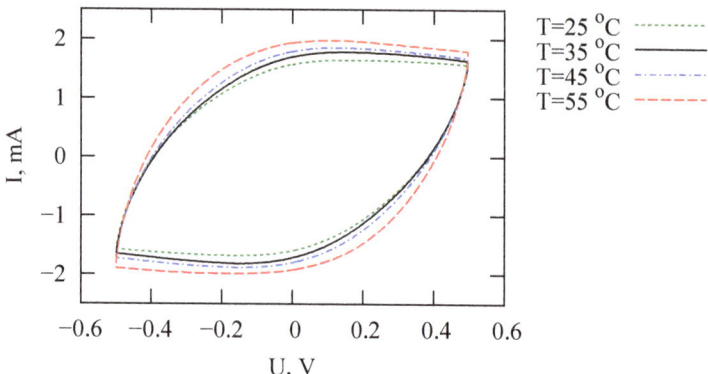

Figure 8. Cyclic voltammograms (10th cycle, scan rate is 50 mV/s) of PANI/VA-MWCNT pseudo-capacitors for different temperatures.

In contrast to the results of reference [17], we observe an increase in the fractional exponent with increasing temperature, which is, in some sense, consistent with the theory of dispersive transport in disordered materials [41,45].

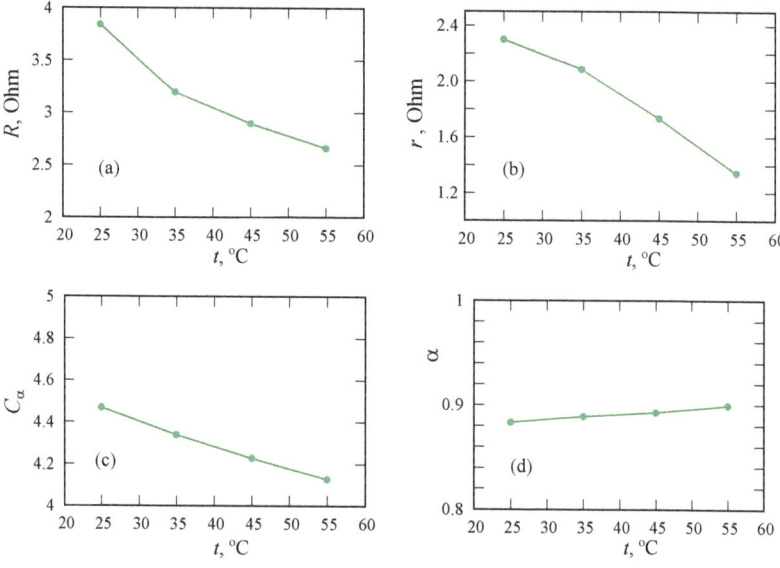

Figure 9. Equivalent circuit model parameters for different temperatures. Resistances R (**a**) and r (**b**), CPE parameter C_α or fractional capacity in 10^{-5} s$^\alpha$·Ohm^{-1} (**c**) and fractional order α (**d**).

5. Conclusions

Pseudo-capacitors with electrodes based on PANI/VA-MWCNT composites have been manufactured and investigated. The measured discharge curves demonstrate the presence of a memory effect in the devices under study, and the impedance spectra are described by a fractional-order equivalent circuit model, which can be justified within the framework of the anomalous diffusion-reaction model or transmission line model. The proposed model is a linearized version of the nonlinear model based on the fractional Cahn–Hilliard equation of phase-field diffusion [16]. The fractional-order equivalent circuit is consistent with the measurements by cyclic voltammetry, impedance spectroscopy, and potentiostatic charging–discharging. We investigate the temperature dependence of the parameters of the fractional model. The resistances R and r associated with the transfer of ions in the PANI/VA-MWCNT structure decrease with increasing temperature. The fractional-order α is weakly sensitive to temperature. This fact indicates that fractional behavior is related to the pore morphology rather than to thermally activated ion-hopping in the PANI/VA-MWCNT composite. In contrast to the results of reference [17], we observe a weak increase in the fractional exponent with increasing temperature, which is consistent with the dispersive transport theory for disordered materials [46].

Author Contributions: Conceptualization, I.O.Y. and R.T.S.; methodology, I.O.Y., M.Y.M.-A. and R.T.S.; resources, E.P.K. and I.O.Y.; data curation, I.O.Y. and R.T.S.; writing—original draft preparation, I.O.Y., M.Y.M.-A. and R.T.S.; supervision, E.P.K. and V.V.S.; project administration, E.P.K. and V.V.S.; funding acquisition, E.P.K. and R.T.S. All authors have read and agreed to the published version of the manuscript.

Funding: This work is supported by the Ministry of Science and Higher Education of the Russian Federation (project FNRM-2021-0002) and the Russian Science Foundation (project 19-71-10063).

Conflicts of Interest: The authors declare no conflict of interest.

References

1. Wu, T.M.; Lin, Y.W.; Liao, C.S. Preparation and characterization of polyaniline/multi-walled carbon nanotube composites. *Carbon* **2005** *43*, 734–740. [CrossRef]
2. Eftekhari, A.; Li, L.; Yang, Y. Polyaniline supercapacitors. *J. Power Sources* **2017**, *347*, 86–107. [CrossRef]
3. Zou, C.; Zhang, L.; Hu, X.; Wang, Z.; Wik, T.; Pecht, M. A review of fractional-order techniques applied to lithium-ion batteries, lead-acid batteries, and supercapacitors.*J. Power Sources* **2018**, *390*, 286–296. [CrossRef]
4. Gupta, V.; Miura, N. Polyaniline/single-wall carbon nanotube (PANI/SWCNT) composites for high performance supercapacitors. *Electrochim. Acta* **2006**, *52*, 1721–1726. [CrossRef]
5. Aydinli, A.; Yuksel, R.; Unalan, H.E. Vertically aligned carbon nanotube–polyaniline nanocomposite supercapacitor electrodes. *Int. J. Hydrog. Energy* **2018**, *43*, 18617–18625. [CrossRef]
6. Kitsyuk, E.P.; Sibatov, R.T.; Svetukhin, V.V. Memory effect and fractional differential dynamics in planar microsupercapacitors based on multiwalled carbon nanotube arrays. *Energies* **2020**, *13*, 213. [CrossRef]
7. Wu, G.; Tan, P.; Wang, D.; Li, Z.; Peng, L.; Hu, Y.; Wang, C.; Zhu, W.; Chen, S.; Chen, W. High-performance supercapacitors based on electrochemical-induced vertical-aligned carbon nanotubes and polyaniline nanocomposite electrodes. *Sci. Rep.* **2017**, *7*, 1–8. [CrossRef]
8. Freeborn, T.J.; Maundy, B.; Elwakil, A.S. Fractional-order models of supercapacitors, batteries and fuel cells: A survey. *Mater. Renew. Sustain. Energy* **2015**, *4*, 9. [CrossRef]
9. Biswas, K.; Bohannan, G.; Caponetto, R.; Lopes, A.M.; Machado, J.A.T. *Fractional-Order Devices*; Springer International Publishing: Cham, Switzerland, 2017.
10. Martynyuk, V.; Ortigueira, M. Fractional model of an electrochemical capacitor. *Signal Process.* **2015**, *107*, 355–360. [CrossRef]
11. Tenreiro Machado, J.A.; Lopes, A.M.; de Camposinhos, R. Fractional-order modelling of epoxy resin. *Philos. Trans. R. Soc. A* **2020**, *378*, 20190292. [CrossRef]
12. Uchaikin, V.V.; Ambrozevich, A.S.; Sibatov, R.T.; Ambrozevich, S.A.; Morozova, E.V. Memory and nonlinear transport effects in charging-discharging of a supercapacitor. *Tech. Phys.* **2016**, *61*, 250–259. [CrossRef]
13. Sabatier, J. Fractional order models for electrochemical devices. In *Fractional Dynamics*; De Gruyter Open: Warsaw, Poland, 2016; pp. 141–160.
14. Allagui, A.; Freeborn, T.J.; Elwakil, A.S.; Fouda, M.E.; Maundy, B.J.; Radwan, A.G.; Said, Z.; Abdelkareem, M.A. Review of fractional-order electrical characterization of supercapacitors. *J. Power Sources* **2018**, *400*, 457–467. [CrossRef]
15. Sibatov, R.T.; Uchaikin, V.V. Fractional kinetics of charge carriers in supercapacitors. In *Volume 8: Applications in Engineering, Life and Social Sciences, Part B*; De Gruyter: Vienna, Austria, 2019; pp. 87–118.
16. L'vov, P.E.; Sibatov, R.T.; Yavtushenko, I.O.; Kitsyuk, E.P. Time-Fractional Phase Field Model of Electrochemical Impedance. *Fractal Fract.* **2021**, *5*, 191. [CrossRef]
17. Kopka, R. Changes in derivative orders for fractional models of supercapacitors as a function of operating temperature. *IEEE Access* **2019**, *7*, 47674–47681. [CrossRef]
18. De Levie, R. On porous electrodes in electrolyte solutions: I. capacitance effects. *Electrochim. Acta* **1963**, *8*, 751–780. [CrossRef]
19. Vicentini, R.; Nunes, W.G.; da Costa, L.H.; Da Silva, L.M.; Freitas, B.; Pascon, A.M.; Vilas-Boas, O.; Zanin, H. Multi-walled carbon nanotubes and activated carbon composite material as electrodes for electrochemical capacitors. *J. Energy Storage* **2021**, *33*, 100738. [CrossRef]
20. Stępień, M.; Handzlik, P.; Fitzner, K. Electrochemical synthesis of oxide nanotubes on Ti6Al7Nb alloy and their interaction with the simulated body fluid. *J. Solid State Electrochem.* **2016**, *20*, 2651–2661. [CrossRef]

21. Kötz,R.; Carlen, M. Principles and applications of electrochemical capacitors. *Electrochim. Acta* **2000**, *45*, 2483–2498. [CrossRef]
22. Mahon, P.J.; Paul,G.L.; Keshishian, S.M.; Vassallo,A.M. Measurement and modelling of the high-power performance of carbon-based supercapacitors. *J. Power Sources* **2000**, *91*, 68–76. [CrossRef]
23. Quintana, J.J.; Ramos, A.; Nuez, I. Identification of the fractional impedance of ultracapacitors. *IFAC Proc. Vol.* **2006**, *39*, 432–436. [CrossRef]
24. Freeborn, T.J.; Maundy, B.; Elwakil, A.S. Measurement of supercapacitor fractional-order model parameters from voltage-excited step response. *IEEE J. Emerg. Sel. Top. Circuits Syst.* **2013**, *3*, 367–376. [CrossRef]
25. Bisquert, J. Theory of the impedance of electron diffusion and recombination in a thin layer. *J. Phys. Chem. B* **2002**, *106*, 325–333. [CrossRef]
26. Augustyn, V.; Simon, P.; Dunn, B. Pseudocapacitive oxide materials for high-rate electrochemical energy storage. *Energy Environ. Sci.* **2014**, *7*, 1597–1614. [CrossRef]
27. Parveen, N.; Ansari, S.A.; Ansari, M.Z.; Ansari, M.O. Manganese oxide as an effective electrode material for energy storage: A review. *Environ. Chem. Lett.* **2022**, *20*, 283–309. [CrossRef]
28. Ansari, S.A.; Parveen, N.; Al-Othoum, M.A.S.; Ansari, M.O. Effect of washing on the electrochemical performance of a three-dimensional current collector for energy storage applications. *Nanomaterials* **2021**, *11*, 1596. [CrossRef] [PubMed]
29. Gil'mutdinov, A.K.; Ushakov, P.A.; El-Khazali, R. *Fractal Elements and Their Applications*; Springer International Publishing: Cham, Switzerland, 2017.
30. Samko, S.G.; Kilbas, A.A.; Marichev, O.I. *Fractional Integrals and Derivatives*; Gordon and Breach Science Publishers: Yverdon, Switzerland, 1993; Volume 1.
31. Bisquert, J.; Compte, A. Theory of the electrochemical impedance of anomalous diffusion. *J. Electroanal. Chem.* **2001**, *499*, 112–120. [CrossRef]
32. Lee, G.J.; Pyun, S.I.; Kim, C.H. Kinetics of double-layer charging/discharging of the activated carbon fiber cloth electrode: Effects of pore length distribution and solution resistance. *J. Solid State Electrochem.* **2004**, *8*, 110–117. [CrossRef]
33. Itagaki, M.; Suzuki, S.; Shitanda, I.; Watanabe, K.; Nakazawa, H. Impedance analysis on electric double layer capacitor with transmission line model. *J. Power Sources* **2007**, *164*, 415–424. [CrossRef]
34. Garcia, B.B.; Feaver, A.M.; Zhang, Q.; Champion, R.D.; Cao, G.; Fister, T.T.; Nagle, K.P.; Seidler, G.T. Effect of pore morphology on the electrochemical properties of electric double layer carbon cryogel supercapacitors. *J. Appl. Phys.* **2008**, *104*, 014305. [CrossRef]
35. Batalla, B.; Sinha, G.; Aliev, F. Dynamics of molecular motion of nematic liquid crystal confined in cylindrical pores. *Mol. Cryst. Liq. Cryst.* **1999**, *331*, 121–128. [CrossRef]
36. Prasad, R.; Mehta, U.; Kothari, K. Various analytical models for supercapacitors: A mathematical study. *Resour.-Effic. Technol.* **2020**, *1*, 1–15.
37. Prabhakar, T.R. A singular integral equation with a generalized Mittag-Leffler function in the kernel. *Yokohama Math. J.* **1971**, *19*, 7–15.
38. Jonscher, A.K. Dielectric relaxation in solids. *J. Phys. D Appl. Phys.* **1999**, *32*, R57. [CrossRef]
39. Nigmatullin, R.R. On the theory of relaxation for systems with "remnant" memory. *Phys. Status Solidi (b)* **1984**, *124*, 389–393. [CrossRef]
40. Weron, K. A probabilistic mechanism hidden behind the universal power law for dielectric relaxation: General relaxation equation. *J. Phys. Condens. Matter* **1991**, *3*, 9151. [CrossRef]
41. Uchaikin, V.V.; Sibatov, R. *Fractional Kinetics in Solids: Anomalous Charge Transport in Semiconductors, Dielectrics, and Nanosystems*; World Scientific: Singapore, 2013.
42. Khamzin, A.A. Trap-controlled fractal diffusion model of the Havriliak–Negami dielectric relaxation. *J. Non-Cryst. Solids* **2019**, *524*, 119636. [CrossRef]
43. Uchaikin, V.; Sibatov, R.; Uchaikin, D. Memory regeneration phenomenon in dielectrics: The fractional derivative approach. *Phys. Scr.* **2009**, *T136*, 014002. [CrossRef]
44. Sibatov, R.T.; Uchaikin, V.V.; Uchaikin, D.V. Fractional wave equation for dielectric medium with Havriliak–Negami response. In *Fractional Dynamics and Control*; Springer: New York, NY, USA, 2012; pp. 293–301.
45. Sibatov, R.; Shulezhko, V.; Svetukhin, V. Fractional derivative phenomenology of percolative phonon-assisted hopping in two-dimensional disordered systems. *Entropy* **2017**, *19*, 463. [CrossRef]
46. Uchaikin, V.V.; Sibatov, R.T. Fractional theory for transport in disordered semiconductors. *Commun. Nonlinear Sci. Numer. Simul.* **2008**, *13*, 715–727. [CrossRef]

Article

Improvement of Lithium Storage Performance of Silica Anode by Using Ketjen Black as Functional Conductive Agent

Guobin Hu [1,†], Xiaohui Sun [1,†], Huigen Liu [1], Yaya Xu [1], Lei Liao [2,*], Donglei Guo [2], Xianming Liu [3] and Aimiao Qin [1,*]

1   Key Lab New Processing Technology for Nonferrous Metal and Materials Ministry of Education, College of Material Science and Engineering, Guilin University of Technology, Guilin 541004, China; kokpinhu@hotmail.com (G.H.); 15564222794@139.com (X.S.); huigenliugz@163.com (H.L.); xuyayachonga@163.com (Y.X.)
2   Guangxi Key Laboratory of Environment Pollution Control Theory and Technology, College of Environmental Science and Engineering, Guilin University of Technology, Guilin 541004, China; gdl0594@163.com
3   Key Laboratory of Function-Oriented Porous Materials, College of Chemistry and Chemical Engineering, Luoyang Normal University, Luoyang 471934, China; myclxm@163.com
*   Correspondence: fangqiu2001@163.com (L.L.); qam2005032@glut.edu.cn (A.Q.)
†   These authors contributed equally to this work.

Abstract: In this paper, SiO_2 aerogels were prepared by a sol–gel method. Using Ketjen Black (KB), Super P (SP) and Acetylene Black (AB) as a conductive agent, respectively, the effects of the structure and morphology of the three conductive agents on the electrochemical performance of SiO_2 gel anode were systematically investigated and compared. The results show that KB provides far better cycling and rate performance than SP and AB for SiO_2 anode electrodes, with a reversible specific capacity of 351.4 mA h g^{-1} at 0.2 A g^{-1} after 200 cycles and a stable 311.7 mA h g^{-1} at 1.0 A g^{-1} after 500 cycles. The enhanced mechanism of the lithium storage performance of SiO_2-KB anode was also proposed.

Keywords: silica-based anode; Ketjen Black; electrochemical properties; lithium-ion battery

Citation: Hu, G.; Sun, X.; Liu, H.; Xu, Y.; Liao, L.; Guo, D.; Liu, X.; Qin, A. Improvement of Lithium Storage Performance of Silica Anode by Using Ketjen Black as Functional Conductive Agent. *Nanomaterials* **2022**, *12*, 692. https://doi.org/10.3390/nano12040692

Academic Editors: Xiang Wu and Jung Woo Lee

Received: 16 January 2022
Accepted: 8 February 2022
Published: 19 February 2022

Publisher's Note: MDPI stays neutral with regard to jurisdictional claims in published maps and institutional affiliations.

1. Introduction

Lithium-ion batteries (LIBs) have attracted much attention due to their high energy density and long cycle life. To meet the demand for scaled-up LIBs, the development of electrode materials with high performance is necessary. Graphite is widely used as anode material for LIBs [1], however, its theoretical lithium storage capacity is relatively low, only 372 mA h g^{-1}. Therefore, silicon-based anode materials with higher theoretical specific capacity (4200 mA h g^{-1}) are considered to be anode materials for next-generation LIBs [2–4]. However, the severe volume expansion (>300%) associated with the various phase transitions during the intercalation and escape of lithium in/out Si particles have been a major disadvantage, as this led to rapid capacity fading and significantly limited commercial application [5]. Although novel silicon anodes with nanosphere [6,7], nanotube [8,9], core-shell structure [10,11] and other new structures could improve the cycling performance [12,13], the complicated process and expensive preparation technology are prohibitive. In addition, the low initial coulombic efficiency and the poor conductivity also limited its application [14–16]. Compared to elemental Si anode, silicon oxides show a smaller volume change during cycling. Furthermore, when using silicon oxides as anodes, the in situ generated Li_2O and lithium silicates during the first lithiation may buffer the large volume change and lead to the improvement of cycling stability. In cutting-edge researches, silica [17] with hollow [18], porous [19,20], and other special structures were composited with carbon [19,21,22], graphite [23], metal [24], and metal oxides [25–28] to improve its conductivity and lithium storage performance. These methods could effectively improve the electrochemical performance of silica anodes.

Carbon conductive agents, which were added during the electrode manufacturing process, played an important role in the impedance and electrode density. However, their functional mechanism still needs further investigation [29,30]. The literature has reported that conductive agents could be used as the mediator [31] to form a conductive network in electrodes, reducing the contact resistance of the electrode and improving the electron transport rate. Commercial carbon black, such as acetylene black (AB) and Super-P (SP), have been used as conductive agents in LIBs [32,33]. Compared with AB and SP, Ketjen Black (KB) has the advantages of large specific surface area, excellent electrical conductivity, and relatively narrow pore size distribution, when used as the conductive agent [34]. However, the systematic study of the effect of KB on silica anodes is sparsely reported.

Herein, a network nanostructure of silica (SiO_2) anode material using KB as a conductive agent with high electrochemical performance was prepared. The effects of KB on the electrochemical performance of silica anode materials were systematically studied. Furthermore, the enhanced storage mechanism of the SiO_2-KB anode materials was proposed. This work revealed that the type of conductive agent played a key role on the electrochemical performance of anode materials.

2. Materials and Methods

2.1. Synthesis

Briefly, SiO_2 aerogels were prepared by the sol–gel method [35]. It was obtained by taking 8 mL of anhydrous ethanol in a beaker, adding ammonia to adjust the pH to 9–10, then slowly adding 0.5 mL of TEOS, and left for 4 h at room temperature before adding 1 mL of deionized water to prepare the gel, and freeze-drying to obtain SiO_2 aerogels.

SiO_2 nanospheres were obtained by first taking 3 mL of $NH_3 \cdot H_2O$ and 60 mL of alcohol to be mixed and stirred thoroughly, then 1.5 mL of TEOS was added into the above solution and continued stirring for 10 h at room temperature to obtain a white emulsion, finally the solid product was collected by centrifugation, washed several times with distilled water and alcohol, and dried at 70 °C for 12 h in a vacuum.

2.2. Materials Characterization

Morphological and compositional analyses for the as-prepared sample were performed with Transmission Electron Microscope (TEM, JEM-2100F, JEOL Inc., Tokyo, Japan) and field emission scanning electron microscopy (SEM, S-4800, HITACHI Inc., Tokyo, Japan), respectively, the crystallographic structure of the obtained SiO_2 were characterized by X-ray diffraction (XRD, X' Pert PRO, PANalytical Inc., Almelo, The Netherlands), the chemical component of the SiO_2 anode was investigated using an X-ray photoelectron spectroscope (XPS, ESCALAB 250Xi, Thermo Fisher Scientific Inc., Waltham, MA, USA) using Al kα radiation, the electrical resistance and electrical resistivity were tested by four-point probe meter (FPM, RTS-2A, 4 PROBES TECH Inc., China).

2.3. Electrochemical Measurements

The working electrodes were fabricated by compressing a mixture of the active materials (SiO_2 nanospheres), a conductive material of KB, AB, or SP, and a binder of polyvinylidene fluoride at the mass ratio of 50:30:20 onto Cu foil current collector (10 μm in thickness), then dried at 110 °C for 12 h. 0.1 mL of 1 M $LiPF_6$ in EC/DMC/DEC (Ethylene carbonate/Dimethyl carbonate/Diethyl carbonate) with volume ratio of 1:1:1 was used as the electrolyte, electrochemical experiments of half cells were carried out in CR2025 coin-type cells, 0.6 mm thick lithium discs are used as counter electrodes, and Polypropylene diaphragm type Celgard 2500 as battery separator. The cells were assembled in an argon-filled glove box (MIKROUNA, LAB2000, Shanghai, China). The specific capacity was measured by a galvanostatic discharge–charge method in the voltage range between 3.0 V and 0.01 V at a current density of 100 mA g^{-1} with SiO_2 as the active material mass on a battery test system (Neware, BTS 5 V 10 mA, Shenzhen, China). Cyclic voltammetry was performed using an electrochemical workstation (CV, CHI 690D, CH Instruments Ins, Wuhan, China) between 3 V and 0.01 V (vs. Li/Li$^+$) at a scan rate of 0.5 mV s^{-1}.

3. Results and Discussion

Figure 1a shows the SEM image of the SiO$_2$ aerogel, and it can be seen that the prepared SiO$_2$ aerogel particles are uniform in size with sphere in shape. Figure 1b shows that the average size of SiO$_2$ aerogel particle is about 100 nm. Figure 1c is the XRD pattern of the SiO$_2$ aerogel, the crystal structure of the prepared silica aerogel only shows a broad diffraction peak at around 23°, indicating an amorphous structure.

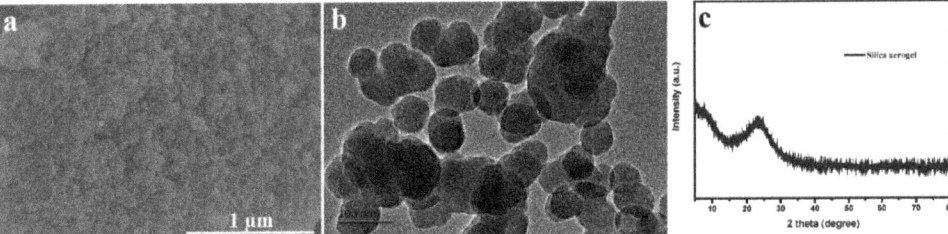

Figure 1. SEM (**a**), TEM (**b**) image, and XRD (**c**) pattern of silica aerogel.

Figure 2 shows the TEM and elemental mapping images of SiO$_2$ aerogel mixed with different conductive agents, respectively. It can be seen that different conductive agents form different structures when combined with silica aerogel. Figure 2a shows that the aerogel mixture with KB (SiO$_2$-KB) has an internet structure with uniformly dispersed nanoparticles, which indicates that a conductive network can be formed to provide a large number of effective conductive pathways and contacts for Li-ions. Furthermore, the elemental distribution shows a cobweb-like carbon chain pathway. For comparison, we also investigated the differences in the composition of the conductive pathways of SiO$_2$-SP and SiO$_2$-AB, respectively, under identical conditions. Figure 2b shows that the SiO$_2$-SP has a branched structure with larger SiO$_2$ particles and more agglomerates than that of SiO$_2$-KB. Figure 2c shows that the SiO$_2$-AB stacks together and has more agglomerate structure than that of SiO$_2$-SP and SiO$_2$-KB. Therefore, it is clear that the SiO$_2$-KB has the best dispersion, indicating it has excellent conductive network channels.

To further confirm the effects of the three conductive agents on the electrochemical performance of SiO$_2$ anode, a four-probe electrical resistance test was carried out and the result is shown in Table 1, the type of conductive agents plays an important role on the electrical resistance of SiO$_2$ anode. KB provides much lower electrical resistance and electrical resistivity than SP and AB for SiO$_2$ anode, which is helpful to improve the rate performance of electrode.

Table 1. The electrical resistance and electrical resistivity of anode electrodes.

Sample	SP	AB	KB
sheet resistance (Ω)	125.0	148.2	64.9
electrical resistivity (Ω·cm)	12.50	14.82	6.49

The cycling performance and coulombic efficiency of the SiO$_2$ anode with different conductive agents are shown in Figure 3a. The first discharge capacity of SiO$_2$-KB reaches 378.2 mA h g^{-1}, the capacity has a slight decrease in the several consequent cycles, and maintains 351.4 mA h g^{-1} after 200th cycles at 0.2 A g^{-1}. In comparison to SiO$_2$-KB, electrodes of SiO$_2$-SP and SiO$_2$-AB exhibit a lower reversible capacity of 139.4 mA h g^{-1} and 118.7 mA h g^{-1} at the first cycle and after 100th cycles display the capacity of 163.9 mA h g^{-1} and 137.7 mA h g^{-1} at 0.2 A g^{-1}, respectively, which indicates that KB is more beneficial in facilitating the silica electrochemical reaction. The specific capacity of SiO$_2$-KB decreased before the first 40 cycles, and then gradually increased, even after more than 100 cycles; the former is mainly due to the gradual lithiation of SiO$_2$ and the generated irreversible products such as lithium silicate and Li$_2$O, and the latter is due to the generated elemental silicon, which can provide the reversible specific capacity by the Si-Li alloy reaction.

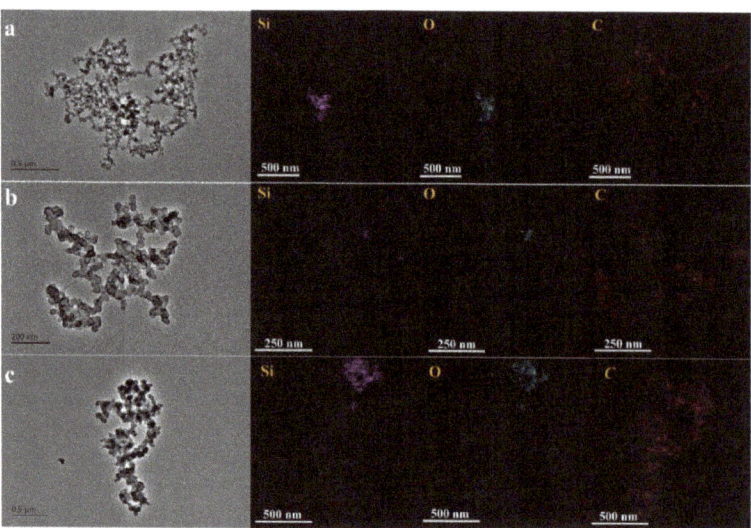

Figure 2. TEM and elemental mapping images of silica aerogels mixed with KB (**a**), SP (**b**), and AB (**c**).

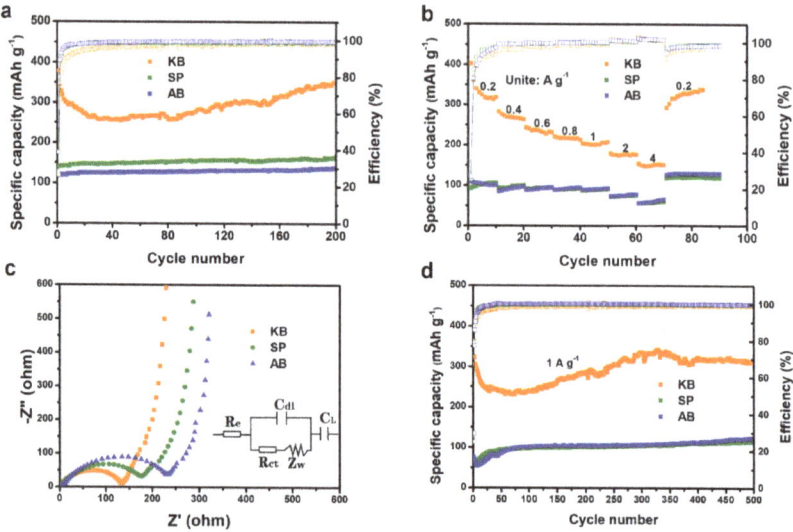

Figure 3. Cycle performance with coulombic efficiency at a current density of 200 mA g^{-1} (**a**), Rate capability at various current densities (**b**), Nyquist plot (**c**), and long-term cycling performance at a current density of 1.0 A g^{-1} (**d**) of silica aerogel anode with three carbon conductive agent.

The rate capability of the electrodes is shown in Figure 3b, with different current densities of 0.2, 0.4, 0.6, 0.8, 1.0, 2.0, and 4.0 A g^{-1}. The SiO$_2$-KB electrode attains an average discharge capacity of 318.2 mA h g^{-1}, 263.6 mA h g^{-1}, 231.9 mA h g^{-1}, 213.4 mA h g^{-1}, 207.7 mA h g^{-1}, 176.7 mA h g^{-1}, and 151.4 mA h g^{-1}, respectively, at the above current densities. The stable high reversible capacity of 331.1 mA h g^{-1} recovered when the current density turned back to 0.2 A g^{-1}. Compared to SiO$_2$-SP and SiO$_2$-AB, SiO$_2$-KB shows more excellent rate properties, indicating that the SiO$_2$-KB has excellent structural stability.

To determine the origin of the electrochemical behavior of SiO$_2$-KB, electrochemical impedance spectra (EIS) test was performed at their open-circuit potential. The equivalent circuits inserted in Figure 3c were employed to analyze the Nyquist plot of the desired anode material. The total impedance could be regarded as the electrolyte resistance R$_e$ and the charge transfer resistance R$_{ct}$, and C$_{dl}$ is the double-layer capacitance. Z$_w$ is the Warburg impedance that reflects the diffusion of lithium-ion in the solid. C$_L$ means the simplified intercalation capacitance. A semicircle was an indication for the charge transfer at high frequency range, while the straight line for the low frequency lithium-ion diffusion in the electrode material [36]. Obviously, the resistance of SiO$_2$-KB (R$_{ct}$ = 130 Ω) is much lower than that of SiO$_2$-SP (R$_{ct}$ = 180 Ω) and SiO$_2$-AB (R$_{ct}$ = 240 Ω), suggesting that KB could remarkably enhance the silica electrical conductivity. Furthermore, SiO$_2$-KB presents an exciting long-term cycling performance and delivers a specific capacity of 311.7 mA h g^{-1} at a current density of 1.0 A g^{-1} after 500 cycles. In comparison to SiO$_2$-KB, SiO$_2$-SP and SiO$_2$-AB exhibit a much lower reversible capacity of 66.4 mA h g^{-1} and 75.9 mA h g^{-1}, respectively, at the first cycle, and after 500th cycles, the capacity only remains 115.5 mA h g^{-1} and 123.9 mA h g^{-1}, respectively, which is shown in Figure 3d.

To confirm the structural integrity of the electrodes after cycle tests, SEM images of the different electrodes with the three conductive agents after 200 cycles were obtained and illustrated in Figure 4. For the electrode of SiO$_2$-KB, the shape of SiO$_2$ remained constant after 200 cycles (Figure 4a,b). For the electrode of SiO$_2$-SP, the silica undergoes a slight agglomeration phenomenon after 200 cycles (Figure 4c,d), while for the electrode of SiO$_2$-AB, the SiO$_2$ particles stick together and form large particles after 200 cycles (Figure 4e,f), which reduces the contact area between the silica, and leads to the reduction in capacity.

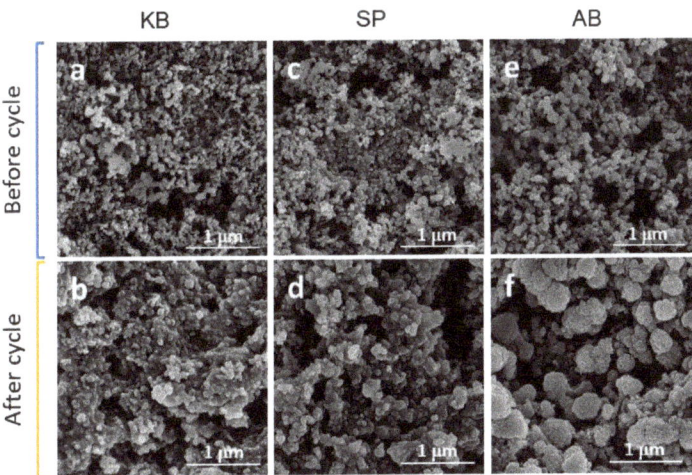

Figure 4. SEM images of SiO$_2$ electrode with KB (**a,b**), SP (**c,d**), and AB (**e,f**) before and after 200 cycles.

Figure 5 shows the CV curves of the SiO$_2$ aerogel electrodes at a scan rate of 0.5 mV s^{-1} with three different conductive agents, respectively. As can be seen in Figure 5, the reduction characteristic peak potential for the reaction of silica to produce lithium silicate and lithium oxide is 0.42 V, 0.65 V, and 0.8 V when using KB, SP, and AB as the conductive agent, respectively. The peak potential is significantly shifted to a smaller voltage for the KB (Figure 5a) compared to the SP (Figure 5b) and AB (Figure 5c). The oxidation characteristic peak potential of 0.25 V for the silicon-lithium alloy when using KB as the conductive agent also shows a significant shift to a smaller voltage compared to the SP (0.28 V) and AB (0.35 V), which is mainly due to the fact that KB has higher conductivity than SP and AB, where KB acts as a microcurrent collector between SiO_2 and the current collector to accelerate the speed of electron movement and also effectively increase the migration rate of Li^+ in the electrode material. Subsequently, the polarization of the silica anode is reduced, which will facilitate the occurrence of electrochemical reactions [37].

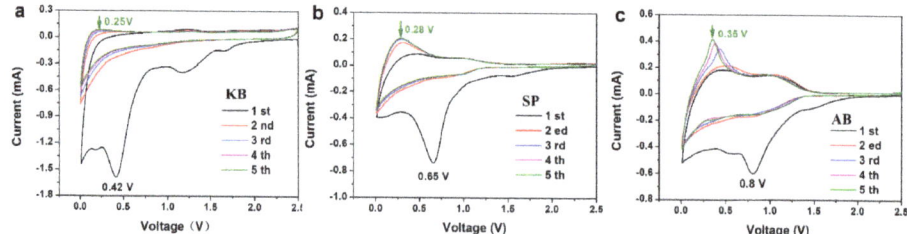

Figure 5. CV curve for the first five cycles of the silica aerogel when KB (**a**), SP (**b**), and AB (**c**) as conductive agents.

The chemical states of Si in SiO_2 anode with KB (Figure 6a), SP (Figure 6b) and AB (Figure 6c) during discharge/charge were identified by XPS, where energy correction for surface contamination was performed using C1s (284.6 eV) as a standard. The Si $2p_{3/2}$ peak shifts from 104.0 to 103.0 eV when discharged to 0.01 V, suggesting the reduction in Si to Li_xSi. When charged to 3 V, the peak shifts back to the original position of the SiO_2 electrode before discharge/charge (blue curve). It is clear that the curve in SiO_2-KB fluctuates more strongly than the smooth curves in SiO_2-SP and SiO_2-AB, indicating that the electrochemical reaction promoted by SiO_2 using KB as a conductive agent produces a higher amount of products containing elemental silicon, which is conducive to the improvement of electrochemical performance.

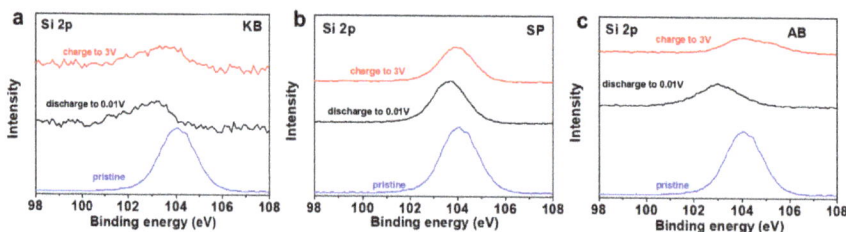

Figure 6. Si 2p XPS spectra of SiO_2 anode with KB (**a**), SP (**b**) and AB (**c**) at different states.

Based on previous researches, the possible electrochemical reaction mechanisms of SiO_2 can be summarized into the following reactions [38,39]:

$$SiO_2 + 4Li^+ + 4e^- \rightarrow 2Li_2O + Si \tag{R1}$$

$$2SiO_2 + 4Li^+ + 4e^- \rightarrow Li_4SiO_4 + Si \tag{R2}$$

$$Si + xLi^+ + xe^- \leftrightarrow Li_xSi \tag{R3}$$

Reactions of R1, and R2 are irreversible (potential of KB-0.25 V, SP-0.28 V, AB-0.35 V) and occur simultaneously although they compete with each other, and the obtained Si is electrochemically active, while Li_2O, and Li_4SiO_4 are electrochemically inactive [22]. The reaction R1 can produce more Si and result in a higher capacity than the reaction R2 [13].

To better understand the storage mechanism and excellent high-rate performance of the SiO_2 aerogel electrode with KB, the CV curves of the SiO_2 aerogel electrode under the three carbon conductive agents at different scan rates from 0.5 to 5 mV s^{-1} were collected and shown in Figure 7a,e,i. Generally, the current obeys a function relationship with the voltage during the sweep [40–43]:

$$i = av^b \tag{1}$$

where a and b are the parameters. The capacity contributions from the diffusion-controlled intercalation process and the surface-induced capacitive process can be qualitatively analyzed by the b-value. For a diffusion-controlled process, the b-value is 0.5, while the b-value near 1 means a totally capacitive-controlled process. According to the fitted line log (v)-log (i) curve depicted in Figure 7b,f,j, the b-value is 0.84, 0.86, 0.66, respectively. Furthermore, in view of the capacity contribution, the current (i) under a certain potential (V) can be divided into two parts [42–44]:

$$i(V) = k_1v + k_2v^{\frac{1}{2}} \tag{2}$$

where k_1v and $k_2v^{\frac{1}{2}}$ present charge stemmed from the surface capacitive charge and diffusion-controlled charge, respectively. The area share of the pseudocapacitive behavior of the silica aerogel electrodes with the three different conductive agents is shown in the paler parts of Figure 7c,g,k. The pseudocapacitance contributions are shown in Figure 7d,h,l. The pseudocapacitance contribution of the SiO_2 aerogel electrode with KB at scan rates of 1.0 to 5.0 mV s^{-1} is 41%, 49%, 54%, 58% and 61%, respectively; 51%, 58%, 63%, 67% and 70%, respectively, for SP and 65%, 73%, 77%, 79% and 80%, respectively, for AB. It can be seen that the use of different forms of conductive agents has a greater effect on the contribution of the pseudocapacitance in the SiO_2 aerogel electrode. The storage mechanism of the SiO_2-KB electrode is dominated by diffusion-controlled intercalation behavior, which is due to the excellent conductive network structure of KB and leads to an accelerated redox reaction. The contribution of surface-driven pseudocapacitance behavior for the SiO_2-KB electrode gradually increases as the scan rate increases, but is still less than that SiO_2-SP and SiO_2-AB. The contribution of the pseudocapacitance behavior of the SiO_2-KB electrode increases with increasing scan rate, however remains smaller than that of the SiO_2-SP and SiO_2-AB electrodes.

Li$^+$ diffusion coefficients during electrochemical charge/discharge for silica aerogels with different conductive agents calculated from the GITT method [45–47] are presented in Figure 8. The voltage change curve of the first charge/discharge under pulse current when using three different carbon conductive agents is shown in Figure 8a, and the Li-ion diffusion coefficient calculated from the pulse charge/discharge curve is shown in Figure 8b. The longer charging and discharging duration of Li$^+$ in the diffusivity test curve of SiO_2-KB is due to the special network structure of KB, which enables the nano-SiO_2 particles to perform electrochemical reactions without agglomeration and less hindering to the transport of Li$^+$. This indicates that the KB conductive agent accelerates the diffusion of Li$^+$ and enables the sufficient electrochemical reaction of SiO_2 [16].

In order to confirm the effect of the KB, the electrochemical properties of SiO_2 nanosphere anode was compared with three different conductive agents, respectively. The morphology of the SiO_2 nanospheres is shown in Figure 9a, which has a nice monodispersity and smooth surface with an average particle size of 100 nm. The inset in Figure 9a shows the XRD pattern of the SiO_2 nanospheres; a broad peak at 23° suggesting that the SiO_2 nanospheres are amorphous in structure similar to the SiO_2 aerogels. Figure 9b–d show the CV curves of the SiO_2 nanosphere anode at a scan rate of 0.5 mV s^{-1} when using the three different conductive agents, respectively. The reduction characteristic peak potential for the reaction of SiO_2 to produce Li_2O and Li_4SiO_4 locates at 0.5 V for the SiO_2 nanospheres-KB anode, which has a significant shift to the left compared to the anodes of SiO_2 nanospheres-SP (0.85 V) and SiO_2 nanospheres-AB (0.9 V), indicating that it also has the similar effect of reducing polarization of SiO$_2$ anodes prepared by different methods when using KB as the conducting agent, and strongly promotes the electrochemical reaction.

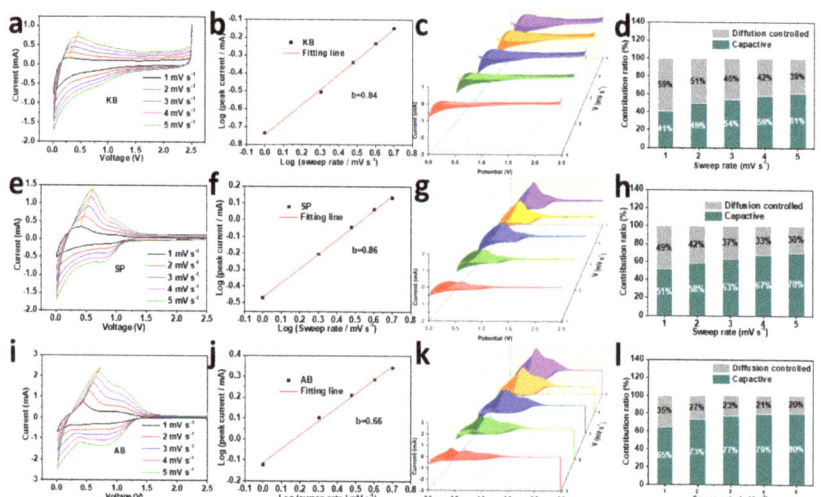

Figure 7. Behaviors of pseudocapacitance in silica aerogels using KB (**a–d**), SP (**e–h**), and AB (**i–l**) as conductive agents.

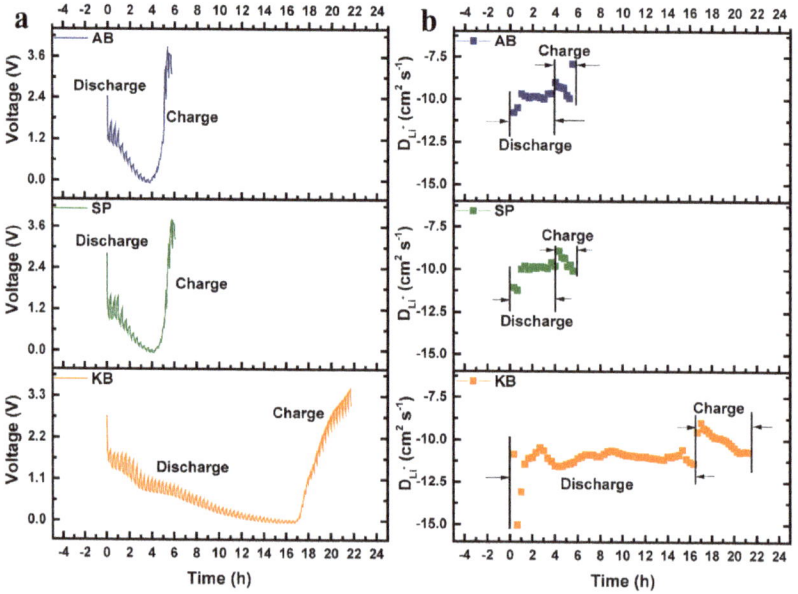

Figure 8. The voltage change curve of the first charge/discharge under pulse current (**a**) and Li-ion diffusion coefficient calculated from the pulse charge/discharge curve (**b**) with different carbon conductive agents.

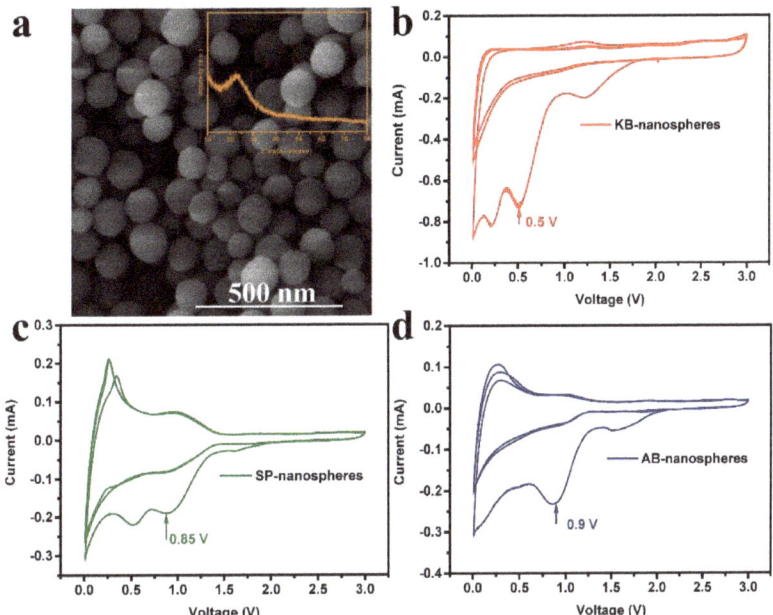

Figure 9. The SEM image and XRD pattern (the inset) (**a**) of silica nanospheres, CV curves of the first three cycles of the silica nanospheres when using KB (**b**), SP (**c**), and AB (**d**) as conductive agents, respectively.

The function of KB can be described as a schematic diagram conductive network, as shown in Figure 10, the special structure of KB connects the SiO_2 aerogel particles together and forms an excellent conductive network of dispersed SiO_2-KB, which provides rich electron transfer channels and improves the electrochemical performance of silica anode.

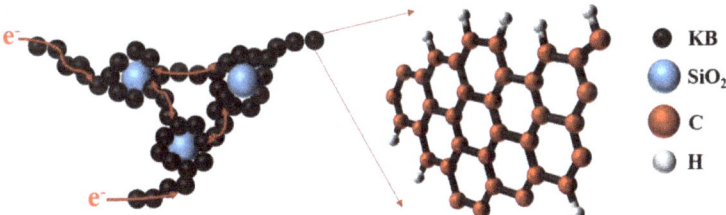

Figure 10. Schematic diagram conductive network of KB.

4. Conclusions

In summary, the effects of the types and the structures of conductive agents on the electrochemical performance of SiO_2 aerogel electrode were investigated, and the results show that the KB as a conductive agent not only can uniformly disperse and wrap the SiO_2 nanoparticles, but also can build a good conductive network to enhance the transport rate of lithium-ions and effectively increase their electrochemical activity. This work proves and verifies that SiO_2 aerogel can be used as a recommended electrode material for high-rate LIBs through choosing appropriate conductive agent.

Author Contributions: Conceptualization, methodology, writing—original draft preparation, investigation, and resources, G.H.; data curation, visualization, X.S.; software, H.L.; validation, D.G.,

L.L. and X.L.; formal analysis, Y.X.; writing—review and editing, supervision, project administration, and funding acquisition, A.Q.; All authors have read and agreed to the published version of the manuscript.

Funding: This work was supported by the Natural Science Foundation of Guangxi province (2018JJA160029, 2018GXNSFAA138041), the Foundation of Guangxi Key Laboratory of Optical and Electronic Materials and Devices (20AA-17), and Innovation Project of Guangxi Graduate Education (YCSW2021199, YCBZ2021062).

Institutional Review Board Statement: Not applicable.

Informed Consent Statement: Not applicable.

Data Availability Statement: The data presented in this study are available from the corresponding author upon request.

Acknowledgments: We acknowledge the support of the Key Laboratory of New Processing Technology for Non-ferrous Metals and Materials of the Ministry of Education, School of Materials Science and Engineering, Guilin University of Technology.

Conflicts of Interest: The authors declare no conflict of interest.

References

1. Zhang, Z.; Wang, F.; An, Q.; Li, W.; Wu, P. Synthesis of graphene@Fe_3O_4@C core–shell nanosheets for high-performance lithium ion batteries. *J. Mater. Chem. A* **2015**, *3*, 7036–7043. [CrossRef]
2. Hou, G.; Cheng, B.; Yang, Y.; Du, Y.; Zhang, Y.; Li, B.; He, J.; Zhou, Y.; Yi, D.; Zhao, N.; et al. Multiscale Buffering Engineering in Silicon–Carbon Anode for Ultrastable Li-Ion Storage. *ACS Nano* **2019**, *13*, 10179–10190. [CrossRef]
3. Choi, J.W.; Aurbach, D. Promise and reality of post-lithium-ion batteries with high energy densities. *Nat. Rev. Mater.* **2016**, *1*, 16013. [CrossRef]
4. Tu, J.; Yuan, Y.; Zhan, P.; Jiao, H.; Wang, X.; Zhu, H.; Jiao, S. Straightforward Approach toward SiO_2 Nanospheres and Their Superior Lithium Storage Performance. *J. Phys. Chem. C* **2014**, *118*, 7357–7362. [CrossRef]
5. Song, T.; Xia, J.; Lee, J.-H.; Lee, D.H.; Kwon, M.-S.; Choi, J.-M.; Wu, J.; Doo, S.K.; Chang, H.; Park, W.I.; et al. Arrays of Sealed Silicon Nanotubes As Anodes for Lithium Ion Batteries. *Nano Lett.* **2010**, *10*, 1710–1716. [CrossRef] [PubMed]
6. Li, J.Y.; Li, G.; Zhang, J.; Yin, Y.X.; Yue, F.S.; Xu, Q.; Guo, Y.G. Rational Design of Robust Si/C Microspheres for High-Tap-Density Anode Materials. *ACS Appl. Mater. Interfaces* **2019**, *11*, 4057–4064. [CrossRef]
7. Xu, Q.; Li, J.Y.; Sun, J.K.; Yin, Y.X.; Wan, L.J.; Guo, Y.G. Watermelon-Inspired Si/C Microspheres with Hierarchical Buffer Structures for Densely Compacted Lithium-Ion Battery Anodes. *Adv. Energy Mater.* **2017**, *7*, 1601481. [CrossRef]
8. Ma, T.; Xu, H.; Yu, X.; Li, H.; Zhang, W.; Cheng, X.; Zhu, W.; Qiu, X. Lithiation Behavior of Coaxial Hollow Nanocables of Carbon–Silicon Composite. *ACS Nano* **2019**, *13*, 2274–2280. [CrossRef]
9. Zhang, C.; Gu, L.; Kaskhedikar, N.; Cui, G.; Maier, J. Preparation of Silicon@Silicon Oxide Core–Shell Nanowires from a Silica Precursor toward a High Energy Density Li-Ion Battery Anode. *ACS Appl. Mater. Interfaces* **2013**, *5*, 12340–12345. [CrossRef]
10. Zhai, W.; Ai, Q.; Chen, L.; Wei, S.; Li, D.; Zhang, L.; Si, P.; Feng, J.; Ci, L. Walnut-inspired microsized porous silicon/graphene core–shell composites for high-performance lithium-ion battery anodes. *Nano Res.* **2017**, *10*, 4274–4283. [CrossRef]
11. Lu, B.; Ma, B.; Deng, X.; Li, W.; Wu, Z.; Shu, H.; Wang, X. Cornlike Ordered Mesoporous Silicon Particles Modified by Nitrogen-Doped Carbon Layer for the Application of Li-Ion Battery. *ACS Appl. Mater. Interfaces* **2017**, *9*, 32829–32839. [CrossRef] [PubMed]
12. Chen, X.; Li, X.; Ding, F.; Xu, W.; Xiao, J.; Cao, Y.; Meduri, P.; Liu, J.; Graff, G.L.; Zhang, J.-G. Conductive Rigid Skeleton Supported Silicon as High-Performance Li-Ion Battery Anodes. *Nano Lett.* **2012**, *12*, 4124–4130. [CrossRef]
13. Gowda, S.R.; Pushparaj, V.; Herle, S.; Girishkumar, G.; Gordon, J.G.; Gullapalli, H.; Zhao, X.; Ajayan, P.M.; Reddy, A.L.M. Three-dimensionally engineered porous silicon electrodes for Li ion batteries. *Nano Lett.* **2012**, *12*, 6060–6065. [CrossRef] [PubMed]
14. Liu, Z.; Yu, Q.; Zhao, Y.; He, R.; Xu, M.; Feng, S.; Li, S.; Zhou, L.; Mai, L. Silicon oxides: A promising family of anode materials for lithium-ion batteries. *Chem. Soc. Rev.* **2018**, *48*, 285–309. [CrossRef]
15. Liang, C.; Zhou, L.; Zhou, C.; Huang, H.; Liang, S.; Xia, Y.; Gan, Y.; Tao, X.; Zhang, J.; Zhang, W. Submicron silica as high–capacity lithium storage material with superior cycling performance. *Mater. Res. Bull.* **2017**, *96*, 347–353. [CrossRef]
16. Zhang, Y.; Li, Y.; Wang, Z.; Zhao, K. Lithiation of SiO_2 in Li-Ion Batteries: In Situ Transmission Electron Microscopy Experiments and Theoretical Studies. *Nano Lett.* **2014**, *14*, 7161–7170. [CrossRef]
17. Chang, W.S.; Park, C.M.; Kim, J.H.; Kim, Y.U.; Jeong, G.; Sohn, H.J. Quartz (SiO_2): A new energy storage anode material for Li-ion batteries. *Energy Environ. Sci.* **2012**, *5*, 6895–6899. [CrossRef]
18. Ma, X.; Wei, Z.; Han, H.; Wang, X.; Cui, K.; Yang, L. Tunable construction of multi-shell hollow SiO_2 microspheres with hierarchically porous structure as high-performance anodes for lithium-ion batteries. *Chem. Eng. J.* **2017**, *323*, 252–259. [CrossRef]

19. Yuan, Z.; Zhao, N.; Shi, C.; Liu, E.; He, C.; He, F. Synthesis of SiO_2/3D porous carbon composite as anode material with enhanced lithium storage performance. *Chem. Phys. Lett.* **2016**, *651*, 19–23. [CrossRef]
20. Li, H.H.; Wu, X.L.; Sun, H.Z.; Wang, K.; Fan, C.Y.; Zhang, L.L.; Yang, F.M.; Zhang, J.P. Dual-Porosity SiO_2/C Nanocomposite with Enhanced Lithium Storage Performance. *J. Phys. Chem. C* **2015**, *119*, 3495–3501. [CrossRef]
21. Xiao, T.; Zhang, W.; Xu, T.; Wu, J.; Wei, M. Hollow SiO_2 microspheres coated with nitrogen doped carbon layer as an anode for high performance lithium-ion batteries. *Electrochim. Acta* **2019**, *306*, 106–112. [CrossRef]
22. Liu, X.; Chen, Y.; Liu, H.; Liu, Z.-Q. SiO_2 @C hollow sphere anodes for lithium-ion batteries. *J. Mater. Sci. Technol.* **2017**, *33*, 239–245. [CrossRef]
23. Li, H.H.; Zhang, L.L.; Fan, C.Y.; Wang, K.; Wu, X.L.; Sun, H.Z.; Zhang, J.P. A plum-pudding like mesoporous SiO_2/flake graphite nanocomposite with superior rate performance for LIB anode materials. *Phys. Chem. Chem. Phys.* **2015**, *17*, 22893–22899. [CrossRef]
24. Tang, C.; Liu, Y.; Xu, C.; Zhu, J.; Wei, X.; Zhou, L.; He, L.; Yang, W.; Mai, L. Ultrafine Nickel-Nanoparticle-Enabled SiO_2 Hierarchical Hollow Spheres for High-Performance Lithium Storage. *Adv. Funct. Mater.* **2017**, *28*, 1704561. [CrossRef]
25. Hou, Y.; Yuan, H.; Chen, H.; Shen, J.; Li, L. The preparation and lithium battery performance of core-shell SiO_2@Fe_3O_4@C composite. *Ceram. Int.* **2017**, *43*, 11505–11510. [CrossRef]
26. Tian, Q.; Chen, Y.; Chen, F.; Zhang, W.; Chen, J.; Yang, L. Etching-free template synthesis of double-shelled hollow SiO_2@SnO_2@C composite as high performance lithium-ion battery anode. *J. Alloy. Compd.* **2019**, *809*, 151793. [CrossRef]
27. Liu, F.; Chen, Z.; Fang, G.; Wang, Z.; Cai, Y.; Tang, B.; Zhou, J.; Liang, S. V_2O_5 Nanospheres with Mixed Vanadium Valences as High Electrochemically Active Aqueous Zinc-Ion Battery Cathode. *Nano Micro Lett.* **2019**, *11*, 1–11. [CrossRef]
28. Khan, Z.; Singh, P.; Ansari, S.A.; Manippady, S.R.; Jaiswal, A.; Saxena, M. VO_2 Nanostructures for Batteries and Supercapacitors: A Review. *Small* **2021**, *17*, 2006651. [CrossRef]
29. Yang, X.; Zhang, R.; Xu, S.; Xu, D.; Ma, J.; Wang, S. Effect of carbon dimensions on the electrochemical performance of $SnSe_2$ anode for Na-ion batteries. *Mater. Lett.* **2020**, *284*, 128989. [CrossRef]
30. Asset, T.; Job, N.; Busby, Y.; Crisci, A.; Martin, V.; Stergiopoulos, V.; Bonnaud, C.; Serov, A.; Atanassov, P.; Chattot, R.; et al. Porous Hollow Ptni/C Electrocatalysts: Carbon Support Considerations to Meet Stability Requirements. In *ECS Meeting Abstracts*; IOP Publishing: Bristol, UK, 2018. [CrossRef]
31. Ilango, P.R.; Gnanamuthu, R.; Jo, Y.N.; Lee, C.W. Design and electrochemical investigation of a novel graphene oxide-silver joint conductive agent on $LiFePO_4$ cathodes in rechargeable lithium-ion batteries. *J. Ind. Eng. Chem.* **2016**, *36*, 121–124. [CrossRef]
32. Liu, W.R.; Guo, Z.Z.; Young, W.S.; Shieh, D.T.; Wu, H.C.; Yang, M.H.; Wu, N.L. Effect of electrode structure on performance of Si anode in Li-ion batteries: Si particle size and conductive additive. *J. Power Sources* **2005**, *140*, 139–144. [CrossRef]
33. Wen, L.; Sun, J.; An, L.; Wang, X.; Ren, X.; Liang, G. Effect of Conductive Material Morphology on Spherical Lithium Iron Phosphate. *Nanomaterials* **2018**, *8*, 904. [CrossRef] [PubMed]
34. Yang, Z.; Kim, C.; Hirata, S.; Fujigaya, T.; Nakashima, N. Facile Enhancement in CO-Tolerance of a Polymer-Coated Pt Electrocatalyst Supported on Carbon Black: Comparison between Vulcan and Ketjenblack. *ACS Appl. Mater. Interfaces* **2015**, *7*, 15885–15891. [CrossRef] [PubMed]
35. Hench, L.L.; West, J.K. The sol-gel process. *Chem. Rev.* **1990**, *90*, 33–72. [CrossRef]
36. Iurilli, P.; Brivio, C.; Wood, V. On the use of electrochemical impedance spectroscopy to characterize and model the aging phenomena of lithium-ion batteries: A critical review. *J. Power Sources* **2021**, *505*, 229860. [CrossRef]
37. Su, F.; Tang, R.; He, Y.; Zhao, Y.; Kang, F.; Yang, Q. Graphene conductive additives for lithium ion batteries: Origin, progress and prospect. *Chin. Sci. Bull.* **2017**, *62*, 3743–3756. [CrossRef]
38. Sivonxay, E.; Aykol, M.; Persson, K.A. The lithiation process and Li diffusion in amorphous SiO_2 and Si from first-principles. *Electrochim. Acta* **2019**, *331*, 135344. [CrossRef]
39. Abate, I.I.; Jia, C.J.; Moritz, B.; Devereaux, T.P. Ab initio molecular dynamics study of SiO_2 lithiation. *Chem. Phys. Lett.* **2020**, *739*, 136933. [CrossRef]
40. Zou, G.; Wang, C.; Hou, H.; Wang, C.; Qiu, X.; Ji, X. Controllable Interlayer Spacing of Sulfur-Doped Graphitic Carbon Nanosheets for Fast Sodium-Ion Batteries. *Small* **2017**, *13*, 1700762. [CrossRef]
41. Wang, J.; Polleux, J.; Lim, J.; Dunn, B. Pseudocapacitive Contributions to Electrochemical Energy Storage in TiO_2 (Anatase) Nanoparticles. *J. Phys. Chem. C* **2007**, *111*, 14925–14931. [CrossRef]
42. Jiang, Y.; Liu, J. Definitions of Pseudocapacitive Materials: A Brief Review. *Energy Environ. Mater.* **2019**, *2*, 30–37. [CrossRef]
43. Fleischmann, S.; Mitchell, J.B.; Wang, R.; Zhan, C.; Jiang, D.-E.; Presser, V.; Augustyn, V. Pseudocapacitance: From Fundamental Understanding to High Power Energy Storage Materials. *Chem. Rev.* **2020**, *120*, 6738–6782. [CrossRef]
44. Liu, T.C.; Pell, W.G.; Conway, B.E.; Roberson, S.L. Behavior of Molybdenum Nitrides as Materials for Electrochemical Capacitors: Comparison with Ruthenium Oxide. *J. Electrochem. Soc.* **1998**, *145*, 1882–1888. [CrossRef]
45. Shen, Z.; Cao, L.; Rahn, C.D.; Wang, C.-Y. Least Squares Galvanostatic Intermittent Titration Technique (LS-GITT) for Accurate Solid Phase Diffusivity Measurement. *J. Electrochem. Soc.* **2013**, *160*, A1842–A1846. [CrossRef]
46. Liu, C.; Wu, X.; Wang, B. Performance modulation of energy storage devices: A case of Ni-Co-S electrode materials. *Chem. Eng. J.* **2020**, *392*, 123651. [CrossRef]
47. Wu, X.; Yao, S. Flexible electrode materials based on WO3 nanotube bundles for high performance energy storage devices. *Nano Energy* **2017**, *42*, 143–150. [CrossRef]

Article

Sea Urchin-like Si@MnO$_2$@rGO as Anodes for High-Performance Lithium-Ion Batteries

Jiajun Liu, Meng Wang, Qi Wang, Xishan Zhao, Yutong Song, Tianming Zhao  and Jing Sun *

College of Sciences, Northeastern University, Shenyang 110819, China; 2000165@stu.neu.edu.cn (J.L.);
2000176@stu.neu.edu.cn (M.W.); wangqi@mail.neu.edu.cn (Q.W.); 2000189@stu.neu.edu.cn (X.Z.);
2000172@stu.neu.edu.cn (Y.S.); zhaotm@stumail.neu.edu.cn (T.Z.)
* Correspondence: sunjing74@mail.neu.edu.cn; Tel.: +86-136-0407-3045

Abstract: Si is a promising material for applications as a high-capacity anode material of lithium-ion batteries. However, volume expansion, poor electrical conductivity, and a short cycle life during the charging/discharging process limit the commercial use. In this paper, new ternary composites of sea urchin-like Si@MnO$_2$@reduced graphene oxide (rGO) prepared by a simple, low-cost chemical method are presented. These can effectively reduce the volume change of Si, extend the cycle life, and increase the lithium-ion battery capacity due to the dual protection of MnO$_2$ and rGO. The sea urchin-like Si@MnO$_2$@rGO anode shows a discharge specific capacity of 1282.72 mAh g^{-1} under a test current of 1 A g^{-1} after 1000 cycles and excellent chemical performance at different current densities. Moreover, the volume expansion of sea urchin-like Si@MnO$_2$@rGO anode material is ~50% after 150 cycles, which is much less than the volume expansion of Si (300%). This anode material is economical and environmentally friendly and this work made efforts to develop efficient methods to store clean energy and achieve carbon neutrality.

Keywords: Si; MnO$_2$; rGO; sea urchin-like structure; lithium-ion battery; high performance

Citation: Liu, J.; Wang, M.; Wang, Q.;
Zhao, X.; Song, Y.; Zhao, T.; Sun, J.
Sea Urchin-like Si@MnO$_2$@rGO as
Anodes for High-Performance
Lithium-Ion Batteries. *Nanomaterials*
2022, *12*, 285. https://doi.org/
10.3390/nano12020285

Academic Editor: Carlos
Miguel Costa

Received: 31 December 2021
Accepted: 14 January 2022
Published: 17 January 2022

Publisher's Note: MDPI stays neutral
with regard to jurisdictional claims in
published maps and institutional affiliations.

1. Introduction

In recent years, environmental pollution caused by carbon emissions has an increasing urgency for developing high-density, long lifetime storage materials or storage devices for clean energy. As a carrier of clean energy, lithium-ion batteries play a pivotal role in the country's goal of achieving carbon neutrality [1–4]. A Silicon-based electrode is one of the most promising candidates as an anode for lithium-ion batteries and is expected to replace the use of a commercial graphite electrode (372 mAh g^{-1}) due to its remarkable theoretical capacity (4200 mAh g^{-1}) [5,6]. It is also widely regarded for its good voltage platform, environmental friendliness, and abundant reserves. Despite these advantages, Si-based lithium-ion batteries still face severe volume expansion during the charging/discharging process, poor electrical conductivity, and a short cycle life [7,8]. The safety hazards and unstable performances of Si materials resist the applications of lithium-ion batteries in commercial use. Therefore, many attempts on modifying Si materials have been made to restrict the volume expansion and enhance electrical conductivity for improving the performance of Si-based lithium-ion batteries [9–12].

The rational design of an anode has been considered as an effective strategy to en-hance Si-based lithium-ion batteries' performances [13,14]. Some studies have reported that coating or doping can effectively reduce the large volume changes and the subsequent accumulation of excessive stress during lithiation–delithiation cycles [15–19]. Especially, Si nanospheres are combined with carbon materials, a strategy that aims to provide ex-tra space to accommodate volumes' expansion and improve the electrical conductivity of the electrode, such as Si/ reduced graphene oxide (rGO) and Si/Carbon Nanotube (CNT) [20–22]. However, the size gap between Si nanospheres and rGO sheets is large, and Si nanospheres tend to agglomerate and do not easily migrate uniformly into the rGO sheets to form a stable structure. It is worth mentioning that transition metal oxides are also good anodes with high theoretical capacity, such as MnO_2 (1233 mAh g^{-1}) [23–28]. In addition, the volume changes of MnO_2 in lithium-ion intercalation and deintercalation is very small. A reasonable coating structure was used to grow MnO_2 evenly outside the Si nanospheres, which can avoid the material crushing and form an unstable solid electrolyte interface (SEI) film due to the volume expansion of Si nanospheres during the circulation process, lowering the capacity. At the same time, rGO has high electrical conductivity and excellent physical properties. The low conductivity of transition metal oxides and Si nanospheres can be compensated by rGO. MnO_2 wraps the Si nanosphere to avoid the agglomeration of the Si nanosphere, increase the contact area with the rGO sheet, ensure the interface strength between MnO_2 and rGO, and improve the structural stability of the material [29–32].

Herein, we propose a simple hydrothermal method to produce sea urchin-like Si@MnO_2@rGO as anodes. In this unique structure, MnO_2 and rGO surrounding Si nanospheres formed a strong armor, relieving the mechanical strain generated by the volume expansion of the Si nanospheres and providing enough space to buffer, which helps provide to a long lifetime. In addition, the sea urchin-like Si@MnO_2@rGO anode showed an initial discharge capacity of 1378.15 mAh g^{-1} at a current density of 0.1 A g^{-1} and a discharge specific capacity was maintained at 1282.72 mAh g^{-1} under a test current of 1 A g^{-1} after more than 1000 cycles, showing excellent chemical performance at different current densities. Therefore, this rational design provides a new route for the development of high-performance Si-based anodes; this work made efforts to store clean energy and achieve carbon neutrality [33,34].

2. Materials and Methods

2.1. Materials

All chemicals were analytical reagent grade and used as received. Graphene oxide (GO) was purchased from XFNANO Materials Tech Co., Ltd, Nanjing, China. Si nanospheres, ethanol, $MnSO_4$, $KMnO_4$, KH550, and $NaBH_4$ were obtained from National Medicines Corporation Ltd, Shanghai, China.

2.2. Preparation of Si@MnO_2@rGO

First, 0.5 g of GO and 0.1 of M $NaBH_4$ were ultrasonically dispersed in 100 mL of deionized water with stirring treatment at 80 °C for 24 h. After cooling to room temperature, the sample was centrifuged to obtain reduced graphene oxide (rGO). The above-prepared rGO of 0.05 g was dispersed in 20 mL of deionized water and sonicated for 30 min, denoted as Solution A.

Then, 0.086 g of Si nanospheres (50 nm) were dispersed in a beaker containing 20 mL of water and 20 mL of alcohol, followed by the addition of 50 μL of KH550 silane coupling reagent. The sample was sonicated for 30 min to form a homogeneous solution, denoted as Solution B. Next, Solution A, Solution B, 0.64 g of $MnSO_4$, and 1 g of $KMnO_4$ were sequentially added in a reaction vessel and stirred with magnetic force for 2 h. The reaction temperatures were set as 20, 50, or 160 °C, respectively. After cooling to room temperature, the products were centrifuged with water and ethanol several times to remove the residual reaction products. A freeze-dried treatment was used to remove the water. Finally, the samples were denoted as Si@MnO_2@rGO-20°C, Si@MnO_2@rGO-50°C, and Si@MnO_2@rGO-160°C, respectively.

The synthesis of Si@MnO_2 was similar to the synthesis Si@MnO_2@rGO, only missing the rGO, denoted as Si@MnO_2-20°C, Si@MnO_2-50°C, and Si@MnO_2-160°C, respectively.

2.3. Electrochemical Measurements

The homogeneous slurry was prepared by mixing Si@MnO_2@rGO, acetylene black, and polyvinylidene fluoride (PVDF) in N-methyl-2-pyrrolidone (NMP) with a mass ratio of

8:1:1. After fully stirring, the prepared slurry was evenly coated on a copper foil by a sputter coater, whose space was 25 μm. The dimeter of electrode was 1.5 cm and the anode material mass load was ~1 mg. The electrolyte of each coin cell was ~45 μL. Then, the electrode sheet was placed into a vacuum drying oven for 12 h at 80 °C. The working electrode was the prepared electrode sheet (Si@MnO$_2$@rGO composites). The electrolyte was a LiPF$_6$ (1.0 mol L^{-1}) in a 1:1:1 v/v/v mixture of ethylene carbonate, dimethyl carbonate, and ethyl methyl carbonate. The separator was Celgard 2400 (Saibo, Beijing, China). The capacity calculation and cycling rate were set by a battery testing system (CT3008, Kejing, Hefei, China). The galvanostatic charge/discharge (GCD) tests were conducted in the voltage window of 0.1–3.2 V. The cyclic voltammetry curve (CV) was carried out with scan rate of 0.1 mV s^{-1} between the voltage range of 0.1–3.2 V using an electrochemical workstation (CHI 600E, Chenhua, Shanghai, China). The electrochemical impedance spectroscopy (EIS) was performed over the frequency range from 100 kHz to 0.1 Hz with an alternating current (AC) impedance of 5 mV and was also recorded by an electrochemical workstation (CHI 600E, Chenhua, Shanghai, China).

2.4. Materials' Characterization

Field-emission scanning electron microscopy (FE-SEM; SU8010, Hitachi High-Tech, Tokyo, Japan) and field-emission transmission electron microscopy (FE-TEM; JEM2100F, JEOL, Tokyo, Japan) were used to characterize the morphology and elemental distribution of the electrode materials. X-ray diffraction (XRD; Rigaku lnc., Tokyo, Japan) was used to characterize the phases, crystallinity, and crystal structures of the samples. The species and chemical composition of the surface elements of the samples were analyzed using X-ray photoelectron spectroscopy (XPS, Thermo AXIS-SUPRA, Kratos, Manchester, UK).

3. Results and Discussion

3.1. Anode Material Design and Morphology Characterization

Figure 1 shows a schematic of the preparation of the sea urchin-like Si@MnO$_2$@rGO composites. First, GO sheets were reduced to rGO by a hydrothermal method with NaBH$_4$. Then, the rGO was collected and dispersed in deionized water (Solution A). Next, Si nanospheres were dispersed in a solution of water, alcohol, and the KH550 silane coupling agent (Solution B). The composites were well grafted by KH550, improving the conductivity. Finally, Solution A, Solution B, KMnO$_{4,}$ and MnSO$_4$ were added in a reaction vessel and stirred with magnetic force for 2 h. The reactions were carried out at different temperatures, resulting in different morphologies. Si nanospheres were coated with MnO$_2$ and rGO, which may have contributed to the KH550. The unique double-layer structure of MnO$_2$ and rGO effectively mitigated the volume expansion of Si nanospheres during the charging/discharging process. In addition, rGO had excellent electrical conductivity, compensating for the low electrical conductivity of Si nanospheres and MnO$_2$. This was the basis of the rational design of the Si-based anode material for enhancing lithium-ion battery performance.

Figure 2 shows the morphologies of the materials grown at three different temperatures. The morphology of Si@MnO$_2$-20°C is shown in Figure 2a. The MnO$_2$ formed disordered and entangled filaments on the surface of Si nanospheres. Figure 2b shows Si@MnO$_2$-20°C at a high magnification. The surface of the material was rough and the diameter of Si@MnO$_2$-20°C particles was 200 nm. Figure 2c is an SEM image of Si@MnO$_2$@rGO-20°C, showing that the rGO covered the Si@MnO$_2$-20°C. However, some MnO$_2$ nanowires appeared on the surface of rGO, which suggested that the material was not stable. Figure 2d,e show the SEM images of Si@MnO$_2$-50°C at different magnifications, respectively. As shown in Figure 2d, the MnO$_2$ formed a sea urchin-like shell coating on the surface of Si nanospheres. Figure 2e shows the sea urchin-like Si@MnO$_2$-50°C. The stings (MnO$_2$) were 100–300 nm in length. Figure 2f shows Si@MnO$_2$@rGO-50°C. The rGO wrapped the Si@MnO$_2$-50°C without obvious damage to the structure. Interestingly, this rGO surface was clean, without any MnO$_2$ nanowires, indicating the sea urchin-like structure was more

Nanomaterials **2022**, *12*, 285

stable. MnO$_2$ nanowires were vertically oriented on the outer layer of the Si nanospheres, imbedding rGO layers, forming a stable structure, which can buffer the excessive expansion of the Si nanospheres during the charging/discharging process and extend the batteries' lifetime. Figure 2g shows the SEM image of Si@MnO$_2$-160°C. The MnO$_2$ nanowires grew longer and coated the surface of the Si nanospheres. Figure 2h is the enlarged view of Figure 2g. The plentiful MnO$_2$ nanowires were interleaved. Figure 2i shows the images of Si@MnO$_2$@rGO-160°C. It can be clearly seen that the MnO$_2$ nanowires were broken and dispersed on the surface of rGO, which indicates that the Si@MnO$_2$@rGO-160°C was not stable enough. The stability of the material is one of the important factors of the lithium-ion batteries' performance. The high stability of the anode material may suggest the long life-time. Therefore, the Si@MnO$_2$@rGO-50°C was chosen for the following test.

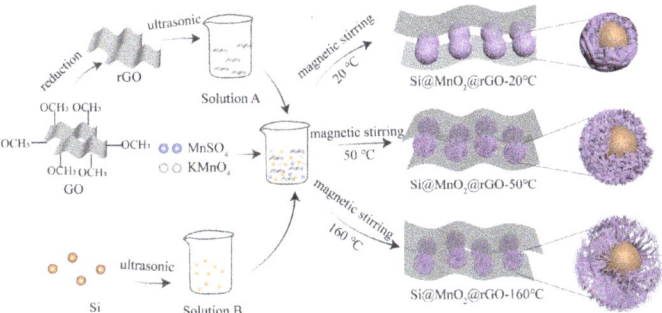

Figure 1. Concise preparation schematics of the Si@MnO$_2$@ reduced graphene oxide (rGO) composites.

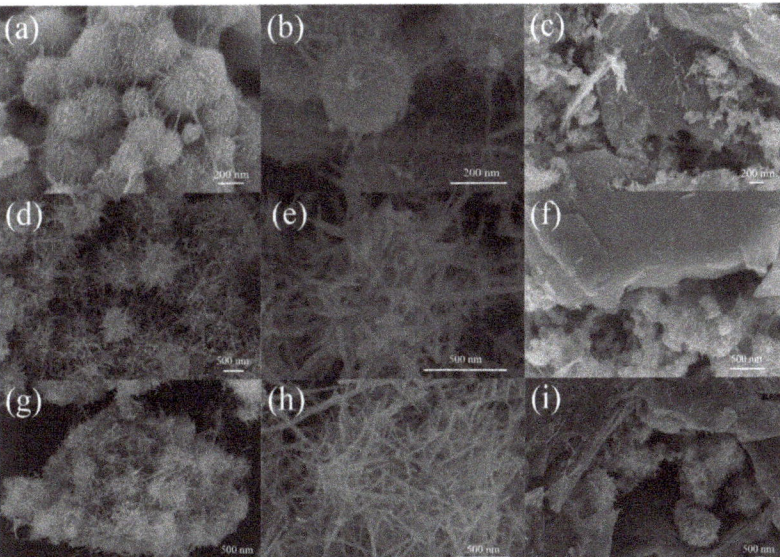

Figure 2. (**a**) SEM image of the Si@MnO$_2$-20°C composite. (**b**) Enlarged view of the SEM image of the Si@MnO$_2$-20°C composite. (**c**) SEM image of the Si@MnO$_2$@rGO-20°C composite. (**d**) SEM image of the Si@MnO$_2$-50°C composite. (**e**) Enlarged view of the SEM image of the Si@MnO$_2$-50°C composite. (**f**) SEM image of the Si@MnO$_2$@rGO-50°C composite. (**g**) SEM image of the Si@MnO$_2$-160°C composite. (**h**) Enlarged view of the SEM image of the Si@MnO$_2$-160°C composite. (**i**) SEM image of the Si@MnO$_2$@rGO-160°C composite.

Figure 3a shows the Energy Dispersive Spectrometer (EDS) analysis of the ball-milled material, using aluminum foil as a substrate, and the inset shows the atomic percentages of elements in the materials. The inset shows that the Si@MnO$_2$@rGO-50°C contained ~48% Si, ~21% rGO, and ~31% MnO$_2$. Table S1 shows the capacity contribution percentages of anode materials (Si, ~63.3%; MnO$_2$, ~35.6%; and rGO, ~1.1%). The existence of MnO$_2$ can not only resist the volume expansion of Si, but also provide the capacity contribution. Figure 3b,c show HRTEM images of the Si@MnO$_2$@rGO-50°C composite at different magnifications. Si nanospheres were coated with rGO and MnO$_2$. The lattice fringes corresponded to the (111) plane of Si, having a separation of 0.31 nm [35,36]. In addition, lattice fringes having an interplanar spacing of 0.47 nm can be seen, and this is consistent with the (200) plane of MnO$_2$ [25,37,38]. The element mapping of the Si@MnO$_2$@rGO-50°C composite is shown in Figure 3d–h. Mn element and O element contributed to MnO$_2$; Si element contributed to Si nanospheres. C element dispersed throughout the image may have contributed to both the substrate and rGO.

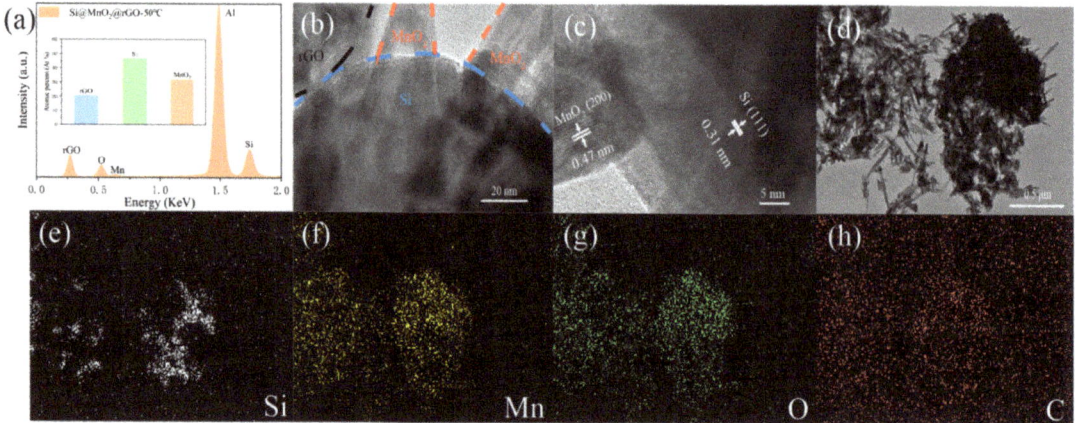

Figure 3. (**a**) Energy Dispersive Spectrometer (EDS) spectra of Si@MnO$_2$@rGO-50°C. Inset showing the respective substance of Si@MnO$_2$@rGO-50°C. (**b,c**) HRTEM images of the Si@MnO$_2$@rGO-50°C composite at different magnifications. (**d–h**) Elemental mapping of the Si@MnO$_2$@rGO-50°C composite.

Figure 4a shows the XRD patterns of rGO, Si, Si@MnO$_2$-50°C, and Si@MnO$_2$@rGO-50°C, respectively. In the XRD pattern of Si@MnO$_2$-50°C, the pronounced peaks at 28.5°, 47.4°, 56.2°, 58.9°, 69.2°, 76.5°, and 88.2° corresponded to the (111), (220), (311), (222), (400), (331), and (422) planes of Si (PDF#77-2108). In addition, the characteristic peaks at 12.7°, 18.0°, 28.7°, 37.6°, 41.1°, 49.8°, 59.5°, 65.5°, 68.5°, and 72.5° corresponded to the (110), (200), (310), (121), (420), (411), (260), (002), (202), and (631) planes of MnO$_2$ (PDF#72-1982). The XRD pattern of the Si@MnO$_2$@rGO-50°C composites contained peaks corresponding to Si and MnO$_2$, as well as the broad peak ranges from 20° to 30°, which were indexed to the standard peaks of rGO [39–42]. Figure 4b shows the XPS spectrum of the Si@MnO$_2$@rGO-50°C composites, which revealed the presence of Si, Mn, O, and C, corresponding to element mapping (Figure 3d–g). Figure 4c,d shows the high-resolution XPS spectra. Figure 4c shows the Si 2p spectrum, whose peak at 99.7 eV was related to Si-Si bonds; two small peaks at 101.9 and 103.6 eV contributed to organic Si and Si-O, respectively, which may have been caused by the slight oxidation of Si in the thermal-treated process. Figure 4d shows the Mn 2p spectrum, which contained two spin–orbit peaks corresponding to Mn 2p$_{3/2}$ (642.5 eV) and Mn 2p$_{1/2}$ (654.1 eV) of MnO$_2$, whose separation between these peaks was 11.6 eV.

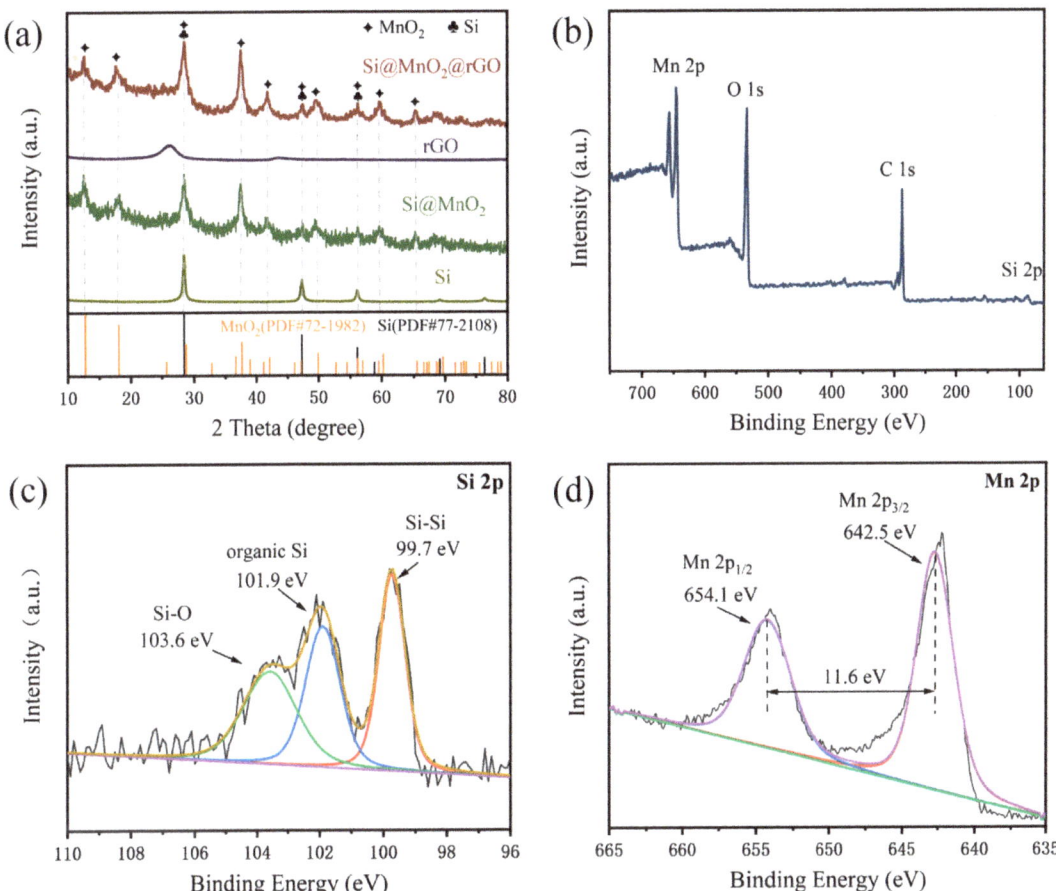

Figure 4. (**a**) X-ray diffraction (XRD) patterns of the Si, Si@MnO$_2$-50°C rGO, and Si@MnO$_2$@rGO-50°C, respectively. (**b–d**) XPS spectra of the Si@MnO$_2$@rGO-50°C.

3.2. Lithium-Ion Battery Performance

A CV experiment carried out to further evaluate the lithium storage behavior is shown in Figure 5. Figure 5a shows the CV curves of Si@MnO$_2$@rGO-50°C as the independent anodes of lithium-ion batteries for the first four cycles at a scan rate of 0.1 mV s^{-1} between 0.1 V and 3.2 V. In the first cycle, a clear cathodic peak at 0.16 V corresponded to the lithium alloying process of crystalline Si and the formation of an amorphous Li$_x$Si phase; a clear cathodic peak at 0.10 V corresponded to the formation of the Li$_{15}$Si$_4$ phase. An anodic peak at 0.24 V was related to the delithiation of Li$_{15}$Si$_4$; an anodic peak at 0.50 V was related to the transition from the Li$_x$Si phase to amorphous Si, according to Equations (1) and (2).

$$xLi^+ + Si + xe^- \leftrightarrow Li_xSi, \tag{1}$$

$$Li_xSi + xLi^+ + xe^- \leftrightarrow Li_{15}Si_4, \tag{2}$$

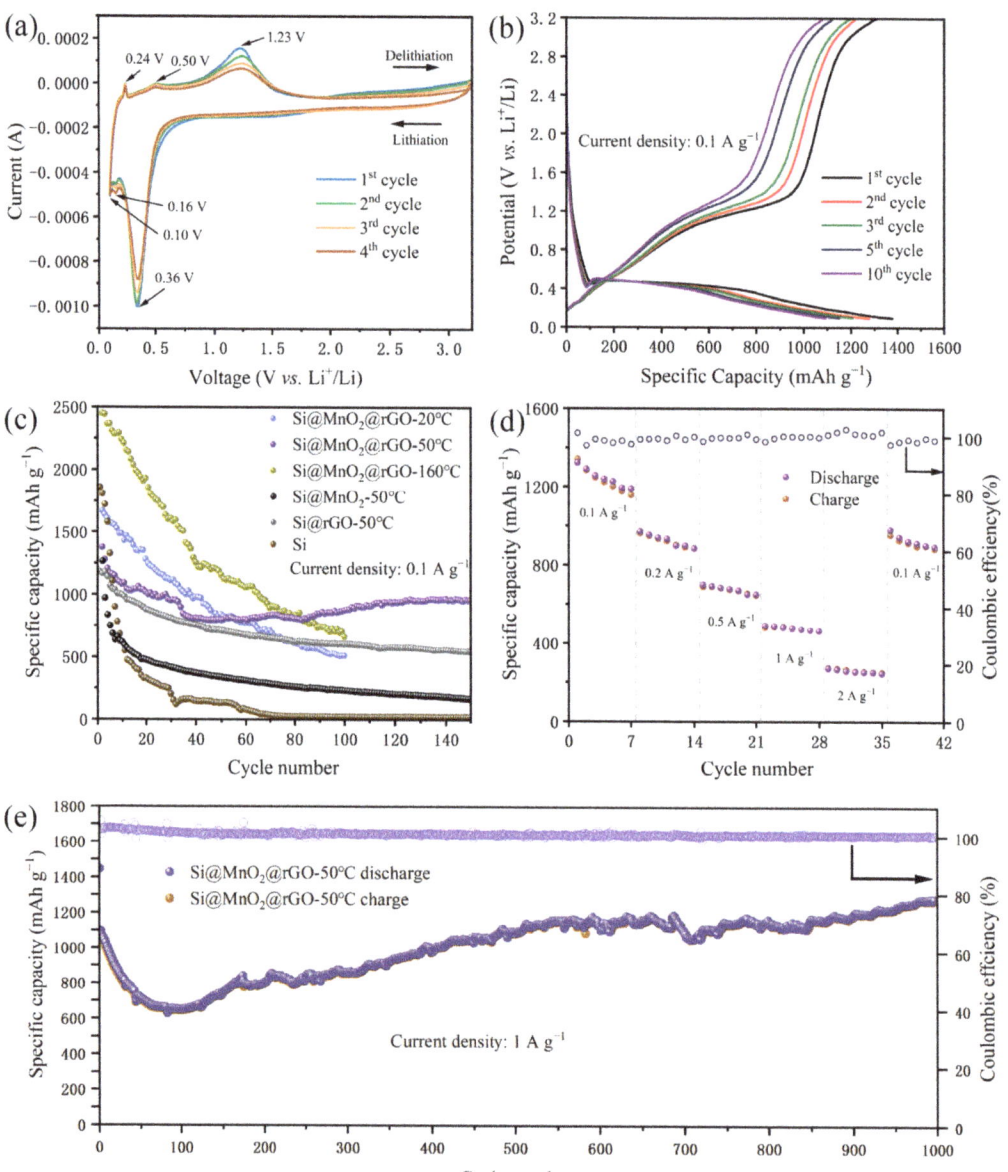

Figure 5. (**a**) CV curves of the Si@MnO$_2$@rGO-50°C at 0.1 mV s^{-1} between 0.1 V and 3.2 V. (**b**) Galvanostatic charge–discharge curves of the Si@MnO$_2$@rGO-50°C at 0.1 A g^{-1}. (**c**) Long-term cycling performance of the Si, Si@MnO$_2$-50°C, Si@rGO-50°C, Si@MnO$_2$@rGO-20°C, Si@MnO$_2$@rGO-50°C, and Si@MnO$_2$@rGO-160°C at 0.1 A g^{-1}. (**d**) Rate performance of the Si@MnO$_2$@rGO-50°C. (**e**) Long-term cycling performance of the Si@MnO$_2$@rGO-50°C at 1 A g^{-1}.

The redox peaks of Si coincided with those reported previously [43–45]. In addition, the reversible redox peaks at 1.23 and 0.36 V, respectively, were consistent with the lithiation and delithiation reactions of MnO$_2$, according to Equation (3). When the cathodic peak was 0.36 V, Li was inserted into the anode to form LiO$_2$ and MnO$_2$ was reduced to Mn. An

Nanomaterials **2022**, *12*, 285

anodic peak at 1.23 V was related to the charging process of the lithium ion battery; Mn can facilitate the decomposition of LiO_2.

$$MnO_2 + 4 Li \leftrightarrow 2 Li_2O + Mn, \tag{3}$$

In the subsequent scans, the redox peaks were largely coincident, which was attributed to the stability of the Si@MnO$_2$@rGO-50°C, indicating the good electrochemical reversibility for lithium-ion batteries.

Figure 5b shows the galvanostatic discharge/charge curves of Si@MnO$_2$@rGO-50°C at 0.1 A g^{-1} over the range between 0.1 V and 3.2 V. In the first cycle, the initial discharge capacity was 1378.15 mAh g^{-1}. In the second, third, fifth, and 10th cycles, the discharge specific capacities were 1279.21, 1208.02, 1150.74, and 1093.70 mAh g^{-1}, respectively. In the subsequent cycles, the charge and discharge curves basically overlapped, which indicated good capacity retention. In addition, the plateau of Si at 0.5 V in the figure was consistent with the CV results.

Figure 5c shows the cycling performances of the anode (Si, Si@MnO$_2$-50°C, Si@MnO$_2$@rGO-20°C, Si@MnO$_2$@rGO-50°C, and Si@MnO$_2$@rGO-160°C) at a current density of 0.1 A g^{-1}. Of these samples, Si@MnO$_2$@rGO-50°C showed the best cycling performance. The initial discharge specific capacity of Si was 1855.62 mAh g^{-1}. However, after 20 cycles, the discharge specific capacity decayed to 330.03 mAh g^{-1}; after 150 cycles, the discharge specific capacity was almost 0. This is because, during the cycling process, the slurry of the active material became dislodged from the collector due to the serious volume expansion of Si nanospheres caused by lithium-ion intercalation and deintercalation. When the Si nanospheres underwent volume changes during the charging/discharging process, the formed SEI film was broken, resulting in new surfaces being exposed in the electrolyte. The exposed surfaces needed external lithium ions to form a stable SEI film, which led to a dramatic decrease in capacity. Although Si@MnO$_2$-50°C had a higher discharge specific capacity than Si nanospheres alone after 150 cycles, it still did not meet current demands for battery energy storage. The initial specific capacity of Si@rGO-50°C was 1180.41 mAh g^{-1}; after 150 cycles, the capacity was maintained at 543.84 mAh g^{-1} (Figure 5c). Compared with Si@rGO-50°C and Si@MnO$_2$@ rGO-50°C, the existence of MnO$_2$ improved the specific capacity, which may have been due to the synergistic effect of MnO$_2$ and rGO. The initial specific capacities of Si@MnO$_2$@rGO-20°C and Si@MnO$_2$@rGO-160°C were 1670.24 and 2450.32 mAh g^{-1}, respectively. Furthermore, after 100 cycles, the specific capacities were 512.14 and 665.19 mAh g^{-1}, respectively, with low capacity retention rates. The initial specific capacity of Si@MnO$_2$@rGO-50°C was 1378.14 mAh g^{-1}; after 150 cycles, the capacity was maintained at 960.21 mAh g^{-1}. Among Si@MnO$_2$@rGO-50°C, Si@MnO$_2$@rGO-20°C, and Si@MnO$_2$@rGO-160°C, although the Si@MnO$_2$@rGO-50°C exhibited a lower capacity, it had the excellent cyclability, which can better meet the commercial need of long lifetime, due to the stability of the sea urchin-like Si@MnO$_2$@rGO-50°C and the dual protection of rGO and MnO$_2$. Interestingly, the structure of rGO encapsulating the sea urchin-like Si@MnO$_2$-50°C mitigated the volume-change effects of Si nanospheres, improving the electrical conductivity and contributing to the high capacity when comparing the Si and Si@MnO$_2$-50°C. A comparison of the rate and cycling performances of the sea urchin-like Si@MnO$_2$@rGO-50°C, at different various current densities, is shown in Figure 5d. At current densities of 0.1, 0.2, 0.5, 1, and 2 A g^{-1}, the specific charging capacities were 1323.87, 971.85, 701.12, 491.85, and 272.47 mAh g^{-1}, respectively. Furthermore, when the current density rose again to 0.1 A g^{-1}, the specific charging capacity recovered to 981.68 mAh g^{-1} and the capacity retention rate was 74.2%. Compared to previous reports of Si and Si@rGO, Si@MnO$_2$@rGO-50°C demonstrated a long lifetime and high capacity retention rates [46,47]. Furthermore, to evaluate the cycling stability at high current densities, Si@MnO$_2$@rGO-50°C was tested at 1 Ah g^{-1} (Figure 5e). In the first charge/discharge cycle, the specific capacity was 1446.85 mAh g^{-1}; after 1000 cycles, the specific capacity was 1282.72 mAh g^{-1} with a coulombic efficiency of 99.4%. The composite material also showed excellent cycling performance at high currents. The decrease in capacity fluctuations at the first 80 cycles was due to the irreversible formation of SEI film on the materials' surface. Interestingly, the capacity exhibited an increasing trend from the ~80th cycle upwards. This was attributed to the reversible growth of a polymeric gel-like film. After a long-cycling charging/discharging process, the polymeric gel-like film gradually degraded. The improved electrode kinetics increased the capacity. This phenomenon has been widely reported among transition metal oxides [48–55]. Table S2 shows the performance comparison among the reported works and indicates the high capacity and long lifetime of Si@MnO$_2$@rGO-50°C. When the cycle reached 150 times, the specific capacity was 960.21 mAh g^{-1} at 0.1 A g^{-1} and 746.13 mA g^{-1} at 1 A g^{-1}. This is because the higher the charge/discharge current density was, the fast the electrode chemical reaction speed became. A large number of lithium ions reacted on the surface of the anode materials instantly, leading to the formation of concentration polarization on the electrode surface. Part of the active materials had no time to react, and the utilization rate of the active materials became smaller, resulting in the decreasing in capacity.

EIS measurements were carried out to further investigate the electrochemical mechanism shown in Figure 6. Figure 6a shows the EIS plots and corresponding fitting plots of Si, Si@MnO$_2$-50°C, and Si@MnO$_2$@rGO-50°C, respectively. Additionally, the inset shows the data equivalent circuit diagrams. The EIS diagram consists of a semicircle in the mid-high-frequency region and a diagonal line in the low-frequency region. Here, R$_{CT}$ in the mid-frequency region corresponds to the charge-transfer impedance and W$_O$ in the low-frequency region corresponds to the diffusion of Li$^+$ inside the electrode material [56,57]. The R$_S$ and R$_{CT}$ values fitted from the equivalent circuit model are summarized in Table S3 for comparison. The Si@MnO$_2$@rGO-50°C exhibited a low value of R$_{CT}$ (146.2 Ω) before cycling, which was the lowest among Si, Si@MnO$_2$@rGO-50°C, and Si@MnO$_2$-50°C. This result was attributed to the rGO networks, prominently improving the electronic conductivity of the electrodes and lowering the charge transfer impedance. Figure 6b shows the EIS plots and corresponding fitting plots of Si@MnO$_2$@rGO-50°C before and after 150 cycles at 0.1 A g^{-1}; cycled Si@MnO$_2$@rGO-50°C was fitted with the same equivalent circuit model in Figure 6a. The R$_{CT}$ (72.7 Ω) was lower after cycling, which implies the charge transfer impedance decreased substantially, demonstrating the stable SEI film and good structure stability during cycling of Si@MnO$_2$@rGO-50°C.

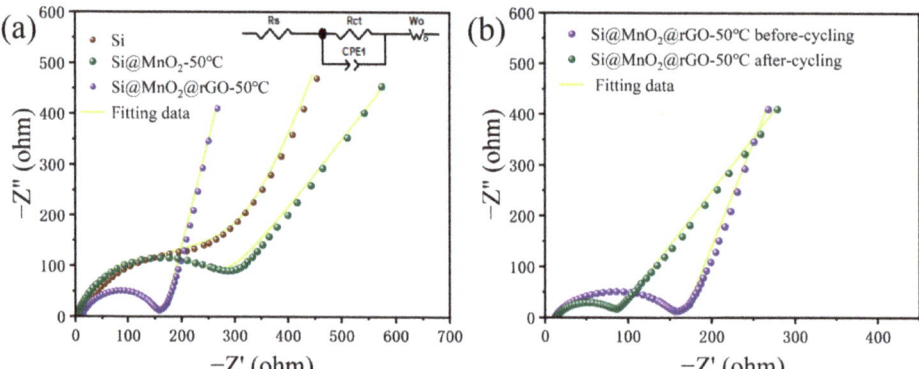

Figure 6. (**a**) The electrochemical impedance spectra of Si, Si@MnO$_2$-50°C, and Si@MnO$_2$@rGO-50°C. The inset shows the equivalent circuit model. (**b**) The electrochemical impedance spectra of the Si@MnO$_2$@rGO-50°C before and after cycling.

Figure 7a,b shows the cross-sectional SEM images of Si before and after 150 cycles at 0.1 A g^{-1}. After cycles, the volume expansion was ~323% (from 5.3 μm to 22.41 μm). The huge change in volume expansion of Si may have contributed to the intercalation and deintercalation of lithium ions during cycles and the exfoliation of the active materials. As shown in Figure S1a,b, the volume expansion of Si@MnO$_2$-50°C electrode after 150 cycles at 0.1 A g^{-1} was ~198% (from 4.29 μm to 12.8 μm). Interestingly, the volume expansion of Si@MnO$_2$@rGO-50°C electrode was ~59% (8.13 μm to 12.9 μm) after 150 cycles at 0.1 A g^{-1}. Such a small volume change in the active materials guarantees long-term cycling stability. Figure S2a shows the top-view SEM image of Si@MnO$_2$@rGO-50°C after 150 cycles at 0.1 A g^{-1}. The electrode sheet was relatively intact without signs of rupture. Figure S2b shows the element mapping of Figure S2a, suggesting Si, Mn, O, and C were evenly distributed. The mechanism of the Si electrode reaction was further illustrated in Figure 7e. The lithiation of Si resulted in the formation of an amorphous silicon–lithium alloy (Li$_x$Si) (0.16 V). Then, the amorphous Li$_x$Si was transformed to crystalline Li$_{15}$Si$_4$ (0.1 V). The delithiation process involved the transformation from crystalline Li$_{15}$Si$_4$ to amorphous Li$_x$Si (0.24 V) and, finally, to amorphous Si (0.5 V) [58–60].

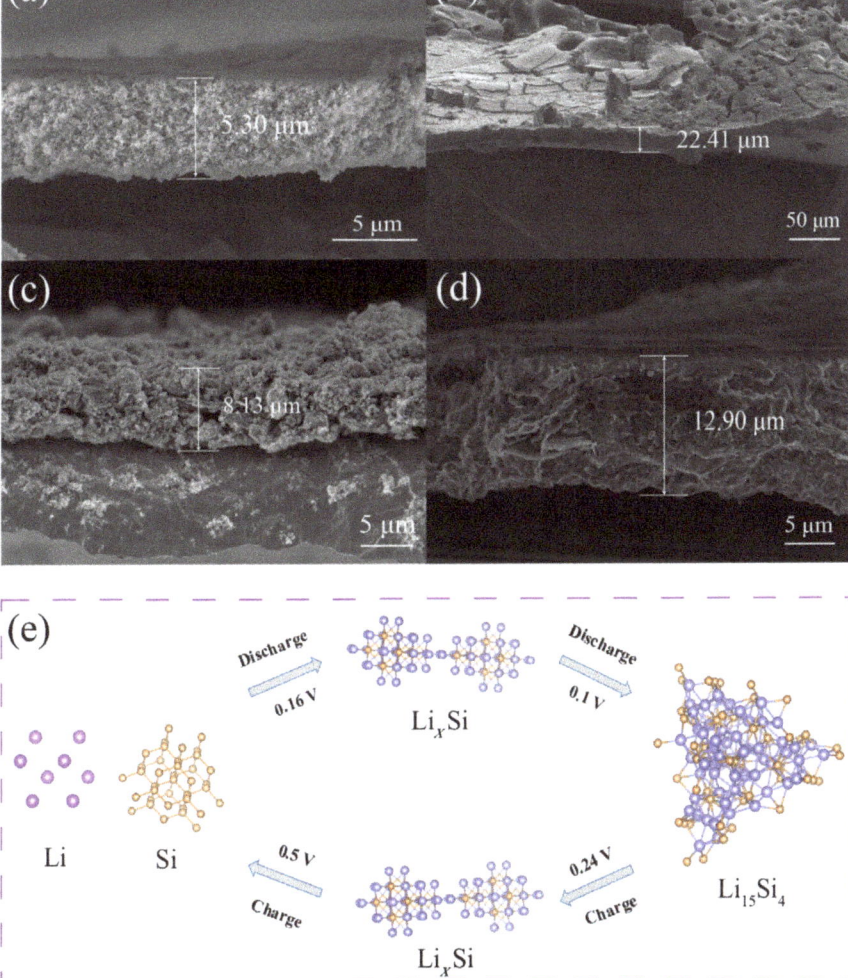

Figure 7. (**a**,**b**) Cross-sectional SEM images of Si before and after 150 cycles. (**c**,**d**) Cross-sectional SEM images of Si@MnO$_2$@rGO-50°C before and after 150 cycles. (**e**) The mechanism of intercalation and deintercalation of lithium ions.

4. Conclusions

In summary, sea urchin-like Si@MnO$_2$@rGO-50°C as an anode for lithium-ion batteries was presented. The reversible capacity, cyclability, and rate capability were very high, which was attributed to the dual protection of rGO and MnO$_2$. The discharge specific capacity was maintained at 1282.72 mAh g^{-1} under a test current of 1 A g^{-1} after more than 1000 cycles. Such high cyclability may be attributed to the sea urchin-like structure reducing the volume expansion of anodes during the charging/discharging process. The present results indicate that the sea urchin-like Si@MnO$_2$@rGO-50°C is a good candidate for high performance anodes of lithium-ion batteries. This work made efforts to develop efficient methods to store clean energy and achieve carbon neutrality.

Supplementary Materials: The following supporting information can be downloaded at: https://www.mdpi.com/article/10.3390/nano12020285/s1. Figure S1: (a,b) Cross-sectional SEM images of Si@MnO$_2$-50°C before and after 150 cycles. Figure S2: (a) Cross-sectional SEM images of the Si@MnO$_2$@rGO-50°C electrode sheet after 150 cycles at 0.1 A g^{-1}. (b) Element mapping images of the Si@MnO$_2$@rGO-50°C electrode sheet after 150 cycles at 0.1 A g^{-1}. Table S1: Capacity contribution of Si, MnO$_2$, and rGO. Table S2: Synthesis strategies and electrochemical performance comparison of Si-based anode materials and MnO$_2$-based anode materials in lithium-ion batteries (References [61–64] are cited in Table S2). Table S3: The R$_S$ and R$_{CT}$ values fitted from the equivalent circuit model are summarized for comparison.

Author Contributions: Conceptualization, J.L. and M.W.; methodology, Q.W.; software, X.Z.; validation, Y.S. and T.Z.; formal analysis, J.L.; investigation, J.L.; resources, J.L.; data curation, J.L.; writing—original draft preparation, J.L.; writing—review and editing, J.L.; visualization, J.S.; supervision, J.S.; project administration, J.S.; funding acquisition, J.S. All authors have read and agreed to the published version of the manuscript.

Funding: This research was funded by the Fundamental Research Funds for the Central Universities, grant number N2105011, and the Natural Science Foundation of Liaoning Province of China, grant number 2021-MS-082.

Data Availability Statement: The data presented in this study are available on request from the corresponding author.

Acknowledgments: Thanks to Siqi Liu from Shiyanjia Lab (www.shiyanjia.com) for the TEM analysis.

Conflicts of Interest: The authors declare no conflict of interest.

References

1. Mo, R.W.; Li, F.; Tan, X.Y.; Xu, P.C.; Tao, R.; Shen, G.R.; Lu, X.; Liu, F.; Shen, L.; Xu, B.; et al. High-quality mesoporous graphene particles as high-energy and fast-charging anodes for lithium-ion batteries. *Nat. Commun.* **2019**, *10*, 1474. [CrossRef]
2. Grey, P.; Hall, D.S. Prospects for lithium-ion batteries and beyond-a 2030 vision. *Nat. Commun.* **2020**, *11*, 6279. [CrossRef] [PubMed]
3. Yang, Y.; McDowell, M.T.; Jackson, A.; Cha, J.J.; Hong, S.S.; Cui, Y. New Nanostructured Li$_2$S/Silicon Rechargeable Battery with High Specific Energy. *Nano Lett.* **2010**, *10*, 1486–1491. [CrossRef] [PubMed]
4. Ge, F.; Li, G.C.; Wang, X.C.; Chen, X.X.; Fu, L.; Liu, X.X.; Mao, E.Y.; Liu, J.; Yang, X.L.; Qian, C.X.; et al. Manipulating Oxidation of Silicon with Fresh Surface Enabling Stable Battery Anode. *Nano Lett.* **2021**, *21*, 3127–3133. [CrossRef] [PubMed]
5. Zheng, G.R.; Xiang, Y.X.; Xu, L.F.; Luo, H.; Wang, B.L.; Liu, Y. Controlling Surface Oxides in Si/C Nanocomposite Anodes for High-Performance Li-Ion Batteries. *Adv. Energy Mater.* **2018**, *8*, 1801718. [CrossRef]
6. Zheng, T.Y.; Jia, Z.; Lin, N.; Langer, T.; Lux, S.; Lund, I.; Gentschev, A.C.; Qiao, J.; Liu, G. Molecular Spring Enabled High-Performance Anode for Lithium Ion Batteries. *Polymers* **2017**, *9*, 657. [CrossRef] [PubMed]
7. Yue, H.W.; Li, Q.; Liu, D.Q.; Hou, X.Y.; Bai, S.; Lin, S.M.; He, D.Y. High-yield fabrication of graphene-wrapped silicon nanoparticles for self-support and binder-free anodes of lithium-ion batteries. *J. Alloy. Compd.* **2018**, *744*, 243–251. [CrossRef]
8. Gao, R.S.; Tang, J.; Tang, S.; Zhang, K.; Ozawa, K.; Qin, L.C. Biomineralization-inspired: Rapid preparation of a silicon-based composite as a high-performance lithium-ion battery anode. *J. Mater. Chem. A* **2021**, *9*, 11614–11622. [CrossRef]
9. Fang, R.; Miao, C.; Mou, H.Y.; Xiao, W. Facile synthesis of Si@TiO$_2$@rGO composite with sandwich-like nanostructure as superior performance anodes for lithium ion batteries. *J. Alloy. Compd.* **2020**, *818*, 152884. [CrossRef]
10. Hassan, F.; Batmaz, R.; Li, J.D.; Wang, X.L.; Xiao, X.C.; Yu, A.P.; Chen, Z.W. Evidence of covalent synergy in silicon–sulfur–graphene yielding highly efficient and long-life lithium-ion batteries. *Nat. Commun.* **2015**, *26*, 8597. [CrossRef] [PubMed]

11. Wang, Z.L.; Mao, Z.M.; Lai, L.F.; Okubo, M.; Song, Y.H.; Zhou, Y.J.; Liu, X.; Huang, W. Sub-micron silicon/pyrolyzed carbon@natural graphite self-assembly composite anode material for lithium-ion batteries. *Chem. Eng. J.* **2017**, *313*, 187–196. [CrossRef]

12. Dai, J.Y.; Liao, J.; He, M.Y.; Yang, M.M.; Wu, K.P.; Yao, W.T. Si@SnS$_2$-rGO composite anode material with high lithium-ion battery capacity. *ChemSusChem* **2019**, *12*, 5092–5098. [CrossRef] [PubMed]

13. Yang, Z.; Song, Y.W.; Zhang, C.F.; He, J.J.; Li, X.D.; Wang, X.; Wang, N.; Li, Y.L.; Huang, C.S. Porous 3D Silicon-Diamondyne Blooms Excellent Storage and Diffusion Properties for Li, Na, and K Ions. *Adv. Energy Mater.* **2021**, *11*, 2101197. [CrossRef]

14. Wu, Y.J.; Chen, Y.A.; Huang, C.L.; Su, J.T.; Hsieh, C.T.; Lu, S.Y. Small highly mesoporous silicon nanoparticles for high performance lithium ion based energy storage. *Chem. Eng. J.* **2020**, *400*, 125958. [CrossRef]

15. Liang, B.R.; Zhu, S.S.; Wang, J.C.; Liang, X.Q.; Huang, H.F.; Huang, D.; Zhou, W.Z.; Xu, S.K.; Guo, J. Silicon-doped FeOOH nanorods@graphene sheets as high-capacity and durable anodes for lithium-ion batteries. *Appl. Surf. Sci.* **2021**, *550*, 149330. [CrossRef]

16. Zhang, Y.L.; Mu, Z.J.; Lai, J.P.; Chao, Y.G.; Yang, Y.; Zhou, P.; Li, Y.J.; Yang, W.X.; Xia, Z.H.; Guo, S.J. MXene/Si@SiO$_X$@C Layer-by-Layer Superstructure with Auto-Adjustable Function for Superior Stable Lithium Storage. *ACS Nano* **2019**, *13*, 2167–2175. [PubMed]

17. Suo, G.Q.; Zhang, J.Q.; Li, D.; Yu, Q.Y.; Wang, W.A.; He, M.; Feng, L.; Hou, X.J.; Yang, Y.L.; Ye, X.H.; et al. N-doped carbon/ultrathin 2D metallic cobalt selenide core/sheath flexible framework bridged by chemical bonds for high-performance potassium storage. *Chem. Eng. J.* **2020**, *388*, 124396. [CrossRef]

18. Kang, Y.; Zhang, Y.H.; Shi, Q.; Shi, H.W.; Xue, D.F.; Shi, F.N. Highly efficient Co$_3$O$_4$/CeO$_2$ heterostructure as anode for lithium-ion batteries. *J. Colloid Interf. Sci.* **2020**, *585*, 705–715. [CrossRef]

19. Li, D.P.; Dai, L.N.; Ren, X.H.; Ji, F.J.; Sun, Q.; Zhang, Y.M.; Ci, L.J. Foldable Potassium-Ion Batteries Enabled by Free-standing and Flexible SnS$_2$@C Nanofibers. *Energy Environ. Sci.* **2021**, *14*, 424–436. [CrossRef]

20. Pham, T.K.; Shin, J.H.; Karima, N.C.; Jun, Y.S.; Jeong, S.K.; Cho, N.; Lim, S.N. Application of recycled Si from industrial waste towards Si/rGO composite material for long lifetime lithium-ion battery. *J. Power Sources* **2021**, *506*, 230244. [CrossRef]

21. Chen, X.F.; Yang, X.F.; Pan, F.L.; Zhang, T.T.; Zhu, X.Y.; Qiu, J.Y.; Li, M.; Mu, Y.; Ming, H. Fluorine-functionalized core-shell Si@C anode for a high-energy lithium-ion full battery. *J. Alloy. Compd.* **2021**, *884*, 160945. [CrossRef]

22. Tu, W.M.; Bai, Z.Y.; Deng, Z.; Zhang, H.N.; Tang, H.L. In-Situ Synthesized Si@C Materials for the Lithium Ion Battery: A Mini Review. *Nanomaterials* **2019**, *9*, 432. [CrossRef]

23. Gou, W.W.; Kong, X.Z.; Wang, Y.P.; Ai, Y.L.; Liang, S.Q.; Pan, A.Q.; Cao, G.Z. Yolk-shell structured V$_2$O$_3$ microspheres wrapped in N, S co-doped carbon as Pea-Pod nanofibers for high-capacity lithium ion batteries. *Chem. Eng. J.* **2019**, *374*, 545–553. [CrossRef]

24. Wu, F.F.; Gao, X.B.; Xu, X.L.; Jiang, Y.N.; Gao, X.L.; Yin, R.L.; Shi, W.H.; Liu, W.X.; Lu, G.; Cao, X.H. MnO$_2$ nanosheet-assembled hollow polyhedron grown on carbon cloth for flexible aqueous zinc-ion batteries. *ChemSusChem* **2020**, *13*, 1537–1545. [CrossRef] [PubMed]

25. Zhao, Y.; He, J.F.; Dai, M.Z.; Zhao, D.P.; Wu, X.; Liu, B.D. Emerging CoMn-LDH@MnO$_2$ electrode materials assembled using nanosheets for flexible and foldable energy storage devices. *J. Energy Chem.* **2020**, *45*, 67–73. [CrossRef]

26. Zhang, Y.; Ren, J.H.; Xu, T.; Feng, A.L.; Hu, K.; Yu, N.F.; Xia, Y.B.; Zhu, Y.S.; Huang, Z.Y.; Wu, G.L. Covalent Bonding of Si Nanoparticles on Graphite Nanosheets as Anodes for Lithium-Ion Batteries Using Diazonium Chemistry. *Nanomaterials* **2019**, *9*, 1741. [CrossRef] [PubMed]

27. Liu, F.S.; Xiao, Y.Y.; Liu, Y.T.; Han, P.Y.; Qin, G.H. Mesoporous MnO$_2$ based composite electrode for efficient alkali-metal-ion storage. *Chem. Eng. J.* **2020**, *380*, 122487. [CrossRef]

28. Wang, B.Y.; Wei, Y.H.; Fang, H.Y.; Qiu, X.L.; Zhang, Q.B.; Wu, H.; Wang, Q.; Zhang, Y.; Ji, X.B. Mn-substituted tunnel-type polyantimonic acid confined in a multidimensional integrated architecture enabling superfast-charging lithium-ion battery anodes. *Adv. Sci.* **2021**, *8*, 2002866. [CrossRef]

29. Pi, W.B.; Mei, T.; Li, J.; Wang, J.Y.; Li, J.H.; Wang, X.B. Durian-like NiS$_2$@rGO nanocomposites and their enhanced rate performance. *Chem. Eng. J.* **2018**, *335*, 275–281. [CrossRef]

30. Mei, L.; Xu, C.; Yang, T.; Ma, J.M.; Chen, L.B.; Li, Q.H.; Wang, T.H. Superior electrochemical performance of ultrasmall SnS$_2$ nanocrystals decorated on flexible RGO in lithium-ion batteries. *J. Mater. Chem. A* **2013**, *1*, 8658–8664. [CrossRef]

31. Rana, M.; He, Q.; Luo, B.; Lin, T.; Ran, L.B.; Li, M.; Gentle, I.; Knibbe, R. Multifunctional Effects of Sulfonyl-Anchored, Dual-Doped Multilayered Graphene for High Areal Capacity Lithium Sulfur Batteries. *ACS Cent. Sci.* **2019**, *5*, 1946–1958. [CrossRef]

32. Park, S.; Jeong, S.Y.; Lee, T.K.; Park, M.W.; Lim, H.Y.; Sung, J.; Cho, J.; Kwak, S.K.; Hong, S.Y.; Choi, N.S. Replacing conventional battery electrolyte additives with dioxolone derivatives for high-energy-density lithium-ion batteries. *Nat. Commun.* **2021**, *12*, 838. [CrossRef] [PubMed]

33. Cui, R.Y.; Hultman, N.; Cui, D.; McJeon, H.; Yu, S.; Edwards, M.R.; Sen, A.; Song, K.; Bowman, C.; Clarke, L.; et al. A plant-by-plant strategy for high-ambition coal power phaseout in China. *Nat. Commun.* **2021**, *12*, 1468. [CrossRef]

34. Zhao, T.M.; Zheng, C.W.; He, H.X.; Guan, H.Y.; Zhong, T.Y.; Xing, L.L.; Xue, X.Y. A self-powered biosensing electronic-skin for real-time sweat Ca^{2+} detection and wireless data transmission. *Smart Mater. Struct.* **2019**, *28*, 085015. [CrossRef]

35. Chen, S.; Chen, Z.; Xu, X.Y.; Cao, C.B.; Xia, M.; Luo, Y.J. Scalable 2D Mesoporous Silicon Nanosheets for High-Performance Lithium-Ion Battery Anode. *Small* **2018**, *14*, 1703361. [CrossRef] [PubMed]

36. Zhao, T.M.; Han, Y.C.; Qin, L.N.; Guan, H.Y.; Xing, L.L.; Li, X.J.; Xue, X.Y.; Li, G.L.; Zhan, Y. Bidirectional modulation of neural plasticity by self-powered neural stimulation. *Nano Energy* **2021**, *85*, 106006. [CrossRef]

37. Liu, H.D.; Hu, Z.L.; Su, Y.Y.; Ruan, H.B.; Hu, R.; Zhang, L. MnO_2 nanorods/3D-rGO composite as high performance anode materials for Li-ion batterie. *Appl. Surf. Sci.* **2017**, *392*, 777–784. [CrossRef]

38. Li, Y.; Ye, D.X.; Liu, W.; Shi, B.; Guo, R.; Pei, H.J.; Xie, J.Y. A three-dimensional core-shell nanostructured composite of polypyrrole wrapped MnO_2/reduced graphene oxide/carbon nanotube for high performance lithium ion batteries. *J. Colloid Interf. Sci.* **2017**, *493*, 241–248. [CrossRef]

39. Xu, P.; Chen, H.Y.; Zhou, X.; Xiang, H.F. Gel polymer electrolyte based on PVDF-HFP matrix composited with rGO-PEG-NH_2 for high-performance lithium ion battery. *J. Membr. Sci.* **2021**, *617*, 118660. [CrossRef]

40. Li, J.B.; Ding, Z.B.; Li, J.L.; Wang, C.Y.; Pan, L.K.; Wang, G.X. Synergistic coupling of $NiS_{1.03}$ nanoparticle with S-doped reduced graphene oxide for enhanced lithium and sodium storage. *Chem. Eng. J.* **2020**, *407*, 127199. [CrossRef]

41. Fang, Y.Z.; Hu, R.; Liu, B.Y.; Zhang, Y.Y.; Zhu, K.; Yan, J.; Ye, K.; Cheng, K.; Wang, G.L.; Cao, D.X. MXene-derived TiO_2/reduced graphene oxide composite with an enhanced capacitive capacity for Li-ion and K-ion batteries. *J. Mater. Chem. A* **2019**, *7*, 5363–5372. [CrossRef]

42. Yao, W.Q.; Wu, S.B.; Zhan, L.; Wang, Y.L. Two-dimensional Porous Carbon-coated Sandwich-like Mesoporous SnO_2/Graphene/Mesoporous SnO_2 Nanosheets towards High-Rate and Long Cycle Life Lithium-ion Batteries. *Chem. Eng. J.* **2018**, *361*, 329–341. [CrossRef]

43. Chae, S.; Choi, S.H.; Kim, N.; Sung, J.; Cho, J. Integration of Graphite and Silicon Anodes for the Commercialization of High-Energy Lithium-Ion Batteries. *Angew. Chem. Int. Ed.* **2019**, *59*, 110–135. [CrossRef] [PubMed]

44. Li, Z.H.; Zhang, Y.P.; Liu, T.F.; Gao, X.H.; Li, S.Y.; Ling, M.; Liang, C.D.; Zheng, J.C.; Lin, Z. Silicon Anode with High Initial Coulombic Efficiency by Modulated Trifunctional Binder for High-Areal-Capacity Lithium-Ion Batteries. *Adv. Energy Mater.* **2020**, *10*, 1903110. [CrossRef]

45. Keller, C.; Desrues, A.; Karuppiah, S.; Martin, E.; Alper, J.; Boismain, F.; Villevieille, C.; Herlin-Boime, N.; Haon, C.; Chenevier, P. Effect of Size and Shape on Electrochemical Performance of Nano-Silicon-Based Lithium Battery. *Nanomaterials* **2021**, *11*, 307. [CrossRef]

46. Zhao, Z.Q.; Cai, X.; Yu, X.Y.; Wang, H.Q.; Li, Q.Y.; Fang, Y.P. Zinc-assisted Mechanochemical Coating of Reduced Graphene Oxide Thin Layer on Silicon Microparticles for Efficient Lithium-ion Battery Anodes. *Sustain. Energy Fuels.* **2019**, *3*, 1258–1268. [CrossRef]

47. Li, X.; Bai, Y.S.; Wang, M.S.; Wang, G.L.; Ma, Y.; Huang, Y.; Zheng, J.M. Dual Carbonaceous Materials Synergetic Protection Silicon as a High-Performance Free-Standing Anode for Lithium-Ion Battery. *Nanomaterials* **2019**, *9*, 650. [CrossRef] [PubMed]

48. Laruelle, S.; Grugeon, S.; Poizot, P.; Dolle, M.; Dupont, L.; Tarascon, J.M. On the Origin of the Extra Electrochemical Capacity Displayed by MO/Li Cells at Low Potential. *J. Electrochem. Soc.* **2002**, *149*, A627–A634. [CrossRef]

49. Hao, Q.; Wang, J.P.; Xu, C.X. Facile preparation of Mn_3O_4 octahedra and their long-term cycle life as an anode material for Li-ion batteries. *J. Mater. Chem. A* **2014**, *2*, 87–93. [CrossRef]

50. Li, L.; Guo, Z.P.; Du, A.J.; Liu, H.K. Rapid microwave-assisted synthesis of Mn_3O_4-graphene nanocomposite and its lithium storage properties. *J. Mater. Chem.* **2012**, *22*, 3600–3605. [CrossRef]

51. Jiang, Y.; Jiang, Z.J.; Chen, B.H.; Jiang, Z.Q.; Cheng, S.; Rong, H.B.; Huang, J.L.; Liu, M.L. Morphology and crystal phase evolution induced performance enhancement of MnO_2 grown on reduced graphene oxide for lithium ion batteries. *J. Mater. Chem. A* **2016**, *4*, 2643–2650. [CrossRef]

52. Kim, K.; Daniel, G.; Kessler, V.G.; Seisenbaeva, G.A.; Pol, V.G. Basic Medium Heterogeneous Solution Synthesis of alpha-MnO_2 Nanoflakes as an Anode or Cathode in Half Cell Configuration (vs. Lithium) of Li-Ion Batteries. *Nanomaterials* **2018**, *8*, 608. [CrossRef]

53. Liu, C.; Wu, X.; Wang, B. Performance modulation of energy storage devices: A case of Ni-Co-S electrode materials. *Chem. Eng. J.* **2020**, *392*, 123651. [CrossRef]

54. Wu, X.; Yao, S.Y. Flexible electrode materials based on WO_3 nanotube bundles for high performance energy storage devices. *Nano Energy* **2017**, *42*, 143–150. [CrossRef]

55. Zhang, D.A.; Wang, Q.; Wang, Q.; Sun, J.; Xing, L.L.; Xue, X.Y. High capacity and cyclability of hierarchical MoS_2/SnO_2 nanocomposites as the cathode of lithium-sulfur battery. *Electrochim. Acta.* **2015**, *173*, 472–482. [CrossRef]

56. Chen, Y.X.; Yan, Y.J.; Liu, X.L.; Zhao, Y.; Wu, X.Y.; Zhou, J.; Wang, Z.F. Porous Si/Fe_2O_3 Dual Network Anode for Lithium–Ion Battery Application. *Nanomaterials* **2020**, *10*, 2331. [CrossRef]

57. Zhang, Y.; Hu, K.; Zhou, Y.L.; Xia, Y.B.; Yu, N.F.; Wu, G.L.; Zhu, Y.S.; Wu, Y.P.; Huang, H.B. A Facile, One-Step Synthesis of Silicon/Silicon Carbide/Carbon Nanotube Nanocomposite as a Cycling-Stable Anode for Lithium Ion Batteries. *Nanomaterials* **2019**, *9*, 1624. [CrossRef] [PubMed]

58. Han, N.; Li, J.J.; Wang, X.C.; Zhang, C.L.; Liu, G.; Li, X.H.; Qu, J.; Peng, Z.; Zhu, X.Y.; Zhang, L. Flexible Carbon Nanotubes Confined Yolk-Shelled Silicon-Based Anode with Superior Conductivity for Lithium Storage. *Nanomaterials* **2021**, *11*, 699. [CrossRef] [PubMed]

59. Kwon, H.J.; Hwang, J.Y.; Shin, H.J.; Jeong, M.G.; Chung, K.Y.; Sun, Y.K.; Jung, H.G. Nano/Microstructured Silicon–Carbon Hybrid Composite Particles Fabricated with Corn Starch Biowaste as Anode Materials for Li-Ion Batteries. *Nano Lett.* **2019**, *20*, 625–635. [CrossRef] [PubMed]

60. Argyropoulos, D.P.; Zardalidis, G.; Giotakos, P.; Daletou, M.; Farmakis, F. Study of the Role of Void and Residual Silicon Dioxide on the Electrochemical Performance of Silicon Nanoparticles Encapsulated by Graphene. *Nano Lett.* **2021**, *11*, 2864. [CrossRef] [PubMed]
61. Gao, X.F.; Li, J.Y.; Xie, Y.Y.; Guan, D.S.; Yuan, C. A Multilayered Silicon-Reduced Graphene Oxide Electrode for High Performance Lithium-Ion Batteries. *ACS Appl. Mater. Interfaces.* **2015**, *7*, 7855–7862. [CrossRef] [PubMed]
62. Wang, T.H.; Ji, X.; Wu, F.Z.; Yang, W.L.; Dai, X.Y.; Xu, X.J.; Wang, J.; Guo, D.; Chen, M.L. Facile fabrication of a three-dimensional coral-like silicon nanostructure coated with a C/rGO double layer by using the magnesiothermic reduction of silica nanotubes for high-performance lithium-ion battery anodes. *J. Alloys Compd.* **2021**, *863*, 158569. [CrossRef]
63. Wang, D.X.; Wang, Y.; Li, Q.Y.; Guo, W.B.; Zhang, F.C.; Niu, S.S. Urchin-like alpha-Fe_2O_3/MnO_2 hierarchical hollow composite microspheres as lithium-ion battery anodes. *J. Power Sources.* **2018**, *393*, 186–192. [CrossRef]
64. Ma, Z.F.; Zhao, T.B. Reduced graphene oxide anchored with MnO_2 nanorods as anode for high rate and long cycle Lithium ion batteries. *Electrochim. Acta.* **2016**, *201*, 165–171. [CrossRef]

MDPI

St. Alban-Anlage 66

4052 Basel

Switzerland

www.mdpi.com

Nanomaterials Editorial Office

E-mail: nanomaterials@mdpi.com

www.mdpi.com/journal/nanomaterials